INTEGRATED BROADBAND COMMUNICATION NETWORKS AND SERVICES

IFIP Transactions C: Communication Systems

International Federation for Information Processing

Technical Committee 6

Communication Systems

IFIP Transactions Abstracted/Indexed in:
INSPEC Information Services
Index to Scientific & Technical Proceedings®
CompuMath Citation Index®, Research Alert™, and SciSearch®

C-18

INTEGRATED BROADBAND COMMUNICATION NETWORKS AND SERVICES

Proceedings of the IFIP TC6/ICCC International Conference on
Integrated Broadband Communication Networks and Services
Copenhagen, Denmark, 20-23 April, 1993

Edited by

VILLY B. IVERSEN
Institute of Telecommunications
Technical University of Denmark
Lyngby, Denmark

1994

NORTH-HOLLAND
AMSTERDAM • LONDON • NEW YORK • TOKYO

ELSEVIER SCIENCE B.V.
Sara Burgerhartstraat 25
P.O. Box 211, 1000 AE Amsterdam, The Netherlands

Keywords are chosen from the ACM Computing Reviews Classification System, ©1991, with permission.
Details of the full classification system are available from
ACM, 11 West 42nd St., New York, NY 10036, USA.

ISBN: 0 444 81584 8
ISSN: 0926-549X

Printed in The Netherlands

PREFACE

IBCN &S — The International Conference on Integrated Broadband Communications and Networks was held in Copenhagen April 20-23 1993. The conference was organized by ICCC and IFIP-TC6. It was fully dedicated to the subject of Broadband Communications, recognizing the importance of this item in shaping the future telecommunication network.

The proceedings contain papers of permanent value, selected among the papers presented at conference. The papers cover key items concerning research, development and applications in Broadband Communications all over the world.

The Conference proceedings present ideas and experience of value, both for operators and users, and for research and development, and both from a technical and a business point of view.

The program was the result of the work of the members of the technical committee and the referees. The conference is made possible by the sponsoring organizations ICCC and IFIP, the host KTAS (Copenhagen Telephone Co.), and the conference sponsors.

Villy Bæk Iversen

Conference Board:

J. Arnbak	ICCC Conference Governor
J. Lindegaard	Copenhagen Telephone Co.
O. Spaniol	Chairman IFIP-TC6

International Advisory Committee

E.J. Boutmy, The Netherlands
K. Boyanov, Bulgaria
F. Eliassen, Norway
F. Kamoun Tunesia
L. Fratta Italy
J. Granado Portugal
P.P. Gupta India
P. Verbeeck Belgium
T. Housley Australia
S.-H. Kyong Korea
X.-J. Liang China
O. Martikainen, Finland
K. Ono, Japan
L. Pouzin, France
J. Puzman, Czechoslovakia
P. Radford, UK
S. Ramani, India
R.P. Uhlig, USA
C. Rindorf, Denmark
H. Rudin, Switzerland
T. Szentivanyi, Hungary
L.M.R. Tarouco, Brazil

Program Committee

J. Arnbak, The Netherlands
A. Casaca, Portugal
J.P. Coudreuse, France
S. Fredslund, Denmark
E. Lorenz-Pedersen, Denmark
R. Haugen, Norway
P. Heuer, Germany
V.B. Iversen, Denmark (Chairman)
E. Nicolaisen, Denmark
J. Johansen, Denmark
D. Khakhar, Sweden
K. Kvols, Denmark
U. Kørner, Sweden
P. Kuehn, Germany
A. Bohn Nielsen, Denmark
P. Schicker, Switzerland
B. Veirø, Denmark
P.K. Verma, USA

Organizing Committee

P. Ibsen (Chairman)
V.B. Iversen
Y. Liu
B. Urbild
K. Andersen
K. Linder
H.H. Stilling
K. Madsen

Sponsors

- IFIP Technical Committee 6 – Communication Systems
- ICCC – International Computer Communication Conference

Supporters

- KTAS – Copenhagen Telephone Company
- Siemens
- IBM
- Ericssoni
- Dansk Dataforening

LIST OF REFEREES

Å. Arvidsson
J. Andrade
S. Blaabjerg
U. Briem
A. Casaca
J. Charzinski
H. Christiansen
P. Davids
L. Dittmann
R. Dittmann
S. Fredslund
A. Gravey
R. Griffiths
O. Gropp
R.B. Haugen
G. Hebuterne
B. Heinrichs
O. Hermanns
P. Heuer
F. Hübner
P.T. Huth
V.B. Iversen
J. Johansen
K. Jakobs
K. Kilkki
U. Körner
U.R. Krieger
P.J. Kühn
Ø. Kure
K. Kvols
K. Lindberger
Y. Liu
J. Lubacz
T. Meuser
A.M. Møller
E. Nicolaisen
A. Bohn Nielsen
A.P. Nielsen
B. Friis Nielsen
K. Nivert
I. Norros
M.S. Nunes
C. Nyberg
J.A.M. Nÿhof
A.R. Pach
S. Møller Pedersen
E. Lorenz Petersen
M. Pioro
L. Reneby
T. Renger
J. Roberts
T. Rydén
J.M. Salles
K.E. Skouby
O. Spaniol
T. Theimer
A. Tonietti
B. Veirø
M. Villén-Altamirano
J.T. Virtamo
B. Wallström
O. Østerbø

TABLE OF CONTENTS

Reviews and Operator Strategies

The Evolution of High Performance Networking: Driving Forces and Handicaps 1
O. Spaniol

Strategies for the Introduction of Broadband Networks and Services in Denmark 11
K.E. Jessen, H. Pedersen, K. Nørgaard and A. Thomsen

Scenarios for Introduction of Broadband Services in Telecommunication Networks 23
R. Haugen and B. Mannsåker

Trends

Visions on the Combination of SDH, ATM and LAN's as seen from the Customer Side 35
J.W.J. baron van Till

Evaluation and projects

Basic Concepts and Performance of High-Speed Protocols 45
T. Stock and P. Tran-Gia

EURESCOM supporting Public Network Operators to Prepare for New Networks and Services 67
O. Baireuther

Standardisation

Activities on traffic engineering for B-ISDN in CCITT, SGII 77
M. Villén-Altamirano

Standardisation of B-ISDN 89
J. Johansen

Developments in Japan and USA

An Evolving Scenario of Communication Network toward B-ISDN 95
T. Saito

Tools and Methodologies for High Speed Network Design 107
A. A. Nilsson, H. Perros and D. Stevenson

Supplier Strategies

Intelligent LAN Concentrators Evolution to ATM 119
H. Sourbes, C. Basso and J. Calvignac

Network Concepts and Siemens Products for ATM-Based Public Networks 131
R. Händel

Business Applications - Supplier's Strategy 139
P. Staxén

New Services

New IBC Services in Support of the Border Crossing Trade in the European Community 151
R.M. Guarneri, L. Resmini and C.J.M. Lanting

Economic Implications of Introduction of New Telecommunication Services 163
M. Falch, A. Henten and K.E. Skouby

Architecture

Frame Synchronized Ring: a Bus Concept for High Speed Processing Applications 177
P. Raatikainen, J. Zidbeck and H. Flinck

The DYCE Concept – Architecture & Implementation Strategies 189
B. Heinrichs, W. Reinhardt and R. Stainov

An Experimental Architecture Supporting Multimedia Applications on Broadband Networks 201
G. Girardi, D. Ercole, M. Ullio, R. Melen and A. Albanese

Connection Acceptance Control

The Poor Gain from Statistical Multiplexing in the Homogeneous and the Heterogeneous Case 213
T.A. Smit

An Effective-Rate Enforcement Algorithm for ATM Traffic and its Hardware Implementation 225
N.M. Mitrou and A.K. Koukos

Effective Bandwidths with Different Quality of Service Requirements 241
N.G. Bean

Usage Parameter Control

Equivalence of Policing Mechanisms 253
A.M. Møller

A Flow Management Architecture for B-ISDN 261
F.A. Aagesen

Test Equipments

Traffic Measurements and Test Equipment in BATMAN 275
L. Dittmann and K. Kvols

The Synthesized Traffic Generator; Objectives, Design and Capabilities 287
B.E. Helvik, O. Melteig and L.Morland

Measurement Techniques and Equipment for B-ISDN 303
R. Grunenfelder

Signalling and Intelligent Networks

B-ISDN Signalling and Intelligent Networks; how two worlds can come together 313
P.F.C. Blankers

Compatibility Aspects of B-ISDN Signalling 325
J. Tuyt

Timer Handling in High-Performance Transport Systems 337
B. Heinrichs and K. Jakobs

Traffic Characterization

On the Limit Behaviour of the ATM Traffic within a Network 349
M. D'Ambrosio and R. Melen

Time Dependent Behaviour and Large Deviations Analysis for a Virtual Path in a B-ISDN 361
S.P. Evans

On the Broadening of Cell Bursts in an ATM Network 373
J. Virtamo and I. Norros

Analysis of ATM Switching

Analysis of ATM Switching Modules with Channel Grouping in a Bursty-Source Environment 383
B. Steyaert and H. Bruneel

A Refined Approximation of a Finite Capacity Polling System under ATM Bursty Arrivals 395
Y.F. Jou, A.A. Nilsson and F. Lai

An Information Model for the QOS Monitoring of an ATM Switch: from Analysis to Implementation 407
S. Arsenis, N. Simoni and S. Znaty

Inter-working

High Speed Data Services for LAN Interconnection in Norway 419
B. Svendsen

Buffering and Packet Loss in the DQDB to ATM Interworking Unit 429
S. Manthorpe

ATM – N-ISDN 2 Mbit/s Interworking Unit 441
M.S. Nunes, P.D.Soerensen, M. Frieden, J. Gonçalves, M. Günay
H. Blaser and V. Tao

Integrated Broadband Communication Networks
and Services (C-18) / V.B. Iversen (Editor)
Elsevier Science B.V (North-Holland)

The Evolution of High Performance Networking: Driving Forces and Handicaps

Otto SPANIOL
Computer Science Department, Aachen University of Technology; Aachen (Germany).
Electronic mail address: spaniol@informatik.rwth-aachen.de

1. The Evolution of High Performance Networking

The Evolution of High Performance Networks may be characterized by the motto of the Olympic games, i.e. by "Citius, Altius, Fortius". The reason for this comparison is that the characteristics of successive types of so called high-performance networks are in good accordance to this motto. In order to justify that claim let's consider some properties of successive network generations:

Generation 1 *High-Speed Networks* *(Citius, i.e. "faster").*

First high-speed networks had very limited applications. There were basically conceived in order to provide a suitable transport facility (subject to throughput requirements) for a few specialized stations, e.g. supercomputers. High-speed networks don't really deserve their name since obviously the transmission speed (per bit) is not increased at all; this parameter is governed by the speed of light and can hardly be improved.

Generation 2 *High-Capacity Networking* *(Altius, i.e. "higher in capacity").*

The second generation of broadband networks concentrated more on application and already to some extent on service integration. Most networks of this kind are intended to be 'backbone' networks, i.e. they served as a backbone facility in order to interconnect many local networks (LAN). Usually, those backbones deliver incoming messages in the same format to the destination network as they have been presented by the source LAN. Thus, communication is mainly limited to stations which are connected to subnetworks of the same kind.

Capacity of connections is the most important parameter: whereas the speed of a single bit may be even lower than for conventional networks the time to deliver a whole message is significantly reduced due to the increased "bits per second" facility of the networks.

Generation 3 *High-Performance Networking* *(Fortius, i.e. "stronger" or "better").*

In a more mature state we will have a coexistence and a cooperation of several types of high capacity networks such as Metropolitan Area Networks or Broadband ISDN systems based on ATM technology (see [HH91]). A variety of services will be offered (voice, video, file transfer, interactive data, mail etc.) which are suitable for different capacity requirements. Multi-media systems will be based on the availability of both new networks and new services.

2. Driving Forces for High Performance Networking

2.1. Technology and Feasability

The continuous progress in networking has ever been and still is technology driven. The networking area in particular takers profit from new components which become available at affordable prices. Such developments include but are not limited to:

- performance imcrease of microprocessors
- availability of fiber optics
- efficiency improvement by increased use of parallelism.

The area of 'high speed' (or better 'high capacity') networking, like many other areas, may be divided into three sections which could be characterized

by:	*"small"*	*"medium"*	*"large"*
or alternatively by:	*"yesterday"*	*"today"*	*"tomorrow"*
or even by:	*"everybodys darling"*	*"risky but feasible"*	*"speculations"*.

The interpretation of 'small', 'medium' and 'large' is rapidly changing; that is to say that 'high speed' is very likely to be considered as 'medium speed' in five years from now and probably as 'very moderate speed' in ten years from now. We have seen a lot of similar changes for example in the area of processors where the performance of a former mainframe is now significantly lower than the performance of a small modern workstation.

As of today the first networking area deals with the **medium speed** range (10 Megabit per secondc) which is now very well established and accepted by everybody and for which a lot of products are beginning to produce the expected results.

The second area is the **high speed** range (i.e. a capacity range of approximately 100 Megabit per second) where the standards and the products are coming and where the integration begins. This will, as usual, last for some years before reaching the intended price figures (if the product 'survives' this period; it may very well be overtaken by new announcements!) and the required stability of operation which will be necessary for reaching a wide acceptance.

From the user point of view it is obvious that experience in the 100 Megabit per second range is still very limited and early efforts of deployment have been sometimes rather painful. But the trend is clear and the stabilization of the standards makes us confident in an evolution which will follow the pace of the Ethernet deployment ten years ago. Indeed, we have good arguments to believe that the following 'equations' are not unrealistic:

FDDI = Ethernet + 10 (years)
or even more speculative:
ATM = FDDI + 10 (years).

This results in a feeling that in the networking area we shall observe some shortened 'Kondratieff cycles' and that developments which happen to appear between two successive peaks take a risk of not being very successful. This might happen for example in the case of DQDB systems which have been in the center of interest a few years ago. Their importance in the research, discussion and installation arena, however, seems to diminish or to disappear (take a look into some recent proceedings volumes!) or to be merely reduced to a mathematical exercise of some unfairness aspects. A rather malicious interpretation could say that the impact of those systems was not much more than the support for some dozens or even hundreds of PhD thesises.

The third area is the **very high speed** range, from 1 Gigabit/sec up, which is today a research area. Even if the 100 Megabit per second range is not yet stabilized, there is a lot of efforts going on already in the Gigabit and the Terabit ranges where many claims have still to be considered as remote dreams.

2.2. Applications (graphics, vireo, multimedia systems etc.)

Apart from 'purely technological reasons' we may identify a driving force towards new networks by a lot of new applications which might be meaningful for a large community but which cannot be supported by conventional networks as of today. In most cases, this situation is valid for requirements which are based on graphics or video. The problem with 'classical' networks is that they offer more than enough capacity for applications which may be characterized by 'data' but that, on the other hand, those networks have a very poor and largely insufficient performance for everything which has to do with image transmission. Thus, there is a need for very powerful broadband network. This driving force will become successful, however, only if the user may be convinced that he really needs and may afford the new facility. Some of the actual demonstrators of multimedia systems are not really convincing in that respect - and some newly established research and development projects have less difficulties for defining and reaching their intended technical aims than for finding enough user groups who need the upcoming system and who are prepared to pay for it.

2.3. Positive experiences gained from local networks (the Ethernet story)

The willingness to consider new networking concept is considerably improved by a lot of positive experiences which have been made by very different groups. This is best demonstrated by the Ethernet story: After a rather difficult start (due to insufficient availability of products, due to unrealistic prices for components, due to missing experience etc.) the Ethernet concept has become the clear winner within the 10 Mbit/sec range. Some important reasons for that success (apart from its technical suitability) have been:

- the relative easiness of installing and operating of the network (even by nonspecialized people)
- the dramatic price reduction of components
- the inclusion of Ethernet interfaces as a standard feature within many workstation types
- the availability of the general concept in several versions (yellow cable, cheapernet, twisted pair, broadband, optical star,...); thus, for almost every need there was a suitable installation strategy which satisfied the requirements
- and, last but not least, the increasing community of users! Indeed, it is always very helpful or even necessary for 'cautious' users to see a lot of others who make a decision for the same concept. Whereas the imitation of strategies made by other institutions might be considered as boring or not really brilliant, we may observe a lot of positive aspects. The imitator will undoubtedly profit from price reduction and may avoid the mistakes made by the early installations. Moreover, the decision to join a large community is always on the safe side: If everything goes clear then the members of the community may congratulate each other for the success; if something goes wrong the unavoidable criticism will be easier to survive by demonstrating that a lot of other people are in the same situation, i.e. that those people made the same mistake and that, therefore, the decision maker cannot be considered as guilty.

3. Handicaps for High Performance Networking

3.1. Insufficient components performance and protocol overhead

The performance of a connection between two endstations is limited by the weakest component of the 'chain'. This may be one of the endstations (or both of them); another reason for a bottleneck may be a 'lousy' design of some intermediate component - e.g. a bridge, a router, an interworking unit etc. It happens very rarely that the bottleneck is due to insufficient network capacity.

Very often, the performance is reduced by a combination of two aspects:

- A limitation of computation performance within one of the endstations.
- An overly complex structure of the underlying protocol stack.

  Protocol stacks and protocol functions have been defined for situations which were radically different from the situations observed in high performance networks:

  For example, these classical protocols did a lot of work for flow control and for error recovery mechanisms. This was necessary due to the relatively high error rate as well as due the limitations of the network capacity, just to mention two major reasons. Meanwhile, however, error rates arevery much lower and, on the other hand, capacities are significantly higher than for networks which have been installed in the early times of networking. Another factor is that for an increasingnumber of applications neither acknowledgments nor retransmissions after errors are meaningful any more; this holds particularly for a lot of real time applications such as voice or imaging.

The consequence of both effects is that a single connection may only use a minor part of the raw network capacity. Even worse than that: if the throughput per connection is increased then it is to be expected that this ends up in a situation where the whole computing capacity of the endstations is wasted for protocol processing - or even bynothing but lower layer protocol handling! One may even extrapolate a situation where several additional workstations would be needed for the only purpose of protocol handling, i.e. without any remaining capacity for useful things!!

A very nice and convincing demonstration of that phenomenon was given by Steenkiste et alii (see [S 93]). They measured (and extrapolated from those measurements) how much host capacity is used by different data processing overheads - and, therefore, how much (or how little) is left over for the application itself. The endstation type under consideration was a DECstation 5000/200. Whereas this may not be the biggest or the most modern workstation the nature of results is independent of the actual workstation type. The obeservation is that already lower layer protocol functions such as 'media DMA', 'packet processing', 'copying and checksum' and 'ACK processing' result in a maximum utilization of no more than 21 Megabit per second per connection of an assumed FDDI network (which has a nominal capacity of 100 Megabit per second!). Thus, if only a fifth of FDDI's capacity is needed for a single connection then this may be done with the given workstation type only at the expense of no remaining capacity for higher layer protocols - and nothing at all remains for applications! The 'dream' or any possible requirement of filling FDDI up to completion by a single connection would result in the necessity to provide more than three additional workstations (at either end!) for the only purpose of lower layer protocol processing (see figure 1 for a corresonding diagram).

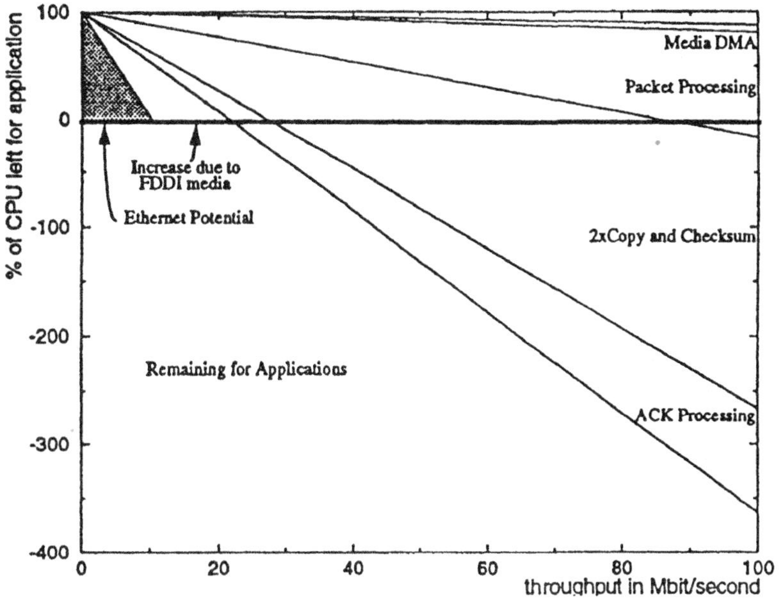

Figure 1: Communication overhead for FDDI networks (from [S 93])

It is more than obvious that such a scenario is unrealistic and far from being desirable. What is to be done in order to avoid such the described effects? We have at least three alternatives:

A. *Increase of workstation performance:*

A better design will considerably improve the situations. There are some reports available which state that some more modern workstation types may have a communication capacity of up to 90 Megabit per second (which would be more than enough in order to fill an FDDI network up to completion by a single connection).

We have to be cautious, however, about the real meaning of such reports. Our experience with respect to announcements and results of own measurements (see [DMS 93]) seem to indicate that sometimes many 'measured' parameters differ significantly from the reality.

Another problem is that the increase of workstation performance will come together with a corresponding increase of network capacity. Regarding the previous example we see that the situation will be completely unchanged if we imagine of having a workstation which has eight times more processing capacity but which is used in a HIPPI network (which has a capacity of 800 instead as of 100 Megabit per second). The discrepancies would become even more serious for the intended Gigabit networks.

B. Reduction of protocol complexity:

It is not difficult at all to observe that good old TCP/IP is far from being the best candidate for high performance networking. Some improvements and optimization procedures have been very helpful in order to make TCP/IP viable or at least acceptable in FDDI environments (see [DMS 92]). However, for real high performance networking we need completely revised protocols - preferably neglecting a lot of protocol options and even omitting some protocol layers. There is an abundant number of proposals for such protocols (e.g. XTP and others). On the other hand, many research, implementation and - last but not least - installation and acceptance problems remain to be solved. Special protocols will result in increased network specialisation. The dream of open communication will become more difficult or even impossible to realize.

C. Better network utilisation by multiplexing of many connections:

It should be obvious that almost all networks take their motivation from the interconnection of <u>many</u> stations. Only a few specialized types of networks (e.g. direct interconnections between supercomputers) may be justified from a return on investment point of view if they are intended only for a few stations.

The strategy of connection multiplexing over shared high capacity media is exactly in accordance with the ATM (Asynchronous Transfer Multiplexing) philosophy. Thus backbone networks for the interconnection of many low capacity networks make sense even if the capacities of both network types are widely different: Most FDDI systems as of today are installed as an interconnection of several or many Ethernet networks. Even if the performance of interconnection units in such structures is far from being perfect and even if the protocol structures are so complicated or outdated that a single connection will be limited to, say, a few Megabit per second (instead as of the nominal capacity of 100 Megabit per second) we may nevertheless observe that the FDDI network in total may become reasonably well utilized. Another positive aspect is contributed by the observation that in such hierarchically interconnected structures the border crossing traffic is much lower that the local traffic. This compares very well with traffic scenarios for telephone usage.

3.2. Acceptance problems

In many countries there are a lot of people who are very critical or negative against the introduction of new technologies. It is often required that <u>all possible consequences</u> of the introduction must be clarified <u>before the first installation</u> and that the absence of any negative impact which might come out of the new technology even in a worst case scenario must be proven beforehand. Another requirement is that only those technologies (or networks) should be investigated which are really needed by the public. This latter statement appears to be particularly dangerous since it happens very rarely that the community of prospective users will know beforehand what might be desired or acceptable. I want to give just two well-known examples:

- The widespread acceptance of fax machines was not foreseen by anybody.
- The telephone was (according to several trustable statements) not intended for interpersonal communications but rather for "bringing the concert to the home": With the latter intention in mind, however, it is highly probable that the installation of the telephone would have been hindered or even forbidden if this would have been preceeded by some committees which deal with possible negative consequences and with acceptance problems.

Data protection and security aspects are other topics which become extremely critical in high performance networks. Here we observe a similar situation as in the case of 'crocodiles in Alaska': It may be possible to prove the existence of some phenomena (if they exist) but it will never be possible to prove the nonexistence (if the phenomenon is puely fiction) since in the latter case no proof at all will be accepted: The opponent would need nothing but to put a question mark about the system of axioms on which the proof must rely on; moreover, it seems to be much more interesting and even profitable to speculate about dangerous scenarios.

3.3. The endless time required for standardisation

Standards are crucial things for networks in order to become successful. However, their creation, the development of products which are in conformance to the standard and finally the acceptance of standards become more and more difficult to achieve. This very unfortunate situation is due to a number of reasons:

1. Standards are mainly created by committees - and the decisions of such a committee are necessarily 'boring'.
2. Finding a consensus needs all too much time. Standards are never completed. This may be explained by the following arguments:
   - 'Remaining topics' are difficult to solve (there must have been a reason for postponing those topics) or only to agree upon the best alternative.
   - A consensus becomes more and more difficult to achieve if more interest groups (with conflicting interests) become involved.
   - Whereas the progress of a standard is relatively fast in the beginning it slows down more and more when it approaches its final stage. We may even state that the progress rate is proportional to the amount of remaining work, i.e. the progress rate finally approaches a zero limit which means that an infinite time is necessary for successfully completing the global standard. This dark scenario is supported by the observation that not a single important network standard has been completely finalized in the recent past.

     The only way out for a decision maker out of this dilemma is that he accepts a standard which is not totally mature. This may be a possible strategy whenever the degree of completion is higher than, say, 80 percent but then the situation remains risky since some (expensive!) investments made may become useless within a rather short period of time. In the case of FDDI networks, for example, the early installations had to live with encapsulation type bridges which were early available but which had a severe drawback: they were non-standardized and vendor-dependent. Therefore, the only thing they had in common with components of other vendors was 'coexistence' but not 'communication'.
- Whenever a consensus is achieved after the first phase (i.e. at a point when a lot of different parties became interested and involved in the game) then this consensus has a high risk to be based on an enormous number of options (in order to 'satisfy the requirements' made by members of different groups). As a consequence, from a practical point of view there are few standards which really deserve their name since nobody is able to install the full version and to pay for all of the options. Thus almost all emerging products will be limited to a subset of a standard and it becomes less and less likely that 'standardized' components may really interoperate with each other.

3.4. Confusion by new announcements

Potential users become more and more confused about several facts:

- Standards are partly finalized (cf. the argumentation of the previous section).
- Products are not available or far from being perfect.
- Announcements are continuously made for new systems. The myth about these new systems is always that they are drastically better than the old ones and that they will make completely obsolete every system which has been installed in the recent past or which is going to be installed in, say, the next twelve months.

This confusion may be shown by a simple diagram (see figure 2). It's up to the reader to fill in some examples for concepts A, B, C; there are plenty of such examples in the networking area!

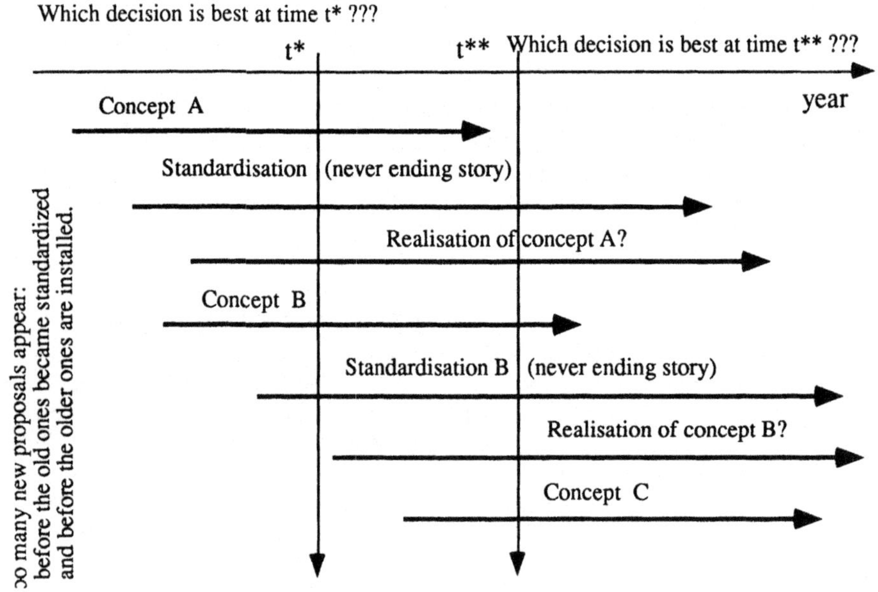

Figure 2: Confusion of users by too many announcements and too many promises

The consequence is that decision makers tend to postpone a planned installation for longer and longer times and that they prefer to live with intermediateor shortsighted solutions. This tendency is supported by the fact that most demonstrators or prototypes of emerging systems are presented in a way which is far from being convincing. Take for example the numerous demonstrators of multimedia systems: Most of them (to my opinion) will lead to a contraproductive effect; they will act in favour of a reduction instead as of an increase for acceptance willingness.

3.5. Infeasability of 'isolated' decisions

It becomes more and more difficult to install and to coordinate a high performance network. This is due to the fact that such a network is very costly and it may only be justified by a global effort of different institutions who want to profit from a shared access to such a network. A corresponding decision has to be made by a large group of people (and decisions of large gremia are ncessarily impossible or - at best - boring due to Parkinssoin's law).Moreover, it is extremely costly and time-consuming to establish the global infrastructure for the high performance network. There are numerous examples for that, e.g. the administrative complexity of installing a backbone network of type FDDI in a large institution. [If something goes wrong then a lot of criticism is readily available and unavoidable; if, on the other hand, everything works perfect - this is a very rare situation! - then this will be considered as completely normal and as trivial].

Another example for the increasing complexity and difficulty of high performance networking is visible from todays expectations for ATM networks: Whereas the ATM concept has been investigated for wide area environments it may be observed that product developments and (what is even more important) the installations will be much earlier in the local environment. The reason is that in this context decisions are easier to achieve and network management will be difficult but (perhaps) affordable. This vision of a creation of different local ATM switches which might be combined together by global (ATM) networks looks reasonable. A significant risk of compatibility problems, however, is not to be neglected since it is far from being guaranteed that all those local ATM'islands' will follow exactly the same design criteria. Thus, interoperability without any problem is not to be expected.

4. Strategical aspects and conclusion

There seem to be (and this impression is probably not unrealistic) more handicaps against than driving forces in favour of both the installation and the acceptance of high performance networks. The majority of those handicaps is largely dominated by nontechnical aspects.

Whenever a fast breakthrough of high performnce networking is envisaged it must be prepared by providing suitable countermeasures against most or all of the cited handicaps. Most important is that - as in the case of Ethernet - a larger community of users has to be convinced to install a commonly agreed type of high performance network. This may be implicitly reached if one of the major vendors of workstations would decide to offer a standard ATM interface for an affordable price as a fixed part of the workstation. If sales numbers are reasonably high then the cost for such a standard add on feature would be acceptable (and it would be accepted by the user - since the user would gain a lot of new functionalities for a relatively small increase in cost).

In the case of Ethernet installations this strategy has worked perfectly well and the corresponding vendors became very successful.

For FDDI, until now nobody really tried a similar strategy - and the sad consequence is that FDDI installations are not very numerous, that even today adaptor units are ridiculously expensive and that FDDI networks might die away if more performant and cheaper network adaptors for other technologies will be offered in the future.

For upcoming high performance networks the situation is still unclear. Let's hope that the desired breakthrough may be reached by following the proposed or a similar strategy.

5. References

[DMS 92] Davids, P., Meuser, T., Spaniol, O.: *FDDI Performance. Measurements and Experiences od the FDDI network at Technical University Aachen.*
In: Tienari, M., Khakhar, D. (eds.): Information Network and Data Communication IV; North-Holland 1992, pp. 439-450.

[HH 91] Händel, R., Huber, M.N.: *Integrated Broadband Networks. An Introduction to ATM based Networks. Addison Wesley, 1991.*

[S 93] Steenkiste, P.A. et alii: *A Host Interface Architecture for High-Speed Networks. Manuscript A3 (pages A3.1-A3.16), in [DS93].*

IFIP WG 6.4 conference series on "High Speed Networks" (published by North-Holland):

[SD 87] Spaniol, O., Danthine, A.: *High Speed Networks.*

[DS 89] Danthine, A., Spaniol, O.: *High Speed Networks II.*

[SD 91] Spaniol, O., Danthine, A.: *High Speed Networksing III.*

[DS 93] Danthine, A., Spaniol, O.: *High Performance Networking IV.*

IFIP WG 6.1/6.4 conference series on "Protocols for High Speed Networks" (published by North-Holland):

[J 91] Johnson, M.: *Protocols for High Speed Networks II.*

[PGP 93] Pehrson, B., Gunningberg, P., Pink, S.: *Protocols for High Speed Networks III.*

Integrated Broadband Communication Networks
and Services (C-18) / V.B. Iversen (Editor)
Elsevier Science B.V (North-Holland)

Strategies for the Introduction of Broadband Networks and Services in Denmark

K.E.Jessen[a], K.Nørgaard[a], H. Pedersen[a] and A. Thomsen[b],

[a]Network Development, Copenhagen Telephone Co.
1 Teglholmsgade, DK–1790 Copenhagen, Denmark

[b]Executive Office, Copenhagen Telephone Co.
21 Noerregade, DK–1199 Copenhagen, Denmark

The paper describes some of the general aspects that a telco (KTAS) must take into account when making a strategy for the introduction of broadband networks and services on the business market. These aspects are: market demand, regulatory issues and different kinds of broadband technologies. Furthermore, the paper includes an overview of the actual Danish broadband activities and a description of a likely Danish broadband strategy for the evolution towards B–ISDN.

Keyword Codes: A.1; H.m; K.6.0
Keywords: General Literature, Introductory and survey; Information Systems, Miscellaneous; Computing Milieux, management of Computing and Information Systems, General

1. BACKGROUND

KTAS (Copenhagen Telephone Co.) is the largest of the regional operating telecom companies in the Tele Denmark Group. KTAS is making strategic plans for the evolution of the existing network infrastructure towards IBCN (Integrated Broadband Communication Network) together with the other regional companies in the Tele Denmark Group.

As a regional operating company KTAS has the following mission:

- KTAS will supply telecommunications services to liberalized markets in open competition.

- KTAS will supply telecommunications services to monopoly market areas on competitive terms, in order to further the continuous development of the Danish infrastructure.

- KTAS will supply communications systems, user equipment, and services related to telecommunications in open competition.

- KTAS shall on a commercial basis provide cable–carried television programmes and deliver end–user equipment and services which naturally relate to a cable TV network

The mission is integrated in KTAS' vision for year 2000, which states among other things:

- The target is to remain an important actor on the liberalized market for telecommunications and to deliver the telecommunication services, which are required in the future with the main emphasis on transport of information of any kind, any volume and any bandwidth.

- The customers are attracted by means of a switched public network with competitive prices and modern techniques.

- Focus will be set on profitable operations through careful and flexible investment planning, which means that networks and network functions are dismantled, when the services can be produced in a better or cheaper way through other networks.

The vision expresses a strong intention to implement IBCN, when the market and the technologies are ready. To do so KTAS participates actively in the process of preparing for the introduction of IBCN. A line of strategically chosen activities within the framework of RACE, ESPRIT, EURESCOM, CCITT and ETSI together with national Danish trials are being carried out. The list of activities focuses on: ATM technology, TMN of broadband networks, Interworking, MAN–networks and markets, Vendor survey, Protocols, AAL (ATM Adaptation Layer) and CBDS.

The company's strategy for the implementation of broadband networks and services must be based on these guidelines and must be in line with the strategy and plans for the other regional companies within the Tele Denmark Group.

The paper describes some general aspects a telco (KTAS) has to take into account, when preparing and choosing a strategy for the implementation of integrated broadband networks and services (IBCN). These aspects are:

- Market demand for broadband applications
- Regulatory issues
- Communication technologies in the broadband area

The paper takes a market–oriented viewpoint, or "user view" as technological issues are treated thoroughly in other papers of the conference.

The paper is to a high degree based on the results of the ESPRIT–project MAXI, where KTAS has been project manager of the market survey and analysis. Activities within ongoing RACE–projects (with Danish participation) and the Danish BATMAN–project are mentioned.

2. MARKET AND INDUSTRIAL EXPLOITATION

2.1 Broadband networks and services

Most of the European telcos have the following long term vision: to supply an integrated broadband service (supporting all kinds of bandwidth) on a single network termination point. Broadband ISDN should be the network infrastructure supporting these service requirements and the network should have no geographical restrictions. The services may vary from telemonitoring demanding only few hundreds of bps (bits/sec) to interactive multimedia services demanding hundreds of Mbps.

This vision represents a goal far beyond year 2000, and there are many steps to decide and to take before this goal is reached. A first step towards a fully integrated B–ISDN goes through the business market, because the potential will most likely be found in this market segment.

In a long term perspective the residential sector will also be important mainly because of CATV–distribution to residential customers. This market, however, is still very premature and will not be dealt with in this paper.

The focus is on the business market, where broadband services may be used to interconnect local area networks.

2.2 The end–users' need for sophisticated broadband applications/services

The 1970's and 1980's have seen heavy investments throughout the world in IT equipment. The use of IT was expected to bring efficiency to operations and save costs in the organizations. However, in the last few years focus has been on the fact that this potential does not seem to have been realized.

This makes the sale of new equipment and related issues such as telecommunications investments rather more difficult: economic demands for investment justifications are becoming tighter. In the 1990's a cost effective budget is a prerequisite for new investments. This change of focus affects the potential for broadband demand.

The computer industry is constantly developing new equipment and new applications to meet a wide varity of needs. But as the budget tightens, the penetration of more far–fetched innovations tends to slow down.

Broadband investments must be justified by cost effectiveness in the end–user organization. Cost effectiveness can be defined in terms of many different factors:

- savings of staff time
- faster response speed or improved cycle/processing time
- ability to develop or offer new products and services
- as a basis for or an aid to the restructuring of operations
- reductions in current operating costs
- improvement in the current use of assets – tangible or intangible
- competitive edge
- improvement in staff morale and loyalty

The factors emphasized vary with the level of decision–making: some factors are strategic, some more short–term. But a trend is clearly observable: short–term cost reduction is becoming more and more important.

Broadband communication opens up a whole new world: new ways of working together, new ways of storing and transmitting data and images, new working procedures, new computer architectures. In this respect it must emphasized that factors such as end user organisation and its traditions may be of crucial importance for the success of a new technology.

A method to foresee and create broadband applications to fulfil the customers' needs is of great importance. The EC–RACE programme and all its different projects aim to develop new applications using the potential offered by higher bandwidth. A general model has been developed for describing the applications at a generic level in order to define "core" applications that could be customized without too much effort – and thus without too much cost. This general model also deals with operational requirements in the generic applications, allowing them to be mapped on to services.

A classification of potential applications is illustrated as follows:

1. Communication needs which could be met by earlier and older infrastructure

2. Bearer service requirements to build private services

3. Data–only applications

4. Complex applications like multimedia.

Looking at the timescale for the four types it is clear that Option 1 is a long–term possibility, whereas Options 2 and 3 are possibilities for short–term demand. The timescale of Option 4 is rather more difficult to judge, but it is unlikely that a high demand for this type is developing on a short–term basis. At any rate, in the RACE programmes and in the user population there is no evidence that the more advanced applications will drive a demand for broadband within the next five years.

2.3 Applications driving broadband development

The future use of broadband communications will be determined by the applications available. In an initial phase, and among early users of broadband services it is expected that simpler applications will drive the demand. As the early users will be among the organizations already using advanced telecommunications, and therefore among users who have a good working knowledge of networks and are able to handle their own needs, a rapid growth in the demand for value–added services is not expected. Instead the demand for bandwidth, protocols and interfaces will dominate.

Telephone interviews have been carried out with 400 large organisations in the 4 largest European countries. These companies were among other things asked to clarify their use of applications requiring high speed communications. The most frequently used applications were transfer of large amounts of data – between LANs or between computers. The main applications used are described in more detail below:

Transfer between LANs is likely to dominate the picture, since the importance of LAN–based architecture is growing in the end user organisations.

Interactive applications over the WAN is very important for a potential demand. Today many customers are not familiar with this facility, since it requires a certain amount of bandwidth, especially when the transferred data is of a substantial volume. It may take half an hour to transmit a CAD file over a 9.6 kbps line, less than five minutes over a 64 kbps line, but only few seconds over a 2 Mbps line. For real–time transfer 2 Mbps is almost a must – even with smaller files, if there are many users on the network.

Large file transfer is another area where 2 Mbps could easily become a bottleneck if interactive transfer is required. Large files vary from 1–2 Mbytes to close to 100 Mbytes. Scientific environments feel strongly that as soon as three or four users try to work with interactive transfer of files of 1–2 Mbytes, the 2 Mbps link becomes a bottleneck, so that seamless interconnection will require broadband.

Access to super–computers is another field where broadband connections are needed. Super–computers are used to process very large amounts of data, often for simulations. Today these masses of data are often transferred by means of magnetic tapes since WANs are too slow for effective transfer. Only very specialized calculations are done on super computers, but this situation may change. The use of high power computing systems is one of the central concerns of the R&D policy in the Community.

Multimedia – combinations of voice, video and data – is an application now entering the market. At present it is primarily appearing on stand–alone systems, but it is likely to grow quickly – first over LANs, then over WANs. While simple video conferencing, thanks to compression algorithms, will have sufficient quality for a long time over 2 Mbps lines, the combinations of the different types will require higher bandwidth.

Table 1
Use and expected use of applications

Applications	Today No. of answers	3–5 years No. of answers
LAN to LAN	131	251
File transfer	463[1]	633
Distributed file sharing	102	169
Remote backup	84	160
Disaster recovery	102	173
Multimedia	32	132
Access to super–computers	53	67
Respondents	**330**	**375**

[1] File transfer: Engineering files, Printing/publishing files, Medical Images and other large data or image files.

3. REGULATORY ASPECTS

From the point of view of an independent consultancy firm (Analysys) the regulation in the 1990's may happen as follows:

3.1 The short–term (1. half of 1990's)

The telecommunication sector is expected to focus increasingly on cost reductions and greater efficiency. Public networks will not be sufficient to sustain major growth in new services, but growth in the demand for basic telephony will in general enable continued growth in revenue for telcos. Where cost–cutting measures are introduced, substantial improvements in profitability and cash flow can be expected. Investments relative to turnover will fall and equipment manufactures will come under increasing pressure to deliver greater functionality at lower overall cost to a slowly growing market.

The increasing financial health of most telcos and the changes in network economies makes it feasible for the Commission of the European Community (CEC) to begin the process of removing existing service and network monopolies and to introduce widespread competition. Competition can be expected to emerge, but it will be focused on specific areas such as long–distance and international services, where larger differences exist between tariffs and cost.

3.2 The medium–term (2. half of 1990's)

The medium–term represents the turning point for European telecommunications. During this period it will become clear, that the long–term growth in the sector depends on new services rather than continued growth of basic telephony.

Competition is expected to increase, but its impact will be limited partly because of the inherent advantages of the telcos and partly because of tariff regions. Telcos which have been able to invest in the network architectures and equipment capable of supporting new services will have an advantage over competitors attempting to gain market-share.

Studies have shown that advanced voice services, narrowband and broadband digitally based network access and mobile services all become increasingly important sources of revenue. Simple voice telephony will therefore become proportionally less significant.

The rate of growth of the market hinges on regulatory reform, particularly the introduction of competition. If regulatory changes are not put in train before 1995 both competition and growth will be held back.

3.3 The regulation on broadband networks

The CEC has made the first initiative to regulate the broadband network market by starting a research programme (using independent consultancy firms) dealing with ONP (Open Network Provision) for broadband networks. To be more exact the research programme deals with ONP for:

- MANs, Frame Relay and B-ISDN
- the Local Loop of public Telephone Networks

The ONP will enable other carriers than the national telco to offer eg. MAN-based services on lines leased form the national telco. To a telco this opens the possibility to enter a foreign market on terms of competition.

4. COMPETING TECHNOLOGIES IN BROADBAND COMMUNICATIONS

4.1 N-ISDN

The Narrowband ISDN has been commercially available in Denmark since the beginning of 1992. This has catered for the long hidden need for switched capabilities with higher speeds than modem communication. The N-ISDN is being deployed throughout Europe and it will allow a unified access to the data services network.

4.2 Frame Relay

Frame Relay is offered either over N-ISDN or over leased lines to the Frame Relay Switch. Equipment is now available to run Frame Relay up to 2 Mbps.

Frame Relay has a low complexity and proves a good price/performance ratio compared with other technologies. Its major importance is the technical ability to support, on a bearer service, data rates in the area 64 kbps to 2 Mbps.

4.3 MAN

The MAN service has been commercially available in KTAS since 1992. It is only offered on special request, however. Many MAN trials have shown good results, and the SMDS service offered in the USA is offered on MAN (DQDB) access lines. The SMDS has a European counterpart in the CBDS–service. This service is standardised by ETSI.

In 1994 fully standardised MANs are available with the CBDS service. By the end of 1994 IWUs will be able to map completely the MAN and B–ISDN standards. This is ensured through the KTAS leadership of the RACE project COMBINE.

4.4 B–ISDN

The B–ISDN uses the ATM switching technique. The cell based access is like the DQDB MAN a new technology, which demands a new type of terminal equipment. The success of this technology is highly dependent on the availability of this equipment.

KTAS estimates that ATM based equipment for local area networks will be commercially introduced in 1993 and it will be installed broadly during 1994. This will replace much of the existing LAN equipment.

4.5 Matching demand and deployment

The demand is a result of a hidden need and the availability of reasonable prices. We estimate that the need has grown and is growing, and this need is far higher than the actual demand, which has shown up so far.

The availability of equipment during 1994 will trigger the demand in this period. So we estimate a demand already in 1994 for services beyond 64 kbps and see Frame Relay as a matching deployment to this demand. Demands beyond 2 Mbps can be catered for by the MAN technology.

This will form a transition to the B–ISDN, which together with our plans to introduce broadband bearer services, will ease the introduction of B–ISDN with its ensured market share at the time of deployment.

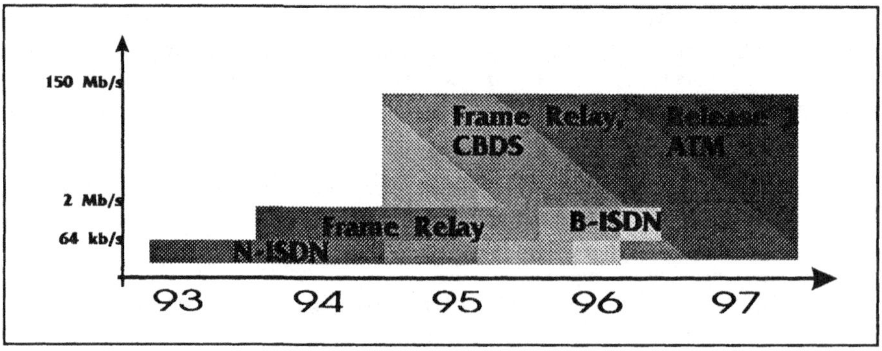

5. DANISH BROADBAND ACTIVITIES, PROJECTS AND TRIALS

In order to build up and consolidate a Danish platform for the introduction of broadband networks the operating companies within the Tele Denmark group are participating in the following main activities (The projects/activities are described in detail later during the conference):

5.1 BATMAN (Broadband ATM Access Network)

A collaboration between Danish operators and industry to gain knowledge and skills about technologies and topologies, which will the basis of a coming B-ISDN broadband network. The project implements a trial network consisting of 3 ATM-nodes.

5.2 PREPARE (Prepilot in Advanced Resource Management, RACE-2004)

The project defines and demonstrates an architecture for management of inhomogeneous broadband network based on TMN-standards and realised net components. The main objective of PREPARE is the definition of an inter-domain communication management architecture required to support the creation and control of end-to-end user services, including the specification of interfaces.

5.3 COMBINE (Composite Broadband Interworking and End-to-end Models, RACE-2032)

COMBINE investigates the interworking of ATM-based broadband networks with other networks foreseen to be present in future broadband network scenarios. Therefore COMBINE's primary objective is to create a conceptual and technological foundation to enable separated broadband islands to work together in providing broadband services, including connectionless data service.

In particular COMBINE will specify interworking strategies with respect to quality of service, level of interworking, routing, addressing, resource allocation and management. The interworking strategy would be validated through design, development and demonstration of heterogeneous networks, functioning on a selected testbed.

5.4 EXPLOIT (Exploitation of an ATM Technology Test-bed for Broadband Experiments and Applications, RACE-2061)

The main goal of the project is concerned with aspects related to the introduction of ATM equipment and services. The project considers the exploitation of an ATM technology test-bed for broadband experiments and applications.

EXPLOIT will develop interworking capabilities between ATM and N-ISDN, Frame Relay and transport of ATM cells over existing transmission systems. The ATM platform in EXPLOIT will be used for traffic experiments on Traffic Control, bandwidth management and performance.

5.5 PRISM (Pan European Reference Configuration for IBC Service Management, RACE-2041)

PRISM's main objective is to investigate TMN reference configurations for the management of IBC services, the associated resources and the user control of those management services.

From its definition of the requirements for service management, PRISM will construct a reference configuration. This is intended to be fully compatible with the overall integrated service engineering (ISE) architecture, and work already carried out in the TMN and intelligent networks (IN) areas by a number of RACE I projects and standardisation bodies. Results will also be validated by a number of case studies.

5.6 Standardization

Furthermore the Danish operators are participating actively in standardization within CCITT (SG XVIII and SG XI) and ETSI (NA5). The activities in NA5 and SGXVIII are concentrated on data communication and service specific aspects in the AAL– and SAS–groups, especially the AAL5 and AAL 3/4 standards and on traffic control and resource management in the TCR–group.

KTAS is active in standardisation of public and private equipment, and the crucial points are to achieve simple standards for equipment as well as for resource utilisation. In this respect KTAS has been successful influencing the AAL5 standard as well as standards for resource management.

6. OUTLINE OF STRATEGIES FOR IMPLEMENTATION OF BROADBAND NETWORKS AND SERVICES

Based on the results of the market research and the other projects and trials in Denmark the likely strategy for the implementation of broadband networks and services could be:

- A MAN DQDB trialnetwork in Copenhagen (the first in Europe) will in 1993 be upgraded to SMDS and opened as a public service.

- MAN networks will migrate towards ATM

- An ATM–based B–ISDN public network will be introduced in 1995 with the possibility of CBDS and Frame Relay.

- From 1997 an integration is expected.

The B–ISDN will be available for deployment in 1995. Since at that time it will be compatible with the MAN equipment, access lines will already be installed to support B–ISDN. In a first phase the existing bearer services will then be supported: CBDS over ATM, and Frame Relay over ATM.

Interworking should be used between MAN and N–ISDN (Frame Relay) and B–ISDN to ensure that the same services will be available to the customers in the future. In this way there will be no sharp transition for the customers.

It should be ensured that there will be no sharp transition for the customers, by interworking so that the will be the same services are available in MAN and N–ISDN (Frame Relay) and B–ISDN.

With the B–ISDN capabilities, the customers will moreover be able to implement integrated multimedia as well as fast connection oriented data transfer.

When release 2 ATM equipment is available – including statistical multiplexing features and separated call/connection control – it is time to use the ATM network for telephony and thereby exploit the full advantages of ATM.

On the user's side the operators should support initiatives to create new applications of the services. Important actors in the first phase will be universities, where an inspiring and creative environment may be the basis from where new ideas spread in society, and also selected large customers with specific needs for broadband services.

7. MAIN REFERENCES AND LITERATURE

A.Thomsen: The European Broadband Market, December 1992 (ESPRIT II 5193: MAXI)

P.M.Nielsen: Broadband ATM Access Network (BATMAN)

C.Gaarn–Larsen: COMposite Broadband INterworking and End–to–end models (COMBINE)

L.B.Sørensen: PRE–Pilot in Advanced REsource management (PREPARE)

Integrated Broadband Communication Networks
and Services (C-18) / V.B. Iversen (Editor)
Elsevier Science B.V (North-Holland)

Scenarios for Introduction of Broadband Services in Telecommunication Networks

Rolf B Haugen and Bente Mannsåker
Norwegian Telecom Research, P.O. Box 83, 2007 Kjeller, Norway

Abstract

There are large activities worldwide to specify and make experiments with broadband networks and services. At the moment most people believe that the broadband network of the future will be based on ATM-technology, and switching manufacturers and research institutions are making detailed plans for how ATM *technically* is to be incorporated into next generations systems and networks. But the crucial question for the PNOs is how to *economically* introduce a B-ISDN concept; it is a fact that, except for CATV, there is really no broadband service that at the moment is attractive enough to pay for the enormous investments necessary. We discuss in this paper two types of broadband services: *video related services* and *data network services*, the former type being the traditionally discussed end user services, whereas the latter are the *aggregation* of services from several users connected to various types of local networks. We argue that the data network services will be the driving force towards a broadband network and discuss some scenarios from broadband islands towards a B-ISDN.

Keyword Codes: C.2.0; C.2.1; C.2.5
Keywords: Computer-Communication Networks; General, Network Architecture and Design, Local Networks

1. Introduction

Broadband services have been very much discussed, and even experimented with, during the last couple of decades. Some services, like cable TV, have found their implementations in dedicated networks. Others, like picture phone and video conferences, have been introduced in the public telephone networks by digital coding (compressing) of the video signals to smaller bandwidths. But, except for this recent breakthrough in compression techniques, there is still no real broadband service offered in public networks that can be said to run on sound economical terms.

In Europe, there has been a tremendous effort the last five years to specify a public broadband network as well as to initiate development of components and subsystems that would make such networks cost effective. Several long term EC project have been established to this end; RACE being the most influential within the telecommunications area.

By now, the evolving broadband network, B-ISDN, based on ATM technique seems to be rather well defined and agreed upon among telecom operators. But one thing is *specification* of a network, another is *cost effective* implementation; i.e. to prices that subscribers can afford. Needless to say, the subscriber loop being the overwhelming bottleneck.

Optical fibers in the subscriber loop are considered to be a prerequisite for broadband services. For *business users* fiber connections are to day, economically spoken, within reach; in fact, medium and large companies might already be connected to the public network via fiber cables. This is by no means the case for *residential* subscribers and small firms. For these categories the cost of installing optical fibers are still an order of magnitude larger than with ordinary copper cables. And for residential users, most PNOs (Public Network Operators) are for political/regulatory reasons excluded from offering the most popular broadband service of today, namely cable TV. An integration of TV and tele services on the same cable might have rendered a fiber optical solution into residential homes cost effective in the near future. The RACE project TITAN is developing computerized tools for calculating (life cycle) costs of optical access networks.With this tool the network planners can extract detailed estimates of total and partial fibering of the local loop as a function of the years up to and beyond year 2000. In the computer program there are forecasting models for service demands and prices of the various network components.

Services that will be driving forces for the PNOs to start implementing broadband networks can be dividing into two classes

- video related services
- data network services

The first class contains services of the type: video telephony, video conferences, TV distribution, multimedia communications, etc. These services typically direct themselves to *one*, or at least a limited number of, end users. To offer video phone or TV distribution services there must be a broadband connection all the way to the end user. That is what makes these types of services hard to offer in the public networks.

The other class of services is of the type: LAN-LAN interconnect, high capacity main frame communication (Cray) etc. These type of services direct themselves towards clusters of users that are hooked on to local networks, hence the "users" are often LANs. And it is the *aggregation* of traffic that makes need of broadband solutions rather than the demand from single users.

To begin with, attention was directed towards the first class of users; in fact, the very notation of B-ISDN somehow leads our minds towards an end user concept. Hence, most broadband services that were experimented with throughout the eighties were of this type, and there was, and perhaps still is, an increasing frustration for not being able to bring these services to the customers to prices they would be willing to pay.

But it is the second class of services that presently "takes off" and requires solutions. Up to now, inter-networking solutions have mostly been provided by *bridges* and *routers* that are interconnected by leased lines. More recently, solutions like Frame Relay, MAN (Metropolitan Area Network), ISDN and even ATM are being discussed.

From a PNO's point of view, the latter type of solutions are by far the more interesting. Could intra-LAN traffic be transported in the switched (public) network? And which type of the above mentioned network solutions have the potential to evolve into the broadband network of the future?

These are the questions many PNOs are discussing today. They all know that there are tremendous revenue potentials in the multimedia revolution that will eventually come, a revolution that is just waiting for cheap network solutions. On the other hand, the data communications area is now an area of competition, and, needless to say, cost effective solutions always have to be found as one proceeds.

2. LAN interconnections

During the last few years we have witnessed a dramatic transition from centralized, proprietary main frame solutions to distributed network solutions with an ever increasing number of interworking PCs or UNIX workstations. The workstations have access to powerful servers over local area networks of the type Ethernet (10 Mbit/s), Token Ring (4-16 Mbit/s) and more recently FDDI (100 Mbit/s).

The number of LANs has thus increased and the traffic has exploded. The number of LANs is about 4 millions world wide and is growing at a very high rate: according to [1] the annual growth rate in Europe is about 40%.The growth in traffic is, however, not only due to the increasing number of users, but also to great extent a result of new applications with their need for transport of large amount of data and short response times.The shift from main frames to network servers enlarge this trend even further. This evolution creates a need for *effective* infrastructures that, in addition to handling the transport of data, satisfy the necessary requirement of *reliability* and *security*. In this multitude of needs and demands proprietary solutions like SNA will be faded out; *open standards* become a must.

We see two trends concerning LANs: Within larger companies and institutions there are need for splitting up large LANs into smaller segments, the reason being both of traffic and security origin. However, certain services, like administrative routines, will still require interworking between the segments. The other trend is to interconnect different LANs between various geographically separated departments or branches of a company, so as to build one large transparent network. Hence, in both cases the result is an ever increasing need for LAN interconnections.

The most common means of interconnecting LANs to day is by *bridges* and *routers*. Bridges, being the simpler of the two, works on the OSI link layer (2. layer) and are limited in routing and load sharing capabilities. Routers, on the other hand, operates on the network layer (3. layer) and works between LANs with the same *network layer*. They support various parallel paths in the network and has better security and load sharing facilities. There are continuous dialogue between the routers in a network. Information of the traffic load in the network will thus be known by all the routers and an optimal routing can be chosen. Needless to say, routers are more expensive than bridges. Figure 1 depicts a typical network with bridges or routers.

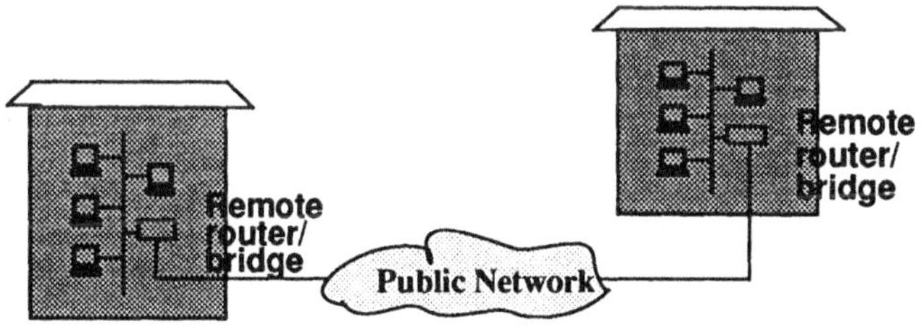

Figure 1. Two LANs interconnected via routers/bridges and leased lines.

Bridges as well as routers have been on the marked for several years and represents to day an important product segment for the data dealers. In [2] is made a survey of sold equipment in Norway up to 1992 and gives thus an indication of the potential marked for the PNOs to attack. The study indicates, however, that the need for capacity for the LAN interconnect is, at the moment, rather modest. This might change as soon as FDDI LANs penetrates the marked.

The traffic out of LANs is typically bursty, i.e outgoing line might be idle for a certain period followed by a period of peak traffic. Hence, leased lines might not be the most cost effective solution. As a result network solutions like, *Frame Relay, MAN* and *ATM* are being considered. In designing these networks, however, attention has to be paid to *delay* and *throughput.* The delay should be of the same order as in the connecting LANs. Usually the requirement is set to be of the same order as the delay between two bridged LANs; this delay admits an absolute maximum of one second and a mean value of 200 ms.

3. Network structures

In this chapter we will describe some of the network candidates for interconnecting LANs:

3.1 Frame Relay

Frame Relay is a new standardized connection-oriented data service where the data is transported in variable length, HDLC-based frames between end users. The service is intended to be a cost effective alternative to leased line for LAN interconnection, and might in the short term be the most attractive way of LAN interconnections. The service was first introduced in US and was followed by several European countries by the end of 1992. Norwegian Telecom will provide frame relay services from mid 1993.

The HDLC (High-level Data Link Control) carries, in addition to the variable data field, Data Link Connection Identifier (DLCI) that identifies the call and Frame Checking Sequence (FCS) that checks for bit errors. Since there is a dynamic allocation of bandwidth, the service is well suited for bursty traffic which is the typical traffic pattern for LAN interconnections.

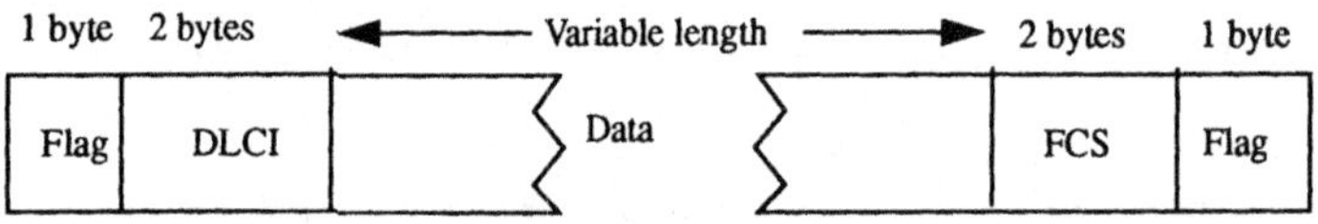

Figure 2. The frame format (HDLC) used for Frame Relay.

A Frame Relay network is made up of network nodes and user terminals, DTE (Data Terminal Equipment). The latter connects the user to the network. DTE can be a PC, bridge, router or host computer that is provided with interface defined for Frame Relay. The standard, which is described in both CCITT and ANSI recommendations, defines signal and data transmission at link level, OSI level 2, in the interface between user equipment and network.

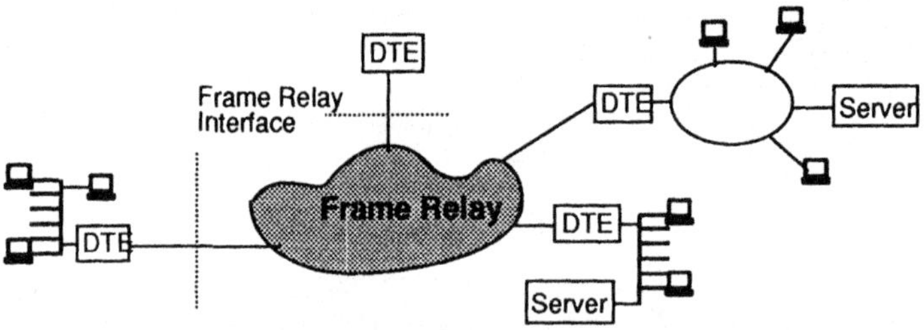

Figure 3. Local area networks interconnected with Frame Relay.

The first versions of Frame Relay work in a Permanent Virtual Connections (PVC) mode which indicates that the paths through the network are predefined by the network operator. When the DTE transmits frames to the network, the nodes will read the identifier, DLCI, look up in tables to find the right outgoing channel and send the frame to the next node (or end terminal). Use of several identification codes permits several parallel sessions in different directions to coexist on the same physical link. In this way a DTE can communicate simultaneously with different destinations over the same access to the network, a prerequisite for LAN-users.

FCS is used to check that the frame has been transferred correctly. Frames indicated as erroneous by the FCS are discarded; means of ensuring an error-free DTE-DTE transmission has to be implemented in the higher-level protocol. If errors occurs and are corrected, data must thus be retransmitted through the entire network. This takes obviously longer time than retransmission only over the link where the error occurred. Hence, the quality of the transmission medium plays an important role for Frame Relay. An approximate limit where Frame Relay is effective is set to a BER $<10^{-6}$ on individual links. This error rate can be met with speeds of several Mbit/s over most transmission links to day; a maximum link transmission speed of 2 Mbit/s is, though, at present defined. Higher speed (2-34 Mbit/s) is considered, but there is a question whether other broadband services, both connectionless and connection oriented, might be more suitable for this part of the marked.

3.2 Metropolitan Area Network (MAN)

MAN is in this context understood as a network consisting of subnetworks that use the IEEE standard 802.6, Distributed Queue Dual Bus (DQDB), as access protocol on a dual bus (see figure 4) which also uses principles defined by ETSI.

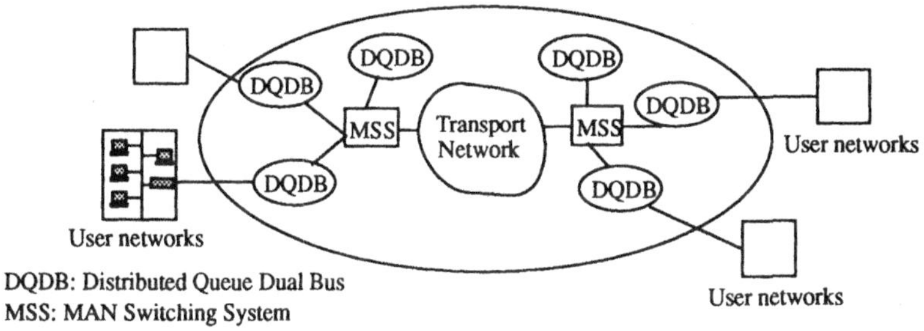

Figure 4. MAN architecture.

The DQDB protocol will eventually support both connectionless and connection oriented (included isochronous) services, at the moment only the connectionless one is specified.

The interest in connectionless service stems from the need to provide cost effective services with high throughput and low delay for bit rates higher than the ones provided by Frame Relay. Bellcore defined thus the Switched Multi-megabit Data Service (SMDS) that operates on US standards for bit rates. ETSI has specified a European counterpart which is named CBDS (Connectionless Broadband Data Service). CBDS offers bit rates of 2 Mbit/s, 34 Mbit/s, 139 Mbit/s and 155 Mbit/s, i.e. bit rates compatible with existing transmission hierarchy in the public network. The connectionless service features makes CBDS particularly suitable for LAN interconnections and might be attractive in a medium and long term perspective. As FDDI, with its nominal rate of 100 Mbit/s, is becoming widespread, the bandwidth requirement for LAN interconnection might come up in the region 30-60 Mbit/s which can easily be handled by CBDS.

CBDS being a service in MAN may also be offered over ATM. Offered in MAN gives regional coverage, offered over ATM gives unlimited coverage depending on the penetration of ATM. The service offers transmission of data units of variable length that tolerates variable delay.

Broadband connectionless services are offered in a number of field experiments in Europe and US. In US commercial service is on the point of starting. However, full exploitation will not be ready until a couple of years.

3.3 ATM (Asynchronous Transfer Mode)

ATM is a connection oriented cell-based technique which provides flexible bandwidth on demand up to 155 (or later 622) Mbit/s and integration of traffic with different quality of service characteristics. The data is transmitted in fixed packets called cells.

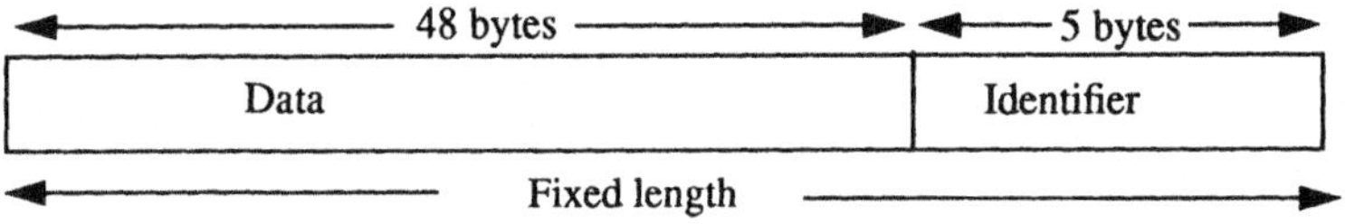

Figure 5. ATM cell.

It is believed to be *the* broadband technique of the future and seems to come quicker than previously believed. An early specification is contained in CCITT recommendation F.811 "Broadband Connection Oriented Bearer Service", but the various RACE-projects and now recently, the ATM Forum, are de facto standardization bodies.

The data manufacturers work actively through the ATM-Forum to have ATM specified according to their needs. Their interest, at present, lie in various LAN interconnects or even substitutions for ordinary LANs. To the largest possible extent, the Forum uses CCITT standards. However, in the case where no standard has been developed, they issue their own recommendation. For the physical medium they have, in addition to the CCITT's, made specifications for multimodus optical cable, shielded twisted pair and the plesiochronous hierarchy.

ATM interface cards for workstations have been on the marked for more than a year. In fact, small companies in the data business were formed to exploit the ATM development, Fore, MPR and Adaptive being three of the forerunners. The initial interfaces make use of the asynchronous byte transfer method supported by the TAXI-chip for the physical transport medium. It uses the same cables as FDDI and the same coding. This allow reuse of components and existing cabling. The manufacturers claim, however, that they also will introduce Sonet/SDH solutions for the physical layer.

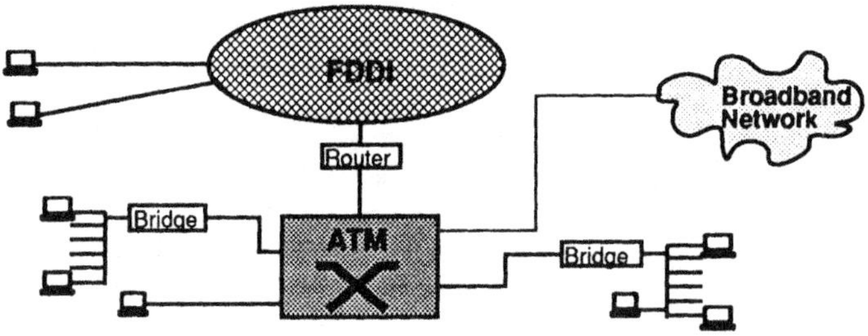

Figure 6. Use of ATM in local environment.

In figure 6 is an ATM switch used partly as a LAN interconnect, partly as a LAN itself. The work stations have inserted an ATM interface card based on the TAXI-chip and can thus directly communicate. The next step would be to deploy an ATM cross connect in the public network to connect ATM islands of the type above.

As an indication of prices: An ATM interface was initially (i.e. about a year ago) priced at 4500 $, this has now dropped to 1900 $ which indicates a take off in the marked. An 8x8 switch costs about 30 000 $.

4. Access Networks

As previously mentioned, the very concept of B-ISDN was based upon the idea of providing broadband services to the end users. Several possible services has been defined and experimented with, most of them related to some form of *image services*. Video Telephony and Video Conferencing are perhaps the most discussed services, but also high quality graphics and bulk data transfer will require large bandwidths. Another important area are services related to entertainment, i.e. TV, or HDTV, distribution, video on demand etc. Many of these services can be combined into the ubiquitous term: *multimedia*. This latter service might very well be the driving force of a B-ISDN of the future.

However, coding techniques have brought forward most of the above mentioned services over ordinary twisted pair that constitute the subscriber loop of today. Video phone services of 64 kbit/s or 128 kbit/s are already offered in ISDN, and even TV signals might be brought several kilometers over twisted pair in the experimental ADSL system of Bellcore. It remains to be seen whether the users will be satisfied with the quality thus provided in the long term.

The subscriber loop of the traditional public network is a multiple, usually triple, star configuration as depicted in figure 7. From the Main HUB (or local exchange) it typically fans out in a tree structure with hierarchical distribution points in form of mechanical racks of cross connects. The more modern network has in addition remote subscriber units (RSU) in form of switching machines, and in larger cities there might even be fiber optical terminating HUBs for providing high capacity links to business users.

Depending on the traffic demand, optical fiber cables might already today be cost effective out to the RSUs. But, except for large business users, fiber is not yet economical neither to the homes nor to the last distribution points (curbs). For real (non compressed) broadband services the transmission medium has to be optical fibers. Hence, fibering the very last mile out to the end users is the challenge for the PNOs the next decade, a challenge that has to be resolved before we can talk about a public broadband network!

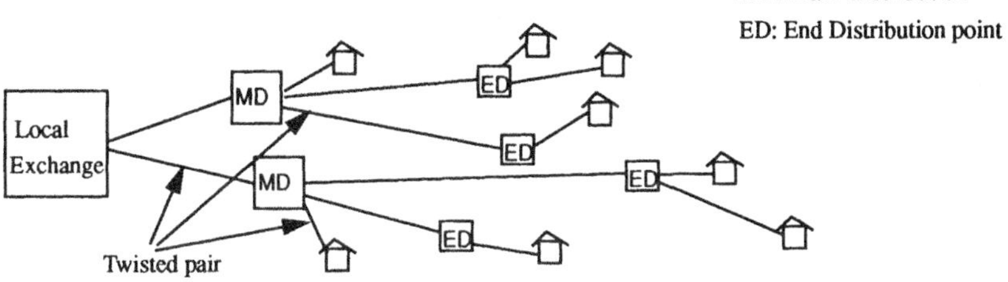

Figure 7. Traditional network structure of subscriber loop.

There are several ways to deploy fiber optics into the subscriber network. The most straight forward way is to replace the distribution points by optical couplers (splitters). A splitter is a pure passive component (crystal) that splits the incoming optical beam into n sub-beams, each with 1/n of the original intensity.The result is a network of the type depicted in figure 8 that is commonly known as PON (Passive Optical Network):

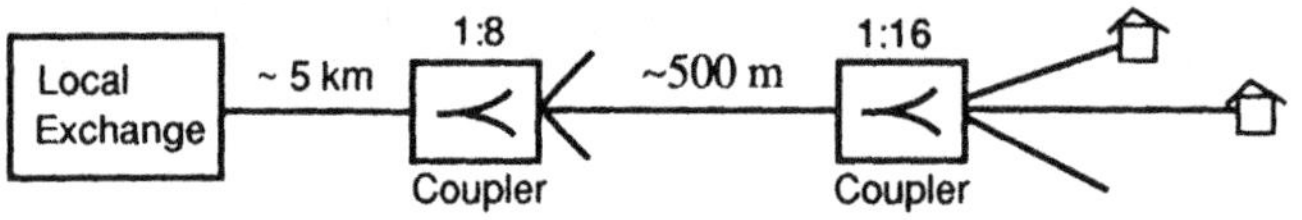

Figure 8. A typical subscriber loop based on PON.

British Telecom designed what they called TPON (Telephony over PON) overlaid the PON structure. TPON aims at being cost effective for ordinary telephone service [3]. Transmission to the subscribers is based on TDMA; the first versions typically had a 20 Mbit/s multiplexer in the local exchange. Two alternatives were considered: Fiber To The Home (FTTH) and Fiber To The Curb (FTTC).

In Europe Deutsche Telekom has several FTTC-projects in the previous East Germany that is very much based on the above mentioned principles [4]. Several FTTH pilot projects do also exist [5], but the cost of the *in-house* optical terminals is presently too high to give hope for even getting close to copper solutions. However, ETSI is now specifying digital interfaces for optical terminations inside local exchanges. This makes the multiplexers superfluous, and the optical subscriber loop more economical interesting; in fact, FTTC now is within reach!

In RACE there is an ongoing project called TITAN (Tool for Introduction strategies and Techno-economic studies for the Access Network) that considers various economical aspects of an optical subscriber loop for residential and small business users. TITAN's scope is the calculation of discount system cost, operation, maintenance and powering costs, life cycle costs and the overall financial budget. Their aim is to compare different optical access networks as well as those of the ordinary twisted pair network of to day. The costs of all network components are calculated using the extensive, integrated cost database developed within RACE, containing data gathered from many European sources. The database includes unit prices for components, installation, civil works etc., and the calculations are performed for various network structures, depending on subscriber density, splitting ratios, number of splitting stages and coupler locations. Assumptions are made concerning demand for different demographic situations, number of users and traffic. The services considered are: N-ISDN, broadcast TV channels and narrowcasting channels offering tele-working, tele-shopping and tele-interaction (tele-learning and tele-entertainment.

TITAN plans to demonstrate cost trends for the various network block elements over the study period 1995-2005. They will also propose a methodology for estimating operation and maintenance costs that allows cost comparisons of different access networks and enables the PNOs to study the effect of various tariff policies on cash flow. TITAN will finally propose a set of recommendations for a reasonable implementation policy.

5. Evolution towards ATM

There are obvious different paths to broadband networks. One thing is the willingness to pay for higher bandwidth, another is the demographical situation in each country or region. Clusters of universities in one and the same urban area will probably require different intermediate term solutions than do small companies scattered sparsely over a wide area. However, the leading star is an ATM-based broadband network, and the PNOs need to make their own strategic way from the present network to B-ISDN of tomorrow, the challenge is to find the most optimal way.

5.1 Scenarios with Frame Relay

Step 1:

We know from [2] that a typical business user in Norway to day with branches in different geographical locations uses X.25 or leased line with capacity less, or equal to, 64 kbit/s to connect their LANs. Hence, there does not seem to be large demand for network solutions with access bandwidths larger than 2 Mbit/s for the next planning period.

Our typical user will furthermore have a PABX connected to the public network and possibly a need for a dedicated 2 Mbit/s service for video conferences. All these services are connected to the access network via a multiplexer (static or statistical). If the total need for bandwidth is high enough he might have a fiber cable to the local exchange where the different services will be directed to the appropriate network.

Since the need for data communication is basically between a limited number of cites (branches) with a bandwidth demand for less than 2 Mbit/s, Frame Relay will be an economical solution. And if a customer wants more random open communication with many other customers, it will be an economical solution to hire a logical channel via a *router service*. This will provide an open connectionless service for the customer. In fact, this may be a suitable solution for mail service.

In figure 9 is depicted a network structure that we believe to be a cost effective solution for the short and medium term for the type of business infrastructure that we have in Norway. And, as we will see, it has a structure that easily evolves into an ATM network.

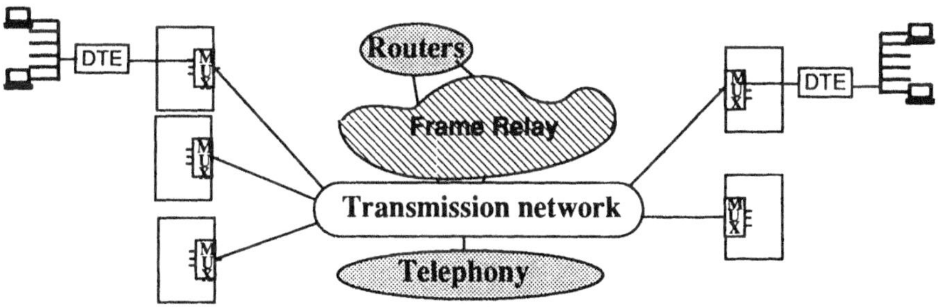

Figure 9. Frame Relay overlay: A first step towards a broadband network.

The traffic in the network will continue to grow as a result of a general increase in data applications but also from the increasing use of graphics and visualization in applications. This will gradually move the customer towards higher bandwidths and drive the networks towards their limiting capacity.

In the core network there will be a continuous upgrading with SDH based optical systems and cross connects for a flexible handling of the various services. SDH based solutions will probably grow further out in the regional networks and also in the local loop for medium and large business users.

Step2:

There will be an overlay of ATM nodes (cross connect and switches) in the core network. This is partly due to a demand for interconnecting FDDIs and local ATM-LANs, and partly a need for interconnecting Frame Relay networks in a flexible way. Figure 10 visualize the next step in the evolution. In this scenario the ATM network is transparent for the Frame Relay services. This means that the various functions for the service have to be performed in the Frame Relay nodes (FR), and ATM is solely responsible for transmission between the FRs. Conflicting situations between the overload controls in the two networks might, however, arise in this scheme.

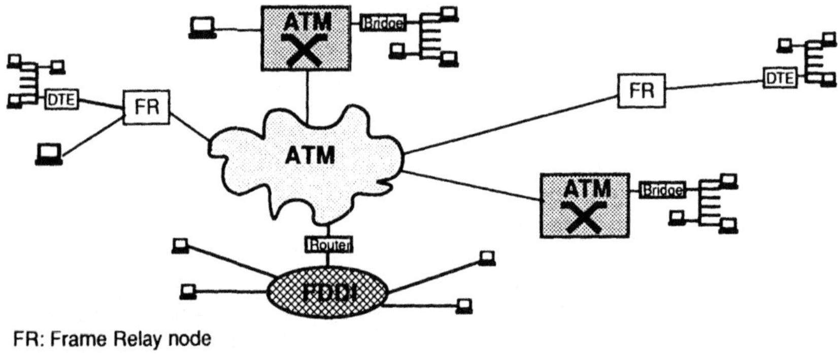

Figure 10. Step 2: Coexistence between Frame Relay and ATM.

Step 3:

The next step would be to introduce InterWorking Units (IWU) between the Frame Relay and ATM networks. This can be done by associate each DLCI with a virtual channel (VCC) in ATM. Then each frame relay connection can also be identified in the ATM network, the individual users can be given the same QoS and overload control can be harmonized.

Step 4:

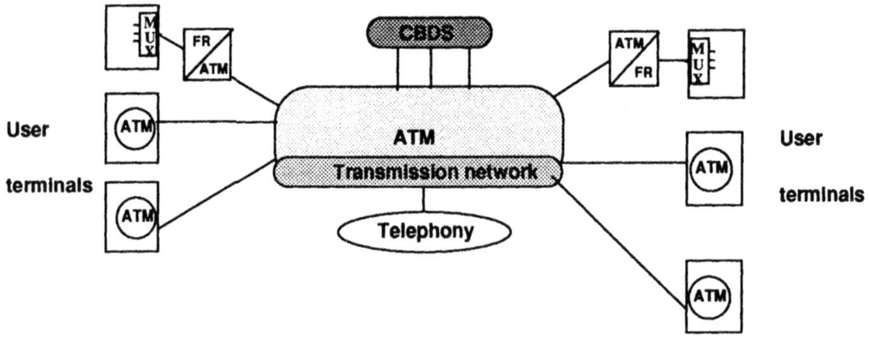

Figure 11. Step 4: Broadband network for business users.

In this stage the frame relay nodes are replaced by ATM-nodes. This situation will arise when the customers start to use new multimedia applications that needs larger functionality from the LANs. ATM based local area network will frequently be the interface to the public network.

Users that want to stick to Frame Relay will be provided with ATM adapters. Since both services are connection oriented and operate in a packet mode, such adapters are assumed to be relatively cheap.

The connectionless router based service can stay in the network until the traffic takes off. It might then be changed into an open CBDS-server. This service will have the same type of functionality as the router-service, but offer higher throughput. The combination of a permanent virtual link service for the main traffic, and an open connectionless service for the more random switched traffic gives a flexible structure for the early broadband networks.

It should be added that this scenario is assumed to be valid for business users. For residential users the situation is more unclear. There is, however, an ongoing development for ATM over PON. Whether such a solution will be allowed to carry TV-channels and whether it prove to be cost effective compared to coax cables remains to be seen.

5.2 Scenarios with MAN

Possible early users of MAN will be universities, high-technology companies, petrochemical companies etc. For these branches there are often use of common resources like super computers and distribution of new video based services from the universities.

Step1:

There are islands of communications using MANs. As the number of this centres in Europe increases they want to interconnect. The first step would be to interconnect the connectionless networks together using routers. Use of SDH based cross connect would belong to this phase. These leased line services can also be connected by SDH-based cross connects. The situation is basically described in figure 4.

Step2:

The next step would be to interconnect the MANs with ATM. In this stage the transport mechanism in ATM is based on permanent connections, i.e. virtual path VP. Use of VP does not necessitate add-on functions in ATM to interconnect the MANs. All the MANs have a logical connection between themselves in a meshed topology:

Figure 12. MAN interconnect via ATM based on virtual path.

Step3:

In this stage a connectionless (CBDS) server is introduced in the ATM network. The CBDS server in ATM has some of the same functions as the switching system in MAN; it can i.e. on the basis of an E.164 address route the connectionless packet through the ATM network to the user.

MAN also support a connection oriented service. This is only a leased transmission line service within a limited distance. MAN can also be used as an access network to ATM. The bus is then used as a concentrator of traffic from different subscribers that only have small traffic interest between them. This may be an early cost effective way to connect subscribers to a

network in high density areas

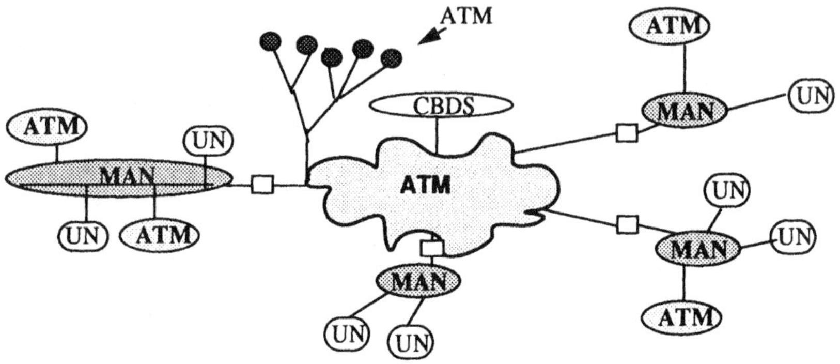

Figure 13. Broadband networks evolving from MANs.

Step 4:

The next stage would be a B-ISDN network based on ATM enhanced with CBDS. Although DQDB and ATM have the same cell structure the network structures and functions are quite different.When the users move to ATM based network with full switching capabilities, the MAN networks will be a bottleneck. It is therefore hard to see how a MAN in a flexible way can evolve into a general access network of B-ISDN. However, if CBDS is the only service requested, the MAN can serve as the access network.

References

1.B. Mannsåker et al, Eurescom P111, Overall Strategic Studies, Task 1.2: Telecomms service trends (1992).

2. B. Svendsen, High speed data services for LAN interconnection in Norway, Proc. IBC&N (1993).

3. D. Faulkner, Journal of Lightwave Technology, 7, No 11 1741 (1989).

4. H. Stegmeier, Progress in optical information technology, Proc. Telecom 91, Technical Symposium (1991).

5. H. Grotjohann, Fibre-to-the-home project with PTT Telecom: OLN-P (AT&T brochure).

Integrated Broadband Communication Networks
and Services (C-18) / V.B. Iversen (Editor)
Elsevier Science B.V (North-Holland)

Visions on the combination of SDH, ATM and LAN's as seen from the customer side

Ir. J.W.J. baron van Till

Consultancy Group Stratix BV, P.O.Box 488, 1180 AL Amstelveen, The Netherlands.
E-mail: vantill@dutetvd.et.tudelft.nl or: C=NL;ADMD=400NET;O=STRATIX; G=JAAP; S=VANTILL; I=JWJ

Abstract

The European telecom carriers have announced not to expect a short term need for the implementation of SDH and ATM. At least a need for mass market broadband applications or a change in public voice traffic is not apparent. Consumers, however, are absorbing IT much faster than the carriers realize. The private networking industry of large and small organizations has an enormous 'reservoir' of Customer Premises Equipment that needs to be linked with high quality broadband switched bandwidth. The three factors driving this urgent need are: informal 'networking' of people, organizational change into the structure of 'networked organizations' and 'digital communication inter-networks'. These three drivers combined are behind the present high business priority for cheap and reliable digital bandwidth which can be managed and scaled according to rapidly changing needs. This private networking is implemented by upgrading, combining and interconnection of present CPE. Analysis of demand trends and recent technical developments shows several critical paths in the voyage towards a broadband IT infrastructure: (A) Workstation-LAN-Ethernet-(Hub+PABX+Router)-Fr.Rl.-ATM-SDH-fiber-, (B) Hundreds of TV channels-fractal compression-Digital Video-ATM- ADSL-TVset, (C) Info services-X.25 network-MTA- city ISDN-dial X.400- Notes- Small business LAN / home PC. The effects on the telecom industry of these 'molecule chains' will parallel those in the computer industry: downsizing, distributed switching and commodity bandwidth.

Keyword Codes: C.2.1; K.4.1; K.4.3
Keywords: Computer-communication Networks, Network Architecture and Design; Computers and Society, Public Policy Issues; Organizational Impacts

1. INTRODUCTION

The 'customer' in the title of this paper can be the prospective user (of applications) of the new Broadband Network services as well as a possible buyer of the broadband equipment.
At present investment planning for telecommunication carriers and PTT's is like a chicken and egg dilemma. New broadband technologies are feasible, like ATM as a new service : non blocking variable bandwith switching, low latency virtual connection and SDH as the new transmission bearer [1]. Very large investments in new infrastructure, crossing public grounds, can however only be made if large scale demand is demonstrated which will eventually pay for its use. This demand is by no means obvious because carriers, especially the European PTT's[2], have a very selective way of looking at demand, experienced as they are with catering for the masses. Network customers *do have* bandwidth needs but behave contrary to plans. For example ISDN is used on premises (digital PABX) for voice and on long distance

networks for data. Just the other way around was forecasted! If we look more closely at the telecommunication demand it turns out that consumers *are* absorbing Information and telecom Technology (IT) on a large scale, although different from the carriers' plans. Fax, PC, modem, LAN, Internet, FidoNet and CableTV *are* very succesful. However, sometimes it takes very long. It should be remembered that also the prospects for telephony looked very slim for the first fifty years. Sometimes it strikes like lightning. Not many people realise themselves that their children live in the 'computer age', like they themselves grew up with pop music. Proof of that are about 130 million Nintendo and Sega game machines sold worldwide: stripped down PC's with a (hidden) datacom connector! So the real problem are the assumptions which were made on client behaviour in the computer and telecom fields. This paper tries to give a reality check to what is happening at the other side of the carrier termination points: the user side. No guarantees however are given . Although independent consultants can look over a lot of fences, Reality behaves very chaotic, non linear and with very complex chains of interrelations, and with many changes taking place at the same time. Simple, total solutions or 'magic boxes' do not apply anymore. *Combinations* of peopleware, software, hardware and interconnections do or don't flourish.

2. DEMAND SIDE TRENDS

First of all, it is not generally known that worldwide the investments in IT (computer and telecom) equipment on Customer Premises (CPE) is five times as large as the total investment in Public Networks. Carriers do not realise themselves that CPE is more than simple terminals: telephones, telex, fax machines. CPE does include computer equipment and software! The second important thing to notice is the long distance telecom traffic volumes. Datacom has yearly growth figures of 20-25 %, voice 4-5%. In some countries international datatraffic, including Fax, already exceeds international voice transmission volume.

Another important aspect is that the benefits of an infrastructure are generated outside the infrastructure itself, that is by the economic applications and uses which are made possible by it. Today we are living in the global networked economy. Unfortunately in Europe the EC did stress the economic importance of digital communication for the telecom industry only, and for a nervous system for its national governments. That is, for the supply side and for administrative uses instead of the economic effect on business and public. Roads are not only made for its builders and tax collectors. Recently the EC telecom policy does acknowledge that 'Telecommunications' are the motor for social and economic activities and competitive edge. As part of the European Growth Initiative the European Investment Fund (EIF) must guarantee infrastructural investments in trans-european networks. This might bring us closer to the United States of Europe: U.S.E.!

In the U.S.A., the emphasis is on the invigourating effect that an improved IT infrastructure can have on schools, education and new knowledge intensive business ventures.

The National Data Highway project plans to digitally connect *all* houses and businesses, within 3- 5 years, by telephone- and cable TV carriers. This must give all USA citizens access to digital telecom, information, knowledge and education and it is within reach because of its obvious competitive advantages for the US society and its economy. Also the Japanese economy will be boosted by investing in telecom infrastructure improvements.

2.1. Focus on organizations

The driver for any technology use is what people actually DO. Therefore it is significant to look at some recent changes in ways people work and in the way they work together. So we look at trends in business needs in the Upperworld of people, tasks, physical processes and movement of tangible objects, as distinct from the Netherworld of information- and communication systems. A recent buzzword is 'business process re-engineering' which is newspeak for streamlining the organization by taking a very hard look at the added value of all tasks to the company objectives. Such reorganizations now anticipate automation so we are observing the impact of organizational changes on IT now, instead of the other way around!

This generally results in segmentation of the company into a number of smaller autonomous business units / teams. Such business units contain all necessary skills to service their own clients and are empowered to make their own decisions. The objective is to be flexible and able to react quickly to turbulent external changes in the marketplace. The new holdings, planetary systems of such small units are called *'networked organizations'*, now held together by an electronic network, replacing several intermediary layers of bureaucracy, where slow vertical reporting took place, which was over-simplifying reality and by making ineffective decisions caused harmful stagnation. Staff is made redundant and non core business services are outsourced or privatised. This does mean that such service groups must keep very close ties with their former mother companies who now hire their services only when needed and if competitive. So electronic networks also bind such satellite service groups to several client companies.

In fact internal and external value chains of small units are woven into networks of companies. Actually we are no longer talking of large companies only but of small businesses or small units where teams, working 24 hours, spanning large global distances work together supported by digital networks, even teleworking from their homes or mobile during trips.The resulting structure has several paradoxial aspects:

* Organizational subsidiarity: some responsibilities are moved higher up & others are brought all the way down to where the work is done.
* Work is brought into the business units & outside the company.
* People, teams and units are independent and empowered to make their own decisions & must report results through the network to the holding every five minutes.
* Units do contain all that is necessary internally & maintain close external coöperation with clients & must be very well informed about the turbulent outside world.

So external (tele)communication is vital: close dialogue relations with emancipated clients are needed to offer flexible services and survive in the new mass-customization. This does also apply to Small Businesses and even shopkeepers. Telecom links are their lifelines! As Peter Drucker long ago [3] wrote: "There is only one valid definition of business purpose: to create a customer. It is the customer who determines what a business is. What the business thinks it produces is not of first importance- especially not to the future of the business and its succes. What the customer thinks he is buying and considers 'value' is decisive- it decides what a business is, what it produces and whether it will prosper".

In other words:

Objects depend on external links. Or:

The succes of an organism or organization depends on the quality of its external relations.

This is similar to Neural Networks: the function resides in the pattern of connections (relations) between nodes, not in the functions of the nodes themselves.

This is also reflected in the way people work. Besides problemsolving structures for decisions and specialisms, emphasis is growing towards peer-to-peer informal *'networking'* by people. This works by asking somebody who might know somebody (etc.) who can solve a particular type of problem. Within a few referals you can find the right person who shares the same interest or knowledge to solve your problem. This 'networking' is very much amplified by tele-communication networks. E-mail, telephone, fax, bulletin boards and computer conferencing systems on for instance Internet are used very succesfully for this type of activity. Not only do people talk and gossip through these networks but they also exchange objects: voicemail, documents, messages, videosequences, business forms. These selfcontained and selfoperative messages are the new type of 'info objects' which are exchanged through electronic networks.

2.2. Network requirements

In the Netherworld of digital networks, information is moved which represent physical movements or human activities in the real Upperworld. So info flow mirrors activity flow. For example airplane passengers or express courier parcels are 'traced and tracked' end-to-end by network computer systems. Along the way in several places the computer status of the passenger or parcel is checked with the actual physical status. A lot of new info flow tools are appearing and are implemented with high priority. Examples are:

* extensions of Mac/PC-LAN functions to WAN interconnection
* extensions of Personal Computing functions to inter-personal functions: collaborative computing, workgroup computing (example:Lotus Notes), document routing, workflow applications (event driven), database replication, smart-e-mail.

The requirement on the network is to give the access possibility for *any to any : person / file / application*. So the demand is to offer high quality ' information logistics'. That is the ability to deliver the right processible info at the right place at the right moment in time, to analyse it, manipulate it and forward it. Various new mechanisms are implemented to fulfil that objective:

* person to person file transfer
* client-client application interconnections like: OLE= Object Linking and Embedding, DDE= Dynamic Data Exchange
* PC - PABX smart connections
* internal and external EDI

Corporate IT policy is redefined away from hardware and software standardisation towards the main emphasis on connectivity between systems, interworking of applications, same meaning of data. In order to be able to 'reach and understand' people and data, corporate standards are enforced for API's and Middle-ware, which resides between application software and networks. The IT paradigm is shifting from System centered to Network centered.

What appears is the 'Enterprise-wide LAN' with network computing: cost effective interconnection of available installed systems. In fact management is very reluctant to invest in totally new large and complicated systems, but wants to protect their investments in present systems and digital networks, so they choose for the combination & upgrading of present equipment. They want the new IT- infrastructure to be OPEN to component changes, and want to mix and match component choices that are non-proprietary. This drive turns computer HW and SW into a commodity and it explains the succes of multiprotocol Routers [4] , TCP/IP and certain ISO protocols. This new *inter-networking* is characterised by the formation of a 'network of networks' and by computer intelligence on both sides of the network and in the network itself. The swiftness and power of this new vogue has surprised many in the IT industry. Actually, in stock value Microsoft is now bigger than IBM , and Cisco is now bigger than Digital Equipment, and still growing.

2.3 Some resulting short term changes in the digital communication landscape

As a consequence of the mentioned driving forces of: 'networking' of people, the internal/external weaving of 'networked organisations / teams' and the digital 'inter-networking' as well as the mutual amplification of these three, strong forces are exerted on the present IT infrastructure. New priorities are expressed in organisations. An example of this can be found in a recent study issued by Infonetics Research Inc. , San Jose, Calif. , quoted by Jayne Wilson [5]. Interviews with 100 firms with at least 1000 employees were held concerning their short term broadband needs. Early ATM uses will not be exotic applications such as live video or multimedia, but more conventional *business critical* ones: Client/server computer applications (rated important by 72% of respondents), File transfer (62%), LAN replacements (62%), Disaster recovery (55%), Visualisation (53%).

So the largest immediate demand for wide area broadband communication will come from the "water reservoir" of the CPE and will start ON Customer Premises: bandwith is nearly free there. The early customers for Wide Area broadband will be those with large CP's: university campuses, airports, Defense establishmens, Railways, Utilities and cities. They are capable to move directly towards a combination of SDH and ATM. The company WAN backbones must behave like LAN's (WAlhaLAN ?) and combine very bursty data traffic with streaming digital voice and video.
Other short term changes come from upgrading of *presently installed* public and private networks and pipes (OSI -1):

* LAN's are upgraded by segmentation and by introducing semi-ATM switching of connected Ethernets
* the new Ethernet switching 'hubs' are a combination of routers, T1/E1 Mux's, PABX's, and LAN servers (ATM in a box)
* 100 Mb/s is introduced for Ethernet on smaller segments of UTP (unshielded twisted pairs)
* IBM is experimenting with ATM (155Mb/s) on STP (shielded twisted pair) cables and has proposed for standardisation the MUNI: Mid-range User Network Interface which will run 25 Mb/s ATM on UTP3.
* for private WAN's, ATM Reverse-multiplexers are being introduced to be able to use a combination of private digital leased lines and public networks.
* In several countries cable TV networks have a very high penetration. For instance in Belgium and the Netherlands more than 90 % of homes are connected with coax cable for video broadcast distribution. This huge one way analog bandwith highway in the ground can be made available for two-way digital signals too. Some cableTV carriers have plans to add switched voice and data services. It is possible now to interconnect LAN's over the cable TV network. Digital Equipment and Zenith have installed packetswitching 4 Mb/s internets for such city-wide LAN's
* PTT's are installing ISDN2 x 64 Kb/s on UTP two wire local loops for subscribers now, which is especially appealing for PC datacom and teleworking. It is now possible to upgrade the use of local loop copper to even higher speeds with HDSL (high bitrate digital subscriber line, 4 w, E1/E1) or ADSL (asymetric digital subscriber line 2w, E1/16kb/s). Compressed digital video can be carried that way into homes on demand over the present copper wires
* another potential source of high volume datacom is mobile computing, which is the extension of LAN's and datanetworks with packet radio crossing the last mile to connect mobile users. TCP/IP and X.400 are used to inter-net these users into the rest of the installed corporate networks and installed base of applications, as if the mobile computers are in the office.
* the interconnection of the above, which will bring benefits for all concerned !

Not only digital bandwidth for voice, data and video is needed in quantity but also the quality (availability, bit error rate, delivery time, service levels) and price are important.
A lot of manufactureres and carriers complain not to hear hear about those user requirements, which indicates a very selective perception. For instance Herman Langhammer, Telecom manager of Du Pont recently [6] gave a very explicit account of business requirements for telecom which are not met today: global open interoperability and low operating costs. The national telecom barriers do protect monopolistic suppliers of services, but also hold back demand. Also the International Telecom User Group (INTUG) and the International Chamber of Commerce (ICC) have very outspoken views on the needs of users with respect to the quality and quantity of telecom services provided by carriers. These customer requirements can be formulated as :

Businesses NEED end-to-end (to the desktop) reliable, flexible and cheap digital bandwidth which can be managed by the customer and is scalable by adding links and switches.

3. MODEL

How will the above needs be met, what will happen if all the mentioned changes take place? To map this diversity and the changes in technology the following model is introduced. THE TILLEVISION MODEL for the IT infrastructure, a tool in practical use since 1988 for design of corporate networks in various variations (clustering of layers) to see in one 'map" the relation between people, telecom, datacom, computers, information and business processes. Also changes can be charted and thus ways can be designed to handle the changes and their impact on the total structure and its management organisation. The user organization and its IT Infrastructure is mapped according to the following layers:

Vertical axis: (extension of the OSI model) functional levels of:
org structure, work group activities, type of info use, user interface, application programs , data storage, processing O.S. / communication protocols, network transport,switching, multiplexing transmission, cabling, physical room/pipes, building/ sites.

Horizontal axis: area's of applicability:
for instance: room floor,building, campus, information service, computer center, local loop, city, country, city, local loop, home.

The resulting IT map of an organization can then be simpyfied by clustering into five vertical levels:

- people, tasks and physical processes
- application services
- end-to-end info exchange facilities, middleware
- switching, transport services
- transmission

and for instance four horizontal levels: headquarters, public grounds, sales office, home

4. CONSEQUENCES

If the needs and possibilities described in this paper are mapped on the introduced Model the analysis shows several immediate consequences.

* Not one uniform structure will prevail. There is not a "best" technique or 'box' which will solve all problems and will meet all requirements. SDH, ATM and DQDB are components which must fit into a designed architecture. In fact the vertical layer functions are complementary: For example ATM= variable bandwith switching, SDH= a transmission bearer. So we will have to live with a number of overlay networks. The crux is to interconnect those overlays well, which is very difficult, as shown by experiences with Intelligent Network services so far. We are faced with cultural differences too, like the old gap between 'switchers' and 'transmission guys'. Can we get the cablecloset dwellers, Macintosh believers, Cobol scratchers, router machinists, Unix freaks and PTT specialists to talk to each other? And there is that little thing called computer power, meaning corporate influence which emanated from computercenter khan's MIPS, UPS, GigaBytes and cool KW's. Now these powerful men want to have the best connections!

* Based on cost, speed and distance considerations the horizontal layer movements are divergent instead of convergent . More techniques appear, shift place and sometimes cluster into boxes like for instance hub's. An upcoming horizontal shift to (parallel) *distributed switching* is apparent, similar to 'distributed processing' which took the computer industry by storm.

* To make a prognosis on what will be successful in broadband growth we have to be aware of one of the law's of agriculture. According to Liebig and Blackman growth depends on the least available element among several factors. Improving plant conditions like adding water, light or fertilizer has a very non-linear effect, with the bottleneck changing place. The driving speed in a traffic jam depends on the speed of the first car. If that one speeds up another car will take over the limiting role. So not one ingredient, but a whole 'critical path' through the model must show promise. Examples of such molecule chains of multimodal network integration are the following combinations which will bootstrap the broadband industry. It are combinations of successful components cutting through and linking several horizontal and vertical layers of our Tillevision Model:

(A) Wkst -*LAN* - Eth.net -(switch Hub+PABX+Router)-FrameRelay-*ATM-SDH* - fiber -, see Fig.1.
Example: Cisco, Stratacom and AT&T seem to have learned from their customers who are applauding the happy marriage of *frame relay adaptation into narrowband ATM*. Such a combination is now also announced by Northern Telecom : the Passport Switch.

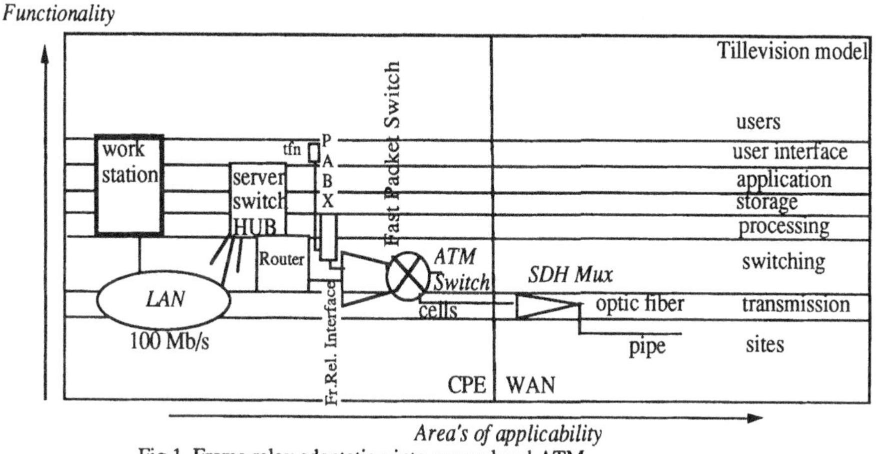

Fig.1 Frame relay adaptation into narrowband ATM ,
examples: Stratacom, NT, Cascade.

(B) Hundreds of TV channels-Fractal compression - Digital Video-ATM- ADSL.- TV set

(C) Info services-X.25 network - MTA - city ISDN-dial X.400- Notes- Small business LAN / home PC
Example: published plans of France Telecom.

This does not look promising for : optical fibers connected to workstations, FTTH, FDDI and DQDB.

4.1 Long term consequences

The interaction of the high priority telecom needs with the availablity of new improved digital broadband switching- and transmission technology will inevitably result sooner or later in the following developments:

* It will make access to people & info independent of time and geographic location. See also De Sola / Noam [7] for more and fundamental changes this will bring. Time indepen-

dance can mean without delay (at the speed of light) or real time-uncoupled by means of store and forward messaging.

* It will do more for trans-national cooperation than embassies and exchanges of ballet groups, as Paul Baran stated years ago, [8].

* It will make the networks that companies start to depend on, for their business and cooperation with their business partners, more available than their component parts , [8].

* Similar to recent history in the computer industry we will see competition on switching (distributed switching ; downsizing switches) and basic transmission infrastructure become inevitable. This will also make high quality bandwidth (driven by price/performance: better services for lower prices) an abundant and valuable commodity instead of a scarce resource. No longer the state will serve out line-provisions to grateful bandwidth-hungry users. A supply and demand market for digital bandwidth is arriving.
WAN tariff is 80% of corporate internetworking cost over 5 year period and as such a bottleneck for business.
Not only the cost levels but even more the economic importance for businesses of the uses of networks is becoming so high that not only the Telecom Managers become more outspoken and less forgiving on these issues but also their company chairmen start to discuss them with heads of state. Those state monopolies or market monopolies who resist this urge will be pushed aside or bypassed.
Those in the industry who are aware of customer requirements and are pro-actively going with the current of partnerships are and will be amazingly succesful as is shown by for instance Finland Post and Telecom [9].

* Actually the carriers are sitting on a broadband goldmine. It is possible if barriers and protections are removed to create a win-win situation where all concerned will benefit. After all, with the Liebig & Blackman law in mind, *interconnection = a synergy business*. It is to be hoped that Mr. Bangemann, the European Commissioner for Industry, will remove the barriers in this IT industry: all parties will benefit. Some more open cross cultural learning might help the European economy a lot : 'good roads create traffic '.

* Strategic telecom policy development on government level is still needed in the public interest to safeguard the following common benefits:

  I. Not to waste scarce resources - if bandwidth is no longer scarce, network management knowhow and flexible configuration skills are.

  II Safeguard end-to-end connectivity in a 'network of networks'.

  III Enforce quality improvement by promoting innovation and market choice and by protecting of civil rights.

If 'Integrated Broadband communication Networks and Services' will be at first used for small- and big 'business critical' applications (the third IT wave of professional network services) then on the horizon we can foresee the fourth wave: consumer / prosumer bandwidth use and demand, but that is material for another paper.

REFERENCES

1. R.J.F. de Vries, Switch Architectures for the Asynchronous Transfer Mode, Thesis Tech. Univ. Twente, Royal PTT Netherlands (1992).
2. P.Heywood, A.Roussel; Europe's PTTs: No Heat, No Hurry; DataCommunications, Dec. 1992, pp. 72-78.
3. P.F. Drucker, The Practice of Management, 1954
4. R.J. Cypser, Communications for Cooperating Systems OSI, SNA and TCP/IP, Addison - Wesley (1992).
5. Jayne Wilson, Users are jumping on the ATM bandwagon, InfoWorld, IDG, Feb.15, 1993
6. H. Langhammer, Du Pont, Germany; Conference paper, 3rd annual Networked Economy Conference, Paris, CommunicationsWeek International, March 1993.
7. I. de Sola Pool, Technologies without Boundaries: on telecommunications in a global age, edited by Eli M.Noam, Harvard University Press (1990).
8. P. Baran, Some Perspectives on Networks, IFIP, Computer Communications Conference, 1977, pp.459-464.
9. J. Heinänen, Experience of Using a Public Frame Relay Network, Finland Post and Telecom, 1993, personal communication, jh@datanet.tele.fi

Integrated Broadband Communication Networks
and Services (C-18) / V.B. Iversen (Editor)
Elsevier Science B.V (North-Holland)

Basic Concepts and Performance of High-Speed Protocols

T. Stock[1] and P. Tran-Gia

Institute of Computer Science
University of Würzburg,
Am Hubland,
W-8700 Würzburg, Germany
Tel.: +49 931/888-5510,
Fax.: +49 931/888-4601,
E-mail: {stock,trangia}@informatik.uni-wuerzburg.de .

Keywords: HSLAN & MAN access protocols, performance evaluation

In the last few years a large number of novel access protocols has been proposed for use in high-speed local and metropolitan area networks. On the one hand those protocols are dedicated for networks operating in gigabit-per-second transmission capacity local networks. On the other hand these protocols are also developed as access protocol to interface users of new high-performance services to broadband ISDN networks. Although normally, the protocol descriptions and operating modes are rather complex and spacious, a number of basic mechanisms can be identified as building blocks to design high-speed protocols. Examples of such mechanisms are 'distributed queueing', 'credit assignment', 'reservation procedures' etc. The main purpose of the paper is to give an overview on these protocol mechanisms and their influence on overall high-speed protocol performance. Further, modelling issues for this class of protocols will be addressed and examples for protocol models and their evaluation methods will be discussed.

1. Introduction

In the current standardization process of high-speed local and wide area networks a large number of protocol proposals have been presented in the literature. The diversity of these protocols can be observed from viewpoints of system structure, control mechanism and signalling schemes. On the one hand, various medium access protocols can be found as candidate for high-speed local and metropolitan area networks (HS-LANs and MANs).

[1]T. Stock is now with Alcatel STR, Zürich, Switzerland

On the other hand, to provide customer access to high-capacity wide area networks like ATM systems (asynchronous transfer mode), generic flow control (GFC) protocols have been proposed in the emerging standardization. Thus, a systematic treatment of these proposals, considering both architecture and performance aspects, is quite complex.

From performance analysis viewpoint various problems arise with this new class of high-speed protocols. Due to the large medium length and the high transmission capacity, packets appear shorter on the medium. An interesting measure to describe this phenomenon is the logical medium length. With C be the transmission capacity, l the medium length and c' the signal propagation speed, the logical medium length Λ is defined as

$$\Lambda = \frac{l}{c'} \cdot C \tag{1}$$

where l/c' is the latency of the medium. The measure Λ represents the number of bits (or rescaled in packets, cells, slots) which can be placed on a row on the medium. In a classical LAN environment with $l = 2.5$ km, $C = 10$ Mbps and $c' = 0.67$ c we yield Λ in the range of a single packet's length. Here classical queueing models and analysis methods can still reasonably be applied [18]. In a high-speed system, e.g. with $l = 100$ km, $C = 100$ Mbps and $c' = 0.67$ c the logical medium length increases by roughly three orders of magnitude. Thus Λ is significantly larger than the typical packet length.

Each attached station would observe another state process, since no time synchronization exists in the system. Therefore new decomposition methods must be developed to cope with this phenomenon, in order to obtain accurate analysis techniques. Due to new requirements in the performance evaluation process, such as delay variation and transfer jitter, analytical methods gain importance, since computer simulations will touch their limits sooner than in conventional systems.

In this paper we approach to present high-speed protocols in a more generic context. We first try to identify a list of mechanisms used as building blocks in the protocol design process. Subsequently we will take a few known protocols as a synthesis of these basic mechanisms and discuss them in more detail. To illustrate performance analyses of such high-speed protocols we will take as examples some DQDB-based protocols (distributed queue dual bus, for an overview see [14], for the standard see [7]) and the CRMA (cyclic reservation multiple access, see [15, 16]) mechanism and outline their performance models.

2. Basic Protocol Mechanisms and their Properties

In this section we will describe the basic protocol mechanisms, some of their basic performance properties and some modifications to overcome the characteristical disadvantages. These descriptions will be rather brief so we refer to the literature for a more detailed description.

2.1 Basic Architecture

We will consider 'linear' topologies in the sense that the transmission media leave the medium access unit into two directions. Two-dimensional topologies like Manhattan street networks are not covered.

The following topologies are mentioned quite often (see Fig. 1):

- ◇ **single directed ring**; perhaps the most basic variant. all nodes are connected to their neighbours via two unidirectional links, an incoming and an outgoing one. Should be one of the cheapest solutions since only one receiver and transmitter are needed.
- ◇ **dual ring**; two of the above ones melted together. The main characteristic is the fact that the second ring points into the opposite direction (therefore: dual, not double).
- ◇ **multiple rings/thorus**; several directed rings operating in parallel and all pointing into the same direction.[2]
- ◇ **folded bus**; since a single unidirectional bus segment would prevent some nodes from sending data to some other nodes, this topology may be considered as the basic bus structure for directed media. It consists of an outbound channel, which leads from the first to the last station and an inbound channel in the reverse direction. There are two variants: the first only provides a single transmitter on the outbound and single receiver on the inbound channel. The second adds a receiver on the outbound channel right in front of the transmitter thus allowing a node to sense the medium ahead of the transmitter.
- ◇ **doubly folded bus/double spiral**; similar to the folded bus including the two variants described above with the exception that outbound and inbound channel are connected through a link segment (see [22] for a description of the Expressnet topology). Therefore outbound and inbound channel point into the same direction so that the data passes the stations on both channels in the same order. The double spiral is topologicaly equivalent but omitts the link segment.
- ◇ **dual bus**; similar to the dual ring configuration where one segment was cut off. Note that in contrast to the dual ring here there is no free choice on which bus segment to transmit data to a specific node.

Architectures which are closed, like all the ring topologies have to be implemented in an active way. This means that the overall topology is based on point-to-point links between the nodes which have to forward data explicitly. In contrast open architectures like all the other topologies mentioned above can be realized passively as well since they provide

[2]Similar parallel structures can be implemented using other topologies, too. But since they are not used very often we will not mention them in the following.

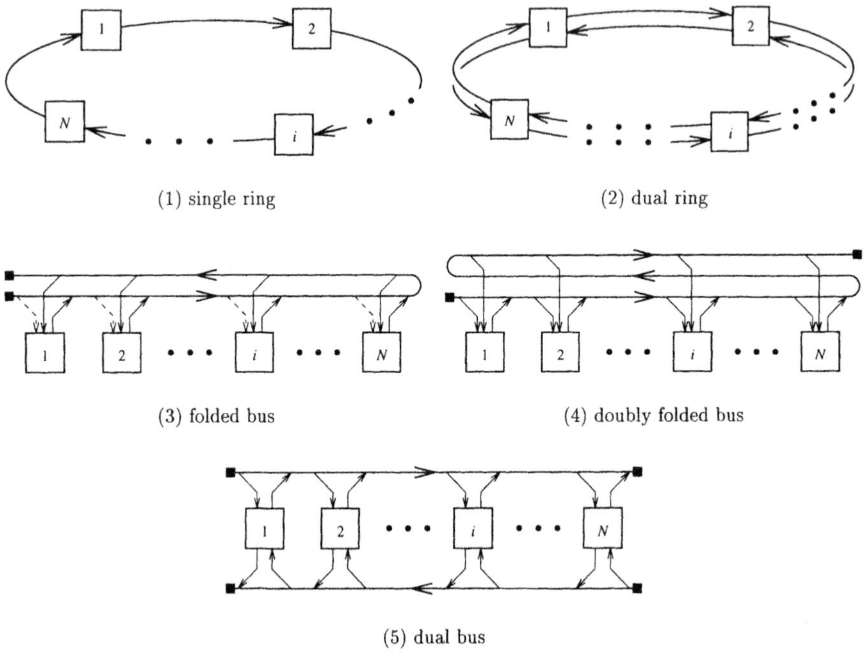

Figure 1: *Basic medium topologies*

an explicit termination at the end of the media. Of course this would require an access protocol that supports such an implementation, i.e. which is based on transmitters that are only capable of OR-writing.

2.2 Access Control

Access protocols will be described on the basis of a — single or dual — ring topology where possible. This is due to the cyclic characteristics of most of the protocols and makes comparing of the concepts easier.

2.2.1 Slot Format and Signalling Issues

The slots consist of two areas (see Fig. 2). The first one is a header, in the following called access control field (ACF), which forms a signalling channel. The second one is a field which contains addressing information and a payload area.

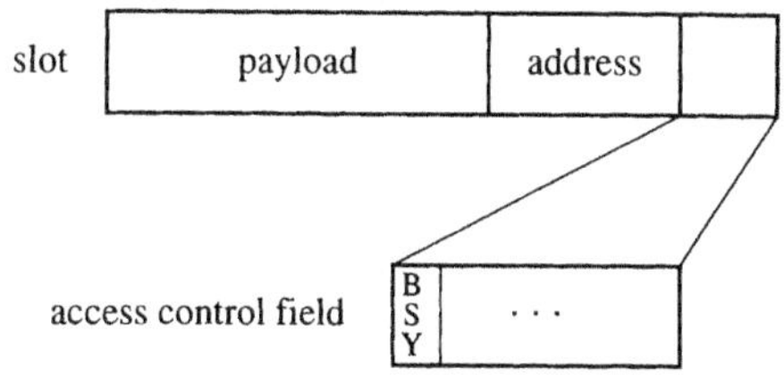

Figure 2: *Slot format*

2.2.2 Token Passing

Mainly there are two variants:

- the first is closely related to the classical token-ring protocol: A single symbol called token is rotating around the ring. A station that wants to transmit some data can take the token off the ring and send a certain amount of data segments before it has to put the token onto the ring again, so that the next station may gain access to the media. Several types of limits may be considered but the most simple one should be a fixed upper bound for the transmission window of each station.
- the second supports multiple tokens: Every empty slot is associated with such a token. This means that a station with some data pending for transmission can utilize the next free slot. This slot travels around the ring and will be cleared again by the sender who has to pass it unused to the next node.

The main disadvantages of these access mechanisms are their disability to support destination release properly thus providing slot reuse as well as the fact that per transmission cycle capacity equivalent to one ring latency is lost due to the rotating token(s). On the other hand they are fairly simple and robust against failures.

Examples of slotted token-based protocols are the Cambridge (fast) ring (cf. [4, 5]) and Fasnet (cf. [10]).

2.2.3 Reservation

The class of reservation based protocols are strongly cyclic, too. Each cycle, which is governed by a master station called *scheduler*, is subdivided into a reservation and a transmission phase.

The main principle is as follows. The scheduler issues a reservation command which passes a stations attached to the ring. Each node that is touched by the command adds the number of segments it wants to transmit in the next cycle (reservation phase). After

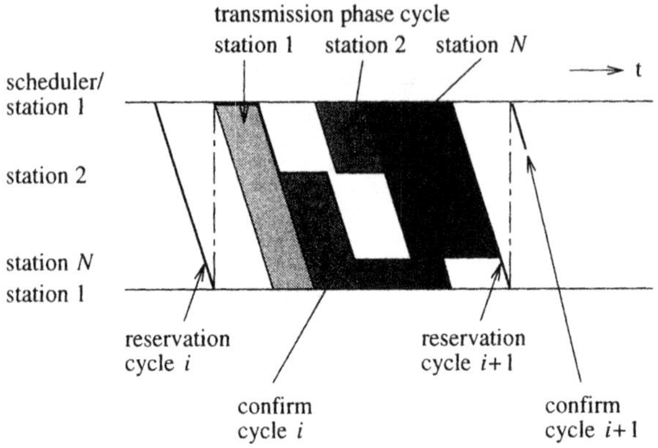

Figure 3: *Space-time diagram of a basic reservation mechanism*

these reservations have travelled around the ring, the scheduler determines how much data each station is allowed to transfer in the corresponding transmission phase. The stations are notified of their quantity by a confirm command also issued by the scheduler. In this simple variant the confirm command may be also be considered as a start signal for the next transmission cycle, in more elaborate version a separate start command may be neccessary.

In the simplest case each station would hold the confirmation command similarly to the token mechanism until it has transferred its quantity in this cycle (see Fig. 3).

After starting the confirmation the scheduler waits for the duration of the transmission phase and the issues the next reservation.

This basic access mechanism still has the same disadvantage as the token concept described previously: one latency of transmission capacity is not used for data transfer while the reservation command rotates around the ring. Thus in improved versions the scheduler should issue the reservation command for the next cycle in advance so that in the ideal case the reservations for that cycle are known to the scheduler before the transmission phase of the current one ends.

One solution could be to issue reservations in realtively short intervals to accumulate a stock of reservations so there are some available when the current transmission phase ends. This is done in the CRMA-system (Cyclic Reservation Multiple Access, cf. [15, 16] and subsection 3.2). The problem with this modification is the fact that because of the potentially large number of reservations on the fly there may be a large accumulation of reservations in the scheduler. CRMA solves this by issueing a cancellation command when a certain level is reached. The cancellation command also passes each stations and notifies them that all reservations not already confirmed are invalid and thus have to be

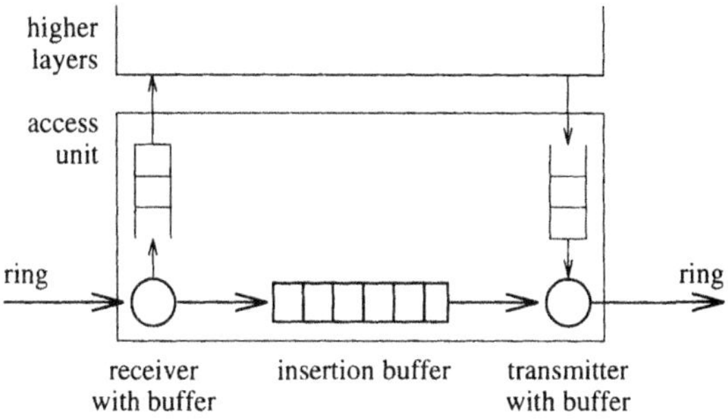

Figure 4: *Station with insertion buffer*

issued again. This mechanism is called backpressure.

Another solution as performed in the Cyclic Ring protocol (cf. [17]) where the reservation for the next cycle is issued one latency before the current transmission phase ends. In case the transmission cycle is shorter than one latency, the reservation command is issued immediately after the last confirmation command. Doing this way there is no unused capacity between successive cycles as long as the transmission phases are longer than one latency.

As described up to now all variants support transmission of longer packets in successive slots. This is not the case anymore when slot reuse is introduced. To implement this in an efficient way an end command has to be introduced. It is sent by the scheduler immediately after the confirm command and forwarded by the individual stations when they have transferred their ammount of data. A further improvement can be achieved by embedding a so called *left count* into the slot stream which allows the scheduler to estimate the number of pending transmissions in the current cycle so that the next reservation can be issued properly.

2.2.4 Buffer Insertion

Another protocol concept which primarily targets to a ring structure is buffer insertion (see [6] for SILK, an early example of a buffer insertion system). These systems implement a fixed sized buffer between receiver and transmitter of a station (see Fig 4). Incoming non-empty slots are accepted by the receiver and checked whether their target address is the station's one. In this case they are taken from the system and stored in the local receiption buffer. This class of protocols therefore is based of destination release although source release would be possible, too. Otherwise they are appended to the data in the insertion buffer or transmitted immedeately if it is empty, respectively. When there is nothing to send or forward empty slots are transmitted.

When a station wants to send some data it has to be distinguished whether the traffic on the ring on the local traffic are of higher priority. In the first case the station has to delay its transmission until the insertion buffer is empty. Once it has started transmitting a packet consisting of several slots it can not be interupted. Therefore the insertion buffer must be big enough to hold the number of slots equivalent to one maximum sized packet. In the second case a station may transmit slots as long as it can be sure that there is no loss of data at the insertion buffer. There is no difference if the buffer is not larger than mentioned above but the station can send more at the same time if the insertion buffer is correspondingly larger.

The problem with both principles is that stations may starve as described in the following. Assume three subsequent stations, the first continuously sending data to the third and the second having a non-empty insertion buffer of minimal size. In this case the intermediate station never will be able to empty the insertion buffer and thus to transmit its data.

Therefore some additional signalling is required. Metaring (cf. [2]) accomplishes this by providing so-called SAT-signals (*SATisfied*) on a second counterrotating ring thus inducing some cyclic behaviour, again. These signals can be held by a particular station as long as it did not already transfer its quantity for the current cycle.

2.2.5 Credit Systems

Credit systems are rather closely related to reservation systems. The main difference between them is that credit systems try to avoid the delays caused by the initial reservation phase.

A cycle is again subdivided into two phases: the crediting and the transmission phase. At the start of a cycle each station gains a fixed amount of credits C_i which is not neccessarily equal for all stations. It is now allowed to transmit up to C_i segments in the current cycle, each transmission decrementing the number of credits by one. Especially under light to moderate traffic conditions this enables a station to transmit its data without extensive delays, provided that there are enough credits available. At the end of the cycle remaining credits are canceled and the stations starts again with C_i.

The question is how to determine the end of a cycle, i.e. the time instant where a station's credits are consumed or there is no data to send. A centralized solution could work as mentioned above by embedding a counter variable into the slot stream. Each station would add the number of outstanding transfers for the current cycle to the counter and thus allow the scheduler to detect the end of the current cycle. The stations are notified of the end of the current and the start of the next cycle by a special *reset* signal which is the sent by the scheduler.

Another fully distributed mechanism was introduced with the Orwell system (cf. [3, 1]). When a station cannot reuse a slot just released it marks the slot as a *trial* slot and enters its own address as target address. This slot is still accessible to other stations. But in case it returns unaltered to the station which marked it as trial this station detected the end of the cycle and thus issues a reset signal.

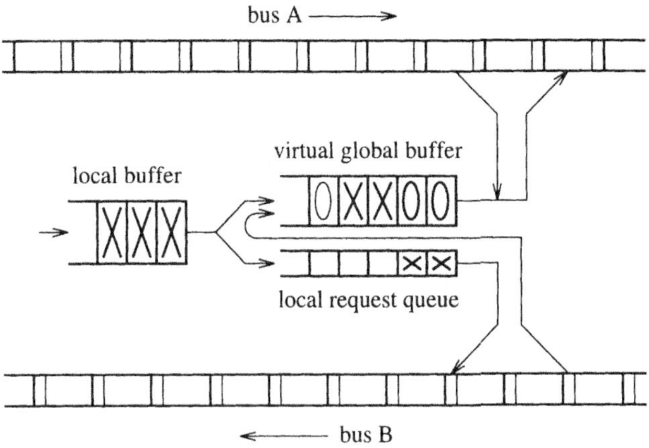

Figure 5: *Data and signalling flow in a distributed queueing system*

In the worst case there is still a loss of capacity equivalent to one latency. The probability for this situation to occur can significantly be reduced by coupling the trial signal to the source instead of the target address so that it can also be connected to busy slots.

2.2.6 Distributed Queueing

Distributed queueing protocols cannot be implemented on a single directed ring since they require a signalling channel pointing into the opposite direction of the payload channel. Most of them were introduced for a dual bus structure (see Fig. 2.1) but some can also be used on a dual ring. For an overview of distributed queueing see [14].

Considering data transfer on the upper and signalling on the lower bus the access procedure works as follows. Seen from a station, the main principles of distributed queueing are

- ◇ broadcast a request to all upstream stations for every slot required
- ◇ keep track of all access requests issued by downstream stations
- ◇ do not access the appropriate bus before all requests of downstream stations prior to its own are satisfied.

The data an signalling flow is depicted in Fig. 5. An incoming packet is segmented and stored in a local buffer. Later a certain number of segments are moved to the station's virtual global buffer. At the same time an equivalent number of requests has been generated and stored in the local request queue. In the virtual global queue the segments find some requests of downstream stations in front of them which must be satisfied by leaving free passing slots unused on the upper bus. Downstream requests are again stored n the virtual queue behind the local segments.

Independently to this procedure the station transfers the requests in the request buffer by utilizing the access control field of the slots on the lower bus. The ACF can either provide single request bits thus allowing one request per slot or a larger request field allowing for the transfer of a bunch of requests in a single ACF.

Using this basic concept several variants can be distinguished. The simplest one forms the basis of the DQDB protocol where only one local segment may be queued in the virtual buffer of a station. Correspondingly only single requests can be issued on the signalling bus where the only provide single request bits. Besides the problem of the potentially large logical lengths of MAN systems, this fact further increases unfairness aspects of DQDB since it does not allow the upstream stations to build a up-to-date record of the downstream queueing situation.

It is possible to improve the situation by allowing to transfer full packets at once from the local to the virtual global buffer (cf. [13] presenting the DQDB/CB protocol). The success of this modification is questionable as will be shown in section 3.1. The main problem is that requests are still carried by single request bits in the access control field of the signalling bus.

This is overcome in the generalized ditributed queueing concept which was introduced in DQMA (cf. [11]). Here the ACF provides an 8-bit request field allowing for up to 255 requests in on slot. Thus the upstream stations have a much more accurate picture of the downstream situation and therefore react faster on sudden changes in load.

Standard DQDB includes an additional mechanism. This is the so-called bandwidth balancing mechanism where each station has to leave every M-th — the bandwidth balancing modulus — passing free slot unused, i.e. may not remove the head of the queue segment in the virtual global buffer.

We want to note that systems which are based on the distributed queueing principle can implement destination-release and thus enable slot reuse. But since this can be done in different ways, which are not as straightforward as in the protocols described in the previous sections, we will not go into detail. Nevertheless it would allow for further classification.

2.2.7 Hybrid Protocols

All previously described protocol concepts follow the same principle: they decide for one single basic mechanism and then try to overcome its drawbacks by introducing additional modifications.

Other proposals try to integrate more than one basic concept from the beginning so that a balanced performance is achieved. As an example we will briefly touch CRMA-II (Cyclic reservation multiple access II, see [25]) which may be implemented on a single directed ring similarly to the reservation systems in 2.2.3.

Mainly CRMA-II is a synthesis of random access and reservation. In the beginning all stations can freely use every empty slot which is released by the destination, again. Embedded into this mechanism there is a reservation process governed by a central scheduler. By utilizing a reservation command which gathers the transmission requests for the next as well as the number of transmissions of all the stations in the current cycle the scheduler can use a confirm command to balance all active stations in the next cycle. The confirm command is dual to the reservation command and grants less requests to stations which were able to transfer more than others in the previous cycle than to those.

The way how this balance is calculated allows for several implementations. But even an algorithm which sums up the requests and cuts at a certain threshold can provide evenly distributed bandwidths among the active stations. As long as too many of them do not send at one time the potential cycle length will not grow excessively (see [21]).

The third mechanism which is optional in CRMA-II is buffer insertion. This may be implemented in cases where it is important to avoid the drawbacks of excessive segmentation. This modification fully integrates into the reservation mechanism mentioned above and thus does not require any additional changes so that stations with and without insertion buffer can operate together on the same system.

3. Performance Issues

Due to the logical length of the media and the large number of on-the-fly unacknowledged data units, which are typical for high-speed systems, the performance investigations of those systems are more complex. In this section we will take two examples to illustrate the problems arised and the complexity of these analyses.

3.1 Distributed Queueing

In the following we will present some simulation based results for three protocols which are based on the concept of distributed queueing. They are DQDB with bandwidth balancing (see [7]), DQDB/CB with some slight modifications [9, 13] and generalized distributed queueing as first introduced with DQMA in [11]. All the protocols were implemented on the dual bus topology shown in Fig. 2.1.

For an approximate performance evaluation of the basic DQDB principle based on classical queueing analysis we refer to [19, 20, 23] where models for different arrival patterns as well as several priorities are considered. A more extensive comparison of this protocol class and other concepts is given in [21].

All three variants were based on a dual bus system of 100 km length with 16 equidistantly located stations. Transmission capacity was assumed at 140 Mbps, signal propagation speed was chosen at 200,000 km/sec and the slot size at 424 bits (432 bits in the case of the DQMA derivate). These parameters result in a logical bus length of about 160 slots. The DQDB bandwidth balancing modulus M was chosen at 8, which is the default in [7].

The arrivals at the stations were given by Poisson streams. The batch size obeyed a bimodal distribution: 50 % of the packets were of maximum length of 80 slots, the other 50 % of the packets had a geometrically distributed number of slots around a mean value of 10 slots.

The amount 20 % of the total load was generated by station 8 the rest of 80 % arrived at the other stations. All target station indices appeared with equal probability. This means that the overall arrival rate matrix is symmetric. But as mentioned in the previous section the access protocol also is fully symmetric so that both directions of data transfer can be decomposed. This way we could restrict ourselves to downstream traffic on the upper bus and thus yield an upper triangular rate matrix.

The figures will show some of the main characteristics of the protocols discussed. This will be done mainly by observing the time from arrival of a packet until it is completely transferred to the bus. The corresponding interval is called T_{14}. In some of the figures we will also present some results for the subintervals of T_{14} that are induced by the time instant when the first segment of a packet enters the virtual global queue. They are denoted by T_{12} and T_{24}, respectively.[3] The delays are given in slots.

Fig. 6 compares the virtual service time a packet experiences when it is transferred at one of the stations 1 to 15. In general this measure increases along the bus since the load along the bus also increases. Thus downstream stations do not see as many free slots as the upstream ones and therefore it takes longer to transfer a single packet. It can be noticed that among all the stations the highly loaded one in the middle of the system shows some exceptional behaviour. The virtual service time of a packet at station 8 is significantly shorter than those at the neighbouring stations. This is due to the fact that station 8 issues more requests and therefore experiences a much smoother pattern of free and accessible slots on the data bus.

As expected the curve for the DQMA-derivate is the flattest since it allows the request for a whole packet to be transferred in a single request field. DQDB/CB shows the largest gradient resulting in a virtual service time which at the station 15 is four times as large as at the first.

Unfortunately the modifications in DQDB/CB (see previous section) — compared to DQDB with bandwidth balancing enabled — further privilege the upstream stations since a packet is transferred to the virtual global buffer earlier. So the unfair characteristics already present in DQDB with bandwidth balancing are further increased.

DQDB with bandwidth balancing more or less behaves as expected. It shows rather constant values up to station 8 significantly above those of DQMA and then grows between the curves for DQDB/CB and DQMA.

[3]The naming conventions of the intervals are due to the approximate analysis which was mentioned above and where one more time instant T_3 played an important role.

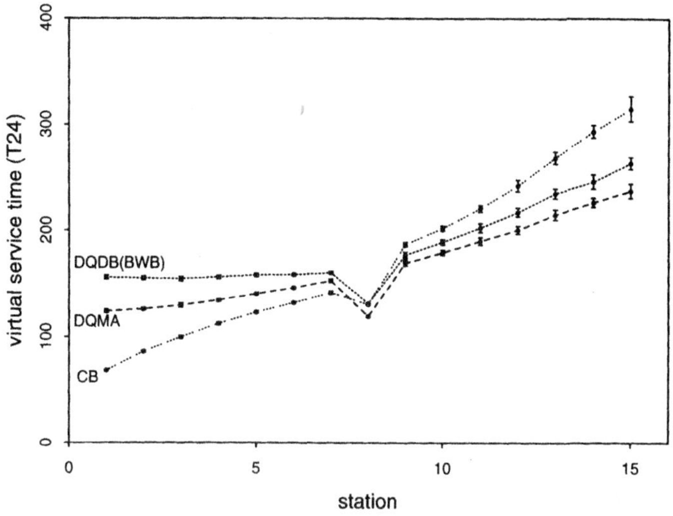

Figure 6: *Virtual service time*

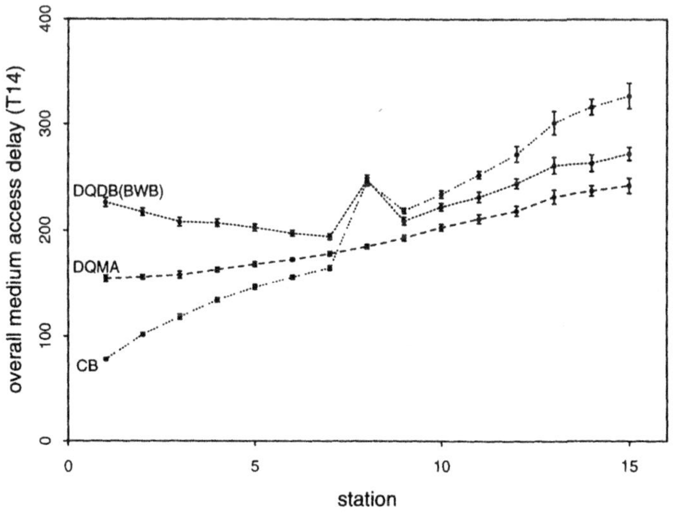

Figure 7: *Overall medium access delay*

Concerning the overall medium access delay T_{14} (see Fig. 7) it can be noticed that DQDB/CB and DQDB with bandwidth balancing have some problems integrating the highly loaded station 8. Again the gradient of the delay of DQDB/CB is excessively large. In contrast DQMA does not show any peaks at all and provides a very flat curve increasing from 160 slots at station 1 to 230 slots at the last active station thus providing the best-balanced behaviour among the distributed queueing protocols considered.

3.2 Cyclic Reservation Multiple Access Protocol

In this subsection we will take the Cyclic-Reservation Multiple-Access (CRMA) protocol as example to illustrate the complexity of a high-speed protocol model with multi-level of information units. The CRMA-protocol has been proposed recently as access scheme for high-speed LANs and MANs, especially in the network capacity region beyond 1 Gbit/sec. Detail description of an early version of this class of protocol, whose prototype has been shown at Telecom'91 (Geneva), can be found in [15] and [16].

We will outline the model and performance analysis as presented in [24]. The model contains two submodels operating on different levels of information units: one deals with packets, the other with slots or cells. To consider the efficiency of overload control mechanisms like the backpressure scheme in CRMA, it is necessary to know the entire state distributions as well as delay and transfer time distribution functions, rather than only the mean values and some higher moments. Thus, in [24] an approximate computational method was derived to obtain distribution functions of performance measures of interest.

The proposed CRMA protocol can basically be used in unidirectional folded-bus or dual-bus topologies. We will summarize only those CRMA properties and features which are relevant in the system modelling context. Since the basic operations are quite analogous for both unidirectional folded-bus and dual-bus structures, we restrict the following description on the folded-bus configuration, as depicted in Fig. 2.1. The headend, which has the function of the scheduler as discussed in the previous section, is a part of node 1 and controls the basic access mechanism (cyclic reservation) and the reservation cancelation (backpressure treatment). Each station is connected to the outbound bus segment, where packets are transmitted, and the inbound bus segment, where they are received. The system operates slot-wise; each slot consists of an access control field (ACF) and a segment field (SF) for payload data. The ACF consists of a busy/free-bit (B/F) to indicate if SF is empty and an access command with its arguments. Examples for the access command format can be found in [15, 16].

There exists a set of basic commands used to control bus access: the *reserve* and *start* commands for the basic access mechanism and the *confirm* and *reject* commands dedicated for the backpressure mechanism. Following the basic access mechanism the scheduler sends periodically *reserve* commands, each has a cycle number and a cycle length as arguments. The interreserve interval Q is set to a constant value ($Q = \Theta$ slots). To reserve slots a node increases the cycle length by the number of required slots on the *reserve* command (R) passing by. For each reservation, a station is allowed to reserve slots for only one packet. A packet consists of a number of X slots and can have a maximal size of X_{MAX} slots. Additionally, the station stores this number as the reserve length in accordance to the cycle number in a local reservation queue. When the *reserve* returns to the scheduler, it is ranged in a global reservation queue. This queue is maintained by the scheduler and operates in a FIFO order. To process a reserve cycle the scheduler sends a *start* command (S) consisting of the cycle number followed by as many empty slots as indicated by the cycle length. A node receiving a *start* will search in its local reservation queue for a entry with the current cycle number. The station will use as many empty slots as indicated by the reserve length.

To protect the scheduler queue from being overloaded a reservation-cancelation backpressure mechanism is introduced, aiming a minimization of the worst-case access delay by varying the length of the interreserve interval according to the traffic demands. It prolongs the interreserve intervals if a scheduler overload condition is detected. As overload indicator, the state U of the scheduler queue is used, i.e. the number of reserved slots waiting in the scheduler queue to be issued. For that purpose the scheduler checks U, after processing a returning *reserve*. If U exceeds an overload threshold L, the backpressure scheme becomes active. In this case no *reserve* is issued until the number of reserved slots is again equal to L. During this phase, all *reserve* commands already issued but still on the bus are cancelled.

The reservation-cancelation backpressure mechanism uses two additional access commands: the *confirm* command having the cycle number as argument and the *reject* command without any argument. Under normal load conditions, the scheduler issues a *confirm* to confirm the reservation after receiving a returning *reserve*, using the same cycle number as argument. If after the reception of a *reserve* the number of waiting slots in the scheduler queue exceeds the threshold L, the scheduler will work in the reservation-cancelation mode. The scheduler issues now a *reject*. Each station receiving a *reject* has to remove all unconfirmed reservations from its local reservation queue. The scheduler will return to the basic access mode when the scheduler queue size drops below the threshold L. Obviously, the first regular *reserve* will return at the scheduler when the scheduler queue becomes empty if the threshold is chosen as the bus latency τ. Therefore τ is a natural choice of the threshold L as discussed in [15].

The major purpose of the performance modelling as presented in [24] is to compute the distribution of the medium access delay and the transfer time of data packets. This is required to investigate fairness issues of the protocol and extreme values of the access delay and its jitter. Thus the performance evaluation has been done analytically using discrete-time methods, since simulations for the range of probability distribution functions required here might be intractable or lead to excessive computation times. Since the system operates slot-wise, analytical methods in time-discrete domain have been chosen.

The performance model was based on an observation of a test packet to be transmitted from a sending station i to a receiving station j. The breakdown of delays and transfer time into subintervals is illustrated in Fig. 8, where the following time instants are marked: (1) packet generation; (2) reservation instant; (3) reservation arrival at scheduler; (4) issuing time of cycle at scheduler; (5) issuing time of test packet at scheduler; (6) access time at

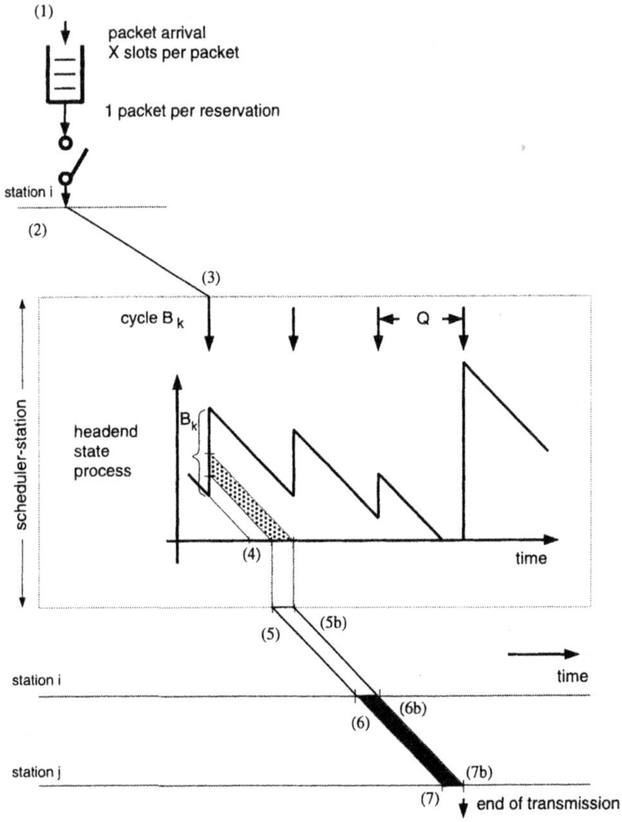

Figure 8: *Components of the medium access delay*

sender; (7) receiving time.

In denoting the time duration between two marked time epochs (x) and (y) with the random variable (r.v.) T_{xy}, T_{12} is the prereservation delay and T_{34} the waiting time at the scheduler. T_{23} and T_{56} represent the propagation delays and form together the turnaround time τ along the bus system. Clearly, the duration τ depends on the bus configuration and the CRMA operation mode under consideration. For the folded bus configuration, τ is just the bus latency, i.e. the propagation delay along the entire length of the bus. The medium access delay is $T_{ACC} = T_{16}$. Its part excluding the bus latency or propagation time is just $T^*_{ACC} = T_{ACC} - \tau$ is more appropriate for comparisons of CRMA performance with different parameters, where the influence of the bus operation mode should be intentionally ignored. From a modelling viewpoint two model levels are considered to investigate the two major delay measures: the prereservation delay T_{12} and the scheduler delay T_{34}. The model for prereservation delay still takes packets as information units while in the scheduler model the state process has to be described on slot level. To cope with this distinction a decomposition approach in conjunction with an independence approximation has been used (cf. [24]. Two submodels arose and have been analysed; a G/G/1 queue with state-dependent feedback and a M/G/1 queue with server vacation.

In the following we will discussed some numerical results out of the study in [24]. Two different types of attached stations are considered: i) stations under normal load conditions (random data traffic) and ii) saturated stations (e.g. stations during a file transfer phase, which can be in a quasi-stationary state). Under normal traffic conditions, the packet arrival process in a station i is assumed to be Poisson with rate λ_i. The packet size X can be arbitrarily distributed with mean EX and the maximal packet length is X_{MAX}. A saturated station is considered to have always a packet to send with the same packet size distribution.

A network with $N = 16$ stations is observed, which are equidistantly located on a folded bus system. The slot length is chosen at 55 Bytes, consisting of 48 Bytes payload segment field, 5 Bytes wide area network header and 2 Bytes CRMA header. Packets are assumed to have a truncated Poisson distribution. The time axis in the diagramms is given in slot transmission time. In the following, a bus of capacity 1.2 Gbps with a length of 10km will be taken, corresponding to a bus latency of 273 slots.

Recall that the purpose of the reservation-cancelation backpressure mechanism is to protect the scheduler queue against overload. The main mechanism is to prolong the interreserve interval artificially in case of scheduler overload. Since a station is allowed to reserve only one packet per reservation command, this mechanism reduces the total load offered to the scheduler under overload conditions. However, since L is chosen to be equal to the bus latency τ, the backpressure mechanism is expected to be active only under very high load levels or in the case where a few stations are sending packets under saturated conditions.

Fig. 9 shows the influence of the backpressure mechanism on the scheduler delay and its distribution function, where a comparison of CRMA with and without the reservation-cancelation backpressure mechanism is illustrated. A normal load level $\rho = 0.5$ is chosen. In this Figure 95% confidence intervals obtained by simulation to validate the accuracy of the approximate analytical method are also depicted. The load-limiting function of the backpressure scheme can clearly be recognized. The distribution function of the scheduler delay is bounded by $L - 1 - \Theta$. This value corresponds to the case where after processing the last reserve command, exactly $L - 1$ slots are in the scheduler queue.

The backpressure mechanism obviously diminishes the influence of a local overload situation, e.g. when a station is heavily overloaded or in saturation, on the global system performance. Due to the decreased *reserve*-rate the packets are delayed in the local queues instead of the scheduler. Thus the traffic is distributed among the stations and not concentrated at the scheduler. Therefore we expect in general a smaller access times for systems running with the backpressure mechanism than without, exspecially under heavy load.

Fig. 10 shows in summary the prereservation delay, the scheduler delay and the medium access delay of station 1, 8, and 16. A position-dependency of the access delay can be

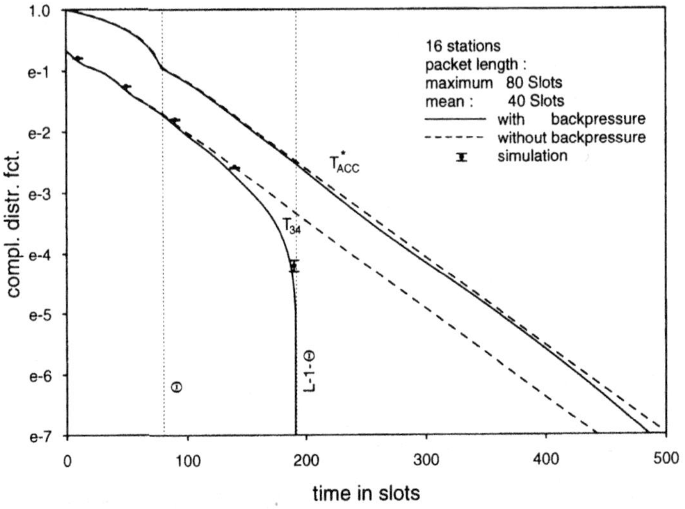

Figure 9: *Comparison of scheduler delay with and without backpressure*

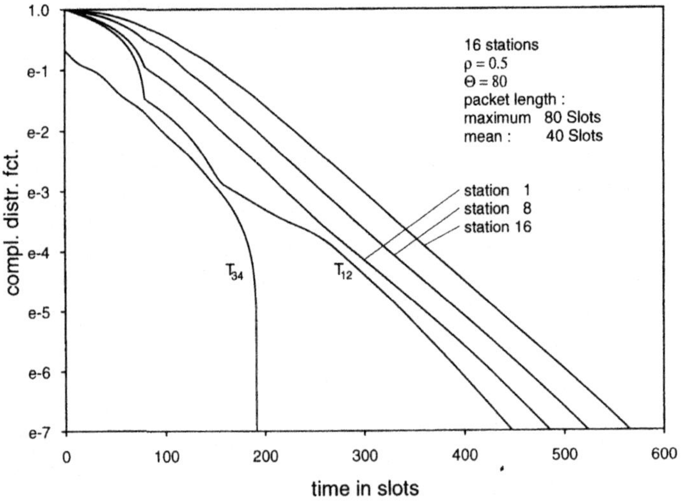

Figure 10: *Delays observed in different stations*

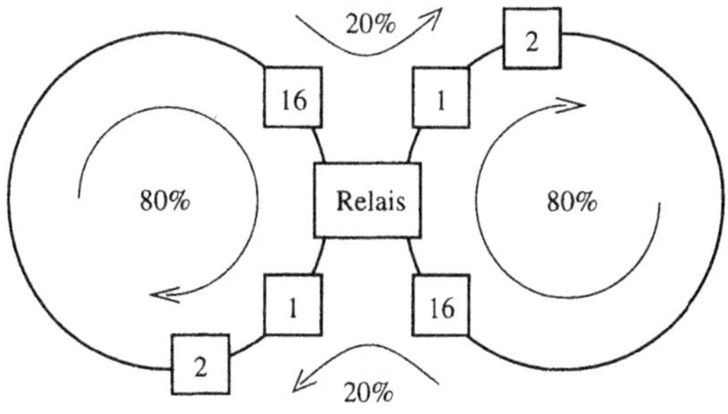

Figure 11: *Configuration of bridged rings*

recognized, which is caused by the different intra-cycle delays of the considered stations. The access delay depends on the sequence of the stations. In Fig. 10 the latency τ (i.e. 273 slots for the configuration b) is substacted, thus the *relative* differences of access delays seen by different stations are smaller than appeared.

4. Internetworking performance

Today a typical bridge between neighbouring LANs consists of an access unit, an error detection/correction-unit (and reassembly and packetizer unit, since e.g. the existing IEEE 802.x LANs and MANs support different maximum packet lengths). It may be very costly to stay with this structure when switching to high-speed environments. Furthermore this may be neither neccessary nor useful since future systems tend to adapt to the ATM-principles of end-to-end flow control, end-to-end error detection/correction etc.

Therefore the bridge functionality should be reduced. Then obviously, some kind of common packet format is required. In the context of MAN and public ATM networks this means a common slot format. As a minimum requirement a common payload format should be introduced so that a complex assembly/reassembly unit can be omitted. Ideally, a common address space is also used, since this would help to avoid potentially large address translation tables. That way only the access protocol specific signalling data has to be converted.

A still remaining problem are the potentially large buffer requirements in the bridge. To give some first insight we observed a system of two rings interconnected by simple bridge as mentioned above. The parameters of the individual rings were identical to those of the dual bus system considered in subsection 3.1, except that 20 % of the traffic generated in one ring is destined for the other ring and therefore crosses the bridge located between stations 16 and 1 (see Fig. 11).

Fig. 12 compares the simulation results for two access mechanisms: a multiple token

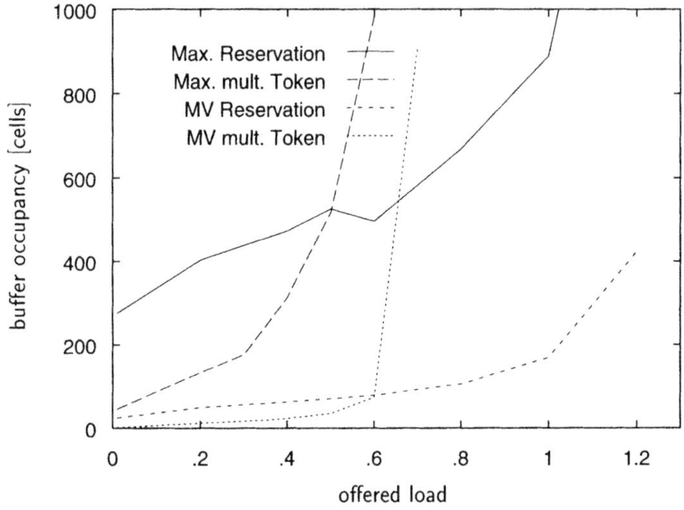

Figure 12: *Buffer occupancy in the bridge*

protocol (cf. subsection 2.2.2) and a reservation based method similar to Cyclic ring (see subsection 2.2.3).

Fig. 12 depicts the maximum (Max.) as well as the mean value (MV) of the buffer occupancy of the bridge during a simulation of symmetrical load (note that only one direction of transfer is observed). As long as it is stable the token based protocol provides significantly smaller mean and maximum values since even in a nearly empty system under the reservation mechanism the bridge typically stores around 25 cells on average.

The advantage of the reservation mechanism is that it supports cell reuse und thus allows for a much higher load than the token based variant. It is stable up to a load of about 120 %.

As a consequence it would be desirable to synthesize a protocol which supports immediate access at low load and equal distribution of bandwidth at high load as the reservation based mechanisms do. This is what e.g. CRMA-II tries to achieve.

5. Conclusion

In this paper we discussed basic concepts and performance issues of the class of high-speed protocol, such as proposed for HS-LANs, MANs or access schemes for B-ISDN. We approached to give a systematic introduction to basic protocol mechanisms in a more generic context, where well-known protocols appeared as specific implementations of these mechanisms. To illustrate the performance evaluation methods used to investigate these protocols, two examples have been outlined: the DQDB protocol class, which is based on the distributed queueing principle, and the CRMA protocol, which represents a reservation-based case. Finally some aspects concerning internetworking and related performance investigation were mentioned.

References

[1] Adams J.L., *Orwell*, to be published in Special Issue of Computer Networks and ISDN Systems, Autumn, 1992.

[2] Cidon I., Ofek Y., *Metaring — A Full-duplex Ring with Fairness and Spatial Reuse*, Proc. INFOCOM '90, San Francisco, 1990.

[3] Falconer R.M., Adams J.L., *Orwell: A Protocol for an Integrated Service Local Network*, Advances in Local Area Networks, Kümmerle et. al. (Ed.), IEEE Press, 1987.

[4] Hopper A., Williamson R.C., *Design and Use of an Integrated Cambridge Ring*, Advances in Local Area Networks, Kümmerle et al. (ed.), IEEE Press, 1987.

[5] Hopper A., Needham R.M., *The Cambridge Fast Ring Networking System*, IEEE Trans. Comp., vol. 37, no. 10, Oct. 1988.

[6] Huber D.E., Steinlin W., Wild P.J., *SILK: An Implementation of a Buffer Insertion Ring*, Advances in Local Area Networks, Kümmerle et al. (ed.), IEEE Press, 1987.

[7] IEEE, *IEEE Standard 802.6-1990 — Distributed Queue Dual Bus (DQDB) Subnetwork of a Metropolitan Area Network (MAN)*, IEEE, 1990.

[8] Kaibel M., *Leistungsvergleich von Multi-Request-Methoden in DQDB-Protokollumgebungen*, Diploma thesis, institute of computer science, University of Würzburg, 1992, *in german*.

[9] Kamal A.E., *Efficient Multi-Segment Message Transmission with Slot Reuse on DQDB*, Proc. INFOCOM '91, Miami, April 1991.

[10] Limb J.O., Flores C., *Description of Fasnet — A Unidirectional Local Area Communication Network*, Advances in Local Area Networks, Kümmerle et al. (ed.), IEEE Press, 1987.

[11] Müller H.R., Nassehi M.M., Wong J.W., Zurfluh E., Bux W., Zafiropulo P., *DQMA and CRMA: New Access Schemes for Gbit/s LANs and MANs*, Proc. IEEE INFOCOM '90, San Francisco, Ca, June 1990.

[12] Mukherjee B., Banerjee S., *Alternative Strategies for Improving the Fairness in and an Analytical Model of DQDB Networks*, Research Report, Division of Computer Science, Department of Electrical Engineering and Computer Science, University of California, Davis, CSE-90-32, July 1990.

[13] Mukherjee B., Kamal A.E., *Scheduling variable-length messages on slotted high-speed fiber optic LANs/MANs using the continuation-bit approach*, Proc. IEEE INFOCOM '91, Bal Harbor, FL, April 1991.

[14] Mukherjee B., Bisdikian C., *A Journey Through the DQDB Literature*, Research Report RC 17016(#75478), IBM Research Division, November 1991.

[15] Nassehi M.M., *CRMA: An Access Scheme for High-Speed LANs and MANs*, Proc. IEEE Supercom '90, Atlanta, Georgia, April 1990.

[16] Nassehi M.M., *Cyclic-Reservation Multiple-Access Scheme for Gbit/s LAN's and MAN's Based on Dual-Bus Configuration*, Proc. EFOC/LAN '90, Munich, F.R.G., June 1990.

[17] Pihlaja J., *Cyclic Ring*, Temporary working document, RACE project 1011, November 1990.

[18] Schwartz M., *Telecommunication Networks*, Addison-Wesley, Reading, MA, 1987.

[19] Stock T., Tran-Gia P., *A Discrete-Time Analysis of the DQDB Access Protocol*, Proc. ITS '90, Rio de Janeiro, September 1990.

[20] Stock T., *Influences of Multiple Priorities on DQDB Protocol Performance*, Proc. 13th ITC, Copenhagen, Denmark, June 1991.

[21] Stock T., *Schnelle lokale und regionale Netze — Konzepte und Fallstudien zur Leistungsbewertung*, Ph.D. thesis, Institute of Computer Science, University of Würzburg, 1992, *in german.*

[22] Tobagi F.A., Borgonovo F., Fratta L. *Expressnet: A High-Performance Integrated-Services Local Area Network*, Advances in Local Area Networks, Kümmerle et al. (ed.), IEEE Press, 1987.

[23] Tran-Gia P., Stock T., *Approximate Performance Analysis of the DQDB Access Protocol*, Proc. ITC Specialist Seminar, Adelaide, Sept. 1989, also in Computer Networks and ISDN Systems, 20 (1990) 231–240.

[24] Tran-Gia P., Dittmann R., *A discrete-time analysis of the cyclic reservation multiple access protocol*, Performance Evaluation, 16, 1992, 185-200.

[25] van As H.R., Lemppenau W.W., Zafiropulo P., Zurfluh E.A., *CRMA-II: A Gbit/s MAC Protocol for Ring and Bus Networks with Immediate Access Capability*, Proc. EFOC/LAN '91, London, UK, June 19–21, 1991.

[26] Wilkes M.V., Wheeler D.J., *The Cambridge digital communication ring*, Proc. Local Area Commun. Symp., Boston, MA, May 1979.

Integrated Broadband Communication Networks
and Services (C-18) / V.B. Iversen (Editor)
Elsevier Science B.V (North-Holland)

EURESCOM supporting Public Network Operators to prepare for new networks and services

Otto Baireuther

EURESCOM GmbH, Schloss-Wolfsbrunnenweg 35, D 6100 Heidelberg/Germany

Abstract

As the telecommunications environment in Europe is facing rapid changes in legal, technical and business terms, a number of Public Network Operators (PNO) felt it appropriate to join research and development forces and settled a in jointly owned organisation in order to provide synergy between the individual R+D efforts. EURESCOM (European Institute for Research and Strategic Studies in Telecommunications) was founded in 1991 to facilitate developments of future advanced telecommunications services and networks. As a framework for the work, visions and possible evolution scenarios towards integrated broad band communication have been developed. The paper presents the Institute, introductes a possible networks and services vision and highlights few selected areas of study serving the evolution to such visions.

Keyword Codes: C. 2.0;C. 2. 1
Keywords: Computer-Communication Networks, General, Network Architecture and Design

Introduction

The past forty years have seen dramatic advances in information and telecommunications technologies (IT&T), and the widespread adoption of these technologies bears the best testimony to the advent of the "information society". Not only the US, also Europe has seen a significant shift from traditional economic sectors like agriculture, manufacturing etc. Services in general and information handling are under the winning sectors from this shift. According to a study launched by the Commission of the European Community (CEC), telecommunications will be the driving force in (Western) Europe economy at the beginning of the next millennium with an annual growth rate of estimated 8% as compared to e.g. the car industry of only 2%.

In 1990, the world telecommunications equipment market totalled to 85 billion ECU, Europe holding a share of about 20 billion ECU. Moreover, the telecommunications services market was estimated for about 230 billion ECU, Europe again taking a share of around a quarter of this, i. e. 75 billion ECU. New and developing countries and states in Middle and Eastern Europe have not been included in the figures and their market opportunities are easily demonstrated by mentioning the average of only seven to eighteen telephone lines per 100 inhabitants.

But European markets are split up in many ways and do not allow for sophisticated exploitation of economy of scale e.g. in production of telecommunications equipment and effective provisioning of services on a pan-European level. Reasons for this are manifold and include historical traditions, regulatory differences, different economical growth and dispersed ways of privatisatising telecommunication services and networks. Recall, a quarter of the world market is split into more than 30 different scenarios whereas another third of that market (the US) is ruled by one unique regime and subsequently provides for one uniform market.

With the Single Act, telecommunication is considered one of the most important facilitators of the unification of Europe. It is assumed to foster the four freedoms, the movement of people, goods, capital and services and thus enable the unification of Europe. In reaction to this, the CEC initiated various activities such as RACE , ESPRIT, Telematic programmes as AIM, DELTA, DRIVE and ENS and TEN. They all strive for the same aim, the advanced telecommunications support within a unified Europe even so they are focusing to different subjects.

Stimulated by the "Green Paper on the Development of a Common Market for Telecommunications Services and Equipment" of June 1987 a number of other bodies have been founded or became active as well:

> **ETSI** has been established for making telecommunications standards with participation form administrations, network operators, manufacturers, service providers, users and research community.
>
> **CEPT** has changed its mission to become the common forum of the regulatory authorities.
>
> **CEC**, under legislation of the Council of Ministers are releasing directives and recommendations for the co-ordinated introduction of services.
>
> **ETNO** forms the platform for the PNOs to prepare for the harmonised implementation of services and political initiatives on a pan-European level.
>
> **EURESCOM**, building a forum for Public Network Operators for joint research and developent of new telecommunications networks and services.

Co-operation of PNOs always needs to ensure that joint decision making and commitment for implementation do not prevent competition of the organisations on the "sales level". Recent experience show that such purpose can well be served by so-called Memoranda of Understanding (MoU).

Most of the known MoU determine the parties involved to implement new networks and services according to unique guidelines in terms of evolution and implementation steps, technical specifications and basic service offerings. The most well-known MoU to-date are:

- for the implementation of pan-European digital cellular mobile communications (GSM=Global System for Mobile communications)
- on the implementation of European ISDN service by 1992 (Euro-ISDN)
- on a European Broad band Interconnection Trial (EBIT)
- on the establishment of a Managed European Transmission Network (METRAN)
- on the co-ordinated implementation of a second generation pan-European paging system (ERMES),
- on the implementation of a pan-European ATM Pilot network.

EURESCOM, the joint undertaking

EURESCOM was founded in March 1991 as a company under German law and holds its seat at Heidelberg/Germany. At that time 24 European Public Network Operators, coming from 19 different countries signed a MoU for that.

Initiative to establish EURESCOM was taken by Deutsche Telekom which made corresponding proposals at a telecommunications committee of CEPT and to a meeting of the EC Council of Postal and Telecommunications Ministers in Autumn 1989.

PNOs' rationale behind the idea was:

- the need to ensure interoperability of national services and to promote the development of pan-European services,
- the need to make the most efficient use of limited R + D resources, taking into account the new role of European PNO and their common identity,
- the wish to stay competitive and take responsibility for the economic success of their investments,
- the wish to take maximum profit of the results of European R + D programmes.

The overall objectives of EURESCOM are to support the development and provision of harmonised pan-European public fixed telecommunications networks and services by furthering science and research in the field. In particular to:

- enable the development of harmonised strategies for the planning and the provision of future public networks and services,
- stimulate and co-ordinate common participation in pre-competitive and pre-normative research projects,
- stimulate and co-ordinate field trials and pilot installations,
- contribute to European and international standardisation,
- contribute to the work of EC, EFTA, and other relevant European R + D programmes.

These objectives will be carried out in accordance with the following principles:

- The activities of the Institute will not hinder or restrict in any way competition among its Members.

- The projects conducted will be carried out by the Members under General Interest or Special Interest categories.

The mission of the Institute and its basic principles of work are set out in the Articles of Association which together with an Intellectual Property Rights agreement (IPR) built the legal formation for the company. Rules for the treatment of IPR are being considered of great importance for such a commercially oriented company as EURESCOM. Voting is done according to shares held by the individual Shareholder (weighted voting).

Furthermore, a Cost Sharing Agreement sets rules for the financing of the work programme and the Institute itself. All Members contribute to the operational budget (e.g. Permanent Staff) and to the cost of the General Interest projects in relation to different classes of shares. Work efforts spent to a particular project are being reimbursed according to actual involvement and contribution of the individual organisation, but based on harmonised rates for labour, travel etc.

Special Interest Projects are funded only by the Members interested in that particular item, without any fixed rule.

European Public Network Operators (PNO) providing national and/or international public fixed telecommunications networks that have been granted rights or obligations for the provisioning of network infrastructures or basic services, or holding companies of such PNOs or associations representing one or more of such PNOs qualify for membership/shareholder ship in EURESCOM. They should have a strong interest in ensuring convergence and integrity of a European network infrastructure, the interoperability of services cross Europe and commit themselves to support advanced collaborative R + D activities.

At present, the following PNOs are Shareholders of EURESCOM: Post and Telecommunications Administration of Iceland, Norwegian Telecom, teleDanmark, Swedish Telecom, Telecom Finland, Association of Telephone Companies in Finland, Telecom Eireann, BT, Mercury Communications, BELGACOM, PTT Nederland, Post and Telecommunications Administration of Luxembourg, Deutsche Telekom, France Telecom, Swiss PTT, PTT Austria, Hungarian Telecommunications Company, Telecom Portugal, Telefones de Lisboa e Porto, Companhia Portugese Radio Marconi, Telefonica de Espana, STET Italy, Yugoslavian PTT (suspended), Hellenic Telecommunications Organisation.

The approach to work is similar to other pan-European research fora, but different to standardisation. After a joint establishment of a work programme (i. e. done with the consent of the Shareholders) and approval of a particular project, a call is established asking the Shareholders for their interest in participating in this project. Each project has been given a budget with its approval by the General Assembly of EURESCOM. Together with interested members and under support by the Permanent Staff, the project will be defined in more detail and subsequently planned. A contract containing the allocation of responsibilities and budgets enables the individual participant to start work for his/her-own. Travelling and meetings should be avoided as much as possible and used only to align individual contributions.

Of great importance is a clear positioning of EURESCOM work in the "European concert". Currently, the Shareholders give priority to short term oriented work delivering results for implementation within about two years after finalisation of a project. In effect, the present programme of work shows only little evidence on long term research work.

EURESCOM projects particularly aim at the establishment of specifications which could be used by its Shareholders for their own operations and/or procurement specifications. Subsequently, work is principally based on international standards but sometimes also supporting the elaboration of such standards. Support for and utilisation of CCITT and ETSI standards is a strong commitment to all the operators and thus for EURESCOM. In relation to research programmes like RACE, EURESCOM's work will clearly benefit from the basic research done on technologies, strategies, service engineering etc. as far as the work results are accessible. In return to such exploitation, PNOs work in EURESCOM may be able to contribute on some strategic and network/service implementation issues for CEC programmes.

Therefore, the general guidelines applied for the definition of the EURESCOM Work Programme can be summarised as follows:

- Focus on the provision of services/networks in terms of pan-European services offered by the PNOs.
- Few identified work areas where PNOs could clearly benefit from close co-operation in order to speed up the evolution of such pan-European services and networks.
- Establishment of PNOs interest and priorities on a pan-European level taking into account existing and proposed pan-European telecommunications initiatives (e.g. RACE), thereby avoiding overlap or duplication of work.

The EURESCOM Work Programme is described in two parts:

- Annual Costed Work Programme (ACWP) which is given a detailed planning for the coming calendar year (incl. a manpower budget), and
- Multi Annual Work Programme (MAWP) which gives an outline for the four following years.

Network and services evolution visions

When starting an activity for R + D in telecommunications it seems to be rather clear that first of all one would need a common understanding on the most likely evolution networks and services will take. Based on this, individual work areas and items can be derived. Applying this logic, one of the significant studies within EURESCOM is collected under the Overall Strategic Studies work. Such work has been planned to first develop an outline and visions for pan-European telecommunications at around 2005, followed by building scenarios and an analysis of possible evolution paths bridging present with these visions. Finally, the results of the study should give input to the definition of a work programme for the Institute.

A telecommunications network vision describes a situation in the future based on current assumptions on customer requirements (in terms of hardware, software and services) and enabling technologies together with policies regarding market issues. Such vision is continuously updated with current changes in the relevant assumptions and is consequently a moving target and any written description is a living document.

Following this, work concentrated on integrating assessments of market trends, business sector trends, technological and institutional/regulatory trends. In order to narrow down the very broad scope of this areas and focus on the needs of Public Network Operators the main interest is clearly a description of the evolution of technology, networks and services, models for system integration, the impact of telecommunications on the way of organising work and techno-economic modelling and assessments.

It should be re-called that the focus of EURESCOM is mainly in providing for the pan-European aspects of the visions rather than in the development of individual PNOs business developments. The above-mentioned mission of the Institute would prevent from planning individual companies businesses as it would restrict competition among the Shareholders.

In the following the most important items contained in the visionary statements will be reported as they are currently under discussion. This does not commit any of the EURESCOM Members to fully support this view of future developments of telecommunications, nor does it contain any commitment for the implementation of such statements:

A *balance need to found between commercial operation and service obligation* as most of the telecommunications companies of Europe have taken a market approach whereas at the same time a sort of universal basic service obligation has been implemented. Nearly every licensed network operator has been given the mandate to provide a basic set of services to commercially less interesting group of customers, sparsely populated areas in the country etc. whereas at the same time their owners/shareholders which is very often the government expects high profits. So, network and service provisioning is and will be influenced by the social consequences it may have to society as a whole and to individuals. As the importance of telecommunications for economy will significantly grow in the future, relationships between economy and telecommunications industry will be even closer than to-day. Equally important is the capability of telecommunications to replace or enhance industrial sectors. Prominent examples in the travel industry (by video conferencing), the data processing industry (through clustering smaller computers temporarily) are well known and will further evolve.

Telecommunications industry will continue to become liberalised in the way that increased competition has already and will in future cause major changes to the industry. Network operators generally have started to restructure their business organisation and will continue to learn when to compete and when to co-operate. So-called global and strategic alliances are being founded and will increase their importance for the industry whereas at the same time the individual companies' re-direction towards customer focus and strengthening the ability to compete are being progressed. One of the most radical changes however is the appearance of independent service providers and agents, both gaining significant importance during years to come. In technical terms this will be supported by greater separation of services from the physical network; Open Network Provision (ONP) is the political initiative promoting this. In parallel, revenues from network transport have decreased but value added service provision revenues have and will continue to increase. Such separation will lead to an increasing number of compatible technical and administrative interfaces between Network Operators, Service Providers, agents and customers/users.

Presentation of personalised services will be harmonised in the sense that:

- easy access to services will be enabled independent of the location of the customer,
- user interfaces will be enhanced by intensive use of speech recognition, synthesising and automatic translation,
- telepresence and virtual reality will provide new dimensions to human communications,
- service presentation and user-friendliness will become an important marketing instrument for service providers,
- customers will have the opportunity to tailor individual services supporting individual applications,
- the trade-off between common presentation and product differentiation for services and terminals will have to be identified and developed,

All in all, such harmonised presentation of services will enhance rapid service provisioning and penetration.

Services will become transparent on a pan-European level as the separation of services and networks will be an important development in the near future. The new feature in this statement is the European level which presumes close co-operation of network providers cross their network boundaries and exhaustive definition of interworking interfaces, procedures and practises. The evolution of service independent platforms (hardware and software) will continue to progress. The migration of networks (physically and logically) is well under way and will foster interworking at all levels. Customers can specify their requirements on network performance features such as stability, availability, robustness etc. Pan-European creation, management and control of services exist at least for big customers in Europe. In order to support this, network operators will have to agree on consistent network architecture and the formal distribution of grade of service and network performance for individual connections. In addition, charging and accounting principles will be consistent and related to incurred costs.

Customer gain greater control on network resources by using flexibility in acquiring bandwidth, controlling cost of usage and determining quality of service. Greater flexibility in resource provisioning by the network provider allow customers to get more resource independent service support and more or less unrestricted service creation is widely available. Network operations managed by the public network control are co-ordinated with the control of the customer network, preferably through direct dialogue of the respective management systems. These improved communications and freedom to configure network resources for their own needs will lead customers to consider the network in part as their own network and will eventually stimulate further use . "Corporate" and "family" networks will be a result of it and will foster support of customer applications in their original form rather than requiring the need to convert it to standardised public services.

The *impact of the residential market will determine the telecommunications industry*. New and enhanced (incl. broad band) services used by the huge number of private users will be dominating revenue growth. The currently existing share of revenue with business and private users (roughly 60 - 70 % against 40 - 30 %) will change dramatically. Reasons for this are mainly the introduction of information processing technology to the home and their request for interconnection. Moreover, the increasing familiarity of users with high tech information processing products in their home will decline the psychological barrier in people's minds for the usage of networking even for applications in their private life. Services like Teleworking, Teleshopping, Telelearning, Telemonitoring, Video Entertainment are just examples of that development.

Of great importance are *information security and privacy on a European level* as more and more sensitive data is being transported. Customers therefor are excessively demanding this as a mandatory feature to all telecommunications, even and especially in applications of personal mobility. Network providers themselves have great concern in protecting their own management and operations information from misuse. A problem which is difficult to solve on the national level but even more complex on the international level. Prerequisite for such precautions are legislation (on a European level) allowing for appropriate implementation.

Telecommunications, information processing and entertainment electronics are more and more converging, and so diminishing technological boundaries. Especially during market penetration phases for each of the three mentioned domains synergy among each other is important and sometimes even the enabler for market success. Customers are using the opportunity for creating their own applications which have a pre-dominant focus on multi-media type of information transfer. This convergence leads to new user service interface definitions using intensively visualisation.

One-stop shop access for all services is a widely used feature based on (contractual) agreements between front offices, service providers and network operators. Such front offices are operated either by one of the traditional actors or by independent third parties acting as an agent for the customer. PNOs are accelerating their co-operation for the pan-European support of at a minimum bearer services while simultaneously competing on the tele services market.

The role of *environmentally friendly services* is gaining importance as the telecommunication share of Gross National Product is still increasing. Telecommunications has either replaced or enhanced other more traditional industries, such as substituting parts of the travel industry both for business and private. The share of the regular income customers have available for telecommunications is continuously increasing and thus enabling new market opportunities for the industry. Another remarkable effect is the sales increase on telecommunications related products in hardware and software, mainly for the residential market.

These visions of future telecommunications exemplify possible targets for evolution. The visions are not to be seen as determining the future of telecommunications in Europe and do not necessarily represent the common position of all PNOs within EURESCOM. However, they will serve for building a framework within the institute to derive areas of interest for joint work and within these areas to define subjects for detailed study work.

Current focal points in the work of EURESCOM

After less than two years of operation of the institute more than 35 projects have been launched under its umbrella, all of them are commonly supported and funded. Rather than introducing individual projects, the following will briefly present the work areas they have been allocated to. This clustering has been done according to nature and objectives of the individual study items.

Infrastructure and Switched Networks

The transport network is a key constituent of service provisioning cost. One of the means to reducing these costs is standardisation of SDH and its introduction which allows for improved provisioning and management of transmission paths. By combing this technology with switching nodes based on ATM technology, the PNOs will have a transport network capable for all kinds of information, incl. broad band, on the medium and even on the long term. Preparations for the implementation of enhanced copper technologies and also fibre optics in the local loop are important short term needs and will also foster the development of all-optical networks (incl. optical switching and signal processing). Signalling for B-ISDN gives another focal point in addition to work on reliability studies of existing and future networks.

Intelligent Networks(IN)

Intelligent networks are represented in an architectural concept allowing new (or changes to existing) services to be introduced (customised) into public telecommunications networks in a rapid and cost effective way. Basically, a separation of service control and transport control is used which supports the opening of the market for new entrants in an Open Network Provision environment. For this, IN is heavily depending on the development of related software. At present, work is concentrated on offering pan-European services, interconnection of national IN (trials) and future IN architecture and service creation environments.

Telecommunications Management Networks(TMN)

TMN architecture will not only allow for effective exploitation of already deployed equipment, but will also result in harmonised and automised interworking between service providers and transport network providers. In their common aim of improving Quality of Service for the customer, TMN and IN are highly complementary in their approach and share the dependence from software. Current TMN projects are focused on management services development, guidelines for the implementation inter-network management functions and specification of TMN applications.

Telecommunication services

Finally, the provisioning of services is the overall aim of EURESCOM work. Subsequently, work is needed to specify /define services before or in association with the specification of the platforms used for the support. In this relatively small area works have been started on usability of existing services, exploration of new applications/services and description of new services (e.g. pan-European directory service).

Software requirements and practices

Telecommunications software will increase its importance for future systems as it will influence all the other domains. It builds a sort of glue for the functioning of telecommunications systems. Rather than actually writing programs, work in EURESCOM is aiming for harmonised software practises, software quality assurance and common platforms to be used in specifications for the procurement of equipment.

Conclusions

EURESCOM, formally founded in 1991 is developing progressively towards a powerful instrument for Public Network Operators in Europe. In its short history it was able to show its ability to:

- provide maximum benefit out of collaborative R + D of its Shareholders and other European R + D, e. g. RACE,
- develop and specify new pan-European networks and services,
- foster the implementation of harmonised service features (on a pan-European level),
- progress the definition of network management functions and principles.

Furthermore, EURESCOM builds a platform for joint contributions to standardisation organisations such as ETSI and WTSC (formerly CCITT). The results of study work is accessible to all its members but only exceptionally to external users. It also an offer to third parties like manufacturers, service providers, users and other organisations working in the domain for a closer co-operation to the benefit of world-wide telecommunications.

Integrated Broadband Communication Networks
and Services (C-18) / V.B. Iversen (Editor)
Elsevier Science B.V (North-Holland)

Activities on traffic engineering for B-ISDN in CCITT, SG II

M. Villén-Altamirano

Telefónica I+D
Emilio Vargas, 6 ; 28043-Madrid, Spain

Abstract

This paper presents the current and planned activities in the CCITT Study Group II to enhace the E.700 series, common framework for Recommendations on traffic engineering for narrowband and broadband ISDN, with specific Recommendations for broadband ISDN.

1. INTRODUCTION

The activity on traffic engineering for ISDN in CCITT began in 1985 when the Plenary Assembly allocated a question on this subject to the Study Group II (Network Operation), which in turn, allocated it to its Working Party on Traffic Engineering (WP II/3). Since then, a new series of Recommendations, the CCITT E.700 series [1], whose objective is to provide detailed guidelines on traffic engineering for ISDN, is being developed. The activity on this question continued in the 1989-1992 CCITT study period and will continue in the starting 1993-1996 CCITT study period, in which the question has been split into two ones, one for narrowband ISDN and the other for broadband ISDN [2].

The E.700 series is structured into four main sections, i.e., traffic modeling, traffic related network performance objectives, dimensioning methods and traffic measurements [3]. These sections cover the major traditional engineering tasks. Figure 1 represents how these tasks are related between them and to the services, architecture and protocols defined in the CCITT I series of Recommendations [4].

The ISDN concept is very broad and includes many very different network structures, services and features. In particular, two major categories of ISDN are considered: narrowband ISDN and broadband ISDN. The structure of the E.700 is comprehensive enough to be a common framework for both of them. Many of the concepts and methodologies used in the Recommendations developed up to now apply to both narrowband and broadband ISDN. Other issued Recommendations are specific for narrowband ISDN and, within this category, for those services and features having the greatest teletraffic impact on early ISDN implementations. Specific work on traffic engineering for

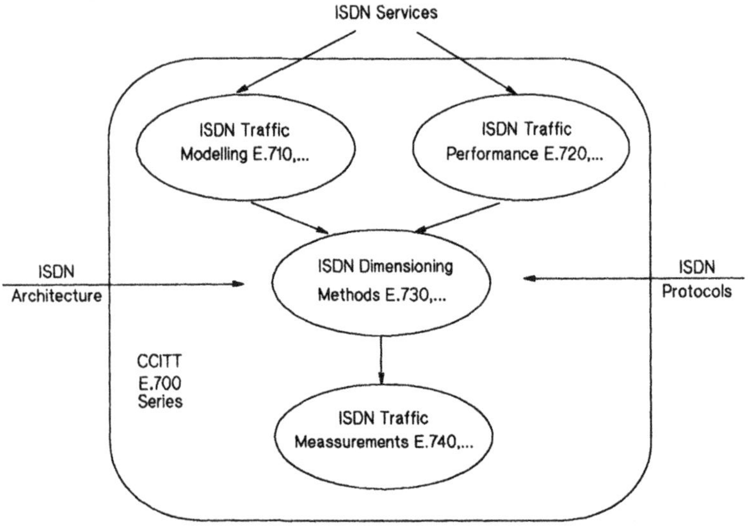

Fig. 1. ISDN traffic engineering tasks.

broadband ISDN started by the end of 1991. The allocation of a specific question on broadband ISDN for the starting study period is a challenge for a quick progress during the four coming years.

The paper is organized following the model of the E.700 series. A section will be devoted to each one of the four main traffic engineering tasks identified in the series. A fifth section will be devoted to the algorithms for traffic controls, given the importance that they will have in broadband ISDN. Each section will give a scope of the Recommendations already issued, with emphasis on those aspects which also apply to broadband ISDN, and will describe the specific current activities on broadband ISDN as well as the activities planned for the starting 1993-1996 CCITT study period. Information taken from CCITT SG II documents and from discussions within the Rapporteur Group will be complemented with personal views of the author.

2. TRAFFIC MODELING

User demand has been traditionally modeled by means of calls. In traditional circuit switched networks, calls were described by a few parameters: relative position of the calling and called users, mean holding time and call disposition were sufficient to describe a call. Users were characterized by their call attempt rate which, in practice, was assumed to be equivalent to the call demand rate.

In ISDN, the large variety of services with different transfer modes and rates, different procedures for establishing a communication, different

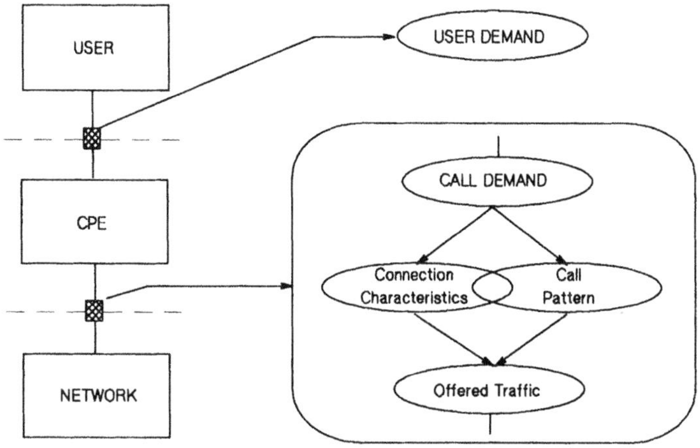

Fig. 2. User demand modeling through call demands.

communication configurations etc., makes the description of a call much more complex. Thus, a systematic approach to describe calls in ISDN has been developed in the E. 700 in order to have a common and comprehensive traffic modeling framework.

Starting with the user demand modeling (Recommendation E.711), the E.700 series develops two separate traffic modes: one for the user plane (Recommendation E.712) and another one for the control plane (Recommendation E.713).

Let describe the approach used to model the user demand by means of *call demands* since this approach may be common for narrowband and broadband ISDN. The user demand, as manifested at the user/CPE (customer premises equipment) interface (see Fig.2), is characterized by an arrival process and by a call duration and/or quantity of information as associated with each arrival. Through the mediation of the customer premises equipment, user demands are translated into call demands for specific telecommunication services at the user/network interface. It is worth nothing that call demand instead of a call attempt is considered as the basic manifestation of the user needs for communications, since, besides the reattempts which can be generated by the user itself, many terminal equipment are able to generate automatic call attempt sequences for each call demand. Thus, the equivalence between call attempt rate and call demand rate, assumed in practice for traditional circuit-switched services, can no longer be retained.

A call demand is defined by a set of connection characteristics and by a call pattern.

The *connection characteristics* are described by the values of the attributes defined in the I series Recommendations which are significant for traffic engineering. Example of these are:

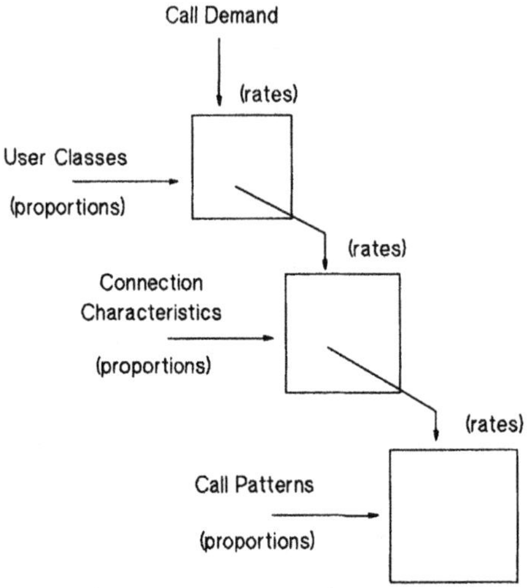

Fig. 3. User caracterization.

- information transfer mode (circuit switched, packet switched, ATM);
- information transfer rate (peak rate in case of packet switched or ATM);
- establishment of communication (reserved, on demand);
- communication configuration (point-to-point, multipoint, broadcast);
- signaling access protocol for layers 1-3;
- supplementary services.

The *call pattern* of a call demand is described in terms of sequences of events at the user/network interface and times between these events. The call pattern is defined by a set of traffic variables. Two kinds of traffic variables are distinguished:

- *Call variables*, which describe events and times between events during the call setup and release phases;
- *Transaction variables,* which describe events and times between events during the information transfer phase.

In particular, call variables are those related to the call attempt arrival process for each call demand and to the call attempt durations while transaction variables are related to the transaction arrival process (packet arrival process in packet-switched, cell arrival process in ATM), and to the transaction lengths (packet lengths) within a call.

Based on the call demand modeling, users are characterized by call demand rates for each type of call demand. This characterization proceeds in three steps, as illustrated in Fig. 3. The user population is divided into classes considered homogeneous with respect to communication needs. In the first step, the proportion of users in each class and the call demand rate per user of each class is evaluated. In a second step the proportion (for each user class) of call demands using each specific set of connection characteristics is considered. In the final step, the proportions of each user class/connection characteristics combination using specific call patterns is introduced to complete the characterization.

The characterization of the users by their call demand rates for each type of call demand also apply to broadband ISDN, as well as the modeling of the call demands by means of the connection characteristics and the call pattern. The connection characteristics to be considered in broadband ISDN are similar to those considered in narrowband ISDN. Among the traffic variables defining the call pattern, call variables are also similar in narrowband and broadband ISDN, but transaction variables require a specific study in broadband ISDN.

In narrowband ISDN, transaction variables are used in packet switched transmission mode. Much of the experience gained in the definition of the packet arrival process in packet switched services can be used in the definition of the cell arrival process. However, this problem takes a special importance in broadband ISDN since the full integration of services has required the definition of certain traffic control mechanisms. CCITT SG XVIII, in Recommendation I.371 [5], has defined the CAC (connection admission control), based on a set of traffic parameters declared by the user at call setup. This set of traffic parameters is called *source traffic descriptor*. Since the CAC is a fundamental control mechanism in ATM, an urgent definition of the parameters of the source traffic descriptor is required; these parameters should be based on the transaction variables which characterize the cell arrival process.

In spite of the close liaison between transaction variables and parameters of the source traffic descriptor, there is a clear distinction between them. In the views of SG II, the following two differences exist:

- Transaction variables are statistical variables which, by its own nature, cannot be surveyed. However both the source and the network should be able to determine through out the duration of the call whether the realized traffic is compliant with the parameters of the source descriptor; thus, instead of a statistical definition, an operational (i.e., algorithmic or ruled-based) definition is much more useful for the parameters of the descriptor. With an operational source traffic descriptor, the parameters negotiated between user and network would be the parameters of the algorithms: each traffic parameter of the descriptor is then defined by an algorithm with its associated parameters.
- Transaction variables represent the expected behavior of the source; meanwhile, the traffic parameters of the descriptor represent the worst possible behavior of the source, since the values of these parameters must never be exceeded, otherwise, actions will be taken against the source.

Keeping these differences in mind, the study of transaction variables and traffic parameters of the descriptor should be linked to each other. The basic requirement of the transaction variables is that they define the impact of the source on network performance; a certain set of transaction variables will be good if two sources with different cell generation process, but with the same values of the transaction variables, have similar impact on network performance.

The traffic parameters of the descriptor must be derived from the chosen transaction variables. Somehow, each parameter of the descriptor will be the result of the "degeneration" of a statistical parameter (the transaction variable) into an operational one; but the faculty of the transaction variables to define the impact of the source on network performance must not be lost in this degeneration. In this sense, a source traffic descriptor will be good if the worst possible source compliant with the parameter values declared by an actual source has not much worse impact on network performance than the actual source. A good source traffic descriptor will lead to a cost effective connection admission control, since it will allow to allocate to the call a bandwidth quite close to the one which it really needs.

Other important requirements of the parameters of the source descriptor are, as stated in Recommendation I.371:

- The user or his/her terminal must be able to predict them at call setup or, at least, to shape the traffic to be compliant with them.
- The network must be able to survey them by means of the UPC/NPC (usage/network parameter control).

Since the definition of the transaction variables and of the source traffic descriptors is crucial for the rest of the traffic engineering task, current SG II work on broadband ISDN has focused on this topic. Contributions received up to now have allowed to clarify the ideas above expressed. A state of the art on transaction variables proposed in the literature has been prepared to have a basis for discussion. Another contribution under discussion proposes a flexible approach which could match with different degrees of evolution of the network or different abilities of the sources to predict or to shape its behavior:

- For the first ATM networks, a descriptor with only one parameter could be used until experience is gained.
- For a more advanced network, a descriptor with two parameters (the previous one and an additional one) could be used. Those sources only able to predict or shape the first parameter could use the one parameter descriptor and the network would assume the worst case for the second parameter.
- In a more comprehensive network, a descriptor with three parameters (including the two previous ones) could be declared, allowing that the less sophisticated sources declare the one parameter or the two parameter descriptor.

The parameters proposed in the contribution are expressed in a statistical way; thus, further study is required to "degenerate" them into operational ones. The contribution proves that a worst case source can be found for each declared descriptor, regardless it is a one, two or three parameter descriptor. The bandwidth allocated to each source will be closer to their needs if it declares a more complete descriptor; thus, with an appropriate tariff policy, sources are encouraged, but not obligated, to declare a more complete descriptor.

Attention is also being paid in SG II to the effectiveness of traffic descriptors to describe particular types of real sources; e.g., a contribution has been presented comparing the ability of source descriptors defined by leaky bucket or sliding window algorithms to describe the cell generation process of VBR video sources.

Work on transaction variables/source descriptor will continue with high priority in SG II. The work plan includes the development of a Recommendation on traffic characterization with 1994 as target date. A close liaison with SG XVIII is mandatory to achieve successful results.

3. ALGORITHMS FOR TRAFFIC CONTROLS

As was said before, the use of the ATM technique emphasizes the importance of the traffic control, due to the integration of a great variety of services with different traffic characteristics, particularly with respect to the cell rate during the information transfer rate, and different QoS requirements. The stringent requirements on cell delay for real time services, such as audio or video services, prevent from using flow control mechanisms within the network. Thus, under statistical multiplexing, the cell flow of a connection may significantly alter the cell flow of other connections. To guarantee the network performance required by the different services a set of traffic control procedures have been defined by CCITT SG XVIII. Thus, traffic engineering Recommendation should be made to give a guidance on algorithms for those controls which guarantee that the network performance provided to the users complies with the specifications.

The controls defined by SG XVIII in Recommendation I.371 are:

- Usage/Network Parameter Control (UPC/NPC)
- Connection Admission Control (CAC)
- Traffic shaping
- Fast Resource Management
- Network Resource Management
- Priority Control and Selective Cell Discarding
- Explicit Forward Congestion Notification

Let's go through all of them to show what could be the activity of SG II with respect to them.

The algorithms for UPC/NPC are closely linked to the definition of the parameters of the source traffic descriptor. As was said in section 2, an operational definition of these parameters is much more useful. An operational definition means that the parameter is defined by means of an algorithm, wich would be the one used by the UPC/NPC to survey the traffic. Thus, the algorithm should be simple enough to be implemented in hardware with a response time smaller than the cell insertion time. A close liaison will be required between SG II and SGVIII to assure, on one hand, that the parameter defined by the algorithm maintain the property of representing the impact of the source on network performance and, on the other hand, that it is implementable.

The algorithms for CAC require as input data the parameters of the source traffic descriptor. Thus, final Recommendations on these algorithms cannot be given until decisions are taken on the source traffic descriptor. Nevertheless, as urgent guidance has to be done on CAC algorithms for the first broadband ISDN implementations, work should progress so fast as the degree of definition of the source traffic descriptor would allow. By now, discussion can start on which type of algorithm can be used: one based on the descriptor declared for all the connections (the already established ones and the one requesting establishment), or one based on the descriptor declared for the new connection and the load measured for the already established ones.

SG XVIII has defined the peak cell rate as a mandatory parameter of the descriptor. Probably, it will be the only parameter in the first network implementations. For this simple descriptor, a CAC algorithm of the first type described above is very obvious. A CAC algorithm of the second type is not so obvious, and studies on it could already start.

Traffic shaping and Fast Resource Management are not considered mandatory but optional by SG XVIII. Traffic shaping may be very useful for the user to assure that its realized traffic complies with the parameters declared. The algorithms of these traffic shapes will be derived straightforwardly from the operational definition of the parameters of the source descriptor. Traffic shaping in the network in conjunction with UPC/NPC as well as fast resource management are mechanisms which could improve the network utilization. Its implementation could be desirable if the previous mechanisms, UPC/NPC and CAC, are not able to achieve a cost effective network. Thus, studies on UPC/NPC and CAC algorithms and evaluation of their efficiency should be done before studying the convenience of traffic shaping in the network and fast resource management.

Network Resource Management, Priority Control and Selective cell Discarding are mechanisms closely linked to the dimensioning practices and will be treated in section 5.

Finally, Explicit Forward Congestion Notification is a mechanism which can be useful to assist the network in case of cell level overload. But since the response of the users to this notification is optional, the best recommendation to the network operators seems to be that they should not rely on this mechanism to control overload, as already stated in Recommendation I.371. No other guidance on this topic from SG II looks necessary, at least by now.

The work plan in SG II on algorithms for traffic controls is to elaborate the provisional text of a Recommendation in 1994 to be circulated to other Study Groups. This text will be the basis for further studies to be finalized in a second issue in 1996.

4. TRAFFIC RELATED NETWORK PERFORMANCE

Standardization of traffic related performance parameters for the ISDN represents a key task because, together with traffic modeling, it forms the basic imput to the dimensioning of an ISDN network and its components. Traffic related performance targets in the ISDN are being developed "top down"; that is, end-to-end performance parameters for major service categories are selected and defined, then targets for these parameters are established to meet user expectations, and finally these targets are subdivided with allocation of performance requirements to individual network components.

It should be noted that many of the delay grade of service parameters in ISDN are determined by the network and protocol architecture of the out-of-band access and network signaling deployed in ISDN. Thus the reference architecture and performance requirements in the user plane of the ISDN are closely linked with those in the control plane.

Presently, Recommendation E.721 proposes and defines end-to-end grade of service parameters for circuit-switched services and establishes target values for them. Recommendation E.723 makes the partitioning of them with allocation of targets to the components of the SS#7 network. Recommendations have to be developed for packet switched services in narrowband ISDN and for broadband ISDN. In particular, in broadband ISDN, traffic related network performance parameters should be characterized (for each service class) for two phases:

- call setup and release phase; and
- information transfer phase.

The grade of service parameters for the call setup and release phases should refer to the permissible delays at these phases as well as to the call blocking probability.

For the information transfer phase the service integrity parameter values should refer to the permissible quality degradation for each service class due to cell delay and cell loss probabilities.

A Recommendation on the subject has been foreseen for 1993.

5. DIMENSIONING METHODS

This section of the E.700 series intends to provide dimensioning rules to assure that sufficient network resources are allocated to handle the offered traffic achieving the traffic related network performance objectives.

The user/control plane perspectives have been used to provide Recommendations dedicated to resources mainly dealing with user plane loads or to resources mainly dealing with control plane loads.

In the user plane, Recommendation E. 731 is dedicated to narrowband ISDN resources operating in circuit switched mode. It provides specific dimensioning techniques to issues such as multi-slot connections, service protection methods, attribute negotiation, service reservation or multipoint connections, which are new in ISDN. Recommendation E.732, not yet developed, will provide dimensioning methods for narrowband ISDN resources operating in packet-switched mode. In the control plane, Recommendation E.733 is dedicated to resources of the SS#7 network. The criteria used to provide dimensioning methods to be applied in the SS#7 network are different from those used in the transport network. In fact a high level performance of signaling network at all times is essential to ensure the survival of the whole ISDN. Thus, the traffic loading efficiency of the signaling devices is not the primary consideration in the dimensioning, but the performance of the network under failure and excessive call demand conditions.

In broadband ISDN, the network signaling will be carried by the transport network also using ATM techniques. Thus, since both user and control loads are carried by the same resources, the user/control plane distinction, although necessary for traffic modeling, is not necessary for dimensioning (except, perhaps, in aspects related to the reliability requirements for signaling).

However, distinction has to be made between cell level dimensioning and call level dimensioning.

In the cell level, guidance should be done on dimensioning of the cell queuing buffers. There is a strong interrelation between buffer dimensioning and CAC, since the number of connections which the CAC can accept for a certain link strongly depends on the buffer length. Thus, studies on CAC algorithms and buffer dimensioning should be carried out simultaneously.

In the call level, for link dimensioning, the CAC algorithms used is an imput data, since CAC determines the bandwidth required by each connection. An urgent task is to give guidance on link dimensioning for the first B-ISDN implementations, when a source traffic descriptor with only peak cell rate will probably be used. In this case, if the CAC is not based on the measured load, but only on the peak cell rate declared by the sources, dimensioning methods for multy-slot connections in circuit switched networks could be easily extended to broadband ISDN. The extension of these methods should be included in the first issue of Recommendation foreseen for 1993.

In the second issue of the Recommendation, scheduled for 1996, dimensioning rules for networks using more sophisticated source traffic descriptors and CAC algorithms should be provided.

Two important issues very related to the dimensioning are the use of cell loss priority (CLP) and the management of virtual paths, in turn related to the fact that different connections can request different QoS or even a same connection can request different QoS for cells with CLP=0 or with CLP=1. SG II must give guidance to the operators on the following questions:

- What is the gain in link utilization obtained with the use of CLP and selective cell discarding ?
- An optimum selective cell discarding mechanism, which assures that a high priority cell (CLP=0) is never lost if a low priority cell (CLP=1) can be lost instead of it, could be difficult to implement. How much does the

effectiveness of the selective cell discarding decrease if simpler mechanisms are used?

- By reserving capacity on virtual path connections (VPC), the processing required to establish individual Virtual Channel Connections (VCC) is reduced, since connection admission decisions on individual VCCs have to be taken only at nodes where VPCs are terminated. On the contrary, this strategy can reduce permissible link utilization. How much can this reduction be?
- Connections requesting different QoS could be aggregated or segregated into different links. What is the gain of each strategy?
- VCCs requesting different QoS could be segregated into different VPCs of a same link. Is there any gain if there is not any mechanism to prevent that the cell flow of a VPC alters the cell flow of other VPCs? What type of mechanic would be necessary? What is the gain obtained?
- Bursty connections and CBR connections could be aggregated or segregated into different links. What is the gain of each strategy?

In order to give guidance on these questions, SG II has planned to issue a pamphlet on the impact of design alternatives on network dimensioning. This pamphlet, which can also illustrate the impact of the use of different source traffic descriptors or different CAC algorithms, will complement the Recommendation on dimensioning rules.

6. TRAFFIC MEASUREMENTS

This section of the E.700 series will describe requirements for traffic measurement and performance monitoring. The objective of these measurements is to provide data which can be used by the Administrations for the following applications, listed from shorter term to longer terms.

- Network management, which requires a real time supervision of the network to take prompt actions to reduce the effect of events such as failures or abnormal increases in traffic demand affecting the network unfavorably.
- Network operation, which requires a traffic and performance monitoring to detect bottlenecks or congestion problems in the network in order to take actions such as rearrangements for load balancing or provision of new resources.
- Network dimensioning, which requires to have a data base allowing to know the offered traffic when a new exchange or route is dimensioned.
- Network planning, which requires a data base for the forecasting of future traffic demands

In broadband ISDN, most measurements required for above applications are call level measurements. Emphasis should be made on classifying these traffic and performance measurements according to the connection characteristics and source traffic descriptors of the calls.

In the cell level, global measurements (i.e., no call-by-call measurements) will be required to check, for example, whether the traffic control algorithms work properly.

The plan in SG II is to develop a Recommendation on measurement requirements in B-ISDN, with two issues: a first one in 1993, including the requirements which can be extrapolated from those of present digital exchanges, and a more comprehensive second issue in 1996.

7. CONCLUSIONS

This paper has presented the CCITT E.700 series, framework for Recommendations on traffic engineering for ISDN, and has described the current and future work in Study Group II to enhance these series with specific Recommendations for broadband ISDN.

Advanced research on traffic engineering for broadband ISDN has been carried out worldwide, leading to fruitful results which constitute an excellent basis for elaborating Recommendations. Nevertheless, a lot of work has yet to be made to achieve a comprehensive set of Recommendations. The author wishes to encourage traffic experts to participate in the work of CCITT SG II in order to achieve the required quick progress on this subject.

REFERENCES

1. CCITT, E.700 Series Recommendations, Recommendations E.700, E.701, E.710, E.711, E.712, E.713, E.720, E.721, E.723, E.730, E.731, E.733, in press.
2. CCITT SG II, Document AP X-6: "Proposed new work program (Questions) for the next study period (1993-1996)", Geneva, February 1992.
3. D. Grillo, A. Lewis, R. Pandya, M. Villén-Altamirano, "CCITT E.700 Recommendation series - A framework for traffic engineering of ISDN", IEEE JSAC, Vol. 9, nº 2, February 1991.
4. CCITT, Blue Book, Vol. III.7, I.100 series Recommendations, Geneva, 1989.
5. CCITT, Recommendation I.371, "Traffic control and congestion control in B-ISDN", in press.

Integrated Broadband Communication Networks
and Services (C-18) / V.B. Iversen (Editor)
Elsevier Science B.V (North-Holland)

Standardisation of B-ISDN

Jørn Johansen
Tele Danmark Research
c/o IT, DTH, Bygning 343
DK-2800, Lyngby, Denmark

Abstract

This paper describes the standardisation process for B-ISDN from the selection of ATM as the transfer method until the present date where most of the ATM functionality's are specified. It states the current status of the B-ISDN standards and identifies still open questions. Traffic issues are not dealt with in detail since other papers treat this subject specifically. The paper finally addresses the work presently done in ETSI/NA5 and discusses the future work to be done. Emphasis is put on the B-ISDN work done and going on in CCITT SGXVIII and ETSI/NA5.

Keyword Codes: H.0; H.1.0
Keywords: Information Systems, General; Models and Principles, General

1. HISTORY

In 1986 Asynchronous Transfer Mode (ATM) was chosen as the basic transfer method for information of all kinds in the Broadband-Integrated Services Digital Network (B-ISDN). During the CCITT study period starting in 1988 most of the characteristics and functionality's of ATM was agreed. Most important was the decision of 1989 where agreement was reached on the cell size (5 byte header and 48 byte payload). By 1991 the signalling requirements for B-ISDN release 1 were finished and in the process the need for more sophisticated signalling systems for future releases was identified. During 1992 the various ATM Adaptation Layer (AAL) types matured and in the beginning of 1993 the earlier identified need for an AAL type 5 led to a specification covering the most important issues. Shortly afterwards this AAL type was chosen to be the AAL type to support signalling both at the User Network Interface (UNI) and at the Network Node Interface (NNI).

Other B-ISDN issues have also been treated during the period: Operation And Maintenance (OAM) methods and procedures, traffic issues, Telecommunication Network Management (TMN) issues, Services to be carried over ATM networks, Evolution, Network Architecture, Interworking and more. Much work still need to be done before B-ISDN is satisfactorily standardised, but by the end of 1993 the protocols for release 1 signalling should be finished.

2. STRUCTURE

The main work in CCITT concerning B-ISDN has been done by:

- Study Group XVIII which is responsible for specifying the concept and the network architecture (I recommendations).
- Study Group XI responsible for signalling protocols (Q recommendations).
- Study Group I responsible for specification of services as seen by the user (F recommendations).

The result of this work is contained in:

- 13 I recommendations covering Vocabulary, Functional characteristics, Service aspects, ATM and ATM Adaptation Layer specifications, Interfaces and Operation, Administration and Management principles.
- A number of (draft) Q recommendations covering Metasignalling, B-ISDN signalling for the first release, the structure of future releases and Q.SAAL (The Service Specific Parts of AAL to be used on top of AAL 5 to support signalling for all releases both at UNI and NNI).
- Some rather vague F recommendations covering descriptions of services for B-ISDN.

3. STATUS

The status of recommendations from the different Study Groups will be described and some of the open points in them will be identified.

3.1 The I recommendations

The presently agreed cell header for the UNI is shown in Figure 1.

<table>
<tr><th>8</th><th>7</th><th>6</th><th>5</th><th>4</th><th>3</th><th>2</th><th>1</th><th></th></tr>
<tr><td colspan="4">GFC</td><td colspan="4">VPI</td><td>1</td></tr>
<tr><td colspan="4">VPI</td><td colspan="4">VCI</td><td>2</td></tr>
<tr><td colspan="8">VCI</td><td>3</td></tr>
<tr><td colspan="4">VCI</td><td colspan="3">PT</td><td>CLP</td><td>4</td></tr>
<tr><td colspan="8">HEC</td><td>5</td></tr>
</table>

Figure 1. ATM Cell Header at UNI.

- GFC, General Flow Control. There are still 2 proposals for the protocol and no final decision is made for the protocol that is intended to assist the user in sharing the available bandwidth between several terminals connected to the same interface.

- VPI and VCI, Virtual Path Identifier and Virtual Channel Identifier. Identifiers to identify cells belonging to the same connection.

- PT, Payload Type. Mainly used for identifying the type of information in the 48 bytes payload field: User Information, OAM Information or Inband Signalling. Also used for indication for information that has passed network elements where congestion has been experienced and to transfer User to User information (poor OSI model!) notably to be used in AAL5 to identify the last cell of a segmented message.

- CLP, Cell Loss Priority. Used to distinguish cells which should be dropped first if a cell stream meets congestion. It is still a possibility to use this bit for violation tagging when a user exceeds his contracted cell rate.

Originally ATM was considered strictly unidirectional and there are still reminiscences of this in I. 311 but from I. 610 and I. 150 it is clear that the VPI/VCI in the opposite direction of an unidirectional user connection only can be used for transfer of data between the same two users and OAM information concerning the user connection.

Concerning the actual measuring of the performance of an ATM connection there are still many open questions before procedures for this are fully specified. Also in the traffic area many issues are unresolved, i.e. Connection Admission Control if statistical multiplexing is used, the User/Network parameter control algorithm, Granularity for the specification of bandwidth, countermeasures against cell clumping effect caused by GFC in the user environment and buffering in ATM switches.

There are also many open points concerning: Charging, procedures for performance measuring and fault management and capabilities for multipoint connections.

3.2 The Q recommendations

The first release for B-ISDN signalling is based on the signalling systems used in N-ISDN, Q931 at UNI and ISDN User Part(ISUP) from Signalling System Number 7 (SS7) at NNI. Open issues in this area are mainly the traffic parameters and the measures to be taken to ease co-operation and evolution to release two and three.

For the next releases it is decided to separate call control and connection control, but it is still not completely clear what this means in detail. The relation between call control and the Intelligent Network (IN) concept is not defined. There are at least two views on which basic assumptions to build the future releases on:

1. Connection Control shall be based on Q931 at UNI and B-ISUP at NNI, and Call Control should be a completely new protocol.

2. The same protocol shall be used at both UNI and NNI.

This means that very little can be said about future releases. Preliminary descriptions can be found in the BAP document (Broadband Access Protocol) and the ISCP Baseline Document (ISDN Signalling Control Part). Covered is the structure, the Association Control and the need for negotiation of the Application Context.

3.3 The F recommendations

This recommendations lack many details and definitions of the many new and exciting services that B-ISDN is supposed to provide are not available. The recommendations treat connection oriented and connectionless broadband bearer services, the distributive services (TV and HDTV) and the teleservices (video telephony, video conference and videotex).

4. ONGOING WORK IN ETSI/NA5

The work in ETSI/NA5 is divided between 8 groups:

- AAL: ATM Adaptation Layer.
- BMA: B-ISDN Management Aspects.
- NAE: Network Architecture and Evolution.
- OAM: Operation, Management and Maintenance.
- PRS: Protocol Reference Model and Signalling.
- SAS: Service Aspects Support.
- TCR: Traffic Control and Resource Management.
- VCM: Video Coding Matters.

The main issues discussed during the last meeting are:

AAL: Support of unidirectional high quality audio and conversational video services. Timing issues for AAL 1, and an ETS for AAL 5.
BMA: Managed Object Model for an ATM crossconnect, and future work on the X interface.
NAE: Interworking with MANs, Interworking with N-ISDN and Point to Multipoint connections at ATM level.
OAM: Policing of OAM cells, loop backs, protection switching and charging.
PRS: Numbering, Addressing and Routeing Aspects. Identification of future work in the areas signalling for release 2 and 3 and IN. Functional Model for an ATM crossconnect.
SAS: Connectionless Broadband Data Services over ATM and access enforcement for these.
TCR: Granularity of the cell rate description parameters, OAM cell policing, policing algorithms, definition of a compliant connection and traffic parameters.

5. CONCLUSION

Even though many unresolved issues are mentioned in this paper, much has been achieved since standardisation of B-ISDN was started. A common agreement on the use of ATM as the transfer mode for B-ISDN has been reached. The cell format is decided and alignment with DQDB has been achieved. The fundamental characteristics for ATM has been specified. ATM has been adopted by the manufactures to be used in LANs to interconnect workstations. The first orders for ATM equipment for public networks has been given and the building of a pan European ATM crossconnect network is in preparation.

REFERENCES

1. I, Q and F recommendations (draft and approved) from CCITT.
2. Report from the Bern meeting marts 1993 of ETSI/NA5.

Integrated Broadband Communication Networks
and Services (C-18) / V.B. Iversen (Editor)
Elsevier Science B.V (North-Holland)

An Evolving Scenario of Communication Network towards B-ISDN

Tadao Saito

Department of Information and Communication
The University of Tokyo
7-3-1, Hongo, Bunkyo-ku, Tokyo 113, Japan

As the component technologies for B-ISDN are now approaching to final development stage. To realize B-ISDN as the infrastructure of the society, various problems are discussed in Japan. This paper explains a scenario to construct B-ISDN network as a infrastructure of future discussed concerning Japanese society.

1.INTRODUCTION

In recent technology, electronics is prominent in their rapid advances and cost reduction. Rapid improvement of cost-performance of various types computers is the benefit of advanced electronics. In recent 30 years, cost performance of computers is evaluated to be improved by the order of 10^3 or 10^4. Although communication technology is also based on electronics, the improvement of cost performance has not been very significant compared with computer.

We have many information handling means. Package type information distribution means represented by VCR. Broadcasting system represented by TV broadcasting. In these information handling means also, a shift from audio signal to video signal experienced in the past 10 ~ 20 years achieved similar cost performance improvement of the order of 10^3 . Communication system which is based on end-to-end connection only has not achieved significant cost performance improvement taking advantage of rapidly developing electronics.

Information handling means surrounding us are in some competitive relationship. Of course some functions of communication systems cannot be replaced by other information handling means, some replaceable portion is now shifting to less costly means. In design of on line computer system, effort to minimize the usage of telecommunication is a common practice. In down loading of computer software, a shift from telecommunication to broadcasting type media is significant.

These tendency is now shrinking the market of telecommunication industry. Frankly observing, the position of telecommunication industry in the whole electronic industry is falling down rapidly.

Broadband ISDN is the technology to recover the position of the communication industry from the present position. By means of B-ISDN technology, end to end telecommunication capacity is expanded by the ratio exceeding 10^3. If the cost of telecommunication can be kept in the same order as the present service, cost performance improvement of 10^3 will be achieved.

This paper summarizes recent discussions to realize B-ISDN as the infrastructure of the 21st century to enhance the importance of communication technology in the electronics industry.

2. BASIC COMPONENTS OF BROADBAND ISDN

In this paper Broadband ISDN is conceived as a network to provide very high speed 150Mb/s a 600Mb/s end to end link to majority of communication user. Needless to say, communication networks are composed of long line transmission systems, subscriber lines and transit and local switches. In realizing Broadband ISDN, major reformation are needed to the basic 3 components. To provide the improvement of cost performance in the order of 10^3, the cost performances of the 3 components should be improved by the same order.

(i) Coherent optical long line transmission systems:

In the history of technical development, the first runner toward broadband ISDN is the long line transmission technology. Especially, coherent optical fiber transmission technology was proposed more than 10 years ago. It is known that using coherent optical fiber communication, transmission speed of 10 Gb/s is realizable by one optical frequency and more than 100 optical frequency can be multiplexed for example into 1.5μm wave length range. This means that transmission capacity of 1 Tb/s is realizable by one optical fiber. If the basic transmission speed is 150 Mb/s, this capacity roughly corresponds to 6000ch. The multiplexity is an appropriate size for a major transmission route. The component technologies to realize high capacity transmission line is near to completion, however, of no use if broadband ISDN is not realized.

(ii) Optical subscriber lines

Application of optical transmission for subscriber access is tried in various countries. Variety of experimental systems were constructed from 1970's. In Japan Hi-OVIS experiment in 1970's and INS Mitaka experiment in 1984 are well known. These early experiments mainly use analog video transmission. Recently, optical subscriber access systems are used for high speed digital line including ISDN primary rate user interface. In Japan majority of subscriber access lines of Tokyo Telecommunication Network (a local new common carrier of Tokyo area) are optical fibers.

Optical subscriber lines economically justifiable are still high capacity office use only. This is called FTTO (fiber to the office). To use optical fibers for home application (FTTH) some cost sharing of optical fiber is considered to be essential at present cost evaluation.

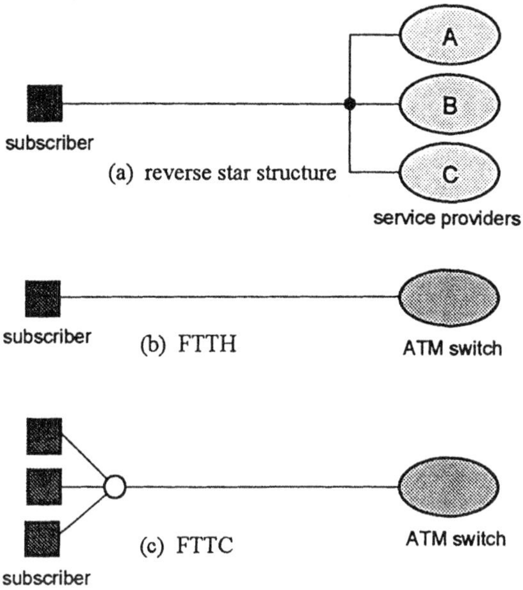

Fig.1 Various optical fiber subscriber plants

Fig.1 shows 3 schemes for optical fiber application for subscriber access for home case. Fig.1(b) is typical FTTH (fiber to the home). The optical subscriber line is used only for telecommunication application. The cost of this type of FTTH is still difficult to justify. One possibility to reduce per subscriber cost of optical subscriber line is to share a optical fiber by a number of line terminals, as shown in Fig.1(c). This style is called FTTC (fiber to the curve). The branching at subscriber side may be passive or active, which is called passive double star or active double star.

Another possibility is to provide one optical fiber for one subscriber and use the fiber by a number of service providers. For example, service providers may be local communication carriers, CATV service providers or other service providers. This is shown in Fig.1(a). If wave length division multiplexing is used in service provider side, service providers can be operated independently.

(iii) ATM local switches

Although the above two components have still difficulties in the cost for realization, key technologies have already been available. Because of this reason, ATM switching technology is considered to be the last obstacle to realize broadband ISDN. Therefore recent studies for broadband ISDN are concentrated on ATM switch development.

Switching technology generation is renewed every 10 years in recent history of switching system development as shown in Table 1. One reason for the cyclic development is the development capability of major research laboratories in the world and also the life cycle of switching system products delivered by manufacturers. To recover the development cost, it is needed to put a product for 10 years. It is also needed to develop new systems taking advantage of new technology to keep competitiveness. Some of key technologies are inherited through switching system generations.

In Japan, basic software technology for switching systems was developed for space division switching systems and inherited through ISDN switches with major enhancement. Basic switch technology of the digital switching system was inherited to narrowband ISDN. In this

Table 1
Development cycle of switching system

1965	Stored Program Switch
1975	Digital Switch
1985	ISDN
1995	ATM
2005	Optical Switch (?)

sense digital TDM switch principle had life of two generations. ATM switch technology is based on a completely different principle. The study of the technology started just after the development of ISDN switches and expected to be generally available in 1995.

When the three components become available broadband ISDN services will be a basic services for public case. How to synchronously deploy the three components to realize broadband ISDN is a major concern in the planning for the new generation telecommunication network based on B-ISDN .

3. INVESTMENT FOR BROADBAND ISDN

One of important concern about the B-ISDN is, how much the B-ISDN communication costs. Since we do not have such facilities installed some predictions are needed to forecast the cost.

The first data to predict the B-ISDN fare is how the revenue of telecommunication carrier is broken down. Table 2 shows one typical value of the breakdown of the revenue taken from the annual report of NTT, Japan 1992. Total earning of NTT of 1992 was about $50 B (US$=120 JYen). This table shows that the amount used for investment for facilities is about 30% of total communication cost. The amount is further broken down into various technical facilities.

Table 2
Typical breakdown of revenue of
a telecommunication business

items	%
Labor	35%
Facility investment	30%
Others	25%
Profit	10%

Table 3 shows composition of technical components in terms of accumulated investment. The data is obtained by analyzing the data reported by NTT and some figures are based on estimation. In this table local line means all the out door facilities for subscriber loops excluding tunnels for feeder cables which are classified into underground facilities.

Table 3
Accumulated investment by NTT
1952-1990

items	US$	%
Local line	42	21
Switch	58	29
Long line	13	6.5
Buildings	25	12.5
Underground facilities	38	19
Others	24	12
total	200	100

Long line includes transmission facilities and electronics at terminal equipments. Therefore all of the facilities are not electronics. Thus, in total investment of $200 B, the electronics for communication is just about 30% composed of switching systems and a part of long line facilities.

Aggregating Tables 2 and 3 it is known that components related to electronics is only about 10% of total telecommunication revenue paid by telecommunication customers. This means that even if cost of electronics come down drastically, the effect just contribute to 10% of total cost. This is the main reason why telecommunication service fare is not essentially decreased in spite of rapid progress of electronics.

Saying in different way, it may be possible to introduce high cost new services in the network, without essential increase in service charge. For example, if the cost of switch is doubled to give new services, total communication cost is increased only 10%. The increase may be covered by increase of traffic stimulated by the new services.

Discussing the B-ISDN in this context, the problem is how much will it cost to deploy nationwide B-ISDN services. In Japan there are several estimations in this subject. Table 4 is the result of estimation by a study group of Ministry of Post and Telecommunication (MPT),

Table 4

Forecast of amount of investment to realize B-ISDN of 50 million subscribers

items	predicted amount B$
Subscriber line	126
Local Switches	63
Long Line facilities	116
Total	305

(based on MPT report)

Japan. The cost estimation is to construct the network for 50 million B-ISDN subscribers. In this estimation, the investment is assumed to start 1995 and continues until 2015. In this estimation, long line facilities includes switches and transmission facilities. Various developments for intelligent network services are also included in the long line facilities.

In this estimation, majority of components of B-ISDN is assumed to cost 3 times compared with these for N-ISDN. For example, local ATM switch for B-ISDN cost about $1250 per subscriber. Optical subscriber line will cost $2500 per subscriber.

If the estimation for investment is correct, total network service cost for B-ISDN will be roughly 50% higher compared with present service cost. It is also noted that the cost increase of only 50% is surprisingly low considering the transmission capacity increase of 1000 times.

4. APPLICATION OF ATM SWITCHING TECHNOLOGY

At present, the effort towards B-ISDN is focused on the development of ATM switches. By this effort, it is quite sure that switching technology for B-ISDN will be completed in the middle of 1990's. However, it is less probable that FTTH will be enough prevailed at that time. Long line facilities have less problems. The facility will be constructed as demand increases. The demand will heavily depend on pricing structure.

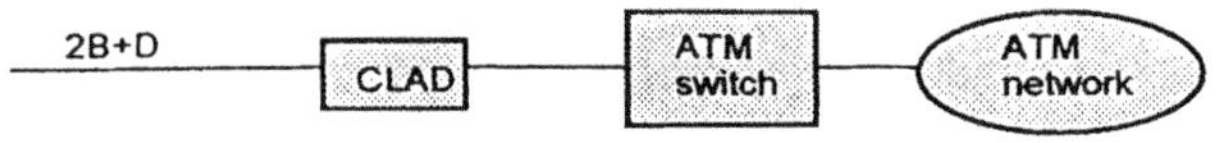

(a) N-ISDM service by ATM local switch

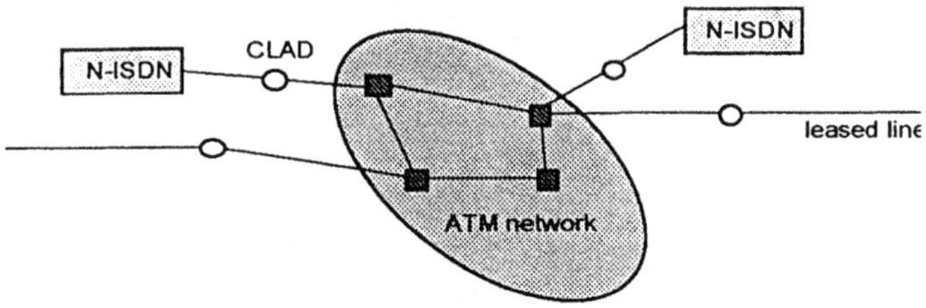

(b) N-ISDN service by ATM long line network

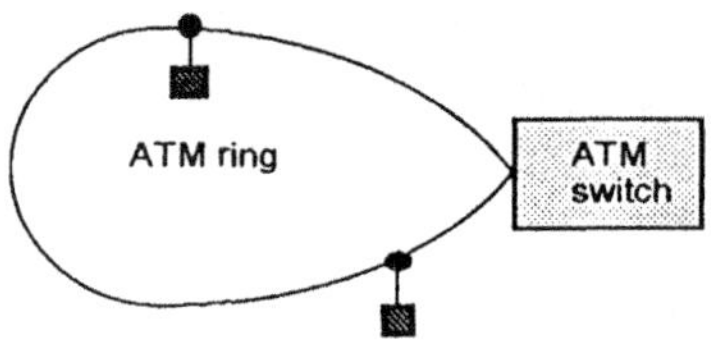

(c) ATM ring

(d) B-ISDN service through FTTH

Fig.2 Various applications of ATM switch

To cope with such a situation where only ATM switches are on market and other components are not available in real field, various applications of ATM switches have been discussed. Definition of B-ISDN by CCITT basically orients to this kind of applications.

Some examples of ATM switch applications are shown in Fig. 2. In the case of Fig.2(a), ATM switches are used as N-ISDN local switches which connects N-ISDN terminals through CLADs. In such application, requested traffic carrying capability is not very large. In this case, all the network including the long line network will be constructed by ATM principle. In the application shown in Fig.2(b), ATM long line network is constructed by ATM principle and local switches are constructed by means of N-ISDN switches. In this network, leased lines can be constructed directly through the ATM long line network. To prevail the ATM interface to more general use, multiplexed ATM network access using bus type or ring type links are under developments. This type of network will be useful to provide broadband services to small number of special users. Finally the network shown in Fig.2(d) is to provide broadband services to majority of communication users.

In the networks shown in Figs.2(a) and (b), majority of communication users are supposed to be N-ISDN or analogue telephone users. If ATM switches become economically competitive with N-ISDN switches, the structures shown in Figs.2(a) and (b) will be available replacing N-ISDN switches. In addition to the application for N-ISDN services, high speed data services can be added by means of the arrangement shown in Fig.2(c).

In traditional definition of ISDN, if a variety of long line networks are used, the unification of access line to variety of services is stressed to clarify the meaning of "integrated services." However if the networks are constructed like the arrangement shown in Figs.2(b) and (c), although the long line network is unified to support variety of services, the access network may be completely disintegrated. In other words different access facilities are needed to use different services provided by long line network.

The network structure shown in Figs.2(a)(b)(c) can be thought as the transient networks to reform to B-ISDN shown in Fig.2(d). The length of transient period will heavily depends on the pricing structure of the B-ISDN services.

5. B-ISDN AND PUBLIC INVESTMENT

Price is the key to prevail the use of B-ISDN. If payload capacity of 135.631 Mb/s is available as the standard interface, if the price is very high, the capacity will not be used by general public. In recent pricing system of digital services in Japan price of the communication line is roughly proportional to $B^{2/3}$ where B is the transmission speed. If this pricing system is used for B-ISDN, price for a speed of 1000 times is 100 times. This price will be prohibitive for general use of B-ISDN.

If the cost estimation of Table 4 is correct, as described in section 3, price estimation for B-ISDN using 150 Mb/s capacity can be 50% higher than present analogue telephone service. In this cost increase, almost half of the increase is to prevail FTTH facilities. As shown in Fig.1(a), the FTTH facility can be shared by a number of service providers including telecommunication service providers and CATV service providers, the cost of FTTH may be shared by a number of service providers. By the sharing the FTTH facilities the B-ISDN price may further be reduced.

To accelerate implementation of B-ISDN, it may be useful to put investment from public sectors into the FTTH. Since all the telecommunication services are privatized in 1985 in

Japan, a new framework for provision of telecommunication is needed to put public fund for B-ISDN. Discussions on the subject is attracting interest related with activation of economy by means of public investment.

6. B-ISDN PRICE STRUCTURE

In the pricing of B-ISDN services, more important problem may be not the absolute value of the price but the relative value of prices between various telecommunication services. The problem is related with traditional conventions on telecommunication pricing as the function of bandwidth, time duration and distance.

(i) Bandwidth: If the price for 150 Mb/s service is only 50% higher than the present telephone service price as described previously, this will be very attractive to promote broadband telecommunication services. However, if traditional pricing system roughly proportional to bandwidth is applied for narrow band services, the price will be too low which may result in insufficient revenue for telecommunication carriers. On the other hand, if price is independent from bandwidth, telecommunication reseller may take excessive profit. To resolve the problem, some new tariff structure for less dependent on bandwidth on information volume is needed.

(ii) Time: Considering the forecast that major use of B-ISDN will shift from point to point communication to video distribution type services, typical holding time is expected to be longer in B-ISDN services. In B-ISDN services based on ATM technology since a number of connection is possible if a few TV program transmissions are continuing, no reason exists to control time duration of local connection. Therefore, telecommunication charge dependence on time is a subject for reconsideration.

(iii) Distance: B-ISDN is a technology of the age when the cost of high capacity long line transmission is extremely low. Using optical transmission using coherent modulation technology, very high capacity long line transmission is realized efficiently. If cost dependent pricing system is applied, dependence of telecommunication charge on distance will become very low.

7. CONCLUSION

Component technologies of B-ISDN are in their final stage of developments. The main issue to realize B-ISDN services prevailed in society is not technical problems any more. The capability which is to be brought about by B-ISDN is so different from the capability of the present day telecommunication system. The capability is expected to be realized without essential increase in service price compared with the present narrowband communication. High capacity service charge with essentially identical to charge of the narrowband service may cause various problems in the transient stage from narrowband services to broadband services. The change must be made under the free competition condition. For smooth the transfer from N-ISDN to B-ISDN, really cost dependent changing system must be devised including all the scope of telecommunication.

REFERENCES

(1) Report of study group on broadband ISDN
Ministry of Post and Telecommunication, Japan (April, 1992)
(2) T. Saito: Broadband ISDN
Journal of Institute of Electronics, Information and Communication
Vol.75, No.11, pp.1193-1199 (Nov., 1992)

Integrated Broadband Communication Networks
and Services (C-18) / V.B. Iversen (Editor)
Elsevier Science B.V (North-Holland)

Tools and Methodologies for High Speed Network Design

A. Nilsson[a] H. Perros[a] and D. Stevenson[b]

[a]Center for Communications and Signal Processing, North Carolina State University, Raleigh, NC, USA

[b]Center for Communications, MCNC, Research Triangle Park, NC, USA

Abstract

The VISTAnet gigabit testbed project features the deployment of a network connecting distant supercomputers and research programs in communications, advanced graphics techniques, and a medical application. It is an environment that has been established to experiment with and evaluate techniques associated with gigabit networking. A description of the testbed goals, infrastructure and driving application are provided. Early insights from the testbed activities are brought out to identify several of the important challenges associated with networked supercomputers and their impact on public broadband integrated services digital networks.

Keyword Codes: B.4.4; C.2.1; C.2.2; C.2.3

Keywords: Performance Analysis and Design Aids, Network Architectures and Design, Network Protocols, Network Operations

1. INTRODUCTION

VISTAnet [1] is a three-year research effort (starting date June of 1990) that involves the deployment of prototype broadband switching systems interconnecting specialized supercomputers. The project partners include BellSouth, GTE, MCNC, North Carolina State University (NCSU), and the University of North Carolina at Chapel Hill (UNC-CH). In addition to the deployment and evaluation of experimental network facilities, the project supports research activities in several areas: gigabit communications, metacomputing, advanced visualization techniques, and medical treatment planning techniques. VISTAnet is a testbed. It is an environment that has been established to experiment with and evaluate tools and methodologies associated with gigabit networking. These experiments include methods for moving data into and out of host computers from the application to the network at gigabit rates, demonstrations and evaluations of an application that requires gigabit networking, and interactive three-dimensional graphics techniques. The overarching goal of the testbed is to advance the state of the art in gigabit networking and its application.

VISTAnet is one of five gigabit testbeds coordinated by Corporation for National Research Initiatives in cooperation with DARPA and NSF [2,3,4], see Fig. 1. Each testbed has a distinct mix of participants, network technologies, applications focus, and research activities. VISTAnet is comparatively geographically compact and focused on a specific set of problems.

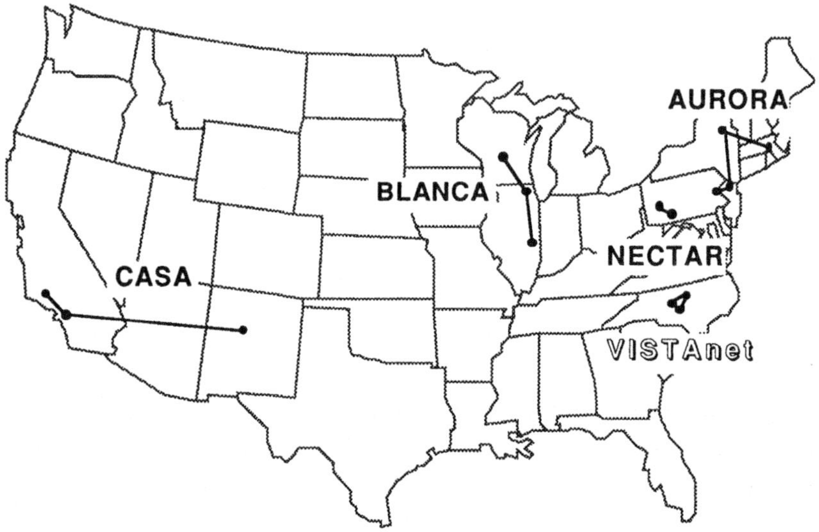

Figure 1. Gigabit testbeds under CNRI.

2. VISTAnet APPLICATION

It has been said that the purpose of computing is insight, not numbers. Images are an excellent method for understanding and gaining insight into the results of scientific computing models. Over the past few years, techniques associated with generating images from computational models have matured significantly, and now the preferred method of getting results from a supercomputer is through animated images. It has often been our experience that individuals associated with broadband communications think of supercomputing network requirements primarily in terms of remote users accessing moving images (two-dimensional data sets) across the network. This type of network usage requires data rates on the order of 150 Mbps to move high-quality images. Deployment of BISDN networks should enable more widespread access to supercomputing resources.

There is an additional kind of networking use for a supercomputing environment, known as metacomputing, that is much more demanding. In our experience metacomputing is poorly understood and not appreciated within the telecommunications industry. The idea of metacomputing is that a computationally difficult problem can be partitioned across multiple, and often specialized, computing machines to increase the speed of execution. In a heterogeneous computing environment, ideally this process is done to take best advantage of the particular strength of each machine. If the job is done properly, computational problems that can not be easily solved in a homogeneous environment may become quite tractable.

The driving problem in VISTAnet —Dynamic Radiation Therapy Planning — is based on metacomputing. In our case, the metacomputer is comprised of a vector pipelined supercomputer, a graphics processor [5], and a massively parallel processor. None of these machines individually can solve the problem at the rates we require.

It is important to note that the role of the network changes in a fundamental way in a metacomputing application. The network is no longer just the means for remotely accessing the computer; rather, the network actually becomes part of the computer. From the networking perspective, one of the most significant consequences is the high data rate required. Rather than moving images (two-dimensional data sets), as in the case of supercomputer access, the networked computers often exchange volume data sets (three-dimensional objects) in a metacomputing situation. Moving three-dimensional objects in real time can easily push the required data rates into the gigabit range. For VISTAnet, the computing machines are interconnected by ATM links operating at 622 Mb/s. In this paper we present some of the tools and methodologies that we have developed for the communication network.

3. TRAFFIC CHARACTERIZATION

3.1. HiPPI link data analyzer (HiLDA)

HiLDA is a custom-designed traffic analyzer for HiPPI [6,7,8] links that was designed primarily to provide real-time data capture for network traffic statistics. The captured data is used to validate statistical traffic models for gigabit networks. Additional features of the HiLDA system provide for programmable error insertion and a host HiPPI interface. The error insertion capabilities is used as part of the protocol performance evaluation activity. The host HiPPI feature is being used to provide a low performance HiPPI to SUN and Silicon Graphics servers.

In the HiPPI host interface mode, HiLDA acts as a network device interface for either a SUN4 or a Silicon Graphics machine. The HiLDA [9] system is installed and measurements have been taken of the HiPPI traffic [10]. We have also used the interface for communication with other HiPPI hosts. In the VISTAnet high-speed network, we use HiLDA extensively for analyzing the HiPPI traffic at all endpoints and as a HiPPI network interface for the medical workstation.

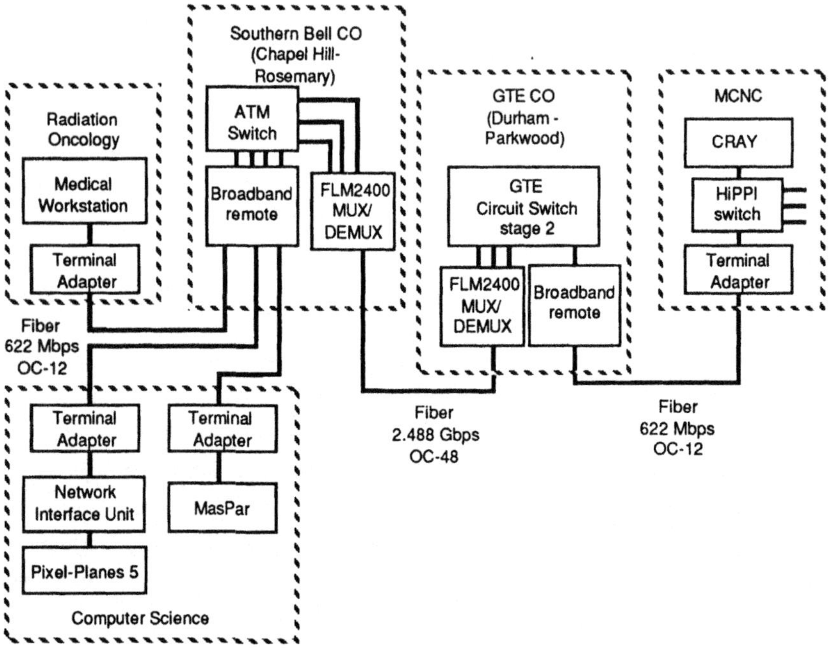

Figure 2. VISTAnet networking equipment

3.2. ATM traffic characterization

The importance of characterizing traffic in high-speed computer communications networks is critical, since many of the proposed congestion control mechanisms and network architectures for high-speed networks contain broad assumptions about the behavior of traffic sources. These assumptions need to be validated through actual traffic measurements in high-speed networks. Very little is known about the actual behavior of traffic in high-speed communication networks, mostly because there are so few experimental testbeds available to perform measurements.The traffic measurements which we have obtained from the VISTAnet testbed are from the medical application Dynamic Radiation Therapy Planning. Our measurements have been taken on traffic which is sent from the Cray Y-MP across the network to the graphics processor, which constitutes the bulk of the traffic which traverses the network. The Cray is used to calculate the radiation dose distribution in response to changes to the treatment plan made by the user at the medical workstation.The Cray is either connected to the graphics processor via the ATM network, or via a point-to-point fiber link which has been used as a temporary substitute for the ATM network for the purposes of testing the application. Our measurements have been taken on a HIPPI channel connecting the Cray to the network, see Fig. 2.

The radiation dose data has a fixed size per session, with observed sizes ranging from roughly 250 KB to 1 MB of data during each session. The dose distribution is recalculated by the Cray upon changes to the treatment plan, which may either result from a single user change at the workstation, or a series of closely- spaced changes to the treatment plan resulting from the user manipulating an interface control tool at the medical workstation.

Between periods when the dose data is being transmitted, no data is sent from the Cray. The duration of these long ``silence periods'' separating the transmission of dose data tends to be highly variable, lasting anywhere from hundreds of milliseconds to several seconds.

4. TRAFFIC POLICING

4.1. Congestion control

Congestion control is required to ensure that the grade of service is met for each connection and that the network's bandwidth is allocated in a fair way. There are two types of control: reactive and preventive. In reactive control schemes, the traffic flow at the access points is based on current traffic levels within the network. In a preventive control scheme, there is an admission control mechanism that is responsible for accepting new connections based on their traffic characteristics. A new connection is accepted if the requested quality of service can be met and if the quality of service for the existing connections is not violated. Due to the bursty nature of a source, it is possible that at times the negotiated traffic parameters of a connection may be exceeded. In view of this, an additional function known as the policing function is required in order to protect the network against congestion due to violation of the negotiated parameters. This policing function is enforced on each connection at the access points of the ATM network. It uses knowledge of the parameters associated with the connection and controls the source by forcing it to conform to these parameters.

Two projects have been carried out within the area of congestion control. The first project dealt with leaky bucket policing mechanisms in preventive congestion control, and the second project dealt with a reactive congestion control mechanism.

4.2. Performance analysis of leaky buckets

The most popular policing mechanism is the leaky bucket. This mechanism consists of a token pool. An incoming cell consumes a token and then enters the network. If there are no tokens, the cell is lost. Tokens are added to the token pool periodically at a fixed rate. The token pool is finite, which puts a ceiling on the maximum burst size of cells allowed into the network. The parameters of the token pool are determined at the call setup. This scheme is referred to as the buffer less leaky bucket. A variation of this scheme is the buffered leaky bucket in which cells that arrive to find the token pool empty, are forced to wait in an input queue. The leaky bucket policing mechanism is intuitively appealing. However, it has been shown that it is not an effective policing scheme [11].

An analytic model of the bufferless and buffered leaky bucket was developed assuming that the arrival process is either a Markov Modulated Bernoulli Process (MMBP) or an on/off process with arbitrarily distributed on and off periods.Using these analytic results, we obtained an insight into the relationship between cell loss and burstiness of the traffic that passes through the leaky bucket when the source attempts to submit excess traffic. A higher loss probability will result in a lower burstiness of the traffic. This is due to the fact that the lost cells thin out the arrival process, thus shifting the distribution of the interdeparture times toward the mean and hence reducing the variance . Generally speaking, larger leaky buckets with a long token interarrival time are more effective at policing the mean arrival rate than smaller leaky buckets with a short token interarrival time. On the other hand, the latter leaky buckets are better at policing the burstiness of the traffic, except under heavy load, than the former ones. In view of this, a dual leaky bucket was proposed which consists of a large and a small single leaky bucket [12].

4.3. A mechanism for reactive congestion control

A reactive mechanism was designed to control congestion in a gateway which receives traffic from other gateways via an ATM network. With reference to the VISTAnet testbed project, this gateway was assumed to provide a HiPPI interface to the CRAY Y-MP. The reactive control scheme allows the sources to violate their negotiated average bandwidth by increasing their average transmission rate by a small increment every period T. This period T was set equal to one round-trip delay. The gateway sends a *slow down* control signal to the sources when the queue length reaches a high threshold. The sources immediately decrease to a lower average rate. The gateway sends a *go ahead* control signal when the queue length reaches a low threshold. The sources wait for a period of one round-trip delay and then they start to violate their negotiated average rate in the above-mentioned fashion.

The idea behind this scheme is that if there is slack capacity in the gateway, then the sources can increase their transmission rate in order to share the slack. (It is assumed that the ATM network can carry the increased traffic.)The efficacy of this scheme was investigated by simulation. The scheme is effective when the bursty period of each source is longer than its round trip delay. The increment by which each source is allowed to increase its average transmission rate also plays a role. If the increment is very small, then it will take a longer time for the sources to take up the slack capacity. On the other hand, if the increments are too big, then the cell loss rate will increase. The distance between a source and the gateway may be different among different sources. This difference is responsible for introducing an unfairness as to how these sources can share the slack capacity.The nearest source receives the *slow down* signal faster than the far away source. It also receives the *go ahead* sooner than the other sources. As a result, it tends to take up a larger percent of the slack.This unfairness problem is quite prevalent in these feedback based schemes, and, it had not been previously identified. For this scheme,it can be easily alleviated by setting the period T for all sources equal to the round trip delay of the further most source.

5. ATM ADAPTATION LAYER

A model has been developed that permit us to determine the optimum convergence sublayer-protocol data unit (CS-PDU) size [13] The cell loss probability was assumed to be a bursty process, i.e., a period of low cell loss probability is followed by a period of high cell loss probability. The general model that describes the interaction among the higher layers, the ATM adaptation layer, the ATM layer, and the physical layer is shown in Figure 3.

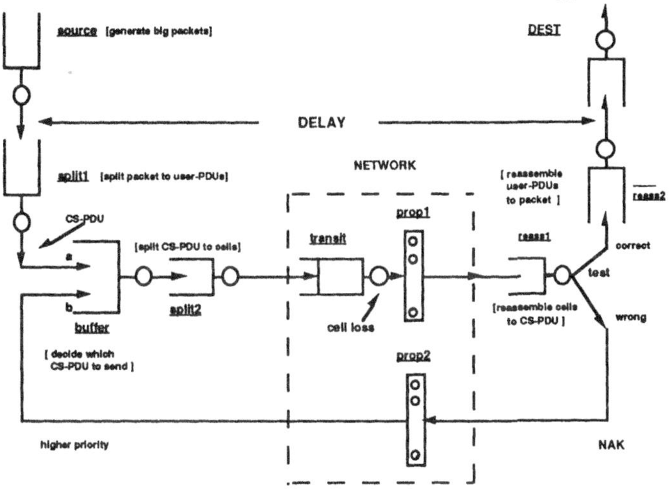

Figure 3. Network and protocol model.

Our results (representative results are shown in Figures 4 and 5) indicate that there is a distinct value of the CS-PDU size that will give us the smallest end-to-end delay. In Figures 22 and 23 the normalized (with respect to transmission time for a cell) average delay and average and maximum delay are shown as a function of the size of the CS-PDU. The parameters used are as follows: network loading, 40%; mean packet size, 63950 B; geometrically distributed message size; network propagation delay 0.1 ms; and transmission time for one ATM cell, 424 ns.

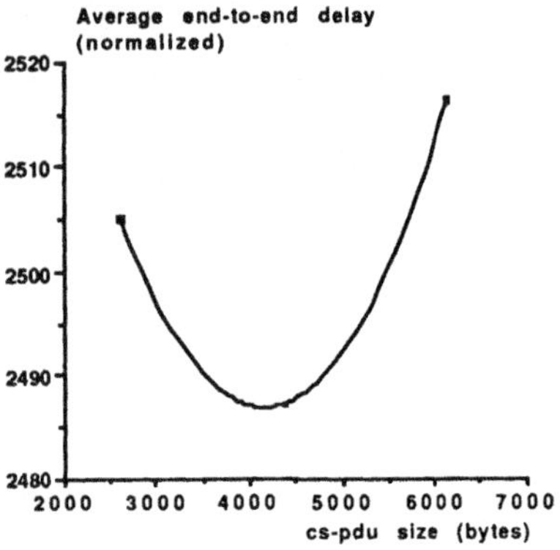

Figure 4. Average end-to-end delay as a function of CS-PDU size.

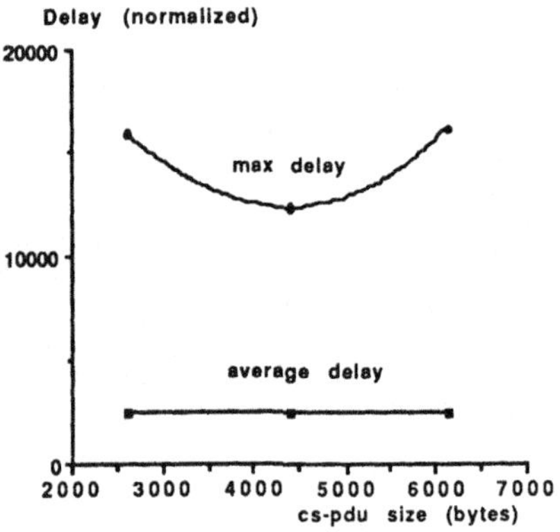

Figure 5 End-to-end delay as a function of CS-PDU size.

The analytical results that have been generated for the network model all deal with the analysis of discrete time queues with bursty arrivals and batch arrivals. Queue length distributions and departure processes have been found, and approximations for the departure processes have been generated.

6. PROTOCOL PERFORMANCE EVALUATION

One of the projects in the VISTAnet communications research program deals with protocol performance evaluation. We have investigated a large number of different transport level protocols. One of the protocols we investigated in detail with respect to its performance was the Xpress Transfer Protocol (XTP) and we will limit the discussion in this paper to this specific protocol. We have developed a detailed simulation program of the XTP protocol operating in a gigabit environment [14]. Note, that we have not modeled the protocol from a finite-state machine point of view, but rather from a performance point of view. Special emphasis has been placed on the ability to modify and experiment with the ARQ (automatic repeat request) mechanism in the protocol in order to be able to change timers and acknowledgement requirements (e.g. immediate acknowledgement or relaxed acknowledgement). This simulation model is part of a bigger simulation in which we also simulate the effects of external interfering traffic. The schematic diagram for the simulation model is presented in Figure 6.

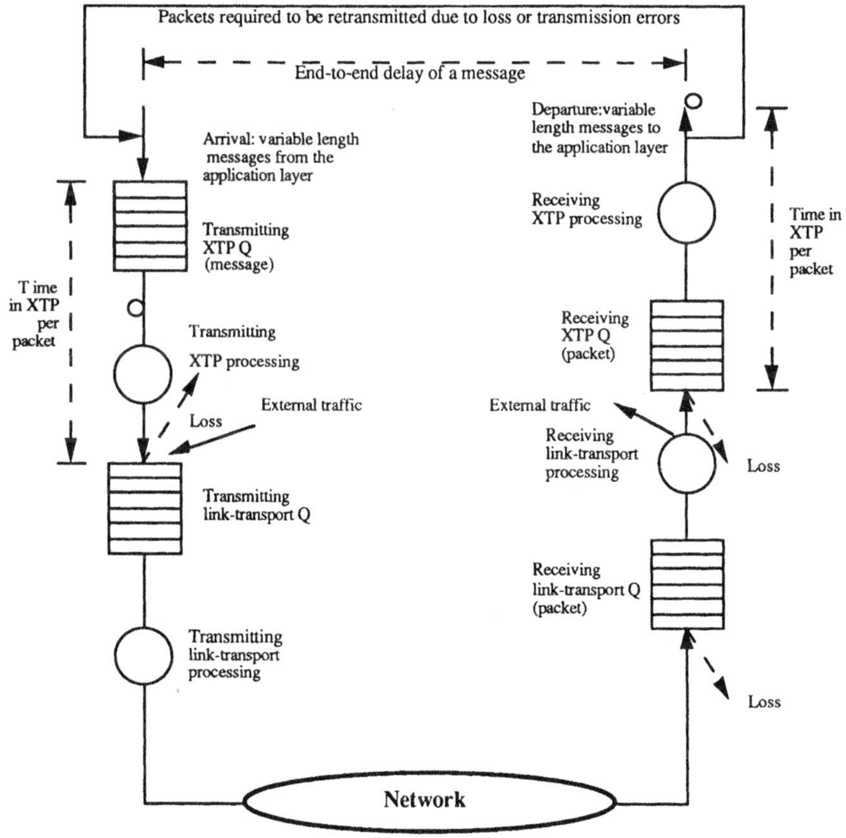

Figure 6. Simulation model.

The arrival processes used are assumed to be Bernoulli and Interrupted Bernoulli both for the virtual circuit traffic and the external traffic. Message lengths are assumed to be geometrically distributed. The service times in the various service modules in the simulation model are assumed to be deterministic. Except for the receiving buffer for the variable length application messages, the buffer capacities are finite. Thus, one source of packet loss is buffer overflow. Loss due to transmission errors is generated randomly. It should be noted that the variable length messages are divided into fixed length packets before transmission. In the simulation, it is assumed that each fixed length packet has a constant probability of being in error. The default value used in the simulation is 8×10^{-6}. The finite buffer size default value was 10 packets. In the simulation results that we present here, it was assumed that the fixed packet length was 1 KB and that the normalized transmission time for such a packet in the network is equal to a time unit. The average number of fixed length packets in a message was assumed to be two. The processing time in the XTP processing and the link layer processing was assumed to be two time units each. The physical propagation delay was assumed to be six time units. The end-to-end delay as a function of the throughput was found for a variety of different transmitter and receiver behaviors. Specifically, two cases for the transmitter were investigated.

1) Go-back- N ARQ

2) Selective Repeat ARQ

For the receiver, three different behaviors were tested. The numbering of these behaviors start at 2 for consistency with the diagram.

2) Do not store out-of-sequence packets and aggressive retransmission (i.e., a NACK is issued as quickly as possible).

3) Store out-of-sequence packets and do not use aggressive retransmission (i.e., wait until the transmitter issues a SREQ).

4) Store out-of-sequence packets and aggressive retransmission.

In Figure 7, the end-to-end delay as a function of throughput is shown for the six possible combinations of transmitter/receiver behavior. The delay throughput results indicate that we get a slightly better behavior for the combinations in which out-of-sequence packets are stored and aggressive retransmission is used.

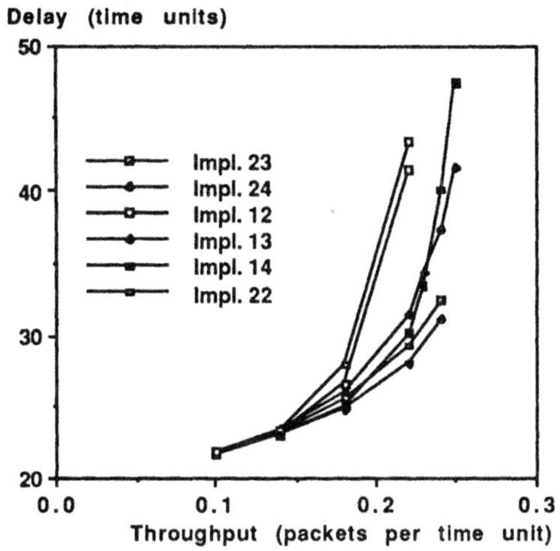

Figure 7. End-to-end delay throughput performance.

The results show that, at the sender side, "SR" yields higher performance than "GBN" except for the case when the receiver discards the out-of-sequence packets, the sender behavior does not matter in that case. At the receiver side, "store" results in smaller end-to-end delay than "discard", and the aggressiveness is helpful only when the system load and the window size are within a certain limit. We expect that this work will give us further insight into how protocols should be selected for gigabit networks and where the trade-offs are with regards to control and buffer management.

7. CONCLUSIONS

The issues associated with high speed networking are many and very diverse. There exists a large number of problems that have to be solved both within the networking area as well as in the metacomputing and application areas. In this paper we have presented some of the insight from the networking area we have gained through our involvement with the VISTAnet testbed activity.

REFERENCES

1. W.L. Bruwer, W.L. Smith, D.R. Spears, “A B-ISDN Field Trial with Medical Applications,” presented at National Fiber Optic Engineers Conference, April 2-5, 1991.

2. “Gigabit Network Testbeds,” IEEE Computer Special Report, September 1990.

3. D. Fisher, "Getting Up to Speed on Stage Three NREN: Gigabit Testbed Lessons on Architectures and Applications," Globecom 91, Phoenix, AZ, 1991.
4. G. Stix, "Gigabit Connection," *Scientific American,* October 1990, 118-120.
5. H. Fuchs, J. Poulton, et al., "Pixel-Planes 5: A Heterogeneous Multiprocessor Graphics System Using Processor Enhanced Memories," *Computer Graphics 23*, 3 (Proceedings of SIGGRAPH '89), 79-88.
6. High-Performance Parallel Interface - Mechanical, Electrical and Signalling Protocol Specification (HIPPI-PH) X3T9/88-127 Rev 7.2 9/1/90
7. High-Performance Parallel Interface - Physical Switch Control (HIPPI-SC) X3T9.3/90- REV 2.6
8. High-Performance Parallel Interface - Framing Protocol (HIPPI-FP) X3T9/89-146, X3T9.3/89-013, REV 2.6
9. D. Winkelstein and D. Stevenson, "HIPPI Link Data Analysis System: Test Equipment for High Speed Network Analysis," presented at Tri-Comm 91, Chapel Hill, NC, April 18-19, 1991.
10. H.C. Kuo, A.A. Nilsson, D. Winkelstein and L. Bottomley "Traffic Measurements on HIPPI Links in a Supercomputing Environment", TRICOMM'93, April 93, Raleigh, NC..
11. D. S. Holtsinger and H. G. Perros, "Performance Analysis of Leaky Bucket Policing Mechanisms," Second ORSA Telecommunications Conference, Boca Raton, FL, March 1992.Submitted to IEEE Trans. on Communications.
12. D. S. Holtsinger, "Analysis of the Dual Leaky Bucket Policing Mechanism for High-Speed Networks," Submitted to the 5th Inter. Conf. on the Performance of Data Networks, Raleigh, Oct. 1993.
13. A.A. Nilsson and Z. Cui, "On the ATM Adaptation Layer", 1992 International Conference on Communication Technology, Beijing, China, September 1992.
14. A.A. Nilsson and Mee-Jeong Lee "A Performance Study of the XTP Error Control", 4th IFIP Conference on high performance networking, Dec 1992, Liege, Belgium.

Integrated Broadband Communication Networks
and Services (C-18) / V.B. Iversen (Editor)
Elsevier Science B.V (North-Holland)

INTELLIGENT LAN CONCENTRATORS EVOLUTION TO ATM

H. Sourbes, C. Basso, J. Calvignac

IBM Networking Systems Development Laboratory, La Gaude (France).

Disclaimer. *The information presented hereunder is a sequence of what is expected to be generally available in the industry. The devices as shown and concepts developed are generic in nature and are not necessarily going to be mapped into IBM products*

Abstract.

The following pages consider the main drivers of the evolution of Local Area Networks (LAN) and try to project the future of current Intelligent Concentrators (I-HUBs). The trend toward ATM is rationalized and explained. Then some possible ways to implement ATM as the LAN technology of the late 90's are explored.

KEY WORDS: ATM LAN, LAN emulation, Connection less service and ATM, ATM at the desktop.

INTRODUCTION - THE EARLY LANS.

In the early days of networking, processing was expensive, as well as bandwidth, and resulting in centralized networking control with hierarchical distribution of functionality. The corresponding topology was a star wired, or a tree structure, from points of control (terminal controllers or network nodes) to end points (workstations or processors). This networking structure presented a maximum efficiency and had inherent capabilities to scale up. We know today multiple implementations (e.g., IBM SNA corporate network, or Airline Reservation Systems),with hundred of thousands of end points, running quite smoothly with a limited number of supporting people.

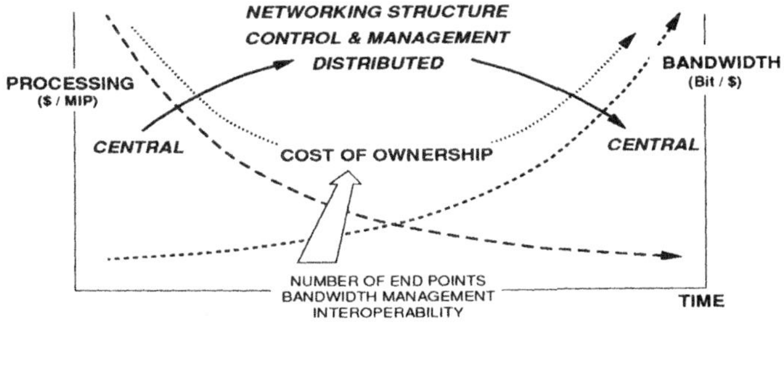

In the early 80's, with the advent of cheaper processing, personal computing became affordable and the networking functions could start to be distributed. This created the possibility to downsize and migrate the communication control of the network at the network periphery. The maximum of this evolution was reached with Ethernet and Token-ring where the media became the network and the control was totally distributed into the network interfaces. This is generally called the **Shared Media** LAN concept.

The early perspectives on shared media were quite attractive:

- The topology and implementation were extremely simple, and the realization cheap: there was no central costly infrastructure, the support staffing was limited to the resources needed for implementation of the physical connections.
- The connectivity was extensive since everybody was a physical neighbor of everybody, as being on the same media.
- The logical interchange was in close reach: one layer away from the hardware, on a single address domain. It was just a question of device driver and protocol transform.

However, shared media carries some inherent constraints:

- Each LAN attached device must align on common physical characteristics, E.G. on a 4Mbps LAN, all stations run at 4Mbps. Thus if certain users need a higher speed, they would have to move on a different LAN or wait until everybody gets upgraded to the new speed, which might get delayed when considering the related costs.
- Everybody has access to everybody's data circulating on the shared media, and this creates exposures to security violations.
- Certain users activities could interfere with other's . For example, business users exchanging small interactive inquiries might be slowed down by the other people exchanging large files, like engineering drawings or complex data sheets.
- As the new facilities attract more users, the lack of management control and central reference makes every network addition or change a problem. This has been the source of multiple competitive network management packages.

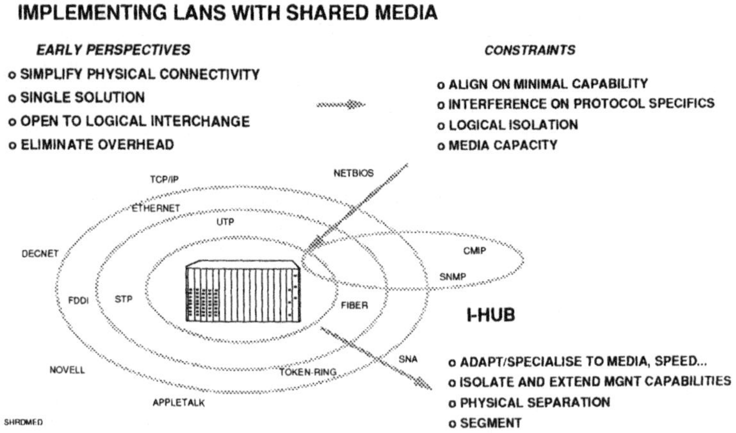

An analysis of the LAN users requirements led in the late 80's to the development of Intelligent-Hubs (I-HUB), as a central point where one could better manage the topology, control the assignment of resources, perform problem determination and address the request for fair access to the media. The Token Ring standard was addressing the problem of arbitrating access to the media, the development of 10BaseT for Ethernet would eventually permit a handling of the collisions since it was now possible to introduce some logic at the point of convergence of the wiring structure. Then the two major LAN standards would converge into a single structure. The I-HUB became a central point to implement the physical interconnections and to insert a minimum level of management of the media. The age of control-less networks was over, and the pendulum started to swing back in the other direction: centralized control and management.

TOPOLOGY MANAGEMENT AND COST OF OWNERSHIP

The structured cabling (or star wiring) and I-Hub implementation provide ways to manage the topology and to respond to end-users' requests for traffic isolation either in case of higher capacity demand or to protect sensible data from unwanted intrusions.

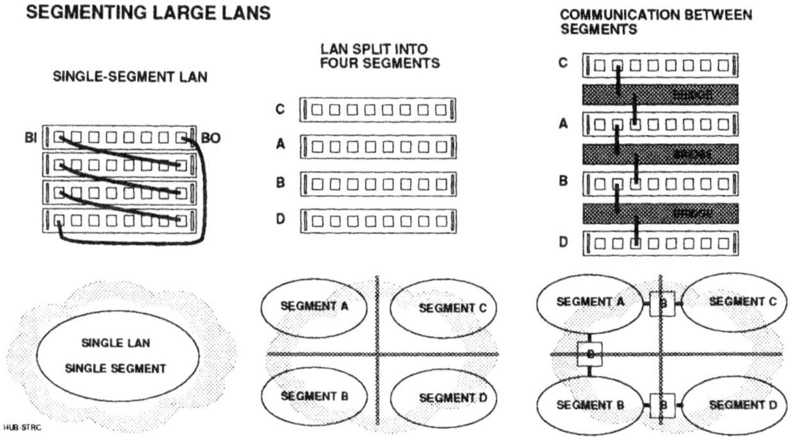

The segmenting of LANs into smaller sub-LANs is the most practical way to answer these requests and hub manufacturers have developed multiple and sophisticated functions to perform this activity with minimum operation resources and delays.

For example, the **IBM 8250** implements in its backplane up to 7 Token-Rings, 3 Ethernets and 4 FDDIs, or a combination of the above.

It is possible to move groups of users from one segment to an other one, (this is called module switching) in order to reconstruct work groups. For Ethernet users, the reassignment is feasible at the port level and so a server could be relocated on a separate segment, when new applications are brought in.

A Hub management program running in a distant maintenance processor can redirect the resources without the need for on site human intervention. It is then possible to remodel groups of users (or even individual ones) according to the evolution of their activities from one segment to another one. This makes it possible to answer quickly and efficiently to activity changes or organizational changes.

The segments are connected together using bridges or routers, with filtering algorithms where this is necessary. When these devices are integrated in the hub, the same remote management program can redirect the bridges and routers physical attachment from one LAN segment to a different one in basically the same way as a user's port.

There are two main ways to structure the LANs for inter-segments connectivity:

- One is based on the use of a common dedicated federating LAN which attaches to the peripheral ones using bridges or routers. A recent consultant survey showed that by mid-92, over 80% of large installations were already using a 16Mbps Token-Ring federating backbone and that many were making plans to migrate the federator to FDDI in order to open new levels of traffic on the backbone. However, there were reports of disappointment because of the performances of the FDDI bridges available on the market then.

- The other one consists in extending peripheral LANs inward (using fiber extensions or Bus-in/Bus-out) to a central hub where local ports are connected to a multi-port bridge or router. This method has some advantages since it concentrates control and topology management on a limited number of places. This is called **collapsed** or router **backbone**.

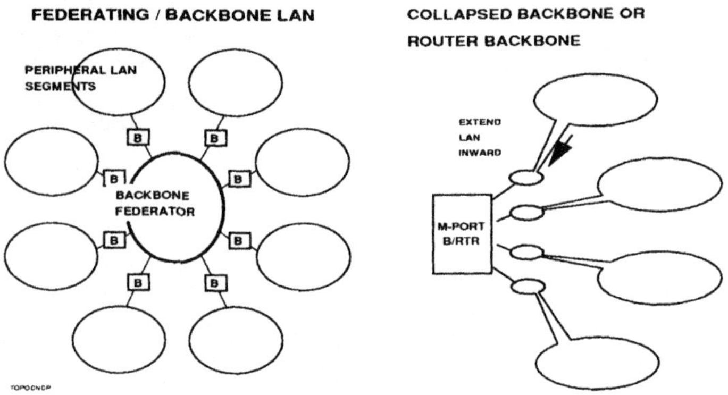

The concept of LAN combined with the concept of personal computing created the concept of workgroup: a set of users isolate themselves on a LAN segment through filtering routers and bridges. There, they can share confidential data, specialized servers for data or applications and remain physically protected from unwanted intrusions or traffic.

All this requires management tools; to monitor the activity, identify the problems, reassign resources, control and modify the topology. Current high-end HUBs offer an extensive set of capabilities in this area. More recently we have heard the requirement to extend this concept of **affinity work group** beyond the physical borders of the establishment or campus. This corresponds to the need for a **seamless integration** between **Local** and **Wide Area Networks**, or at a minimum from LAN backbone to LAN backbone across WAN.

As new applications develop, their bandwidth requirement augments and the capacity needed on each workstation has to be adjusted. One way is to reduce the number of workstations sharing the same segment

This induces more segmenting and increases the complexity of the LAN topology.

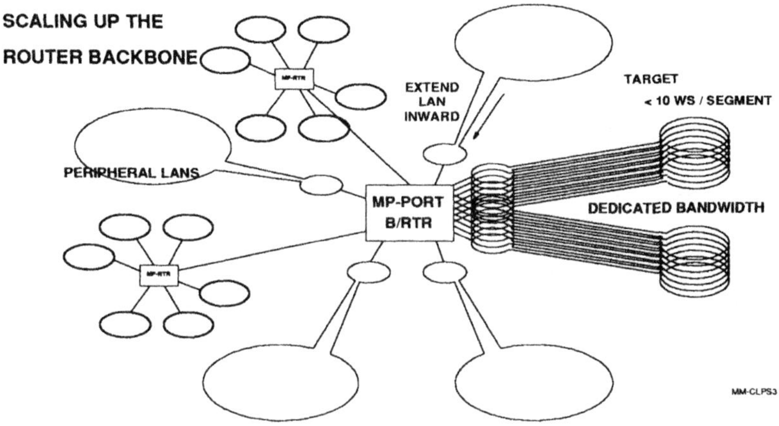

In certain extreme cases the workstation will be left alone on a segment: this is called **dedicated bandwidth**. The relation with other LAN partners is done using bridging techniques based on MAC addresses. This dedicated topology has a lot in common with switching, and since ATM has all the ingredients to implement switching in a networks, one could consider ATM as a natural evolution to reach higher bandwidth per user and better segment isolation and topology management.

ENVIRONMENT AND TECHNOLOGY EVOLUTION

The LANs (combined with cost of processing and speed of transmission) permitted the deployment of Client/Server computing were the various partners in a work group specialize to what they do best. Independent computing and human interface for workstations, intensive computing or data repository for larger processors. This has a direct effect on the need to control access to the media: a large data base server should have a higher priority, a wider and faster path to the media.

With the deployment of intelligent workstation networked over LANs, the role of central computing has changed: the large processors become specialized either for intensive computing or for large data repository. However, it is understandable that the network demand from this high performance processing points will differ from a regular single user's whoever he is, and that the network should give it a priority access. However, this is not feasible with current shared media without constructing complex hierarchical structures.

With ATM, each device can negotiate its specific **bandwidth needs** and **peak data rate** at the call set up.

The end user's workstation applications can now offer better presentation services in the form of graphical interfaces, or can prepare complex requests to other servers. This drives the bandwidth at the station level to new highs. Certain complex visualization functions using video compression techniques might require dedicated bandwidth in the ranges of 1 to 6Mbps per workstation.

APPLICATION BANDWIDTH DEMAND EVOLUTION

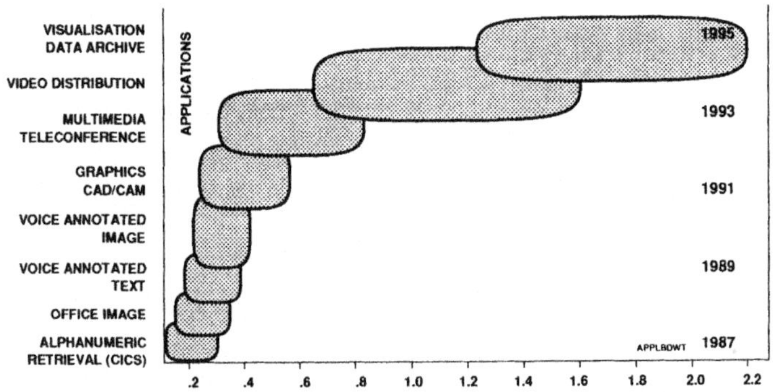

As the data processing evolves into information processing, the network must now enable the deployment of **multimedia applications**. These applications combine audio, video and data on a single device:

MULTIMEDIA APPLICATIONS

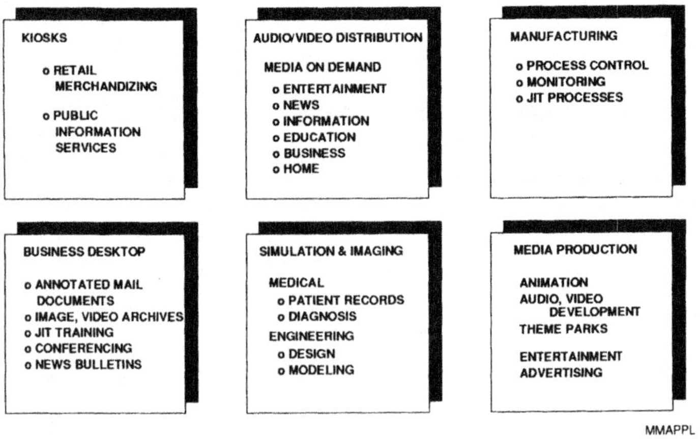

Today, these applications are in their infant stage and are often restricted to small privileged work groups. However, we expect multimedia to reach its real dimension when it will be possible to efficiently interchange audio/video/data not only on a local campus but across the corporate network using private and public communication services. Most consultants predict multimedia communication to be the fastest growing sector of the information processing starting 1994.

The key multimedia business applications are generally considered to be the following:

1. Multimedia Electronic mail
2. Video clip store and forward,
3. Video Servers / Kiosk applications (interactive digital full motion video)
4. Video conferencing and group ware.

The last two have a special significance to the network: as they require guaranteed bandwidth, synchronous transmission and short network latency (less than 150ms).

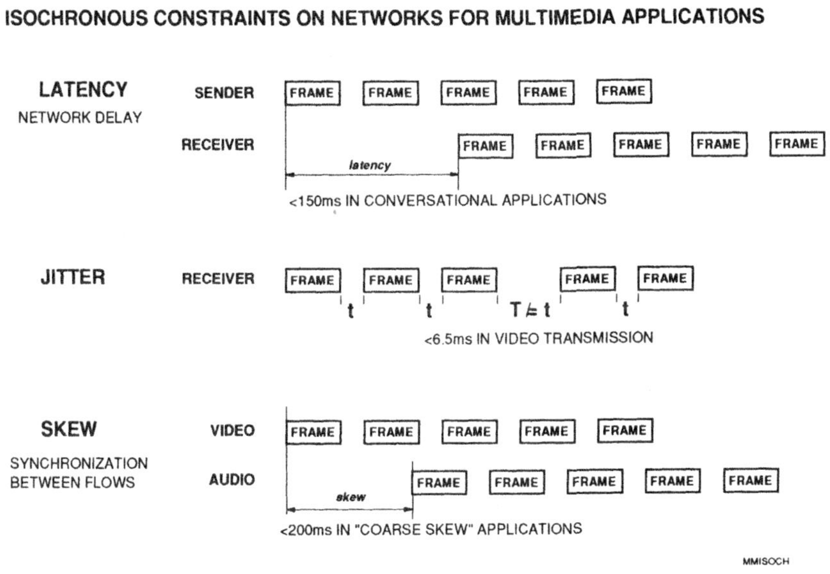

DRIVERS OF ATM LANS

ATM appears to match the current most pressing requirements of the LAN environment both in terms of controlling networks topologies and opening to new applications:

1 .It brings better ways to manage the **cost of network ownership**:
- By streamlining the topology: it is based on a central switching fabric from where every device gets the connectivity,

• ATM is scalable at speeds from 25Mpbs to 1.2 Gbps, and has inherent segmentation capability.
• Each attached device can run at its own pace and has a full dedicated bandwidth to the switch.
• ATM being issued from the CCITT's work on Broad-Band ISDN, initially a WAN architecture, there should be **seamless integration between the LAN and the WAN.**
• ATM is a centrally controlled and managed system, where operating resources can be optimized

2 It opens the door to a new generation of applications (**multimedia**)
• Thanks to its **scalability** to higher bandwidths, ATM systems can support the bandwidth required by these applications, without permanently impacting the network. each session is tailored to the needs of the particular application. This is defined at call setup . bandwidth on demand.
• ATM is **full duplex** and will permit interactive video conferencing.
• ATM is designed to support **isochronous** transmission which is mandatory for videoconferencing (CCITT H-Series Compliance).

ATM AS A LAN TECHNOLOGY

Several products are now available and various manufacturers have described their preferred scenario for the introduction of ATM in the LAN environment.

Considering that most LAN system operations today are based on connectionless services, and that ATM is connection oriented by nature, various consultants and analysts expect to see an initial deployment of ATM as a backbone system, interconnecting routers. The routers would then "shield" the LAN addressing chores from the installed base until a time where a sizable migration to ATM of the end user population has occurred.

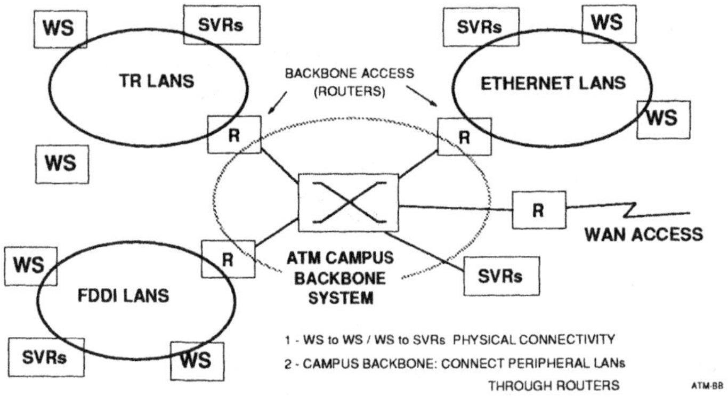

With this scenario, we are faced with a "chicken and egg" problem: ATM adapters will remain expensive until applications using native ATM interfaces are popular, but there will not be many of these since the potential candidates will deterred by the price of the adapters.

This problem is aggravated by the fact that the only interfaces that have been defined probably require a cabling upgrade to support speeds of 100 or 155Mbps. These speeds are probably appropriate to attach high capacity servers or network routers but are certainly not commensurate with the average workstation processor's speed or with the applications that the majority of users want to run, considering that ATM offers dedicated bandwidth by nature.

In IBM, we think that this approach has all the ingredients to delay the acceptance of ATM to connect workstations and thus reduce the possibilities of enabling multimedia applications in the business community. Additionally, the need for higher capacity LANs is here and we start seeing a flurry of alternate technologies competing to higher throughput to the desktop. To name a few:

- 100Mbps Fast Ethernet,
- Iso-Ethernet (demonstrated at Comdex-Fall 92 by NSC and IBM)
- 20Mbps Full Duplex Switched Ethernet (Cabletron)
- Priority-Token-Ring,
- FDDI/CDDI aggressively priced ...

The risk is that all these technologies competing for the same market will further delay the acceptance of a more general solution and will increase networks complexity, incompatibility, and cost.

ATM FORUM CONTRIBUTIONS

Taking into consideration the size of the current installed base of LAN products and the corresponding investments, IBM has introduced three contributions at the March 93 meeting of the LAN Forum with the purpose of easing the migration to ATM and promoting the use of this technology for desktop applications

1 A mid-range speed UNI for desktop access and support.

This data rate should be consistent with proven available, cost effective technology for operation over 100 meters of Category-3 unshielded twisted pair (UTP-3). The proposed speed is 25Mbps of dedicated ATM bandwidth to the workstation.

2 A LAN Emulation service interface.

Since the majority of installed protocol stacks rely on facilities provided by today's LANs (connection-less access, and multicasting), we asked the Forum to initiate a study that would propose a LAN Emulation service on top of ATM in terms of UNI extensions and traffic management. The objective is to provide a similar service to what exists for Token-Ring and

Ethernet. Then, End-Systems would connect to the "ATM-LAN" and interact to other ATM attached or TR/Ethernet attached systems as if they were attached to a traditional LAN (i.e. without call set-up).

3 An ATM/LAN comprehensive management system.

This addresses the elements of a LAN management technique that could be used in an ATM environment. It is essential that network administrators be provided with a set of tools similar in capabilities to what exists today on Ethernet and Token-Ring to perform the Problem Determination and Management of topology

LAN EMULATION AND INTER-OPERABILITY CONCEPTS

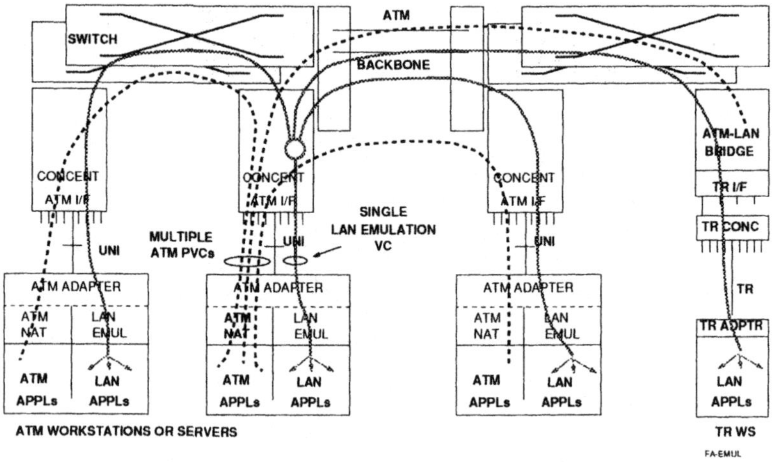

ATM ROLL OUT

In the first stage, ATM should be introduced as a backbone technology since the number of potential end users will be limited. The ATM switch should be integral part of the LAN HUB to permit the deployment of ATM as an additional LAN technology. The switch must be supported by connection modules that would be used for two purpose:

1 To build a HUB-TO-HUB backbone with links that could use the same fibers as the one used for FDDI,

2 To attach high performance servers in native mode.

Immediately following this introduction, a Mid-Range speed concentration should be offered to enable the deployment of new generation applications.

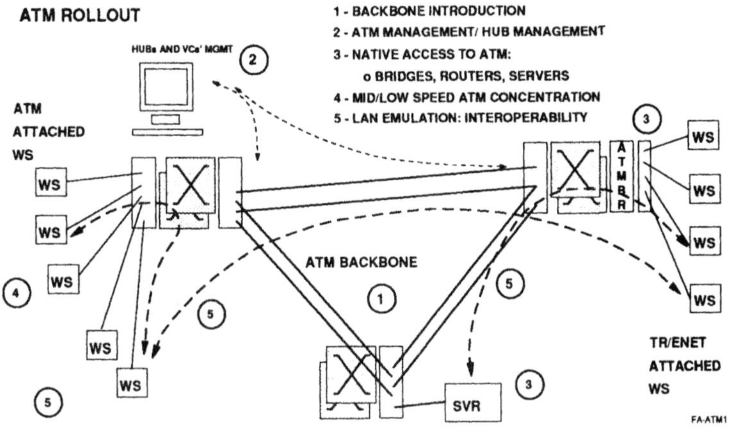

A bridge between ATM and current LANs (Ethernet/ TR) must also be offered in the hub to permit data interchange with ATM attached servers and workstations.

The LAN emulation support must be offered in this stage since that base of ATM native application will be very small. The bulk of the traffic will be migration traffic.

Later on an interface with WAN products could become available This will be the key to extend campus applications to the Enterprise level. This LAN-WAN interface will drive the need for extended, end to end networking interfaces, where problems like generalized directory, best route calculation can be processed.

As a result of this feature being available, it should be possible now to see new application offerings in the area of videoconferencing or groupware in the local environment extended outside the corporation thanks to the connection with B-ISDN or equivalent services either public or private.

CONCLUSION

We think that ATM has the potential to replace the current LAN technologies, and offer possibilities that would open up a new world of application. However, current LANs have over ten years of experience and a tremendous installed base. The technology is cheap and well known and except for specific places they have the right capacity. This is a big handicap for ATM and it could slow its acceptance to a point where it would end up like FDDI and ISDN before: never make it.

The challenges are identified:

- address interoperability and migration of installed infrastructure
- develop a strong and reliable networking fabric
- develop a comprehensive way to interoperate with LANs and WANs to implement a real seamless appearance
- Integrate a management structure equivalent to other LANs.

If properly addressed, then ATM will rally enough support to provide competitively priced hardware and software and will deploy fast.

ACKNOWLEDGMENT

Special thanks go to Jean Yves Le Boudec and Hong Linh Truong who proposed the concept of LAN emulation and to Kip Potter and Jim Harford who generalized it. The Mid-Range ATM speed proposal was refined by Lee Haas and Franc Noel with the help from Dan Ater from Chipcom. Daniel Abensour shared with us his views of ATM and High Speed Interconnect in Enterprise Environment and Jim Russell provided the study on Multimedia Networking Performance Requirements.

REFERENCES

1. Le Boudec J.Y., The Asynchronous Transfer Mode: a tutorial . Computer Networks and ISDN Systems (1992)279-309.
2. Le Boudec J.Y., Erich Port, Linh Truong. Flight of the FALCON. IEEE Communication Magazine. February 1993
3. Boyer P., Guillemin F., Servel M., Coudreuse J.P., Spacing cells protects and Enhance Utilization of ATM Network Links. IEEE Network. September 1992.
4. John. MacQuillan. ATM as a LAN. Business Communications Review-Columns. 92/2, 92/3, 92/6
5. John MacQuillan. LAN in a Box. Business Communications Review. 91/8, 91/12.
6. John MacQuillan. The end of Shared Media LAN. Business Communications Review. 92/10

Integrated Broadband Communication Networks
and Services (C-18) / V.B. Iversen (Editor)
Elsevier Science B.V (North-Holland)

Network concepts and Siemens products for ATM-based public networks

R. Händel

Public Communication Networks Group, Siemens AG, Hofmannstrasse 71,
8000 München 70, Germany

Abstract:
Near-term requirements of public ATM networks are discussed and the matching products are introduced. These include ATM switches, adapters / interworking units which allow interconnection of existing LANs and MANs with ATM networks, and connectionless servers. The foreseeable network evolution will call for enhanced ATM product features in the medium term such as customer-controlled ATM connection establishment, support of bursty traffic, improved OAM capabilities.

Keyword Codes: C.2.1, C.2.3
Keywords: Network Architecture and Design, Network Operations

1. THE CURRENT SITUATION CONCERNING ATM

At present ATM is in the focus of many people's interest (as can be seen for instance from the high frequency of ATM conferences being held). Somewhat in contrast to the first CCITT discussions of ATM as the technology for public B-ISDNs, nowadays mainly the data communication world is pushing the concept and implementation of ATM networks and products. However, due to the inherent potential of ATM to support all types of services - real-time and delay-insensitive, constant and variable bit rate (bursty traffic) -, the evolution of ATM networks and equipment towards multi-media capability has to be considered for the planning of products.

On the other hand, whenever it comes to the assessment of ATM markets and applications and the required equipment, considerable uncertainty is realized in terms of

- availability of standards
- acceptance by potential customers
- commitment of network providers
- competition by other technologies / services such as SMDS, Frame Relay, X.25, SDH leased lines etc.

From a manufacturer's perspective it is essentual both to enter the ATM market on time (short-term requirement) and to plan the ATM products for the medium / longer term that will meet the assumed market needs.

SIEMENS

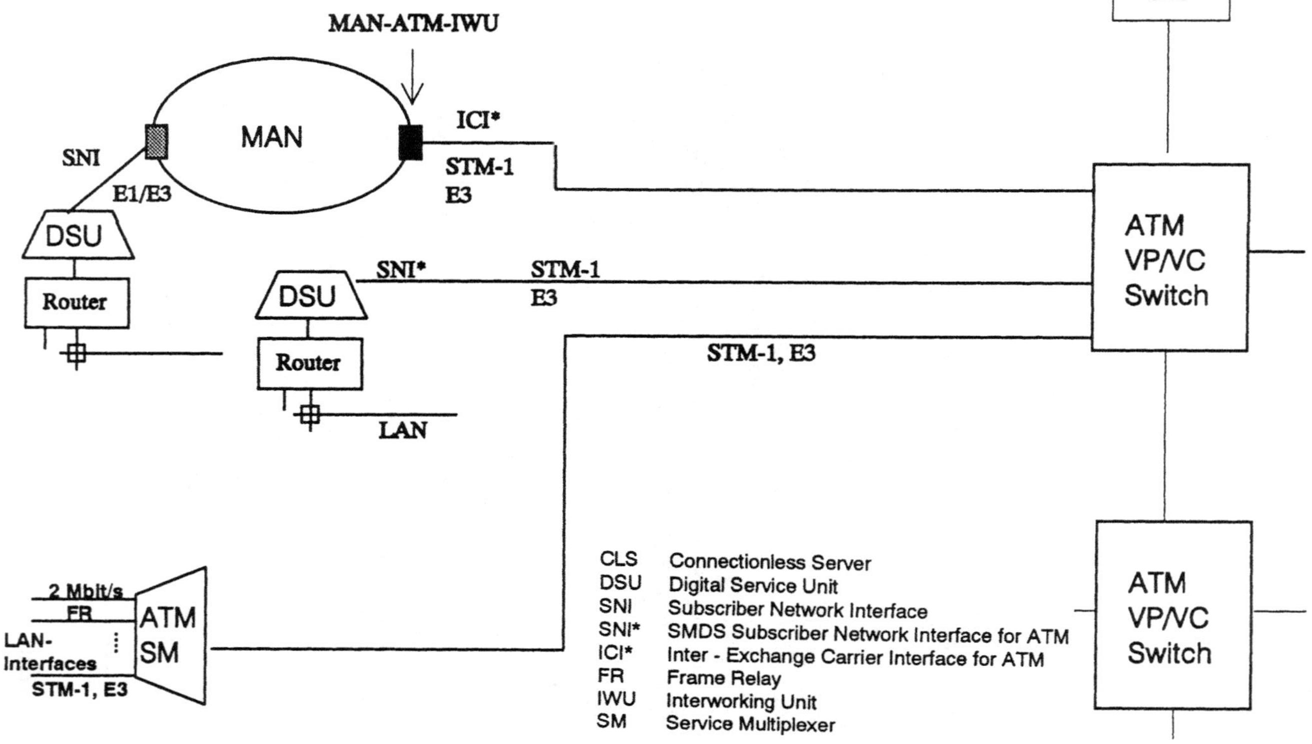

Fig. 1. ATM network architecture to support mainly data services

2. THE ATM MARKET AS SEEN BY A PUBLIC NETWORK EQUIPMENT MANUFACTURER

What does a public network provider need ATM for?

Quite generally, deployment of a new technology is advantageous if it can provide

- new features which meet customer needs
- better performance
- lower costs / tariffs.

ATM obviously exhibits new features such as flexible bit rate allocation per connection, support of virtual private networks by means of the virtual path concept, economising on transmission capacity (less bit rate wastage owing to cell multiplexing), support of burst-type traffic (statistical multiplexing gain).

ATM can be introduced in data backbone networks that are suitable for the transport of X.25, frame relay, connectionsless data etc. with the potential to be upgrated later on towards a universal network (B-ISDN). ATM networks will generate new types of applications (e.g. high-speed multi-media services between work-stations). This evolution will most probably start in customer premises networks which will allow quick implementation and testing of ATM-based equipment and services. Interconnecting private ATM islands via ATM wide area networks will soon provide the means for end-to-end connectivity which is essential for making optimum use of the ATM capabilities.

The wide-area market for ATM is not necessarily identical to the public ATM network market. The first ATM users will be business customers, mainly big companies which are familiar with corporate networking and basically able to implement their own ATM-based network requesting from the public network provider only a (physical layer) cell transport facility but no cell-switching. Such private ATM networks can be commonly used by several companies (network-sharing), or one company that owns the network offers spare capacity to other customers thus becoming a competitor to public network providers (provided the regulatory situation permits this).

While in today's conventional public networks emphasis is put on high reliability, unique services that can be offered to all customers, the need for embedding new network features into the established operation and maintenance base, the future markets may above all require quick (and therefore somewhat "dirty") and cheap solutions. The acceptance of public ATM will be determined by its timely provisioning and, what is most crucial, customer-friendly tariffing. This of course impacts ATM product development.

3. SHORT AND MEDIUM TERM PRODUCT REQUIREMENTS

3.1 . ATM network architecture

The assumed short-term ATM network architecture is shown in Fig. 1.

The ATM backbone network is built up by ATM virtual path / virtual channel (VP/VC) switches which are interconnected via SDH links (STM-1 with 155 Mbit/s or PDH links (E3 with 34 Mbit/s). For countries using the 1.5 Mbit/s-based hierarchical levels E3 is to be replaced with DS3 (45 Mbit/s). The term "ATM switch" in Fig. 1 is used generically; in the introduction phase of ATM only semipermanent connections established and released through OAM actions will be provided. Attached to the ATM switch can be connectionless servers (CLS) to support the connectionless services SMDS and CBDS on top of ATM. The number of CLSs in an ATM network depends on the expected amount of connectionless traffic; it is possible to have a CLS with each of the ATM switches.

Though the end-to-end usage of the ATM transfer mode combined with unique higher layer protocols would be highly desirable as this avoids adaptation and protocol conversion, for a couple of years one will have to live with the pre-established data communication infrastructure such as DQDB-based MANs, frame relay interfaces, X.25, and all sorts of LAN interfaces and protocols (Ethernet, FDDI etc.). Of course those customers also want to be connected with the ATM backbone network, and network providers want to offer such connectivity.

The architecture shown in Fig. 1 allows to retain existing LAN equipment including routers. Conversion to cell format is performed at the edge of the ATM network. LANs can be connected directly to the ATM network or via MANs. Users can communicate with each other without having to take care of the internal network structure.

In cases where several different interfaces are required by customers, e.g. LAN interfaces and 2 Mbit/s interfaces according to CCITT Rec. G.703, ATM service multiplexers will be used.

Fig. 1 doesn't explicitly show the underlying transmission networks, i.e. the PDH or SDH line and cross-connect equipment nor ATM multiplexers / concentrators that could possibly be implemented.

3.2. ATM product requirements

The basic *services, interfaces and interworking functions* to be supported by ATM networks products are listed in Table 1.

Table 1
Required ATM services and interfaces

Services:	- ATM bearer service
	- Constant bit rate service (circuit emulation)
	- Frame Relay
	- SMDS/CBDS
Interfaces:	- 1.5 / 2 Mbit/s
	- 34 / 45 Mbit/s (ATM)
	- 155 Mbit/s SDH / SONET (ATM)
Interworking functions:	- LAN to (DQDB/SMDS-based) MAN
	- LAN to ATM
	- MAN to ATM

The ATM bearer service merely performs cell relaying. No adaptation layer (AAL) functions are deployed, i.e. the 48 byte information field is fully usable.

In the very first phase of ATM implementation only peak bit rate reservation will be implemented due to the lack of available standards and practical experience with burst-type traffic handling.

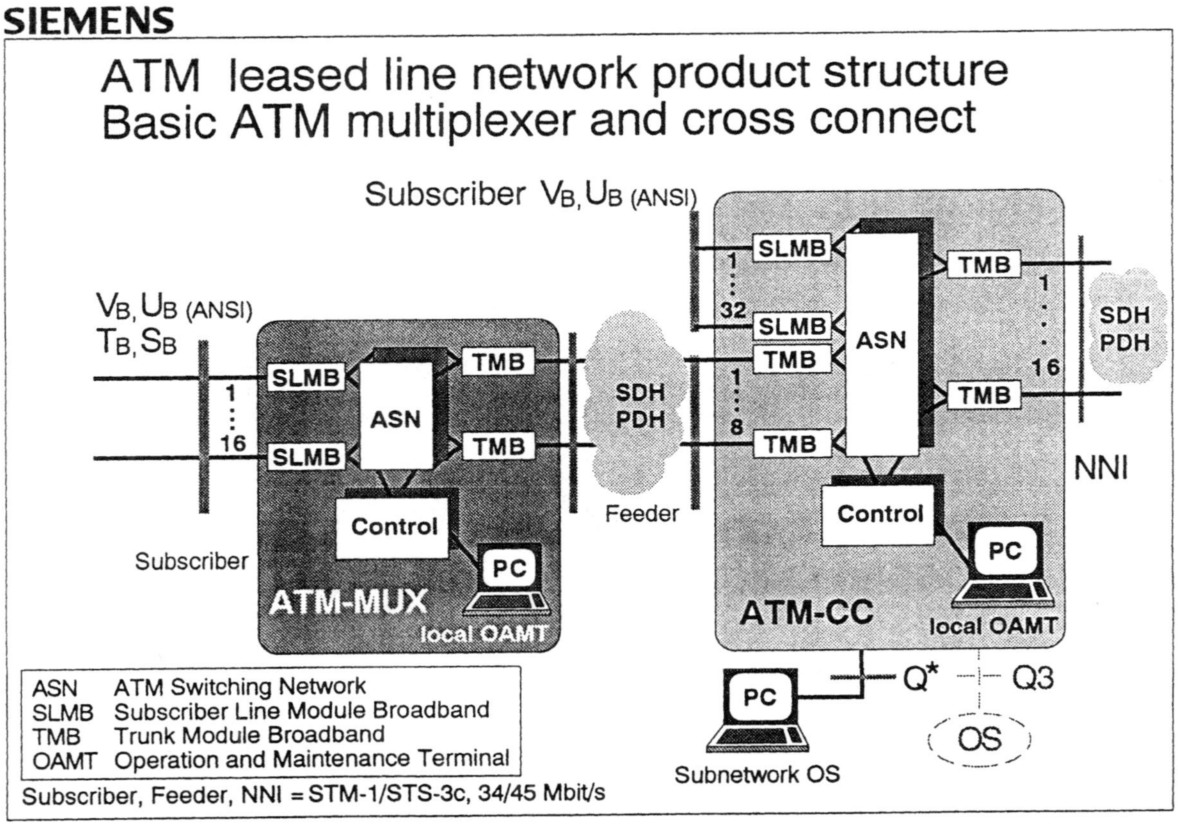

Fig. 2. Siemens' ATM VP/VC switch design

Constant bit rate services must provide n times 64 kbit/s and 2 Mbit/s (G. 703) circuit emulation using AAL type 1 according to CCITT Rec. I. 363.

The connectionsless service will initially be based on Bellcore's SMDS standard. When a settled standard for a connectionless ATM data service (CBDS) is available, it will be implemented in future product releases.

The required interfaces for the ATM cell transport are the plesiochronous E3/DS3 (34/45 Mbit/s) and the synchronous STM-1 or STS-3c / OC-3c, resp., with 155 Mbit/s. These interfaces will be deployed both as user-network interfaces and network-node interfaces.

The ATM service multiplexer will provide in addition to the above interfaces non-ATM interfaces such as 1.5/2 Mbit/s with integrated circuit emulation function (complete interface signals or n x 64 kbit/s) and frame relay adaptation to ATM.

As the first ATM networks will be rather small overlay networks, products must be modularly tailored to suit such an environment. In terms of numbers of interfaces, from only a couple of them up to a maximum of about 50 should be appropriate.

4. SIEMENS' ATM PROJECTS AND PRODUCTS

Siemens has been awarded contracts for ATM projects in Germany (DBP-T), Japan (NTT) and USA (US-West and MCI). Other national and international ATM projects, e.g. in UK, Italy, Switzerland are underway and a pan-European ATM network is announced for 1994 through the MoU of several European PTOs. Siemens is in contact will all the major players and has offered or yet will offer ATM products.

The Siemens products cover the whole range of ATM-related equipment as shown in Fig. 1:

- ATM VP/VC switch
- Connectionless server
- MAN/ATM interworking unit
- DSU for LAN connection
- ATM service multiplexer.

This equipment will be available for the planned ATM projects in 1993 / 1994. In addition to these products an ATM add-drop multiplexer for ATM-based dual rings with 2.5 Gbit/s (STM-16) ring interfaces will be provided for the NTT trial in 1993.

In the following the Siemens VP/VC switch will be presented in some detail. As Fig. 2 exhibits, the ATM switch consists of an ATM switching network, interface modules, control and OAM facilities. All crucial parts are doubled (switching network ASN, control). The ASN is a single-stage, non-blocking switching network composed of 48 switching elements with 16 inputs and 8 outputs each. The switching elements perform serial / parallel conversion, central buffering and output queuing. The ASN is expandable without service interruption from 16 to 64 ports. (Two ports are used internally to connect the doubled control). The available external ports (up to 62) may accomodate any mixture of UNIs, NNIs and feeder lines to connect remote ATM units (concentrators). The interface bit rate can be 34/45 Mbit/s or 155 Mbit/s.

For OAM purposes both a local terminal and connection (via a proprietary Q*-interface) to a remote terminal are provided.

The VP/VC switch can be deployed as ATM cross-connect and switch or as remote multiplexing and concentrating unit (see Fig. 2).

Other essential ATM switch parameters are listed in Table 2.

Table 2
ATM switch parameters

VP and VC switching:	yes
Number of virtual connections that can be handled per port:	4000
Maximum link load:	85%
Mean cell delay:	$< 250\mu s$
Cell delay variation:	$< 300\mu s$ (10^{-10} quantile)
Cell less ratio:	$< 10^{-10}$
Cell misinsertion ratio:	$< 10^{-13}$
Usage parameter control (policing):	Leaky bucket algorithm per virtual connection at the UNI
- bit rate monitoring accuracy:	< 640 bit/s
- probability of unjustified cell discarding:	$< 10^{-10}$
NNI control:	Control of total load
Connection admission control based on:	- Peak bit rate allocation - Peak bit rate allocation with overbooking - "Sigma Rule" for support of statistical multiplexing using *mean and peak* bit rates

5. ATM NETWORK AND PRODUCT EVOLUTION

The main requirement on medium-term ATM networks is the customer-controlled access, i.e. immediate provisioning of ATM connections on a customer's demand. For a end-user it doesn't matter whether this is accomplished via user-to-network signalling or via a dedicated user-to-network-management channel provided the latter method is able to establish / release connections quickly (preferably within seconds).

Signalling specifications for ATM networks are expected soon (at least for the features according to CCITT release 1), so switched ATM complying with international standards will probably be implementable around 1995.

The Siemens product roll-out plan foresees switched ATM based on stable CCITT standards. The ATM switch will work as stand-alone equipment or coupled to EWSD via an interworking unit. Thus 64 kbit/s-ISDN services will be accessible.

6. CONCLUSION

As a manufacturer of equipment for public telecommunication networks, we at present see the following market needs to be met in the near and medium term:

- now deliver first ATM-based products to enable network providers to implement trial networks and enter soon after (from about 1994 on) the phase of commercial exploitation of this new technology. The products will support OAM-controlled ATM-connections with peak bit rate reservation.
- The period of about 1995 to 1997 will require enhanced ATM functionality such as adequate support of bursty traffic (statistical multiplexing), improved OAM and network management, customer-controlled provisioning of ATM-connections.

The requirements for the end of this decade or even for the years beyond 2000 are difficult to figure out at the time being. Up to now, according to my knowledge, no public network provider has yet committed himself to establishing a universal ATM network integrating all services. Feasibility, cost and user-acceptance of this promising vision must still be assessed. The many ongoing ATM activities - both theoretical work and practical experience - will help resolving these issues.

Integrated Broadband Communication Networks
and Services (C-18) / V.B. Iversen (Editor)
Elsevier Science B.V (North-Holland)

Business Applications - Supplier´s Strategy

Peter Staxén

Ericsson Telecom AB, Strategic Network Development, S-223 70 LUND, Sweden

Keyword Codes: C.2.1; C.2.4
Keywords: Computer Communication Networks, Network Architecture and Design; Distributed Systems

ABSTRACT

This paper outlines Ericsson´s current view on business communications and indicates the strategy for introduction of ATM-based broadband communications for both intra-premises and inter-premises networks.

1. INTRODUCTION

Telecommunications have always been vital for business operations. The latest example of this is the fast deployment of the communication infrastructure in the eastern parts of Germany in order to attract business. Business communications is also the target of a multi-billion dollar industry supplying various types of equipment. Within virtually every major enterprise, separate communication departments have emerged to take care of corporate communications. Wide area communications have been achieved by leasing links from the public telecommunication operators.

However, liberalization of the telecommunications market, the increased competition in all trades and technical evolution, open new possibilities. New telecommunication providers, targeted at business communications, offer one-stop-shopping to business customers. Simultaneously, many enterprises outsource their communications department while focusing on their core business. Finally, new technologies offer opportunities to enhance the effectiveness of operations.

In this paper, business communications are seen from a telecommunication system supplier´s perspective. Ericsson has been delivering communication systems to both public telecom operators and corporations for a long time. We are also addressing those actors who are focused at providing wide area communications to the business community. Our strategy is to provide an appropriate solution for every situation.

First of all, this paper will highlight some trends in business communications. Next, the market for business communications will be described from a supplier´s perspective. Ericsson´s position in this context will be given in the following section. Finally, given the trends, the market and the position, some network solutions will be described.

2. TRENDS IN BUSINESS COMMUNICATIONS

The nature of business communications changes with the business climate and the technology that enables these communications. The climate for conducting business is characterized by an increasing globalization of operations as a strategy for achieving higher market shares and higher profits.

This globalization leads to an increased competition on most markets. While trying to expand on foreign markets, enterprises witness how foreign competitors threaten their home markets. The liberalization trend, striving towards introduction of competition and larger consumer choice, enforces the trend of increased competition.

In addressing this increased competition, many organisations adopt the strategy of reducing the operational costs, improving of existing services and introducing new services. The overall consequence of these strategies is that communications are seen as a part of the strategy - not just a generic need.

The reduction of operational costs is often performed by streamlining the organization and, more importantly, effectivisation of information flows. Two important techniques support the latter: one of them is Electronic Data Interchange (EDI) and the second is the integration of computer systems and PABXs.

2.1. PABX and computer integration

The emerging standards for PABX-to-computer interfaces will accelerate the integration of telecommunications and information processing. Although such proprietary interfaces, e.g. Computer PABX Interface (CPI), have existed for years, there are several advantages to the standardization of this interface. First of all, it creates a "safe and familiar" platform for independent software vendors to develop new applications that will take advantage of the PABX and telecommunications. Secondly, as a consequence, these new applications, integrating communication and information processing, will improve the efficiency in business operation. Furthermore, standardization of the interface will also lead to emergence of intelligent PABX peripherals with advanced information processing capabilities such as media translation, e.g. fax-to-voice mail, video telephony, etc.

PABX-to-computer standards are developed both in Europe and the USA. In Europe, ECMA´s Computer Supported Telecom Applications (CSTA) standards are in the implementation phase. In the USA, ANSI's Switch Computer Application Interface (SCAI) specifications are just approved.

So far, the synergy of PABXs and computers is manifested by so called 'Call Centres'. These systems, developed from the first Automatic Call Distribution (ACD) systems, are primarily used in organisations with large volumes of transaction oriented business.

The standardization of PABX-to-computer interface will lead to growth of Call Centre market, even in smaller organisations, since the applications will run on inexpensive PCs or LANs rather than on mini- or mainframe computers. It is expected that the Call Centre technology will be used for:

- automatic identification of incoming calls
- call tracking and handover between agents
- handling of complete calls by employing voice prompting systems

2.2. Telecommuting

The essence of all communications is the ability to reach and to be reached. This accessibility is diversified, depending on the purpose of communications: accessibility to humans, accessibility to information and accessibility to processes. Each of the above needs imposes requirements on the communication systems and networks.

Access to humans is enhanced by introduction of mobility. So far, the concept of mobility has been associated with the ability to reach a person 'on the move' both locally, i.e. within office premises, and generally, i.e. outside office premises. However, the meaning of mobility should be expanded to cover telecommuting and arbitrary workplaces.

Telecommuting is a new trend which results from our social behaviour. Traffic jams, air pollution and increased focus on family life result in incentives for working remotely, i.e. from home. Telecommuting requires several important things, some of them mentioned below:

- an efficient communication link between the home and the office
- appropriate terminal types
- rigid user authentication in corporate networks
- tele-conferencing

Arbitrary workplace means that one should be able to access all necessary functions and information regardless of the location within the organisation. This can only be achieved by having a uniform, broadband communications infrastructure which is transparent to the services provided over this infrastructure.

2.3. Broadband

The concept of broadband communications is not new at all. In fact, broadband communications emerged in LAN environment long before they attracted the attention of the telecom community. In this environment, interface electronics and communication protocols were available in the 1980's supporting 4, 10 and later 16 Mbps data transfer rates.

However, traditional LANs are shared media networks, thereby limiting the effective bandwidth available to each station attached to the LAN. Furthermore, these LANs cannot support isochronous communications required for audio and video services. The emergence of the Asynchronous Transfer Mode (ATM) protocol and its ability to handle both bursty data and isochronous traffic, gives new meaning to broadband communications in the corporate communication context.

Broadband will not only relate to the "backbone" capacity but also to the user-network access. This will also change the LAN topology. We see clearly that future LANs will be star shaped rather than ring or bus shaped as in the existing Ethernet and Token Ring LANs. Advantages of the star topology are guaranteed access rate and network robustness and reliability.

The ATM technology, the trend for client-server computing, distribution of service functionality to terminals, need for high-speed wide-area communications like LAN interconnect and emergence of multimedia communications are just a few factors which contribute to the need of a broadband corporate communications infrastructure.

Such an infrastructure, will allow transport of audio, data and video information in a cost-efficient and reliable way. Cost efficiency will result from lower maintenance costs like simplified cabling and user management. Reliability will be achieved thanks to fibre optic transmission and redundant network structures.

Introduction of broadband capabilities in the corporate infrastructure will also promote development of advanced terminal equipment and applications.

3. MARKET STRUCTURE

The market for business communications has hitherto had a very rigid structure, stemming from the traditional separation of voice and data communication. Voice communication, being legislated or *de facto* monopoly of the TOs, was a public utility - even within a company's premises. The market for PABXs was liberated just a few years ago.

In contrast, data processing and communication has been regarded as a strategic asset of a corporation and handled as such. Internal organisations were built up to handle data processing. Inter-site communication was handled via leased lines from public operators.

We have thus experienced a situation where the TO handled corporate voice traffic, though some large corporations established own voice networks, while data traffic was confined to the business customers themselves. The gradual liberalization of the telecom market and technical opportunities to effectively handle voice, data and video in a common infrastructure are changing the scene for business communications.

Public TOs are trying to regain the confidence of the business users by providing Virtual Private Network (VPN) services. TOs strive for a share of the valuable business data traffic as well. Services like Frame Relay and SMDS and network concepts like MAN are supposed to provide the vehicle for making this ambition come true. Of course, TO would be more than eager to get into intra-corporate communications for other services than telephony as well.

On the other hand, Customer Premises Equipment (CPE) vendors use the latest technology to integrate various types of communication media within a corporate network infrastructure. Media separation will only be necessary for traffic external to the corporate network, at least as long as the public infrastructure will handle voice, data and image separately. CPE vendors lack, however, the experience of wide area networking in the TO sense of the word.

There is, however, a third force on the market, so called service providers (SPs) of different shapes, which aim at providing complete solutions to business users. SPs position themselves on the market between the TOs and CPE vendors. The business idea of the SP is quite simple:

1) buy, or invest in, bulk transmission capacity
2) invest in network equipment which refines bandwidth granularity, improves quality of the basic service and is able to support multiple customers

By acting in this way, the an SP is able to provide to the business customer a service of higher value or lower cost, than would be possible to achieve by a single customer leasing basic transmission capacity. Other types of service providers aim at providing new services.

Since SP aims at serving several customers, it implies than extensive administrative functions (service management), operation and maintenance functions (network management), charging/billing functions and VPN functions are required for supporting such a business idea. It is also natural for an SP to support the international communication needs of his customer.

There are several examples of already existing SPs and their number is expected to grow. SPs can be formed by separation from the TO or by new organisations. As examples,

UniSource, MFS Datanet and SITA can be mentioned.

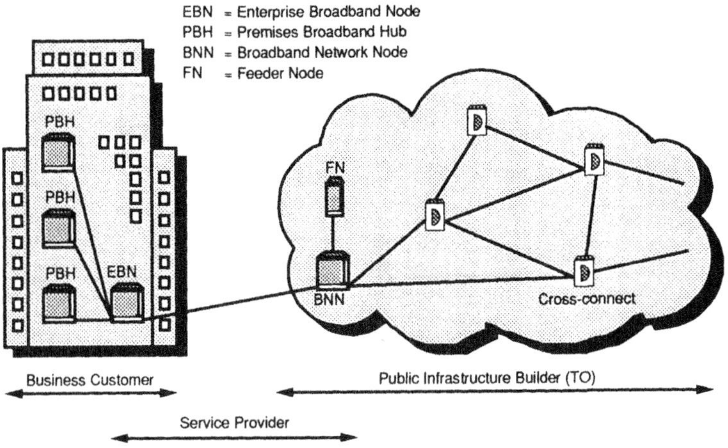

Figure 1. The structure of business communications market. Service providers are focused at providing communications to business customers at a lower price and/or higher quality than offered by the Telecom Operators (TOs).

4. ERICSSON POSITION

Ericsson has been supplier to the telecommunications market for over 100 years. We are represented in over 100 markets worldwide. Our competence and products encompass all major areas: transmission, switching and management for voice and data communication, mobile communication and Intelligent Networks. We are also developing products to meet the emerging demand for broadband communications.

The AXE system has proved to be successful, operating in both fixed and cellular networks. Its modular architecture has been vanguarding and it is fundamental in many national communication infrastructures. Business communications are supported by Centrex and ISDN-BRA functionality integrated in the switch.. The TMOS management platform is becoming a *de facto* standard for operator´s interface to the management functions. Ericsson´s approach to IN has significantly influenced the standardization of IN. The ETNA family provides a complete product range for building of modern SDH-based transport networks. With more than 40% share of the world market for cellular communication systems, Ericsson has proved its leadership and competence in wireless communications.

All these elements support the TOs in providing efficient communications and services to business users. Virtual Private Networks (VPN) can be successfully realized with IN technology.

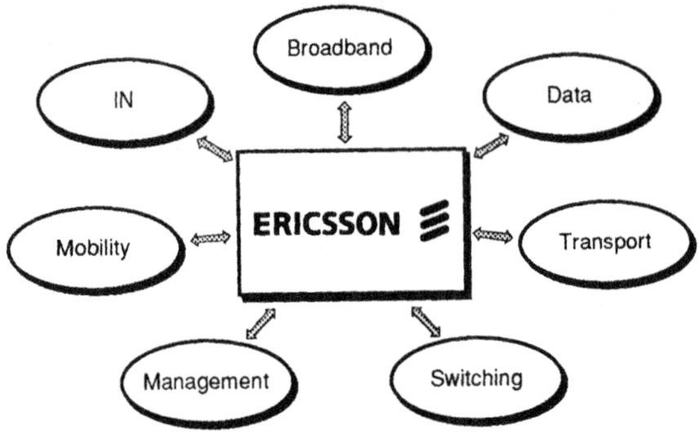

Figure 2. Ericsson has long experience in all key technologies to provide flexible and efficient business communication solutions.

However, Ericsson's offerings are not confined to national and international infrastructure builders. Ericsson has also successfully supplied business network solutions for both voice and data communications. The corner-stones of the Ericsson Business Network are the MD110 digital PABX system, the ERIPAX family of products for data networks, the IDNX bandwidth manager and the NM400 network management system capable of handling both MD100 and ERIPAX systems.

The MD110 is capable of handling both voice and data traffic. Its modular and distributed architecture makes it capable of handling a wide range of extensions in campus, metropolitan or national networks. Specialised applications of MD110 include financial dealing systems, air traffic control, high-security private networks, Call Centre systems and CSTA platforms. The MD110 family has recently been extended with a local broadband communications infrastructure, the MD110-BPN, based on fibre optic transmission at 100 Mbps and wiring hubs for voice, data and video communication.

ERIPAX is a family of products and systems for building corporate or public data networks on local, national or international scale. It implements the X.25 and Frame Relay standards for wide area communications. ERIPAX also supports SNA and TCP/IP networking protocols.

MD110 and ERIPAX networks can be integrated at the physical level, providing premises and wide area networks that combine the transport of both voice and data on the same physical media, by use of the IDNX transmission resource manager in WAN and MD110 Broadband Premises Network within premises.

NM400 is the complete management system for the complete Business Network addressing both MD110 and ERIPAX networks. NM400 provides also a gateway to NetView - IBM's LAN management system. A graphical workstation provides a user friendly interface to the NM400 management functions. The network can by viewed with different levels of detail using a "zoom" function. Independent assessment by Datapro International rate NM400 as the most advanced network management system for voice-and-data networks.

Furthermore, the Business Network concept includes a wide range of wired and wireless access methods and interfaces to all types of public networks: PSTN, ISDN, PSPDN, cellular and mobile data networks.

Despite its success, Ericsson is not resting. On the contrary, during recent years Ericsson has spent a substantial share of its turnover on research and development in key technologies. Several alliances have been entered in order to strengthen Ericsson´s position and enable the company to apply its know-how to new areas.

The coming years will see the introduction of ATM technology in both local and wide area networks to support high-speed data and multi-media communications. In fact, there are already operational ATM-based LANs, and CPE vendors will queue to make new product announcements related to ATM-based LAN and WAN networking.

4.1. The Ericsson approach to broadband communications

The primary objective for Ericsson's broadband development program is a timely and cost-effective introduction of broadband capabilities, helping the network and service providers to timely respond to the market demands.

With these objectives in mind, Ericsson has developed two concepts that form the pillars of the broadband program: the ATM Pipe Switch and the Generic Broadband Module (GBM). Together, these two offer a unique flexibility that will considerably increase the robustness of the network and reduce network management costs. They also facilitate the ability to upgrade and reconfigure the network elements easily, to meet changes in the demand.

The broadband network architecture, as envisaged by Ericsson, is based on several key building blocks. These building blocks are all based on the same technological GBM platform.

Different GBMs provide different functions, but the basic configuration includes an ATM switch core, a processor board and device boards such as access boards, server boards or signalling termination boards.

All the boards are packaged in a GBM subrack. By changing or adding circuit boards and software packages, the broadband functionality of the GBM can be varied. The GBM concept can therefore evolve in line with progress in technology and changing market requirements.

The GBM uses a well-defined internal structure and interfaces to create a modular system with a flexible use of hardware units. GBM-based products with different switching rates can be interconnected directly. This is also the case with different generations of switch equipment. All boards in a GBM are self-identifying and can be plugged into any available slot. The system is self-configuring, which makes it easy to handle and improves the robustness.

A single GBM can be used as a small broadband node, or it can be combined with other GBMs to implement larger broadband nodes of almost any size. Notably, the node concept is distributed and a single can consist of several GBMs spread over a geographical area. The GBM concept is equally applicable in corporate networking.

4.2. Node Types for Intra-Premises Communications and private WANs

The GBM forms the platform for two node types intended for the intra-premises communications: the Enterprise Broadband Node (EBN) and the Premises Broadband Hub (PBH).

The EBN constitutes the central switching resource of the Broadband Business Network. It may act as a central hub in a building or campus network. Sometimes, the EBN will belong to the TO or service provider as part of "Centrex" offerings.

EBN incorporates ATM switching, servers for higher order protocols such as voice, video, data routers, functions for least cost service/routing, etc, and interfaces to existing CPE.

Existing equipment can consist of MD110 PABXs, ERIPAX nodes, LANs, routers, etc. Progressively, more and more of these functions will be implemented within the EBN, obviating the need for physically separate servers.

EBN acts also as a gateway between the intra-premises and inter-premises communication networks. EBN will handle PDH, SDH or ATM interfaces depending on the situation.

The PBH acts as a concentrator towards the EBN and can be located on each floor of an office building or each building in a campus network. End-user equipment, potentially utilizing a variety of protocols, is connected directly to the PBH. Apart from concentrating the traffic, the PBH also performs switching. During migration from MD110-BPN based infrastructure to ATM-based infrastructure, PBH will be able to co-exist and interwork with Broadband Interface Modules (BIMs) which perform the same task as PBH. The PBH user interfaces encompass 2 Mbps PCM circuits serving MD110 Line Interface Modules (LIMs) and cordless base stations for fixed and cordless telephony; direct ATM connections for computer equipment with ATM interfaces, packet switching equipment (e.g. ERIPAX) and LANs.

4.3. Node Types for Wide Area Communications provided by operators and service providers

The Broadband Network Node (BNN) constitutes the heart of wide area networking. Its main function is to terminate various interfaces from customer premises and switch the traffic. The BNN is operated and controlled by the network operator - a TO or Service Provider.

The traffic from business customers is normally multiplexed and concentrated by a Feeder Node (FN), which functionally corresponds to the Premises Broadband Hub.

The BNNs are interconnected by a broadband transport network. This network consists of transmission links, based on either PDH or SDH/SONET, and cross-connect points. The cross connect-points, based on the GBM platform, will initially perform ATM-VP cross-connection. As the demand emerges, selected nodes may be equipped with various type of servers, e.g. a connectionless server for the support of SMDS/CBDS. The cross-connect nodes will be managed through the FMAS functionality of the TMOS management system.

In the next evolution step, signalling and routing capabilities will be added to the broadband transport nodes. This will enhance the flexibility of the network by allowing the customers to set up connections on demand.

By providing a complete range of products for CPE, network infrastructure, switching, transport, services and management, Ericsson is capable of providing business telecommunication solutions. The virtue of these solutions is a uniform operation and maintenance for the operator of networks and services, resulting in lower OA&M costs, robustness and reliability.

5. NETWORK SOLUTIONS

In this section, two network solution examples will be shown. These example are by no means exhaustive, they are simply intended to demonstrate the functionality that can be achieved by combining various products and concepts described above.

Figure 3 outlines the stratified view of a current network structure for intra-premises communications.

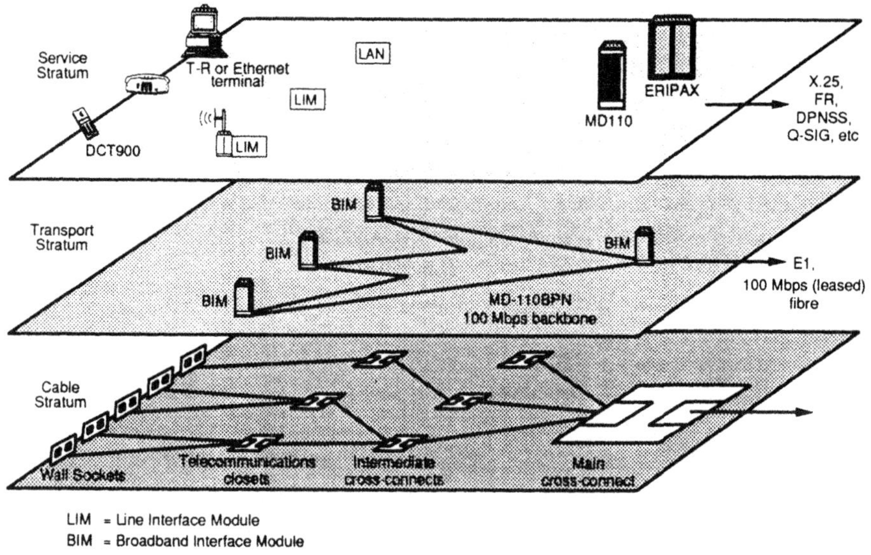

Figure 3. Stratified view of an intra-premises communication infrastructure

In figure 3, the communications network is decomposed into three strata to provide a comprehensive view of the various infrastructure components. On Cable stratum, structured cabling system according to EIA or ISO standards provides the physical infrastructure that has a hierarchical star topology. Fibre is used from the main cross-connect point to the telecommunication closets. Coax, UTP or STP cables are used between the closets and the wall sockets.

The MD110-BPN provides the transport infrastructure consisting of the Broadband Interface Modules (BIM) which are connected to the 100 Mbps fibre ring. External communications can be provided over E1 (2 Mbps) links or over 100 Mbps (leased) fibre.

Each BIM can be equipped with a number of interface to interconnect various types of user terminals. In the central switching room of an office, the MD110 PABX central node and the ERIPAX node are attached to the BIM. The MD110 has the CSTA interface which enables provision of ACD and other advanced transaction oriented services. At the telecommunication closets, the BIMs are equipped with interface modules for data communication (LAN) and voice communication (LIM). LIMs can handle both wired and cordless telephony via the Freeset system that served as a base for the DECT standardization. X.25 and Frame Relay, protocols are offered for interfacing the public networks. DPNSS and Q-SIG protocols enable signalling between PABXs.

The migration to ATM is easily accommodated without affecting the Cable Stratum. The changes take place in the Transport and Service Strata which is depicted in figure 4 below. The central BIM is replaced by the EBN which uses ATM switching. With the introduction of the EBN the intra-premises network starts a transition to star topology. ATM-based hub, the PBH, is interconnected via a dedicated 155 Mbps (or higher) ATM link to the EBN. The "old" BIMs can also be connected to the EBN via a dedicated 100 Mbps link. In this way the bandwidth that is offered to the terminals can be gradually increased. EBN can also handle the MD110-BPN ring-based backbone in the migration phase.

Service functionality is increased by adding, special telephony, video or data servers. The variety and capacity of external transmission links is also significantly increased with the introduction of the EBN.

Since the PBH basically performs the same functions as the BIM, installed communication equipment can be retained. The ATM connection capability of the PBH allows direct interfaces to ATM based work station when they emerge.

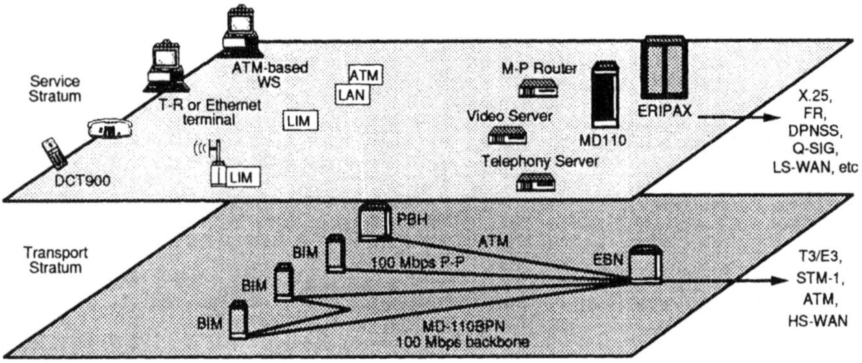

Figure 4. Migration to ATM-based intra-premises communications infrastructure

The evolution of inter-premises wide-area networking can be illustrated in a similar way, see figure 5. The optical fibres, depicted in the Cable stratum, are deployed and cross-connection points are located in strategic points. Above this physical infrastructure, which has numerous redundancy paths, a PDH/SDH stratum is established.

Various SDH-based network elements from the ETNA product family provide a robust and efficient communications infrastructure. The Multiplexers, SMUX, can be configured in several ways (terminal multiplexer, add/drop multiplexer and intermediate regenerators) and are capable of terminating both PDH and SDH transmission rates. The digital cross-connects, DXC, perform cross-connect functionality of PDH paths and act as gateways to the PDH transmission systems. All ETNA products conform to TMN management principles and are remotely controlled, configured and operated via the FMAS part of the TMOS management system.

Entities performing specific services are attached to the interconnection network. The common platform for these entities is the AXE-10 system. From the business users point of view AXE-10 is capable of providing several telephony-oriented services: telephony, Centrex, ISDN, mobility and various types of IN-based services such as VPN. Data communications are provided on the ERIPAX platform. As mentioned earlier, ERIPAX implements the X.25

and Frame Relay standards for wide area communications and supports SNA and TCP/IP net-

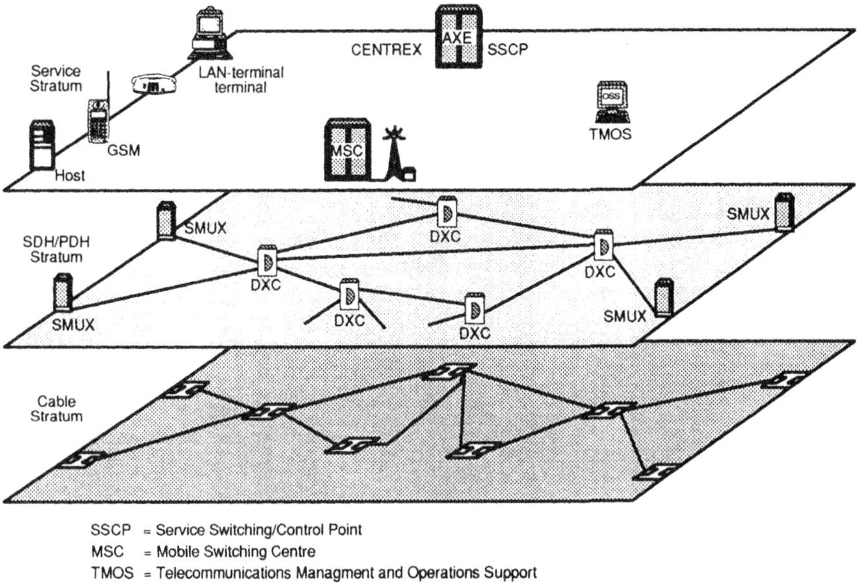

Figure 5. Stratified view of the present and near-term wide-area network structure

working protocols. Management of the ERIPAX network will be done from a TMOS application.

In figure 5 above, the SSCP (Service Switching/Control Point) functionality of AXE-10 denotes the IN capabilities of AXE-10. Ericsson provides a graphical Service Creation Environment (SCE), being a subset of TMOS, that enables a fast creation of customized services.

The most significant aspect of the migration towards ATM, see figure 6, is the property of ATM as an enhancement to the basic PDH/SDH infrastructure. Ericsson considers ATM being a complement to SDH, by allowing a more efficient resource utilization and provision of a common multiplexing and switching platform for several services.

On the ATM stratum in figure 6 we recognize several network elements mentioned previously. One of them is the Feeder Node (FN) which is an access node for business premises. The FN can interface to FDDI, Frame Relay, Token-Ring and Ethernet LANs as well as n*64 kbps lines.

The FN may be connected to the BNN using different bearer services: G.703/704 2 Mbps lines, Frame Relay or ATM interfaces. All these interfaces are provided by the BNN node which is based on the GBM concept. BNN has an ATM network interface and a number of BNNs are interconnected on ATM-VP level via ATM cross-connect systems. These are also based on GBM and management of the ATM cross-connect units is performed via FMAS.

The ATM cross-connect systems can be equipped with various types of servers, such as Frame Relay and SMDS servers as indicated on the Service stratum in figure 6. The transition to ATM is smooth, without disruption to the services offered to the business community. ATM offers a platform to be built upon.

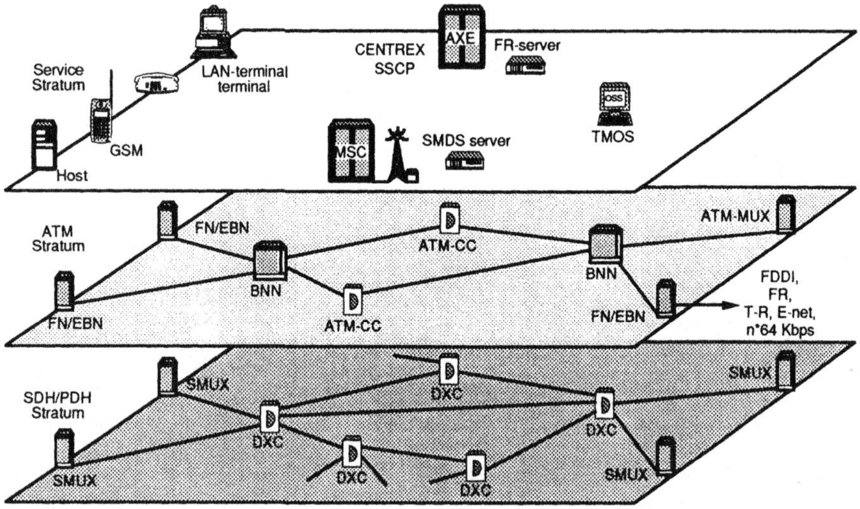

Figure 6. Stratified view of wide-area network structure during migration to ATM

6. CONCLUSIONS

In this paper, Ericsson's view of the market for business communications has been developed. It is foreseen that more and more actors will try to enter this market segment with different type of service offerings.

Ericsson´s global market presence and experience and competence in all relevant technologies makes the company well suited for provision of sophisticated communication solutions to all those actors who address the business communications market. Ericsson can provide complete networking solutions, with integrated handling of voice, data and image, for both premises, campus and wide-area networking.

The advantage of these solutions lies in a coherent and cost-efficient operation and maintenance and a networking concept which is flexible to accommodate future, new services.

Integrated Broadband Communication Networks
and Services (C-18) / V.B. Iversen (Editor)
Elsevier Science B.V (North-Holland)

New IBC Services in support of the border crossing trade in the European Community

R.M. Guarneri[a], L. Resmini[b], C.J.M. Lanting[c]

Abstract:

The 1993 internal market generates opportunities for increased international trade for all trade partners, including small and medium size companies. To take advantage of these opportunities, the latter need efficient and cost effective means for exchanging information, e.g. publicity, marketing, product requirements, product specifications, legal obligations, etc.

The emerging IBC network and services have the potential for becoming the infrastructure able to meet the requirements of this information exchange, including those of EDI, for use by medium and small size companies.

This creates important opportunities for dynamic small and medium size companies as users as well as providers of specialised value added services.

To contribute to the overall competitiveness of the European internal market, and in particular to that of the weaker regions, effective promotion programmes by the CEC would be important tools to speed up and broaden the introduction and use of these IBC services.

Keyword Codes: H.3.5; H.4.3; H.5.1
Keywords: Online Information Services; Communications Applications; Multimedia Information Systems

---

a Italtel SIT c/o RACE Industrial Consortium, 165 Boulevard du Souverain - B-1160 Brussels - Belgium
Tel: +32.2.6748532 - Fax: +32.2.6748538
Email X.400: C=BE;A=RTT;P=Y-NET;ON=SP1;S=GUARNERI;G=RENATA

b Universita L. Bocconi, Via Sarfatti 25 - I-20100 Milan - Italy
Tel: +39.2.5836.5324 - Fax: +39.2.5836.5318
Email: ubec27@imicilea.it

c DATSA Belgium, Wielewaallaan 16-2 - B-3010 Kessel-Lo - Belgium
Tel: +32.16.350282 - Fax: +32.16.350030
Email X.400: C=FR;A=ATLAS;P=EDT;ON=EDT GRENOBLE;S=DATSA

INTRODUCTION

This paper intends to give an overview of the new European economic scenario in view of the Single Market in 1993, and outlines the role that the new advanced telecommunication services based on IBC networks can play in this situation with particular regard to the case of small and medium size companies.

The paper is structured in seven sections.

The first section briefly describes the intents of the European Single Market, and identifies the opportunities that this may offer for the growth of the European economy. Then, in the second section, an overview is given of the new IBC services which may play important role in the support of the international trade. Specific requirements coming from the small and medium size companies are identified in the third section, while in the fourth section the case of EDI is looked at in more detail. The European Single Market will present opportunities not only for the users of the telecommunication services, but also and even more for the providers; this is analysed in section five. Finally section six looks at the possible introduction scenarios for IBC services to promote their usability by small and medium size companies, and section seven briefly outlines the promotion programmes of the CEC in this field.

1. EUROPE 1993: AN OPPORTUNITY FOR MULTINATIONAL TRADE

For the last seven years, since the issue of the White Book of the Commission in June 1985, the European Community has been involved in a process of economic integration, with the objective to create a new Europe without frontiers. The deadline for the completion of the process is fixed to December 31, 1992.

By that date the unification of the EC internal market will be completed; what this unification actually means is often summarised by listing the "four freedoms" within Europe (the EC) starting January 1st, 1993.

After the celebrations for the new year, 324 million EC citizens (although only a few sober people will realise it at that moment), will be able to enjoy:

- Free movement of people;
- Free movement of goods;
- Free movement of services;
- Free movement of money.

What does this mean, and how will be reflected in the European society.

The first freedom implies that EC citizens as well as foreign tourists will be able to move within the EC without being subject to border controls. Specifically EC citizens will be able to work or study in any country of the EC.

With the opening of the borders within the EC, new opportunities will be created for companies that wish to enlarge their business outside the national space. Goods will circulate within the EC without waste of time at the borders, and without any constraint of fiscal, administrative or bureaucratic nature.

The third freedom will allow companies to offer their services within the EC, and consumers will be able to choose the best offer at the best price. Particularly in the telecommunication sector it is expected that new services will be offered using modern technologies.

As for the fourth freedom, while waiting for the introduction of the ECU, as the common currency, EC citizens will be able to travel in the EC using the currency they prefer. Private citizens as well as companies will be able to make any money transfer in any of the member states, or invest money wherever they believe is profitable.

Although it has to be remarked that the latest monetary facts in Europe may make this fourth freedom of somewhat difficult implementation, the principles are certainly promising, and there will be significant advantages for the overall European economy, as competitiveness will be stimulated, and initiative and creativity encouraged.

But how to take advantage of the new situation. Certainly access to information (information on legal aspects, for example, will be of significant importance in multinational trade) is the key issue.

Information is the basic input of the modern economy, linking industry, services and markets together. New information technology and telecommunication services are being developed daily. Modern economy depends increasingly on telecommunications as its basic infrastructure, providing the means to transport and trade a range of other services, as well as the means to support the trade of goods.

The economic benefit for Europe 1993 will therefore depend in large measure on completing the internal market for the telecommunications sector as well. Developing Europe's telecommunication infrastructure will prove central to the achieving the full benefits from completing the internal market, and to improving Europe's position in the global economic competition.

2. IBC SERVICES IN SUPPORT OF MULTINATIONAL TRADE, OVERVIEW

In the recent past a large number of reports, communications and opinions from different sources have been published on the economic situation in Europe and its future development. The Commission's Green Paper on the development of the Common Market for the Telecommunication services and equipment outlines the common understanding that the opportunities presented by the new telecommunication technologies will have a very important impact on the future economic growth in the European Community, as improved access to better information raises productivity throughout the economy.

The quality of telecommunications will therefore be crucial for the future economic growth, since it will determine the capacity of the economy both to generate, and to use the most important factor of modern economy: information and knowledge. Services already account for nearly two thirds of the EC output. By the year 2000, two thirds of the GNP of the EC will be generated in information related activities.

However, transition to the new service-driven and information-based economy requires very large public and private investment in a new infrastructure for the development of value added telecommunication services.

The current studies in the field of telecommunications have identified the ATM (Asynchronous Transfer Mode) technique as the most promising technique for the support of advanced services requiring large amount of data to be exchanged among different parties. While the ATM technique is being consolidated from the technical side, the possible applications of the high speed communication capabilities provided are also investigated under the concept of IBC (Integrated Broadband Communication). IBC systems will be able to carry mixtures of voice, data, text, image and video in a way that integration of services, as seen by the user, is possible. This will enable the creation of application packages which will tailor the capabilities of the systems to the requirements of the users. Specifically application pilots are being investigated in order to identify areas where advanced telecommunications can be of value in supporting existing processes, and in offering new opportunities. In addition trigger areas, where the market for advanced telecommunication services will develop, can be identified.

Various economic sectors have considered in this investigation ranging from banking, to advertising, health care, manufacturing, and so on. In this paper however we will only consider aspects where trade is involved and specifically how multinational trade can have a new push from the development of IBC services and applications.

In order to provide an overview of the possible services to be exploited in the field of international trade, some basic requirements will be listed first. First of all the product will have to be made known to the international market, then it will have to be sold in an international environment, possibly after sales service will be needed.

With these three basic requirements, some value added telecommunication services belonging to three different categories may be already identified. To give some names they are, e.g. Electronic Catalogue, Electronic Product Selection and Electronic Trading Post, Remote Delivery of Expertise (e.g. on legal/contractual matters related to the various specific countries), Remote Maintenance and Training.

Particular importance will have the first two types of services; with the Electronic Catalogue, for example, by means of interactive terminals it will be possible to provide information on the articles available by a certain distributor together with payment instructions, while, on the other hand, the customer could also provide delivery instructions. A similar service is the Electronic Product Selection, where the purchasing staff of any store would have access to the live view of articles referenced by the central purchasing service of a distributor. This allows a dynamic, informative selection with detailed marketing information, while the products can be ordered at the place where they are expected to be sold, avoiding loss of time and money. With the Electronic Trading Post the capability would be provided to establish a presence in a market place for the products of a geographically remote entity, by making promotional material available via the telecommunication system. This could be linked to the activities of Chambers of Commerce. In addition to all the services aiming at promoting products in a foreign environment, significant importance will have all the services aiming at providing information on the different legal and contractual aspects of sales in different countries. Although the Single Market aims at harmonising the different requirement in different countries, still a lot of effort needs to be done, and in the mean time it is important to know whether a certain product can be sold in a certain country, and which modifications are required to make the product marketable.

Another example of application of IBC services to international trade is the well known home shopping. Home shopping is a service that is growing both in US and Europe. However, this growth is limited because either an image is transmitted without the possibility of interaction (TV), or there is an interactive dialogue but no moving images (videotext). IBC will bring interactive image transmission, together with textual and/or voice information and enable the customer to direct the dialogue, select his choice and modality of payment. This will considerable help the development of home shopping, also across different countries, and will extend the categories of goods that might be sold in this way.

A number of examples of services aimed at the development of border crossing trade can be given in general terms, however before going further it is appropriate to have a look at what are the requirement in this specific field, and in particular at the case of the small and medium size companies.

3. REQUIREMENTS AND USE OF INFORMATION SERVICES FOR TRADE BY SMALL AND MEDIUM SIZE COMPANIES

The complexity of the international business and the lack of know-how can place a firm in a competitive disadvantage. This is particularly true for small and medium size firms which constitute an important and often major part in the economic structure of several countries.

It is well recognised that a critical factor in competing on foreign markets is knowledge and expertise, and that a key element in the construction of the competitive competence of companies is information. If national producers receive the adequate training and information, the exporting process itself becomes easier and less expensive because firms know how to profit from this activity.

Exporters need a lot of information concerning the economic situation of the foreign markets they wish to exploit. This information ranges from economic and financial indicators, business practices, tariff and non-tariff barriers, import regulations, existing commercial and co-operation agreements, public contracts, exhibitions and trade fairs, standards, export prices, transportation, lists of potential customers or distributors, etc.

The gathering and dissemination of commercial information are now undertaken by several institutions, such as governmental trade information offices, export boards, trade funds, Chambers of Commerce, management institutes, etc. Most of them are state-owned or state-controlled. Export related information, in fact, can be considered as "public goods", because they confer the same benefits simultaneously on a number of people. Thus, governments have an interest in the dissemination of information because of the spillover effects they generate.

Commercial information is available to exporters in a variety of formats: publications, such as market guides, market opportunity studies, "how-to" manuals on the mechanics of exporting, newsletters, bulletins or specialised magazines for particular markets or sectors; export information telephone services, computerised data banks of national products, services and companies interested in selling their products abroad. Information available on-line is generally obtained by direct link or through a "re-distributor" or, upon request, by telex.

This kind of services, however, does not satisfy the users. Information they receive is often not correct, up to date, timely or costless. In those conditions, information is of little utility or worthless, and they are hardly able to confer a competitive advantage in terms of knowledge and expertise to the exporting firms.

IBC services may constitute the right answer to the need of the providers and the users of export-related information.

4. USE OF EDI BY MEDIUM AND SMALL SIZE COMPANIES

One of the means to exchange and distribute trade information is Electronic Data Interchange, EDI. EDI describes an expandable set of standardised messages, supported by a set of procedures, that in turn can make use of a variety of services to exchange these messages. EDI is therefore itself not an IBC service, but rather a potential user of IBC services.

Two links can be established between the development of EDI and that of IBC services. Although it already is gaining popularity, the main deployment of EDI will coincide with the introduction of IBC services. Secondly, the exchange of messages including graphical, photographic, or even sound or moving image information, requires the availability of wide area high bandwidth services provided only by IBC.

The use of EDI brings a number of important advantages within the reach of small and medium size companies:

- reduction of administrative delays, overheads and costs associated with dealing with customers and suppliers;

- extension or removal of the geographical limits on the area that can be served with products or services;

- a standardised business communication method between small and medium size companies to support strong co-operation, for example in the form of consortia;

- a standardised access and transfer mechanism for interchange of trade information with information and service suppliers.

It is often argued that EDI brings advantages to all industry, and that the big, rich companies will be in a better position to take advantage of it then smaller companies.

However, it can easily be shown that the functions supported by EDI are already commonly available to big companies in the form of proprietary, intra-company information exchange processes and flows. Further, that when big organisations impose their proprietary information flows onto smaller business partners, that the latter suffer rather than profit from it: best case the dependency on one big partner increases, worst case the smaller partner is buried with numerous different procedures by its bigger partners. The standardised approach of EDI gives small and medium size companies not only equal rights but also equal chances in exploiting information technology benefits to effectively control overheads and costs.

Whereas some services can be provided without geographical restrictions, the provision of other services and the distribution of products will impose the necessity for additional distribution and support points. EDI here can support the communication with these support and distribution points efficiently even when other companies are involved, which again is of particular importance for small and medium size companies.

In case the EDI procedures and messages are not only based on voluntary agreements between business partners, but in addition reflect legal requirements and legally accepted business practices, an additional advantage can be obtained. For, the use of these EDI procedures and messages then implies operating within a legally accepted and sound framework. Ideally, the same legal framework should govern all local and international trade.

5. OPPORTUNITIES FOR PROVIDERS OF TRADE RELATED VALUE ADDED SERVICES

As stated before, the continuous increase in demand for trade related information and communication on the one hand, and the introduction of IBC services on the other, provide a fruitful base for the improvement of existing and the development of new value added services, thereby creating new opportunities for value added service providers.

A wide ranging variety of services can be offered, either for a closed or specialised audience, or as a publicly available service. It is useful here to note that many value added services already exist in some form, often for a closed community. Current services typically fall short of user requirements in two areas:

a. bandwidth limitations, often resulting in poor performance, poor user interface, and limitations in understandability and completeness of the information provided;

b. a lack of standardisation of user interfaces, access procedures, protocols and the use of underlying services.

The first issue will be resolved effectively by IBC's ability to deliver high bandwidth at only marginally increased or even decreased communications costs.

The second issue leads to great annoyance for users: access to multiple value added services means working with different and incompatible user interfaces best case, different protocols or even different equipment otherwise. It also leads to unnecessary costs and to slow market acceptance of value added services. This further affects the possibilities for small and medium size companies to seize opportunities for value added service provision.

Taking up the latter point, a set of requirements for the provision of value added services by small and medium size companies can be formulated as follows:

i the supporting (IBC) infrastructure should be available "everywhere", and at reasonable and equal conditions, for both users and suppliers of value added services;

ii also, more general value added services that can be used as infrastructure for specialised services should be available everywhere and at reasonable and equal conditions for both users and suppliers: Message Handling and Directory Services, and EDI transfer services are examples;

iii standardisation of user interfaces, application interfaces, access procedures, protocols and use of underlying services; this standardisation can either take the form of industry agreements, or of formal standardisation; but what really counts is the market acceptance and stability of the standards;

iv possibility of charging for the use of services via the tariff and billing facilities of the provider of the (IBC) infrastructure or supportive value added services.

Can these conditions be met, then the provision of value added services in the IBC environment by small and medium size companies will only depend on the imaginativeness and quality of the services offered.

6. DEVELOPMENT SCENARIOS FOR IBC SERVICES FOR TRADE

The variety and the quality of the IBC services make them particularly appealing for those operators who daily handle large amounts of different pieces of information. Therefore, governmental trade information offices are a potential user of IBC services.

As we have explained before, this kind of public agencies must treat a large amount of different data that need to be quickly processed and made available to users as soon as possible in different formats. Moreover, users are often very different in structure and far from each other.

By the IBC networks and services, export information offices can exploit the links with their peripheral agencies or other EDP centres and commercial libraries in order to ensure a better and faster circulation of the information. Consequently, even the effectiveness of the services, which is the utility perceived by the users, must increase.

Exporters will find easier the management of the foreign markets because it is easier, faster and less expensive to get useful information. Needless to say, the impact at the firm level of the increased effectiveness of the information services, will probably not appear as sales, but rather as an increase in knowledge and competence applied to the export strategies. This effect is not less important than the sale effect, since it is able to change the organisational structure of the companies themselves.

This is of particular importance for small and medium size firms, whose organisation, strongly oriented to the production, lacks of marketing oriented services, such as post-sale assistance, marketing, advertising, legal department, etc. Through an integrated terminal, each firm can get qualified services to support and complement its productive activity. To give an example, through a "tele-assistance" service, exporters can obtain professional advises speaking simultaneously to several consultants who are able to send a firm specific documents, or to show catalogues or products, to get in touch with potential buyers and sellers, etc. The advantage is that nobody has to move to meet other people, without losing the advantage of a "personal contact", very important in the counselling activity. Users, in fact, can make use of all the non-verbal languages (gestures, expressions, positions) they would use during a personal meeting.

This scenario applies to a world where information technologies are costless and freely available. A more realistic perspective is that of a gradual introduction of these new information technologies supported by single governments or by the EC itself. Pilot experiments in the export related services would be useful for many reasons. On the one hand, they would contribute to the rationalisation and the efficiency of the supply side; on the other hand, they would surely stimulate the demand for new integrated services, not only related to commercial aspects.

7. OPPORTUNITIES FOR PROMOTION PROGRAMMES BY THE CEC

The CEC is playing an important role in the development of IBC. The different facets of its role are:

- the role of promoter of both the idea and the realisation;

- that of enabler with respect to the development of technical specifications and standards, notably in the RACE programme;

- a regulatory role, where the CEC moves to put in place legislation and procedures to ensure open competition and equal rights and conditions for access by users and providers of IBC services;

- the role of promoter of the use of the IBC services;

- the role of a user itself of the IBC services.

The CEC has already very effectively developed activities to fulfil its responsibilities in the first two roles.

For the regulatory aspects it is still a bit early, but taking into consideration what has been stated before here, it may be concluded that new regulations are required, also for areas that previous had been considered as "de-regulated".

Opportunities exist for the CEC to further promote the use of IBC services. One possibility is the promotion and funding of demonstrations' projects and pilot services. A second possibility is to promote IBC services by using them to offer public information services on EC activities, in particular CEC activities and CEC sponsored activities. In fact, extending existing programmes to use of IBC services would already provide a solid basis for this. Finally, the use of IBC services by the CEC for its internal needs can be a show-case, the creation of a market as well as the endorsement and promotion of the necessary standards.

It is of key importance that in all cases demonstration and pilot projects and services provided are based on sound economical considerations. Although this should not be misinterpreted as requiring all these activities to be profitable from the first day. Instead, all activities should be justifiable as (longer term) investments. In this way not only money is saved, but also the credibility and the influence of these projects is greatly enhanced.

SUMMARY AND CONCLUSIONS

The European Single Market is presenting every company with opportunities and threats at the same time. While a larger market will be available to all the companies, competition will be stronger, and firms will have to get ready to face the new challenges. Information will play an important role in placing a firm in a competitive position, so that the internal market can be protected while, at the same time, expansion outside the national borders can be sought. In this scenario an efficient telecommunication infrastructure able to support high speed advanced services is a must.

Numerous studies have shown that the telecommunication infrastructure has a decisive influence on the economic growth of a country. Just by looking at the telephony service, it can be seen that there is a major correlation between telephone density and GNP; with growth in telephone density, the GNP per head of population also increases. This correlation will show to be even stronger with the use of IBC services, as telecommunications will represent even more a production factor for all industries of the economy.

Specifically small and medium size firms will benefit from these new services as it will possible for them to get timely and effective information at lower costs.

Specific promotion programmes however are required to spread the knowledge about these new services and to promote their use by the companies. The CEC has started already, within the RACE programme, to promote activities aimed at the evaluation of the impact and the usability of the new advanced telecommunication services on the various economic sectors. Additional actions may be taken by governmental agencies to actually promote the use of the new services by the small and medium size companies that may not have the possibility to get access to these services directly, but that, at the same time, would certainly benefit from their use.

BIBLIOGRAPHY

Commission of the European Communities - Directorate General XIII; XIII Magazine; July 1992 Issue No 7

Commission of the European Communities; Green Paper on the development of the Common Market for Telecommunication services and equipment; June 1987

Documentation Europeénne - Office des publication officielles des Communautés Europeénnes; Un grand marché sans frontières; Luxembourg 1989

European Documentation - Office for official publications of the European communities; Research and technological development policy; Luxembourg 1988

"International perspective on trade promotion and assistance" edited by S. Tamer Cavusgil, and Michael R. Czinkota; Quorum books 1990

Seringhaus F.R.H., Rosson P.J. "Government export promotion: a global perspective"; Routledge, New York 1990

Integrated Broadband Communication Networks
and Services (C-18) / V.B. Iversen (Editor)
Elsevier Science B.V (North-Holland)

Economic Implications of Introduction of new Telecommunication Services

M. Falch, A. Henten and K. E. Skouby

Institute of Social Sciences, Technical University of Denmark, Bldg. 301, DK-2800 Lyngby, Denmark

1. INTRODUCTION

The focus in this paper is on the economic implications of introduction of broadband and advanced telecommunication services. It is based on research work performed within the RACE-programme[1], first of all the project R1044[2] and R2085. The economic implications are analyzed by comparing the estimated costs with the expected benefits and are used to forecast the demand for broadband connections. It has been possible to make qualified and rather detailed estimates of costs - even if some reservations are necessary as the subject is a not fully developed technology. Expected benefits have proven more difficult to estimate. The potential benefits are expected to be realized mainly through a reorganized economic environment - which presently is vaguely conceived. To approach these potential benefits we discuss the results from application pilots within RACE and other existing studies of the financial sector and the transport sector. These studies will serve as illustrative examples of the potential benefits rather than forecasts of the total benefits of IBCN.

2. BACKGROUND

The development of new services based on telecommunications is increasingly seen as a decisive strategic relation for the economic growth in the industrialized countries. Indicators are pointing to the fact that the convergence of telecommunications and computer technology into telematics is generating a forceful technological and economic growth. The rate of innovation in as well transmission as switching equipment has accelerated. This again has inspired to new forms for delivering traditional telecom services and to entirely new services. An indicator of this development can be seen in the fact, that the growth rate of the total revenue of telecommunication services in fixed prices has almost been twice the growth rate of GNP in most OECD-countries from 1974-85.

It is this combination of technological and economic growth which has inspired, e.g., the

---

1 RACE (= Research and development in Advanced Communications technologies in Europe) is an EC funded research programme.

2 Results from this project are presented in [1].

EC-Commission to see telecommunications as a decisive structuring input for the future economy (see e.g. [2]). In accordance with this the EC has been pushing to create the best possible conditions for the growth of new telecom **services** as it is especially in this area the growth potential is seen. The traditional services are expected to continue their rather solid growth, but new services are expected to grow by 25-30% p.a. [3].

It has been argued that one of the most important obstacles to the growth in new services has been the regulation of the industry and the national telecom monopolies. A thorough deregulation and deconstruction of the monopolies is partly realized by now. To assure sustained growth in the services, the next condition is seen as heavy investments in telecommunications: At the EC level 300 billion Ecu is to be invested during the coming 15-20 years. Such heavy investments call for analysis of the economic impact. In this paper the first steps in such an analysis are carried out using figures for the Danish economy. This is not justified by Denmark being representative for the EC-economy, but by the purpose of the paper - to give illustrative examples - that can be met in this way as relevant Danish data and economic relations are available whereas they are non-existent or non-available at the EC level.

3. TWO DIFFERENT KINDS OF IMPACT

The economic impact of introduction of new telecommunications can be categorized in two different types of impact:

1) Impact related to the **costs** of providing telecommunications services.

2) Impact related to the economic **benefits** related to usage of new telecommunication services.

3.1. Costs

The costs are related both to increased network investments and to the development and provision of the new services itself. Most of the investments are expected to be implemented by network operators. An analysis of these costs demands not only knowledge on future prices for telecommunications equipment and the future number of connections for each type of service. The choice of future network-configurations and evolution-strategies, will also affect the costs. Neither of these factors are purely technical matters but depend on the service-requirements of users and the demography of the market. Studies of the future demand for advanced communication services are thus essential for an estimation of the costs.

3.2. Benefits

While the costs are mostly related to the network operators' production of bearer-services, the benefits are related to the use of tele-services. From an economic viewpoint benefits will basically be realized in the form of increased productivity and competitiveness. The benefits will vary from sector to sector and must be expected to be reaped at a different point of time in different sectors. Therefore, the impact will both involve a general rise in productivity as well as structural changes in the economy.

To make a forecast on the benefits it is necessary both to study the particular services in order to estimate their potential benefits **and** to identify the economic sectors that are expected to benefit from them. In this context the demand for new telecommunication services is a central issue.

3.3. Economic implications

To address these questions it can be concluded that a study of the economic implications of new telecommunication services must include different levels of analysis:

* Microstudies on the effect of particular services
* Mesostudies on the sectoral impact
* Macrostudies on the impact on the overall economy

In this paper we will combine all three levels by studying a limited number of services, analyze their sectoral impact and discuss the overall economic implications contrasted with the macroeconomic implications of the investment needed.

Application pilots within the financial sector and the transport sector will be used as case studies on the benefits that can be obtained by use of new telecommunications services. These case studies will form the basis for a more general discussion of the impacts on the sectoral and societal levels.

4. THE COSTS OF IBCN

It is difficult to make exact cost estimates of equipment based on technologies not yet matured. The RACE programme however, has come out with some broad cost calculations for different network-architectures, based on certain assumptions on the level of demand. Once the costs are known, the economic implications at the societal level can be calculated using a traditional macroeconomic model.

The other type of costs mentioned above is system development, maintenance of the tele-services to be provided and costs related to creation of information delivered by these services. These costs are much more difficult to estimate and will depend on both the level and type of demand for new telecommunications services. Knowledge of the communication needs, future tele-services are going to fulfil, is therefore necessary, if this type of costs is to be included. In our analysis, we have focused on the pure network costs, as they are the most essential for network-operators in their planning.

A cost model has been developed by use of a dedicated software tool STEM[3]. Assuming a demand scenario for each type of service, the needs for new telecommunications equipment and the related costs are calculated. According to our analysis, provision of 10Mb broadband services by optical fiber to residential customers will cost in the area of ECU 2.000 per year per household. This result is supported by similar American calculations, but the price is considered to exceed what residential customers are willing to pay. This may, of course, change if new services, that can provide completely new kinds of benefits to the consumers, are developed in the future.

---

[3] STEM = Strategic Telecommunications Evaluation Model, developed by Analysys Ltd., Cambridge for techno-economic modelling.

The key cost-component is the optical fiber in the local loop, including the laying of cables. While other cost components are envisaged to be reduced over time, the cost of laying optical fibers will remain quite stable. Two implications arise from this: First, a substantial cost reduction will hardly occur in the near future. Second, substantial savings can be obtained by applying hybrid technical solutions, which do not require fiber-to-the-home whereby the cabling work is reduced (e.g. fiber-to-the-curb architectures).

The cost of broadband to business customers only has been calculated to be in the area of ECU 4-5,000 per year for a 140Mbps connection. This price is not prohibitive - provided that the benefits will justify the investment. In view of this, it seems that provision of broadband to business customers is a more viable strategy, than the full scale introduction of broadband, which was envisaged in the early days of the RACE programme.

5. BENEFITS OF IBCN

5.1 Demand for IBCN

The demand for IBCN is a crucial factor in estimating the benefits. Demand forecasting of telecommunication services is a well-established discipline, building on statistical analysis of time-series for demand and demographic data. However, in the case of forecasting of demand for new services, no time-series are available and this approach is therefore difficult to apply.

We will address the issue from two different aspects: First, we will, by use of Input-Output tables based on national account data, map the existing demand by sectors. Second, we will study potential applications in particular sectors more carefully. In this subsection we will concentrate on the first aspect.

In order to specify the demand by sectors, calculations on the Danish input-output tables have been made. These tables provide information on the demand for goods from one industry to another. The distribution of demand for communication between different economic sectors is given in Table 1.

The consumption of communication includes the demand for both tele- and postal services. In a separation of tele- and postal services, telecommunications accounts for the major share of the expenses (approximately 2/3). As new telecommunication services are expected not only to substitute some of the existing tele-services, but also some of the existing postal services, the demand for communications altogether is a measure just as relevant as the demand for telecommunications alone.

It follows from table 1 that the service sectors dominate the telecommunications market. Financial and Business Services are both separated from the rest of the service sector due to their significant consumption of communications.The two sectors have further grown considerably in significance while the distribution between the remaining sectors has been fairly stable.

This is reflected in the development of the communications intensity calculated as the consumption of communications as percentage of total input (Table 2). The intensity has increased over time in all sectors excluding transport services where it has been stagnating[4].

---

[4] Information for the last year in the time series, 1988, is not available for transport services. Based on qualitative information it is estimated that the stagnation in communications intensity in the later years has been replaced by the increase known from other sectors.

Table 1
Distribution of consumption of communication by sectors (%)

SECTOR	1978	1985	1987	1988
Primary Sector	2.47	2.15	1.91	2.07
Manuf. & Constr.	26.07	27.07	28.03	26.83
Transport	6.23	4.29	3.51	n.a.
Financial Services	7.82	8.55	13.21	14.51
Business Services	8.85	12.80	13.85	14.28
Other Market services	16.90	17.58	17.03	19.07*)
Public Agencies	31.66	27.57	22.47	23.04
Total	100.00	100.00	100.00	100.00

*n.a. = not available, *) including transport*
Source: Figures calculated on basis of Danish Input-Output tables.

Table 2
Consumption of communication as percentage of total input (1980-prices)

SECTOR	1978	1985	1987	1988
Primary Sector	0.22	0.21	0.28	0.27
Manuf. & Constr.	0.45	0.48	0.66	0.67
Market Services Total	0.72	0.91	1.30	1.27
- Transport Services	0.58	0.49	0.52	n.a.
- Financial Services	1.88	2.60	4.58	4.69
- Business Services	1.86	2.30	3.00	2.81
Public Agencies	1.22	1.20	1.34	1.28
Total	0.63	0.74	0.99	0.96

n.a. = not available
Source: Figures calculated on basis of Danish Input-Output tables.

However, it is in the financial and business service sectors that the development has been most pronounced.

It follows, that the communications intensity, measured as consumption of telecommunications compared to the total consumption of inputs, in the manufacturing sector is about 2-3 times that of the primary sector, the intensity in the service sector in total is about twice as high as that of the manufacturing sector. For the private service sector this is mainly due to the intensive use of communications in the financial and business service sectors. The financial sector has an intensity of more than three times the rest of the service sector. Also in absolute terms, the service sector is the most significant accounting in total for more than half the communication consumption.

The consumption of telecommunications includes expenses paid to network operators, i.e., expenses paid for line charges and basic equipment. For more advanced users of telecommunications expenses for advanced equipment and system development exceed line charges. In a typical financial service company line charges are less than 40% of the total amount spent on telecommunications [5]. The overall economic significance of telecommunications is thus more unevenly distributed than indicated in the table.

Another way to approach the demand for broadband services is to analyze the present demand for other advanced telecommunication services, which serve similar communication needs.

The RACE project PACE (Perspectives for Advanced Communications in Europe) has made a four-nation survey of business data communication [6]. In this survey, based on a sample of firms in Germany, France, Italy and UK, the datacom expenditures paid to public carriers are estimated for different economic sectors (Table 3). It follows that, apart from the financial sector, the sectoral distribution for datacom expenditures and the overall expenditures for communications derived above is quite similar. The financial sector's share of data communications is 32.1% compared to 14.51% of overall communications. This can be interpreted as an indication of the fact, that the financial sector has gone further in the direction of using telematic services to solve its communication needs, than any other sector, and the potential for further growth could therefore be smaller than in other sectors.

If we look at the demand for 2Mb connections (Table 4) the sectoral distribution is quite similar to that of the total expenses for communications.

The expected sectoral distribution of broadband services will depend on the kind of communication needs broadband is expected to serve in the future. If broadband services primarily are seen as fast data communications (e.g. LAN-to-LAN connections) table 3 should be used as a starting point. However, if broadband will serve a wider range of communication needs the distribution indicated in table 1 is more likely. By all means the differences are not very big. This can be taken as an evidence for the fact, that the sectoral distribution is quite robust to changes in the kind of services provided.

Table 3
Distribution of Datacom Expenditures to Public Carriers (1989, %, MECU)

Industry	Germa-ny	France	Italy	UK	Total
Manufacturing	23.3	24.7	25.2	23.8	24.7
Bank/Ins.	22.2	33.2	32.7	28.8	32.1
Gov.agencies	26.0	22.2	154	13.7	20.1
Utilities	7.6	4.3	20.0	8.1	7.4
Distrib./ other serv.	15.3	15.5	12.7	18.1	15.8
Total (%)	100	100	100	100	100
Total (MECU)	1170	967	771	811	3712

Source: Calculated on basis of PACE 90. Perspectives on Advanced Communications for Europe, vol. VI: Business Data Communication Services, 1990, p. 29, Exh. 3-8.

Table 4
2Mbit connections by sector in KTAS's concession area

SECTOR	NO. OF CONNECTIONS	In %
Agriculture	0	0
Manufacturing	61	14.7
Financial Service	48	11.5
Transport	24	5.8
Tele	33	7.9
Other Services	107	25.7
Public Agencies	91	21.9
Others	52	12.5
Total	416	100

Source: Calculations based on data from the network operator, KTAS

5.2. Benefits in the banking sector

The banking sector is often seen as a vanguard user of telecommunications. This view is supported by the above analysis showing the most intensive use of telecommunications within the financial sector. The most important reasons for the wide spread use of telecommunications in the banking sector are:

1. The entire product can be delivered in electronic form.
2. Information is available in a digitalized form.
3. Human and financial resources have been sufficient to invest in information and telecommunications technologies.

Thus the banking sector has a favorable point of departure for a telematization, compared to other sectors.

In the RACE programme three pilots for applications within the financial sector has been adopted: R1055 MERCHANT (electronic retail cash handling), R1059 DIVIDEND (Dealer Interactive Video) and R2027 BANK (banking applications). The latter has just started in 1992, so few results are available.

If we look at the type of tasks fulfilled by telecommunications and discussed in the pilots they can be categorized as:

* Payment transactions
* EDI
* Financial databases
* Communication with customers
* Dealer systems

Payment transactions are probably the most telematized of the services. In some countries, the overwhelming majority of the payment transactions by banks are performed by use of telecommunications. This has drastically reduced the clerical work and the number of bank branches. Also, transactions can be carried out easier and faster. This process has been developed mainly by use of the existing narrowband network facilities.

EDI - electronic document transfer has developed at a somewhat slower pace due to the lack of a common standard. The benefits related to electronic payment systems have been hampered by the fact that information related to the payment often has to be delivered manually.

Financial databases contain financial news as well as information on bad payers, economic-statistical data and others. With regard to financial news, speed can be essential to the usability. Therefore, there will always be a market for systems, which are able to provide faster transfer of data. The benefits at the social level are however, not obvious.

Communication with customers can today be made electronically by use of either homebanking or Automatic Teller Machines (ATM's). Both communication channels are actually used. While homebanking still is dominated by business customers, ATM's are mostly used in retail banking. In communication with customers, multimedia terminals would be a way to increase the possibilities for self-service - especially for retail customers.

Dealer systems are used for dealing with a wide range of financial assets all over the world. In respect of the huge amounts involved, a maximum of speed and usability is essential.

The benefits of telecommunications are obvious in the financial sector. However, it is less obvious - also from the pilots - what kind of benefits use of broadband will imply. Full telematization of the payment system can easily be implemented by use of narrowband only, although broadband in some cases might be more cost efficient.

The most obvious areas for the use of broadband services are customer related functions and dealer systems. These are exactly the functions that have been dealt with in the application pilots. In both cases multimedia terminals could be an improvement. However, the lack of existing broadband facilities in the networks has made it difficult to make banks willing to really invest resources in development and testing of such systems. Even with funding from EC it has not been too easy to make banks interested. However, if it is possible to further automatize customer related communications by use of terminals requiring broadband there is clearly a potential for substantial savings. In this case, the communication needs will be just as dispersed as the customers of the bank: A broadband homebank system demands provision of broadband to residentials. As noted before penetration of the residential sector is hampered by the costs of broadband. Another solution could be terminals placed at central locations. The maximum number of such terminals would probably be within the same range as the present stand-alone payment terminals (about 300 in a country like Denmark). Terminals may also be placed within bank branches. This would increase the demand to about 3,500 if every branch is provided with one connection each.

The benefits here will first of all be a reduction in the number of branches needed in the future accompanied by staff reductions. It might be a counter acting factor, that the number of branches already is decreasing and it will probably be difficult to speed up the closing of branches without major organizational disruptions. The benefits of broadband are thus difficult to identify in the short term, and the system might only be viable if combined with other broadband needs coming from other sectors.

5.3. Benefits in the transportation sector

DRIVE and EURET are the two major EC R&D programmes in the field of transportation. DRIVE was initiated in 1988 and is concentrated around road traffic. EURET, which is a new programme, is dealing with other means of transportation, rail, air and sea.

Removal of goods or persons is a transportation of physical entities while communication may be considered as a transportation of information. Benefits at two different levels may accrue from the implementation of new telecommunication services and improvements of existing services. The first level has to do with the possible substitution between transportation and telecommunication. The second level is dealing with optimisation of efficiency in transportation.

5.3.1. Substitution

The substitutional effects have been widely discussed and have often been opposed with the argument that telecommunication will not substitute but supplement transportation. However, with the advent of new and more advanced telecommunication services, substitution may become possible. In a report from Arthur D.Little [7] the benefits from a substitution between transportation and telecommunication are calculated to be 23 billion dollars (1988 dollars) in the USA when estimating a level of substitution between 10 and 20%. This level of substitution will allow 6 million automobile commuters to work at home, 3 billion shopping trips to be replaced annually, 13 million business trips to be eliminated annually through teleconferencing, and 600 million trucks and airplane delivery miles to be eliminated annually through electronic transfer of documents. These types of substitution will - according to the report from Arthur D.Little - require a nationwide fiber-optic network, including fiber-to-the-home with broadband capacity. The costs of such a network are estimated to be around 300 billion dollars including user equipment.

5.3.2. Optimisation of efficiency

The second level of benefits dealing with productivity gains is a prerequisite for some of the abovementioned benefits but will not necessarily lead to decreasing transportation needs. A DRIVE project - FLEET - has contributed with a structuring of information flows in relation to transportation activities. DRIVE is a programme for R&D in the uses of telematics in **road transportation** that is reflected in the following table. But the structure of information flows is just as relevant for other forms of transportation: railways, airplanes and seaways.

Table 5
Transportation information flows

	Planning	Execution	Controlling	Administration
Freight & logistic management	*Transport order *Transport offer calculation	*Transport documents *Customs clearance	*Cargo tracking *Shipment status	*Invoices and payment *Cost & perform. follow-up
Fleet management	*Fleet Monitoring *Tactical route planning	*Load plan *Transfer of documents	*Route guidance *Vehicle/cargo tracking	*Cost & perform. follow-up
Vehicle management	*Trip preparation *Documentation control	*On-road vehicle/ cargo Monitoring *General communication	*Recording of data *General information	*Vehicle cost perform. follow-up

Source: FLEET

In the FLEET project, additional information flows were structured. In this table, though, we have only included a sample of these flows illustrating the multitude of different types of information required for improving the efficiency of transportation.

The planning phase involves a whole range of documentation work that may be improved by telematics. Electronic Data Interchange (EDI) is an important tool in this area. In several business sectors, EDI has reached a take-off stage, but in the important road and rail transportation areas, EDI is mostly in the preparatory phase. In the Danish railroad system, though, some 20% of all consignment notes in part-load traffic are handled in an EDI system, and a 40% level is planned for by the end of 1993.

It has been calculated that the costs of trade procedures in the EC, including handling of documentation, amount to approximately 35-40 billion ECU each year. In the case of Sweden, a similar figure for the year of 1983 was approximately 2 billion ECU, and Swepro (the Swedish office for trade procedures) has estimated that 20% of these costs may be eliminated by implementing electronic communication of documents [8]. With such a level of reduced costs, some 7-8 billion ECUs could be saved each year in the EC area utilising EDI services. This corresponds to app. 0.2 billion ECU in Denmark.

During the execution and controlling phases, it is e.g. important to track both the transportation unit and the cargo transported. For the sellers and buyers of goods, tracking of cargo is essential to secure Just-in-Time delivery allowing for an optimised production management. Transportation units, to an increasing extent, become parts of the extended stock of production firms.

In the administrative phase, a processing of statistical data concerning both documentation and fleet management will be relevant. Such different types of information may be used in the strategic management of the transportation firm.

6. CONCLUSION

6.1. The Banking Sector

In the banking sector, one method to measure the benefits is in terms of staff reductions. These are related to a reduction in the number of branches and a higher degree of self service by customers.

The early experiences from introduction of automatic teller machines (ATMs) showed a potential of a reduction of staff by one for every two terminals installed. As the costs per employee is 30,-40,000 ECU per year, there is potentially a substantial saving from the installation of broadband terminals - even if we must include the costs of broadband ATMs and connection charges in the calculations to justify the investment in a broadband. Therefore, it is likely that banks, with the current level of costs, will introduce broadband ATMs. Furthermore, they may wish to use broadband at locations where the communication need in total is of a considerable size in terms of traffic. A good estimate of the number of such locations will be the current number of 2Mbit connections.

The total demand from the financial sector given the existing level of services will thus be app. 100 connections (cf. Table 4 transformed from the KTAS-area to national level) with potential to 3,500 connections - based on the number of branches and existing ATMs - depending on the level of costs for all elements.

6.2. The Transport sector

If the estimations of Swedish and EC-level benefits are translated to a Danish environment, the benefits related to EDI will be in the range of 0.2 billion ECU. If we assume that these benefits are related to introduction of broadband and that they will be reaped by the bigger firms (more than 50 employees), app. 700 connections are demanded. Also in this example the estimated benefits will by far outweigh the investment costs.

If we, as above, take the current number of 2Mbit connections in the sector as a point of departure, the total demand from the sector will be 50 connections with a potential of 700 - given the actual structure of the sector.

6.3. Generalisation to other sectors

Assuming that the results from the two sectors discussed can be used as 'rules' for the total economy, we have a demand of 900 broadband connections in Denmark with a potential demand of 20.-25.000 connections. Transformed to the EC level, this implies a demand for 40.000 connections with a potential of 1 million connections in the EC.

The method of comparing benefits from advanced communication services with costs of investment in a broadband network indicates the investment in broadband makes sense and has potentials. However, we have neither shown that the potential benefits are conditioned by the implementation of broadband, nor that relevant services will necessarily be developed if investments are made in broadband. Whether these potentials will be realized thus also depends on tariff structures and other connections costs and on whether relevant services will be available. If new services are developed, estimates above are likely to be conservative. There are indications that substantial demand for broadband connections will rise in sectors, that are not heavy communication users today, e.g. manufaturing and printing industries.

The generel conclusions in the RACE application pilots, that no known, single application can justify the introduction of broadband and that most applications easily can be run on 2Mbit connections or even ISDN, does not seem to comply well with the results discussed above. The conclusions of the pilots are not necessarily that broadband is economic unrealistic. It is rather that no services will be developed until the bearer service, broadband is established. As a consequence of this, the approach taken in this paper is different. We have looked into the economic implications in the sectors, given that certain applications are using broadband. This implies e.g. that no single application has to justify the introduction of broadband and this seems to be a reasonable assumption when we are discussing **integrated** services.

REFERENCES

1. M. Bonatti a.o. (ed.): Integrated Broadband Communications: Views from RACE, Network and Engineering Aspects. North-Holland (1991).

2. Herbert Ungerer: Telecommunications in Europe. Brussels (1988), p. 21.

3. Ibid p. 55.

4. D. P. Reed & M. A. Sirbu: An Engineering Cost and Policy of Proposed Fiber Optic Telephone Networks in the Residential Subscriber Loop. M.C.J. Elton (ed.): Integrated Broadband Networks, North-Holland (1991).

5. R. McGahey a.o.: Financial services, Financial centers, Westview Press (1990), p. 51.

6. PACE: Volume VI: IDS 3 - Business Data Communications Services (1990).

7. Ashok B.Boghani et al: "Can Telecommunications Help Solve America's Transportation Problems", Arthur D.Little, Cambridge, USA, 1991.

8. Teldok-Info no. 9, Stockholm, June 1991, page 3.

Integrated Broadband Communication Networks
and Services (C-18) / V.B. Iversen (Editor)
Elsevier Science B.V (North-Holland)

FRAME SYNCHRONIZED RING

a bus concept for high speed processing applications

Pertti Raatikainen[a], Juha Zidbeck[a], and Hannu Flinck[b]

[a]Telecommunications Laboratory, Technical Research Centre of Finland (VTT),
P.O.Box 34, SF-02151 Espoo, Finland

[b]Switching Systems, Nokia Telecommunications,
P.O.Box 33, SF-02601 Espoo, Finland

ABSTRACT

This paper introduces a high throughput bus concept, named "Frame Synchronized Ring" (FSR-bus). The bus is based on a ring topology, which is an economical way to construct an efficient interconnection network for high speed multiprocessor applications. The paper describes the bus structure, its timing arrangements, frame structure, medium access control algorithm, and explains its operation.

The performance of the bus is estimated using the models developed for slotted rings. To evaluate its efficiency, simulated throughput and transfer delays were compared to those of the traditional multidrop bus and the Orwell ring. Simulation models for each of the three buses were developed and run in cases when the traffic in the ring was uniformly and geometrically distributed. The results show that, in both traffic cases, the FSR-bus performs distinctly better than the multidrop bus. Compared to the Orwell ring, the FSR-bus is slightly better when the traffic is uniformly distributed and performs clearly better when the traffic is geometrically distributed.

Keyword codes: B.4.1, B.4.4, C.1.2
Keywords: Data Communication Devices, Performance Analysis and Design Aids, Multiple Data Stream Architectures (Multiprocessors)

1 INTRODUCTION

The forthcoming broadband networks and real time multimedia services pose new and challenging demands on hardware and software solutions. The data transfer rates of many high speed networks (MAN, FDDI, or BISDN) are today in the range of 100 to 600 Mbps. If, for example, packet routing or protocol processing is performed in real time, transmit rates of these high speed networks address a throughput demand of at least several hundred Mbps between the processing and I/O-components. The transfer rates in the future networks will be higher, and consequently the data processing rates will be in the order of 1 Gbps. Examples of the future high throughput applications are high speed packet switching, real time video compression, real time multimedia coding, and real time processing of security algorithms.

To remove the processing bottleneck, new high throughput solutions for hardware and software must be developed. Since the processing power of individual processors is limited, tasks have to be parallelized. Allocation of tasks to a number of processors raises the problem of transfering data between processing components. Since traditional multidrop buses are incapable of handling high data rates between processors in real time, new processor architectures as well as more sophisticated protocols have to be developed.

The Technical Research Centre of Finland started a research program in 1990 to develop an efficient and cost-effective interconnection network for high speed multiprocessor applications. The network was to fulfil technical as well as production related requirements, set by the parties involved with the project. Having gone through a number of candidate topologies and comparing their theoretical efficiencies as well as construction complexities, the ring topology was chosen to be studied in more detail. Several different ring structures and medium access control schemes were developed. This paper introduces a method, named Frame Synchronized Ring, which was found to be the best one of the candidates.

The concept of the Frame Synchronized Ring is introduced in Chapter 2, and in Chapter 3 a description of the performance of the bus is given. In Chapter 4, throughput and transfer delay of the bus are analyzed using models developed for the slotted rings. Chapter 5 presents simulation principles and summarizes the results obtained in the simulations.

2 FRAME SYNCHRONIZED RING

The Frame Synchronized Ring, FSR-bus in the sequel, consists of a number of Nodes, connected together with an *N*-bit wide parallel data bus. By the side of the data bus is a transmit request line and a frame synchronization line. The request line is used by the Nodes to require access to the bus. Its transmit direction is opposite to that of the data bus. The frame synchronization line is used to carry frame synchronization pulses from Node to Node. Its transmit direction is the same as that of the data bus. Figure 1 illustrates the FSR-bus structure.

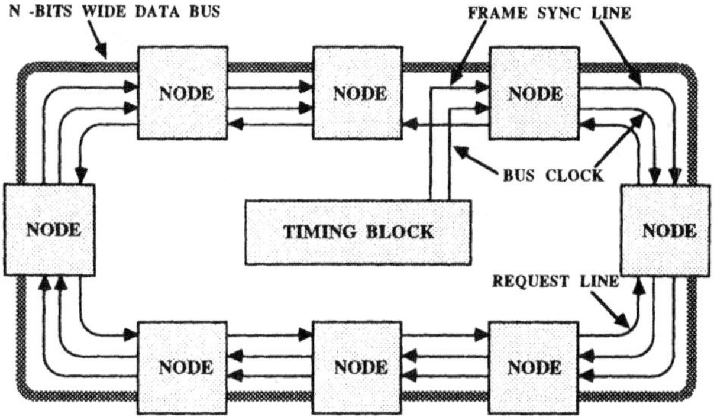

Fig. 1 Block diagram of the FSR-bus

All the mentioned signals are synchronized to the bus clock. The clock signal is fed either parallel to the data bus, as illustrated in Fig. 1, or it can be fed directly from the timing block to each of the Nodes. The parallel clock line is used if the clock rate of the system is close to the maximum attainable clock rate of the Nodes, because clock skew becomes a serious problem.

If the clock rate is reasonably low, compared to the Nodes' maximum clock rate, the clock signal can be directed to the Nodes using star or tree structured lines.

Data is transmitted in parallel form in frames. The frames consist of *L* consecutive time slots, each of which are *N* bits wide. The optimum frame length *L* equals to the number of Nodes *M* in the ring. The frames are transfered time slot by time slot controlled by the bus clock. The period, during which the time slot traverses from the output of a Node to the output of the next Node, is called a bus cycle.

The start of a frame is indicated by a frame synchronization pulse, the duration of which is one bus cycle. It circulates around the ring side by side with the first time slot of the frame and is sent from a Node to the next at the beginning of every bus cycle. The use of a separate frame synchronization line eases the implementation of the Node. A more general solution would be to carry the synchronization information along with the frame. In this case, all information in the frame, excluding the synchronization information, would have to be scrambled to avoid faulty frame alignment. This in turn would necessitate extra hardware in the Nodes to execute the necessary scrambling and descrambling functions. Including the synchronization information in the frame would also contribute to an increase of overhead. Fig. 2 illustrates the bus timing signals.

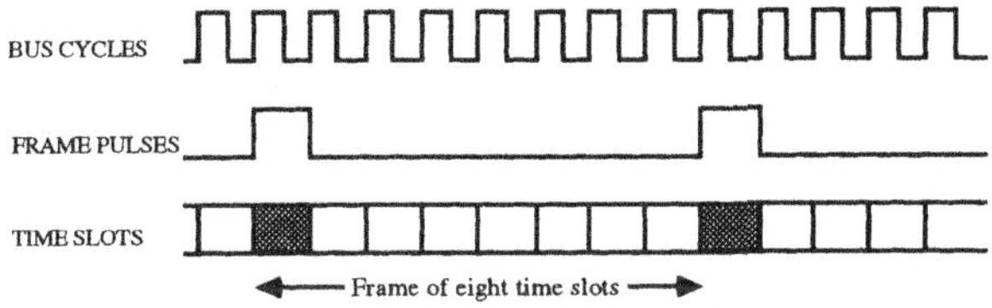

Fig. 2 Bus timing signals

Frame structure in the FSR-bus

The frames of the FSR-bus are divided into fields, as shown in Fig. 3a. The length of each field, measured in time slots, depends on the application. The *header*, the first field of the frame, includes addressing, control, and status information. If the frame length L is small and the bus width N is large, it is reasonable to have a header of only one time slot long. The *data* field includes the actual data, which is to be conveyed from the source Node to the destination Node. The length of this field is normally larger than the sum of the other two fields. The last field in the frame is the *trailer*, which includes the frame check sequence. The *trailer* is an optional field, and it is used if the frame length is long. When the frame length is small, the use of the *trailer* contributes to an excessive amount of overhead.

The *header* is divided into several subfields, which are illustrated in Fig. 3b. The first subfield is one bit wide, and it indicates whether the frame is occupied or idle. The second subfield, also one bit wide, is linked to the *destination address* field, and it indicates whether the address is a group address or an individual Node address. Group addressing can be used, when the frame is to be delivered to several Nodes. Single addressing is used when the frame is addressed only to one Node. The optimum size of the *destination address* field is equal to $\log_2(M)$ bits, where M is the number of Nodes in the ring.

The third subfield is the *ring address* field, which is in use when several rings are connected together to form a network of rings. Individual rings can be connected to form for example a star-structured, a ring-structured, or a tree-structured topology. The Nodes, connecting rings together, have to be able to identify the addresses of the other rings. The optimum size of this field is equal to $\log_2(K)$, where K is the number of rings connected together.

The fifth subfield is called the *source address* field, and it indicates the address of the transmitting Node. The optimum size of this field is equal to $\log_2(M)$. If the frame traverses from a ring to another, the address in this field is replaced by that of the connecting Node in the receiving ring. The field is utilized, when the destination Node has not been able to receive a frame, addressed to it. The Node, which receives frames that have its own source address, interprets that the frames originated by itself have not been received and removes them from the ring. This mechanism is important to avoid blocking, especially in situations, when any one of the Nodes is permanently incapable of receiving frames.

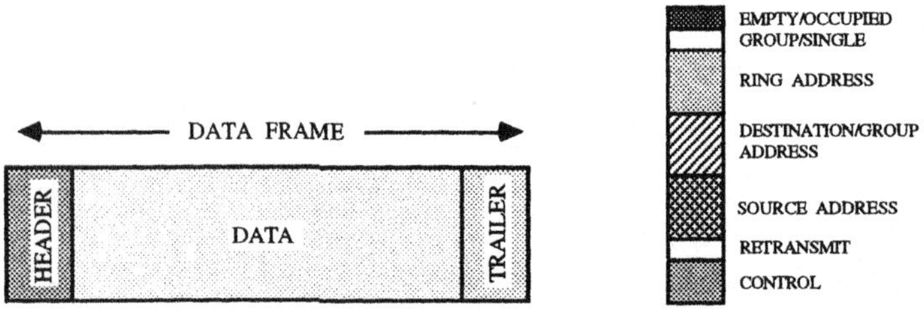

a.) Fields of the data frame *b.) Subfields of the header*

Fig. 3 Frame structure of the FSR-bus

The next one bit field is called the *retransmit* subfield, which is closely related to decoding of the *source address* field. The *retransmit* bit is used by the destination and source Nodes in a situation, when the destination Node is not able to receive a frame addressed to it. The destination Node activates the *retransmit* bit, which indicates to the source Node that this frame must not be deleted. When a Node receives a frame that is originated by itself and the *retransmit bit* is active, the Node deactivates the bit and passes the frame back to the ring or retransmits the frame after a predefined delay. The use of this subfield is optional.

In addition to the aforementioned subfields, there is a field called the *control* field. This field includes extra bits reserved for future needs, some of which may be application dependent. One such requirement is related to the transfer of long data packets. Long packets have to be divided into several mini-packets, which are conveyed in the frames. The source must indicate the beginning as well as the end of the packet to the destination. One bit, in the control field, can be reserved to indicate the first frame and another bit the last frame of the packet. Extra bits can also be used to indicate what connection a particular mini-packet belongs to.

3 OPERATION OF THE FSR-BUS

An essential feature of any bus is the way in which nodes receive data, transmit data, and require service from the bus. In the Frame Synchronized Ring, a peripheral attached to a Node transmits data by setting the data frames into the Node's transmit buffer, and the Node controls the transmission to the ring. Long data packets are divided into a number of mini-packets that are conveyed in the frames from the source Node to the destination Node. The received mini-packets are combined to form packets that are stored into the Node's receive buffer, from which the peripheral reads them. If the traffic is congested, a specialized request line is used to require service.

Data flow control is distributed among the Nodes. Each Node monitors the by-passing traffic and detects the header of every by-passing frame to make the required decisions according to the information in the header subfields. The Node checks the address and control fields to detect whether the incoming frame is empty, if it is addressed to the Node, if it is to be deleted, or if the frame is passing the Node.

The incoming frame is empty if the *empty/occupied* field in the header is active. The incoming frame is destined to the Node if the *destination/group address* in the header matches that of the Node. The Node receives the frame time slot by time slot and sends out an empty frame, a frame of its own, or passes the incoming frame depending on the status of the bus control signals and the state of the *group/single* field. If the address is a group address, the incoming frame is passed to the next node. In case the retransmit bit is in use and the Node is not able to receive a frame addressed to it, the Node activates the retransmit bit. The activation of this bit indicates to the source Node that the destination Node was not able to receive the frame. It is up to the application whether the frame, not received, is deleted or recycled by the source.

If the *destination address* in the header is not identical to that of the Node, the frame is either passed or deleted. The frame is deleted if the *source address field* equals to the address of the receiving Node. If the *retransmit bit* is used, and it is active, the frame is passed to the next Node or stored and retransmitted after a predefined delay. During these actions, the *retransmit bit* is deactivated by the source. If, on the other hand, the retransmit bit is inactive, the frame is deleted. When the frame is deleted, the Node sends out an empty frame or a frame of its own depending on the status of the bus control signals.

When the Node has data to send, it decides, according to the address and control information along with the request line information, whether it can start to send or not. The transmission can begin if the outgoing bus is idle and if it is the Node's turn to transmit. The outgoing bus is expected to be idle if there is no incoming data, which is passing the Node. On the other hand, the outgoing bus is considered to be occupied if the precently incoming frame is passing the Node, i.e. the frame is not addressed to the Node and it is not deleted or the destination address is a group address. The Node has a turn to transmit if there are no on-going requests from the downstream data bus or, in the case of on-going requests, the Node was ready to transmit at least one bus cycle earlier than it discovered downstream requests. The transmission of a frame begins at the frame synchronization pulse.

If the bus is congested and the Node cannot get access to the bus, the request line is used to require service. The request is indicated by setting the request line to active state, i.e. sending continuous binary ones onto the request line in the direction of the upstream data bus. The Node is allowed to require service, if the Node has data to send and it is the Node's turn to transmit.

If the Node does not have the turn, it must pass the downstream requests to upstream and wait until the requests from downstream have been removed. Immediately after detecting that the on-going downstream requests have been removed, the Node is allowed to enter its own request. When the Node has been prevented from entering its request, it is not allowed to utilize empty incoming data frames. It must pass them to downstream as long as there are incoming downstream requests.

If the Node, after entering its request, detects requests from downstream, it prevents them passing to upstream as long as it keeps sending its own request. The Node removes its request, when it discovers that it can have access to the bus. The removal takes place at the same bus cycle as the Node starts to transmit its data frame. When removing its request, the Node does not immediately pass the possible pending requests from downstream to upstream. The downstream requests remain pending until the Node receives the first frame that passes by. From the access control's point of view, it is of vital importance that after the removal of the Node's own request, the next upstream Node detects request discontinuity of at least one bus cycle.

After gaining access to the bus, the Node is allowed to send at least one frame. In case it has more frames to send, it can send them one after another as long as the outgoing bus is idle and there are no requests coming from downstream. When the Node detects requests from downstream, it must stop sending after the on-going frame or after a fixed number of frames. After finishing the transmission, the Node passes the downstream requests to upstream if the currently incoming frame is passing the Node. Otherwise (i.e. the incoming frame is addressed to the Node itself, the frame is to be deleted, or the frame is empty) it suspends the requests and sends out an empty frame.

In case the Node, when not requiring service, detects requests from downstream, it passes them upstream if it simultaneously receives a frame which is passing the Node. After the Node has passed requests from downstream to upstream, it cannot prevent them passing until there is a request discontinuity of at least one bus cycle. After the discontinuity, the incoming requests are considered to be new ones.

4 PERFORMANCE ANALYSIS

Since the FSR-bus is a slotted ring, its throughput and packet transfer delays can be estimated using mathematical models, developed for the slotted rings. Several basic models have been developed of which processor sharing and priority queuing models seem to be the best ones according to the published articles.

The transfer delay in the FSR-bus can be estimated utilizing processor sharing models, originally developed by Kleinrock [Kle76]. The basic formula is given in Eq. (1), where ρ represents the load in the ring, x is the average service time of a packet, and τ is the average delay on the transmission line. Kleinrock's calculations assume that the loads into the ring are Poisson processes.

$$T = \frac{x}{[1 - \rho]} + \tau \qquad (1)$$

Since the equation is developed for a single bus, and the FSR-bus, which consists of a number of bus segments (i.e. single buses), utilizes destination release policy, Eq. (1) must be further developed. If the average path length in the ring is r, then the ring can be considered to correspond to M/r single buses, where M is the number of Nodes in the ring. Taking into account the above, a coefficient α=r/M has to be added to Eq. (1) giving the following expression.

$$T = \frac{x}{[1 - \alpha\rho]} + \tau \qquad (2)$$

In the FSR-bus, long data packets are segmented into several mini-packets that are then conveyed in the frames. Besides data, the frames also include control information, which can be considered as overhead. If the packets include D time slots of data and H time slots of overhead, the overhead factor is defined as h = (D+H)/D. To take into account the overhead, Eq. (2) must be multiplied by h as is done in Eq. (3). The formula does not account for swap time, which in some publications is called switchover time.

$$T = h\frac{x}{[1 - \alpha\rho]} + \tau \tag{3}$$

Zafirovi-Vukovic and Niemegeers have developed in [Zaf87] a formula for the average transfer delay in the Orwell ring. This formula, which is based on a priority queuing model with two priority classes, may be used to allow a comparison between the transfer delay in the FSR-bus and the well-known Orwell ring. The FSR-bus utilizes the destination release policy, just like the Orwell ring, for which reason a model developed for the Orwell ring should also give reasonably good estimates for the FSR-bus. The model is given in Eq. (4), where ρ_t represents the load of the peripheral behind the Node, ρ_r represents the load in the ring, x is the average service time of a packet, and τ is the average delay on the transmission line.

$$T = \frac{[\rho_r + \rho_t]\,x}{2[(1 - \rho_t - \rho_r)(1 - \rho_r)]} + \frac{x}{[1 - \rho_r]} + \tau \tag{4}$$

Equation (4) does obviously not give exact transfer delay estimates for the FSR-ring, because its bus access method differs from that of the Orwell ring. Eq. (3), on the other hand, can be expected to give better estimates, because it takes into consideration path length and overhead.

Curves that represent Equations (3) and (4) are shown in Figures 4 and 5. The system size is set to 10 Nodes, and the average packet length is 100 frames. Frame length is 10 time slots, and the header length is one time slot. Fig. 4 depicts the transfer delays, when the Node transmits packets to the other Nodes with equal probability 1/M (i.e. uniform distribution). Fig. 5 shows the corresponding curves as the Node sends packets to the other Nodes with the probability 2^{-d}, where d represents the distance (i.e. the number of bus segments) between the transmitting and receiving Nodes. In this case, the traffic is said to be geometrically distributed.

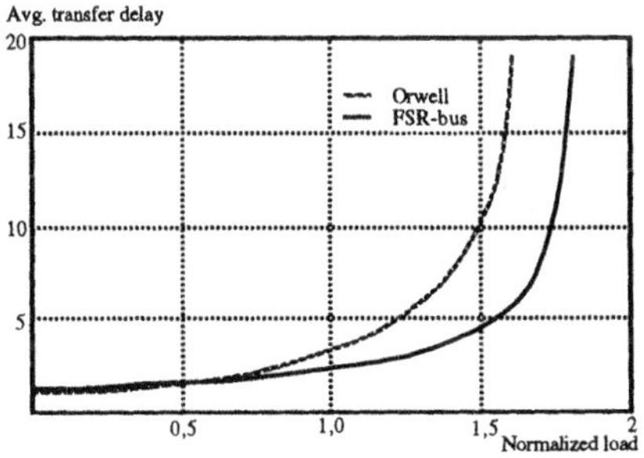

Fig. 4 Average transfer delay as a function of the normalized load with uniformly distributed traffic

According to the figures, the FSR-bus can be expected to function more efficiently than the Orwell ring, especially under geometrically distributed traffic. Since the swap time is not included in Eq. (3), the real transfer delays can be assumed to be longer than the estimated ones. More exact mathematical transfer delay expressions for the FSR-bus are for further study.

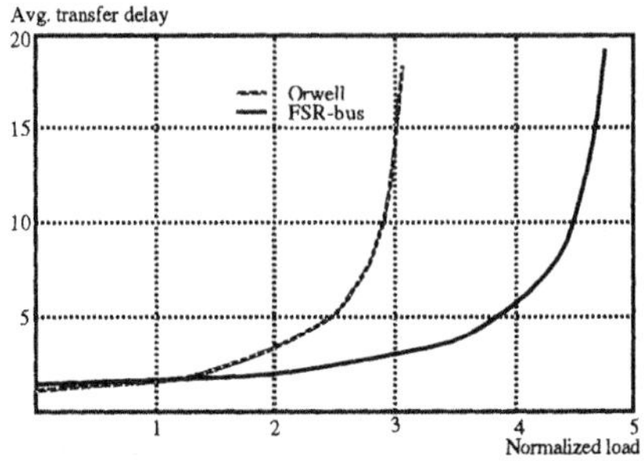

Fig. 5 Average transfer delay as a function of the normalized load with geometrically distributed traffic

5 PERFORMANCE SIMULATIONS

The logical functions of the FSR-bus and its bus access method were tested using a C++ based simulation program. The program, which simulated throughput and average transfer delay, was developed to certify fairness and reliability of the access method in variable loading situations. The throughput was simulated as a function of the system size (i.e. number of Nodes in the ring bus) and the average transfer delay as a function of the load. Similar programs were also developed for the multidrop bus and the modified Orwell ring. Short descriptions of the bus access schemes for these two buses are given below.

The simulations assumed an ideal physical layer, i.e. error free transmission and no overflows in receive buffers. The maximum transmit rate of the peripherals were assumed to be equal to the transfer rate of the ring. The simulated ring sizes were 5, 10, 20 and 40 Nodes. The simulations were carried out using exponentially distributed packet lengths. The average packet length was chosen to be 1,10, and 100 frames. Each simulation case, with constant parameters, was run for 10 million bus cycles.

Traditional multidrop bus

In the traditional multidrop bus, a peripheral holding the bus is expected to transmit one slot onto the bus during one bus cycle, and the destination peripheral is expected to receive the slot during the same cycle. The peripherals can transmit one frame at a time, after which the transmission turn is granted to the next peripheral in line. No arbitration time is assumed. The traffic generator and the other features of the bus model are the same as those for the FSR-bus.

Modified Orwell ring

In the modified Orwell ring, each node has a counter as in the original method [Fal85]. The counter counts the frames sent from the node. The main differences, compared to the original version, are that the counters are reset by a separate reset signal, parallel to the data bus, and an additional handshaking signal is introduced. The handshaking signal circulates opposite to the data flow and guarantees that all possible simultaneous reset sources are synchronized. The node asserts its reset signal when its counter reaches the maximum value. The reset signal is removed when the resetting node observes a reset from its upstream neighbour. The maximum value for the counter is set equal to the number of nodes in the system.

Throughput

Figures 6 and 7 illustrates the average throughput of a node as a function of the number of nodes in the system, in which the average packet length is 10 frames. The throughput is normalized to the transmission rate of one bus segment. Therefore, the effective data transmit rate of the node is divided by its maximum transmit rate, which, in this case, equals to the ring transmit rate. The interarrival times between packets is set to 0, which means that the nodes always have something to send. Since the load in a ring bus, which uses the destination release technique, depends on the average path length, the throughput is calculated separately for uniformly and geometrically distributed traffic. For comparison, the throughputs of the multidrop bus and the modified Orwell ring are also depicted in the figures. Solid lines illustrate the FSR-bus, dotted lines the Orwell ring, and dashed lines the multidrop bus.

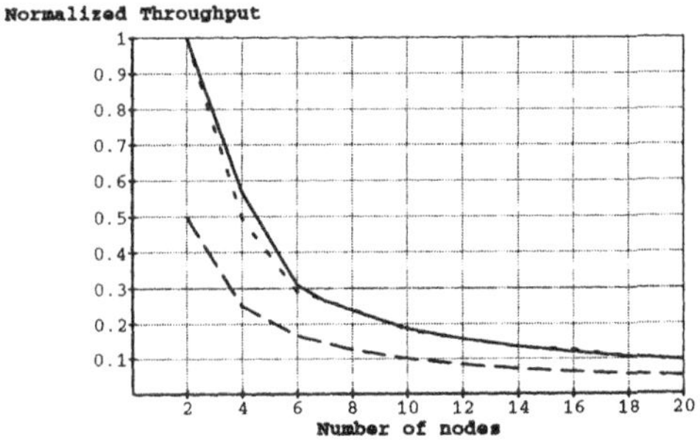

Fig. 6 Average throughput of a node in case of uniformly distributed traffic

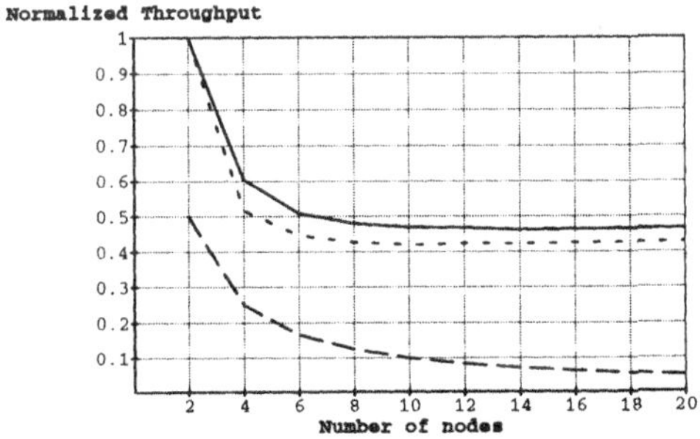

Fig. 7 Average throughput of a node in case of geometrically distributed traffic

The throughput curves represent the amount of bus capacity that one node can utilize. As shown in Fig. 6, which illustrates the case of uniformly distributed traffic, the throughput of a peripheral, attached to the node, is at most twice that of the node in the multidrop bus, when the number of nodes is small. As the number of nodes increases, the throughput curves approach each other. In the case of geometrically distributed traffic, the throughput in the ring bus is at least twice that in the multidrop bus (Fig. 7). As the number of nodes increases, the ratio of the ring's throughput to that of the multidrop bus also increases. The FSR-bus is roughly 10% more efficient than the modified Orwell ring. Increase of the packet size favours the FSR-bus, and decrease of the packet size reduces the differences.

Transfer delay

Figures 8 and 9 illustrate the average transfer delay as a function of the normalized load. The load is normalized to the maximum capacity of a single bus segment. Due to this, the maximum normalized capacity of the ring is equal to the number of nodes, but for the multidrop bus it is equal to 1 irrespective of the system size. The transfer delay is the sum of the queueing time in the transmit buffer, service time of the packet, and the packet transfer time on the line. The interarrival times of the packets, coming into the system, are Poisson distributed. The packet sizes for the depicted cases are exponentially distributed with an average packet length of 10 frames. The size of the system is 10 nodes.

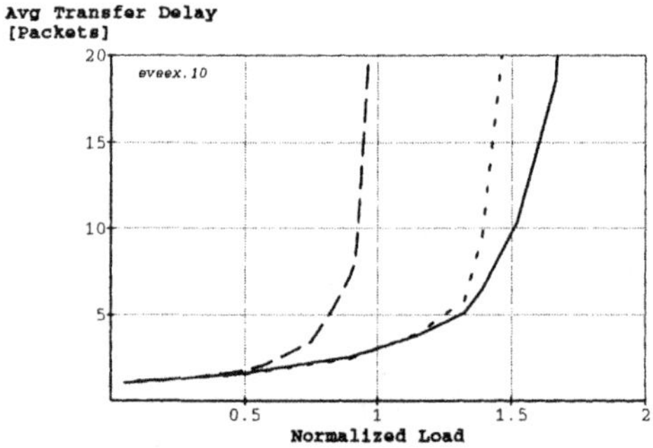

Fig. 8 Average transfer delay as a function of the normalized load with uniformly distributed traffic.

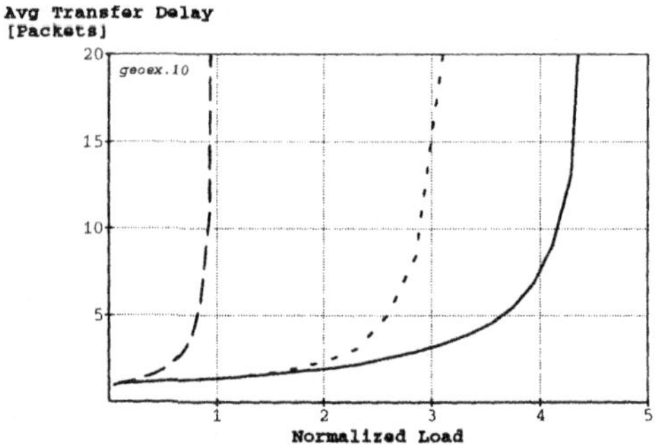

Fig. 9 Average transfer delay as a function of the normalized load with geometrically distributed traffic.

As can be seen from the figures, the multidrop bus generates a delay of 5 packets already at the normalized load of 0.7. The result is independent of the traffic distribution. For the FSR-bus, the same delay is reached at the load of 1.33, when the traffic is uniformly distributed (Fig. 8), and at the load of 3.56, when the traffic is geometrically distributed (Fig. 9). In case of the Orwell ring, a 5 packet delay is reached at the load of 1.23, when the traffic is uniformly distributed, and at the load of 2.59, when the traffic is geometrically distributed. In both traffic cases, the FSR-bus provides lower average transfer delay than either of the two other buses. When the packet size or the number of nodes in the system was increased, the simulation results were more in favour of the FSR-bus. Decreasing the packet size or the number of the nodes had an opposite effect.

Variance of the transfer delay gives some idea of how fairly bus resources are allocated between nodes in the ring. As an example, Figures 10 and 11 illustrate variances of the transfer delays (95 percent confidence level) in the FSR-bus and in the Orwell ring respectively, when the load is uniformly distributed. The figures show that the variance in case of the FSR-bus is larger than in case of the Orwell ring, which means that the bus access method in the FSR-bus is not as fair as in the modified Orwell ring. In the case of geometrically distributed load, the situation turned out to be the opposite, and the variance of the transfer delay in the Orwell ring is slightly larger than in the FSR-bus. This means that the fairness of the FSR-bus access method is more sensitive to load distribution than that of the Orwell ring.

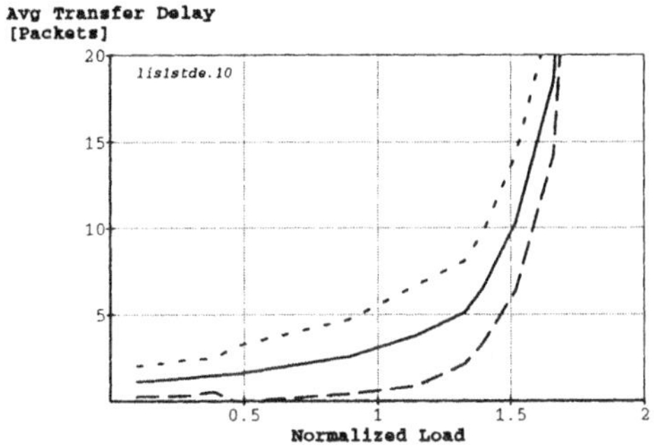

Fig. 10 Transfer delay variances in the FSR-bus as a function of the normalized load with uniformly distributed traffic.

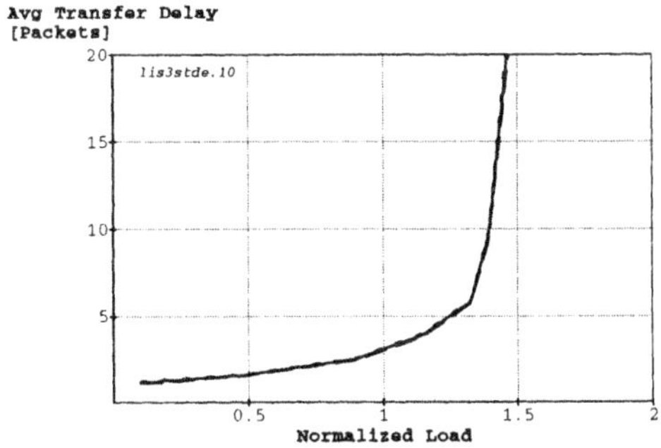

Fig. 11 Transfer delay variances in the Modified Orwell ring as a function of the normalized load with uniformly distributed traffic.

Maximum transfer delays were recorded only for the FSR-bus. The results showed that the maximum delays were about ten times longer than the average delays. The maximum delays were finite in all of the simulated cases, which indicates that the access method of the FSR-bus is fair at least in the loose sense. No one of the transmitting Nodes was prevented from transmission for an infinitely long time. However, the simulations did not cover all possible traffic conditions, for which reason the results still do not prove that the access method is fair. The proof of the fairness is for further study.

6 CONCLUSIONS

A new parallel bus concept, named Frame Synchronized Ring (FSR-bus), for high throughput solutions has been introduced. The bus is based on a ring topology, which is an economical way to construct an efficient interconnection network for multiprocessor applications. The bus structure, its timing arrangements, frame structure, medium access control algorithm, and operation of the FSR-bus have been explained.

The performance of the bus was estimated utilizing queuing models, developed for slotted rings. The models showed that the FSR-bus is more efficient than the well-known Orwell ring. In order to verify the theoretical results, simulated throughput and transfer delays were compared to those of the multidrop bus and the Orwell ring. Simulation results for uniformly and geometrically distributed traffic are introduced for each of the three buses. In the uniformly distributed case, the FSR-bus performs distinctly better than the multidrop bus. In the geometrically distributed case, the throughput of the FSR-bus is always more than twice that of the multidrop bus. The more Nodes, i.e. bus segments, in the ring the larger is the difference.

Compared to the Orwell ring, the FSR-bus is slightly better when the traffic is uniformly distributed. In the case of geometrically distributed traffic, the FSR-bus performs clearly better than the Orwell ring. As the packet length or the number of the nodes is increased, the difference in performance (i.e. throughput and transfer delay) is more in favour of the FSR-bus. Variance of the transfer delay, when traffic is uniformly distributed, is larger for the FSR-bus, which means that the FSR-bus does not have as fair an access method as the modified Orwell ring. When the traffic is geometrically distributed, the result is just the opposite. The transfer delay results on the whole indicate that the FSR-bus access method is sensitive to traffic distribution. However, the recorded maximum transfer delays were finite, which supports the view that this bus access method is fair. The proof of the fairness is for further study.

Transmit credit, i.e. the number of frames that a Node is allowed to transmit during a transmission turn, was fixed to one in the simulations. In the future, it will be necessary to also simulate the impact of the transmit credit on the throughput and transfer delay. The developed mathematical formula for the transfer delay, based on the processor sharing model, overestimated the delays in the FSR-ring under light load. Under heavy load, the model underestimated the delays giving too positive results. More exact theoretical models for the FSR-bus will be developed in the future.

REFERENCES

[Fal85] R. M. Falconer, J. L. Adams: "Orwell: A Protocol for an integrated services local network," Brit. Telecom. Tech. J., vol 3, no. 4, pp. 27-35, Oct. 1985.

[Kle76] L.Kleinrock: Queuing Systems, vol. II, Computer Applications, John Wiley and Sons, New York, 1976.

[Zaf87] M.Zafirovic-Vukotic, I.G.Niemegeers: Performance Modelling of the Orwell Basic Access Mechanism, Computer Communications Rev 17, Special Issue, pp. 35-48, 1987.

Integrated Broadband Communication Networks
and Services (C-18) / V.B. Iversen (Editor)
Elsevier Science B.V (North-Holland)

The DYCE Concept - Architecture & Implementation Strategies

Bernd Heinrichs[1], Wilko Reinhardt[1], Rumen Stainov[2]

[1]Lehrstuhl für Informatik IV, RWTH Aachen; Ahornstr. 55; 52074 Aachen, Germany;
Tel.: +49 241/80-21410; FAX: -21429; e-mail: heibelwilko@informatik.rwth-aachen.de

[2]Fachbereich "Angewandte Informatik und Mathematik", FH Fulda; Marquardstr. 35
36039 Fulda, Germany; Tel.: +49 661 9640 300; FAX: +49 661 9640 349

ABSTRACT

This paper presents an approach to design a powerful and flexible controller architecture in order to support new applications across various heterogeneous networks. The approach allows a run-time configuration of the protocol modules. In contrast to existing systems, the developer has the choice to apply different implementation strategies to a module, changing in this way its encapsulation and execution behaviour. Thus, one and the same module can meet different configuration requirements depending on its implementation form as: a typed dynamic library procedure, a typed unmovable thread, a typed movable thread, or a composed protocol graph. The proposed environment is a light-weight framework, called DYCE-2, which represents an implementation of the Dynamic Configuration Entity (DYCE) [Hein93]. DYCE is the central part of our flexible protocol architecture. To build such a flexible and also high-performance system we have to find out and uncover protocol functionalities which can operate concurrently. To achieve results on protocol inherent functional parallelism we have done extensive simulation studies using a petri net based specification technique.

Keyword Codes: C.2.2, D.4.4
Keywords: Network Protocols, Communication Management

1. INTRODUCTION

During the last years a new generation of applications has been developed, providing distributed multimedia communications. The rise of multimedia applications dramatically changes the demands on the underlying communication systems. Different and even contrary requirements have to be fulfilled. On the one hand isochronous data streams from live video and audio signals request high bandwidth (approximately up to 100 Mbit/s), in-time delivery of the data and low delay variation (jitter). Isochronous media can accept some bit errors or packet loss depending on the used compression scheme. On the other hand asynchronous streams resulting from the transfer of still images or textual documents need a reliable transmission but can accept small delays. Besides the evolution in the application area, high

---

Work described in this paper has been partly funded by the RACE II project EUROBRIDGE. The EUROBRIDGE Consortium consists of the following members: Ericsson EUROLAB, D; RWTH Aachen, D; Teles, D; Detecon, D; Broadcom, IRL; CTS, IRL; CET, P; Intracom, GR; VTT, FIN; Siemens-Nixdorf, IRL speed networks with increasing transmission capacities and decreasing error rates have been come up in parallel (e.g. FDDI [FDDI89], FDDI II [FDDI90] with 100 MBit/s, DQDB with 45 and 150 MBit/s [DQDB90], HIPPI with 800 MBit/s [HIPP92] or CRMA II [CRMA91] and Metaring with Gbit/s rates [CiOf91]).

The trend of developing multimedia applications and high speed networks leads to a performance gap within the common protocol structure. Three major problems of today's protocol architecture can be identified and will be illustrated in the following paragraphs.

1.1 Insufficient Protocol Functionality to Support Multimedia Data Exchange

One of the weakest items of today's communication systems is the missing capability for the application to define Quality of Service (QOS) parameters. The application should be able to request functional and performance QOS from the underlying protocol architecture. Functional QOS parameters are concerned with features that provide services to support useful mechanisms for the data flow throughout the various layers. These mechanisms help to handle the data streams in an efficient manner. Performance QOS define values which are requested to be fulfilled by the system.

Currently available protocols from the ISO and Internet communities are not designed to support the requested functionality and performance of multimedia applications. Synchronisation mechanisms of related data streams, efficient security mechanisms, reservation of resources in advance, multipoint connections and support of QOS are missing. Synchronisation is needed to keep the interrelation between multiple data streams and to ensure that documents will be distributed to all participants of a multipeer session in time. Security mechanisms, like data encryption or passwords are desirable for the transmission of secure information over local and public networks. This item becomes more and more relevant on account of the increasing conference facilities and database queries with sensible data (e.g. medical information of a patient).

Today's communication protocols support only point to point connections. Within multimedia applications like video- or teleconferencing, multipeer connections are indispensable to achieve an efficient information exchange. All kind of media can be transmitted from one entity to many (distributed teaching), many to many (video conference) or many to one (distributed database access).

Multimedia conferencing tools need a high bandwidth over a limited period of time. At peak periods the necessary resources might not be available and the conference has to start or continue with reduced quality. To avoid this, the participants should be able to reserve resources a particular time period in advance. At the scheduled time all requested QOS parameters are guaranteed by the network, independent of the actual traffic situation.

1.2 Inflexible Protocol Architectures

Multimedia applications require versatile QOS support from the communication system. Today's protocols are designed for unreliable and low speed networks. The architecture of the protocol stacks follows a very strict layering. There is no flexibility to react on different reliability and speed of transmission media. E.g. if the transmission medium is very reliable (fibre optic media provide an error rate lower than 10^{-12}) error correction mechanisms can be switched off or at least the effort could be reduced. It is well understood that no single protocol stack is able to support the adaptation from the application to the network efficiently, because a single stack would be very inflexible in handling data streams with opposite demands and much protocol overhead is needed to support the different functionalities. Therefore a flexible configuration and linking of necessary protocol blocks [PoSc91], [Haas91], [Hein92], [Zitt91] according to the specific actual demands is suggested. The common goal is to design application-tailored, configurable protocols, established by a run time binding. Our concept to realize such a protocol architecture is illustrated in chapter 2 of this paper.

1.3 Inefficient Implementation of Protocols

Protocol implementation strategies have a major impact on the performance. An inefficient implementation can destroy the benefits of a correct and efficient protocol specification. During the implementation phase the interoperability of protocol execution with the underlying operating system as well as the efficient use of OS facilities play an important role. Studies found out that in nearly all single protocols many instructions can be processed in parallel [Hein92]. With the support of suitable hardware the processing time of packets inside a protocol element can be decreased.

The remainder of this paper suggests an approach to solve the listed problems. We introduce a flexible architecture supporting the communication requirements of various multimedia applications by dynamic protocol configuration in accordance to the requested QOS parameters. The core part of our architecture is called **DYCE** (**D**ynamic **C**onfiguration **E**ntity) which will be introduced in chapter 2. Appropriate implementation strategies are also depicted. Chapter 3 covers first results on the amount of functional parallelism achievable when doing a careful analysis of single protocol functionalities. As an example we evaluate the connection establishment phase of an XTP session.

2. THE DYCE CONCEPT

Many research groups are working in the area of mulimedia communication support. DYCE is mainly influenced by the ADAPTIVE architecture [SBS92], the x-kernel/AVOCA[Stil92] project and the Function-based Communication SubSystem F-CSS [HuBe91]. Experience indicates that it is difficult to specify one protocol optimal for all application/transport/network combinations. Rather than developing a single complex all-emcompassing protocol, it is more feasible to construct an architecture that permits fine grain selection and configuration of various protocol mechanisms.

2.1 QOS Support

As mentioned in the introduction an efficient support of multimedia communications requires a flexible design of the communication system. The base of a flexible design is the exact specification of the demands from the applications. The requirements are described in terms of QOS parameters. In accordance with the OSI 95 [LeB192] approach we suggest to introduce threshold and compulsory QOS beside the OSI concept of best effort QOS. Threshold QOS parameters describe performance values which will be accepted by the application in the worst case. During the data transfer phase the valid QOS values can vary. If these QOS values go beyond or below the threshold, actions are requested from the application. Compulsory or guaranteed QOS have a more legal flavour than Ferrari points out in [Ferr90]. All parts of the communication systems negotiate a fixed value or a range of values and it is obliged to guarantee the value during the connection.

2.2 Architecture Model

For the implementation of specific protocol parts we use a strategy employing a high modularization of the protocol software. That means the protocol function will be subdivided into a number of smaller functionality protocols, each capable of achieving a single functionality (a specific QOS). This implementation will allow to react dynamically to changing environment conditions and makes it possible to use protocol inherent parallelism (cf. chapter 3). To make the protocol stacks configurable and re-configurable, run-time binding procedures will be provided to build suitable and dedicated communication stacks. In the following figure 2.1 the proposed concept is depicted.

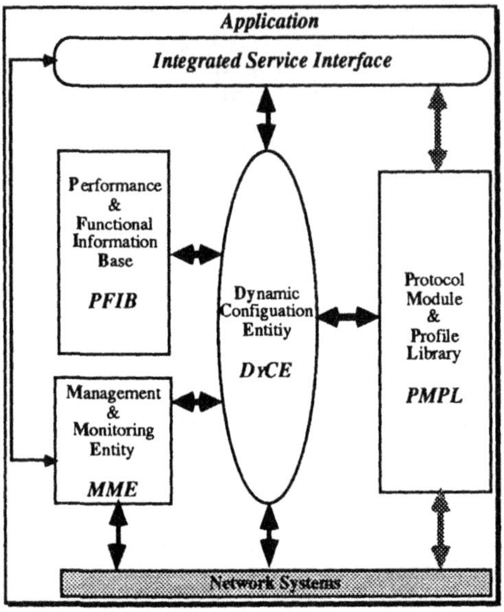

Figure 2.1: DYCE Architecture

In the centre of interest is the ***Dynamic Configuration Entity*** (DYCE). DYCE is responsible for selecting suitable protocol units according to the QOS demands from the applications. It has to map the required QOS on a combination of protocol elements that fit the required characteristics. The information needed for the mapping process is stored in the ***Performance & Functional Information Base*** (PFIB). PFIB contains the mapping strategy, information on local available protocol elements, their functional and performance properties as well as information on available protocol elements on most frequented receivers. It provides the information on characteristics of the several units. This includes functional properties like e.g. error correction schemes, synchronization facilities, security mechanisms or conference management tools and performance properties e.g. throughput, jitter and delay behaviour. Data describing the capabilities of remote sites are useful to speed up the connection establishment phase, since the failure probability during the connection set-up phase will be reduced. If information about the remote site is not available, we propose to use the X.500 directory service for extracting the necessary data. Using a suitable cache strategy reduces the overhead of the search for desired data. If a suitable combination of protocol units is detected that fulfils the requirements from the application, the selected units are launched together dynamically. Thus the connection to the remote site can be established via the actual protocol stack.

DYCE should include information on the actual traffic situation during the selection procedure. This information can e.g. influence the routing decisions or is useful to check in advance if the desired bandwidth is available. To achive this a ***monitor component*** has to be integrated into DYCE. Another task of the monitor entity is the control of the offered QOS values. This is a necessary item if the user requires threshold QOS. The user specifies an extreme value that describes what is acceptable in the worst case. If the monitor entity measures that the threshold value is not supported anymore it informs the application and requests a decision if lower QOS are acceptable or if the connection should be aborted. In case lower QOS are accepted by the application DYCE initiates a re-negotiation of the stack configuration.

Most existing applications are not able to specify their exact QOS requirements. Since the precise information on the functional and performance requirements are indispensable for a flexible and adaptive communication system, we have interviewed various application providers, developers and projects to provide us with quantitative and qualitative service requirements of their applications. From the results [ReTi93] we derived ***QOS profiles*** that describe various classes of related QOS requirements.

Based on our experience and on ongoing research we are able to allocate best suited protocol stacks to a specific application at connection set-up time in accordance with the requested QOS profiles. In the first realisation step pre-configured protocol stacks will save the effort of dynamic stack configuration. But in further implementation we are thinking about the possibility of free defined sets of QOS parameters. We regard this as an additional feature besides the pre-defined profiles.

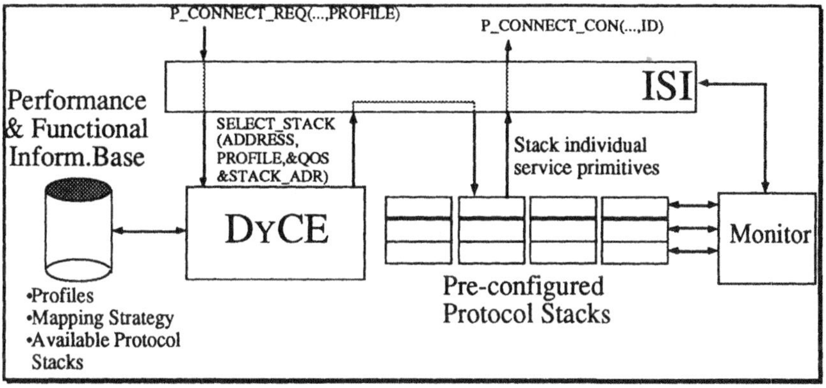

Figure 2.2: Co-operation of the DyCE Architecture Components

Fig. 2.2 depicts cooperation of the various units of the DYCE concept. The application layer protocols access the underlying protocols via the integrated service interface (ISI). This interface is needed, since it hides the diversity of the service interfaces so that the application has a general view of the underlying communication system. Whereas the ISI is above the presentation layer we use service calls similar to the presentation layer function calls (`P_CONNECT(<traditional Parameters>,PROFILE)`). During connection set-up the application has to define within the parameter PROFILE the desired QOS class. After a connection set-up request DYCE will be activated. It is called by the function `SELECT_STACK (ADDRESS,PROFILE,&STACK_ADR,&QOS)`. The return values of this function are a set of protocol specific QOS and an address structure to access the selected protocol stack. In case that the desired connection is available, ISI initiates the connection set-up via the selected protocol stack. Otherwise the application will be informed that the connection set-up is rejected. Now the connection can be set up via the selected protocol stack (cf. fig. 2.2).

2.3 Implementation strategies for DYCE

The different implementation strategies provide different degree of modularity, encapsulation and effectiveness. An object-oriented implementation for example, offers high encapsulation of both, the data and the operation on them. Furthermore, the object classes and the inheritance build a good base for flexibility and varying modularity. The cost paid, however, is the decreased effectiveness and performance. A hardware implementation is a usual way to achieve a better performance. In contrast to the object-oriented approach, the so called threads offer a very efficient implementation.

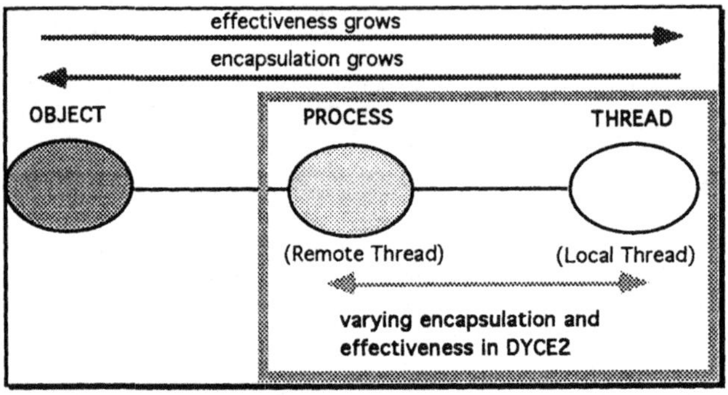

Figure 2.3: Possible Implementation Strategies

Threads exist inside a light-weight process and they share the common data and environment. The effectiveness of a thread is comparable with the effectiveness of a monolithic implementation [DPH92]. In the same time both, the shareable data and the operations on them (the threads), are not protected, so that a failure or even a bad coordination can lead to unpredictable process states. Figure 2.3 shows the trade-off between the different implementation strategies.

In order to deal with those problems we apply the principle *devide et impera* in DYCE by meeting the competing requirements at different design layers:

(L1) *Modularity and effectiveness* of the run-time system by dynamic varying the granularity and encapsulation of the manageable protocol modules in DYCE2 (Figure 2.3)

(L2) *Effectiveness and generality* by a formal definition system allowing to uncover the protocol inherent parallelism in order to specify an optimum for the application functionality and granularity of the parallel protocol modules.

The *first principle* behind (L1) for dynamic varying the granularity and encapsulation of the MPMs was introduced by the DYCE2-implementation [Stai92] and is characterised by following features (Figure 2.4):

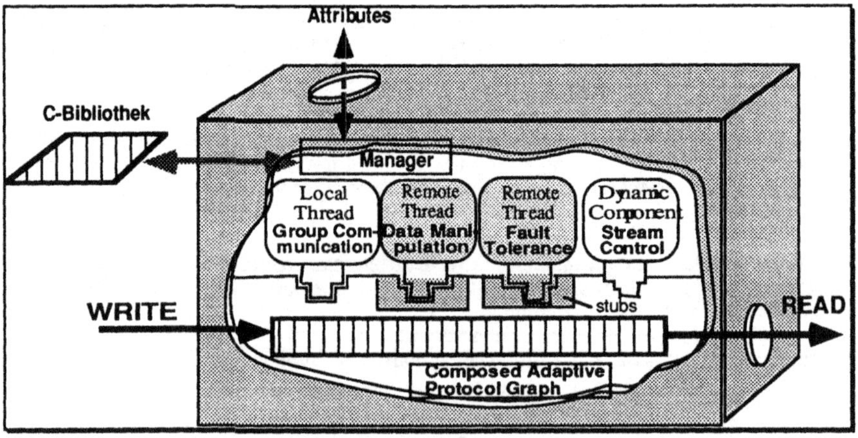

Figure 2.4: The DYCE2 Architecture [Stai92]

(F1) *Light-weight implementation.* DYCE2 represents a light-weight facility for dynamic specification and binding of threads. Our goal was to provide the same flexibility as a fine-grained modular system without compromising the performance and the effectiveness of the monolithic solutions. To do this, DYCE2 ensures incremental growth of the operating system modules at two levels: (1) by threads providing a low cost shared memory communication and (2) by processes providing a high degree of encapsulation and parallelism. Additionally, DYCE2 allows one and the same module to be implemented as a thread or as a process, so that a run-time varying of the implementation strategy and therefore the granularity and the encapsulation of the manageable protocol modules is possible (Figure 2.3). For example, different modules of an adaptive transport system source, written in a conventional language like C, can be implemented in DYCE2 by four different methods: a Reusable Software Component Library Approach, a Thread Library Approach, a Remote Threads Approach and the Implementation of a Complex Adaptive Protocol Graph. Properly speaking the DYCE designer has a choice to apply different implementation strategies to a MPM-module, changing in this way his encapsulation and execution behaviour - in form of: a typed library procedure, a typed unmovable thread, a typed movable thread or a complex protocol graph. Table 2.1 gives an idea of the characteristics of the different implementation strategies in DYCE2.

	Dynamic Component Library	Local Thread Library	Remote Thread Library	Complex Adaptive Protocol Graph
Built-in Communication Overhead	**very low**	**low**	**high**	**depends** on implementation technique
Configuration Granularity	**restricted** limited number of typed modules	**restricted** limited number of typed modules	**restricted** limited number of typed modules	**flexible** hand-made module dependencies
Negotiation Basis	**restricted** choice of existing library modules	**restricted** choice of existing library modules	**flexible** allows remote implementation	**predefined** included in protocol graph
Parallel Execution	**not possible**	**restricted** on shared memory multiprocessor	**Yes**	**predefined** included in protocol graph
Fault-Tolerance support	**not possible**	**restricted**	**Yes**	**predefined** in protocol graph
Heterogeneity	**not possible**	**not possible**	**possible** allows remote implementation	**not possible**

Table 2.1: Comparison of the DYCE2 Implementation Strategies

(F2) *Convenient interfaces.* Suitable abstractions should be provided for the *user* to ensure dynamic negotiation and initialisation of the desired functionality, as well as for the *system designer* to configure and extend the DYCE2. The *user level* access and the invocation of the MPM, belonging to a manageable library group, are provided by attributes implemented in form of DYCE2 capabilities. In the current implementation we adopted the capability principles of AMOEBA. The *designer interface* is provided at the language level, by introducing a new "remote" type data structure shared by the light-weight DYCE2 kernel and the MPMs. Furthermore, the "remote" type of the interface variables together with the "remote" type of the procedures using them defines the implementation strategy of the corresponding MPMs without special language-level constructs.

(F3) *Remote configuration and execution.* This feature allows a MPM to be exchanged between the communicating parties, as well as to be executed in parallel on a different machine. The Remote Thread covers those characteristics. It interacts with the kernel by a specialised mechanism supporting shared remote variables and providing a high-level encapsulation of the remote objects by RPC-similar stubs, so that a part of the user program can be moved and executed on a remote machine.

(F4) *Support for multiversion programming.* This is a technique ensuring high availability and performance. The idea is to start several threads or processes executing in parallel different versions of a module and to accept the first valid result. The power of this approach is that it masks (1) the hardware and software failures, and (2) the non-deterministic nature of the transport system.

The *second principle* behind (L2) provides a high level support for the specification of the parallel protocol modules and of the dependencies between them. By contrast to [SBS92] providing a predefined set of module types, we represent a petri-net-based approach, language and machine allowing the DYCE-implementation to be optimal tailored to the specific protocol. This supports the implementation of a DYCE run-time system (for example DYCE2 in fig. 2.3) for a given protocol as

- a *specialised complex protocol graph* deciding step-by-step about the manageable modules and their mapping on the parallel machine
- a *protocol stack* allowing dynamic configuration of the protocol modules and specifying their possible implementation strategy.

Next section deals with the main aspects of the second principle. The proposed formal specification approach and language is supported by a generic MDMA-controller architecture model for parallel implementation of the protocol modules. This model, and therefore also a DYCE-implementation can be mapped on an arbitrary computer architecture.

3. ACHIEVING HIGH PERFORMANCE USING PROTOCOL INHERENT PARALLELISM

As pointed out in section 1 we need an efficient implementation strategy to realize sophisticated communication mechanisms like XTP multicasting. Implementations typically aim at minimizing protocol processing overhead (e.g. the number of packet exchanges, layer interactions, buffer operations, heavy usage of timers, interrupts and memory read/writes). Before introducing our approach it is necessary to define the term *efficiency of an implementation.* Efficiency depends on the instruction path length, the number of operations at a processing node, delays due to copying of data between adjacent protocol layers and delays resulting from poor synchronization between peer entities. Efficiency of parallel protocol processing can be defined as the quotient of the worst-case processing time of the fastest sequential implementation and the product of (parallel running time x number of processors).

Our concept is based on a flexible protocol structure allowing selection of and access to preconfigured protocol profiles or protocol modules. In addition to the integration of existing protocol implementations we use a strategy employing a high modularization of the protocol software [HeRu92]. Fig. 3.1 illustrates the design process.

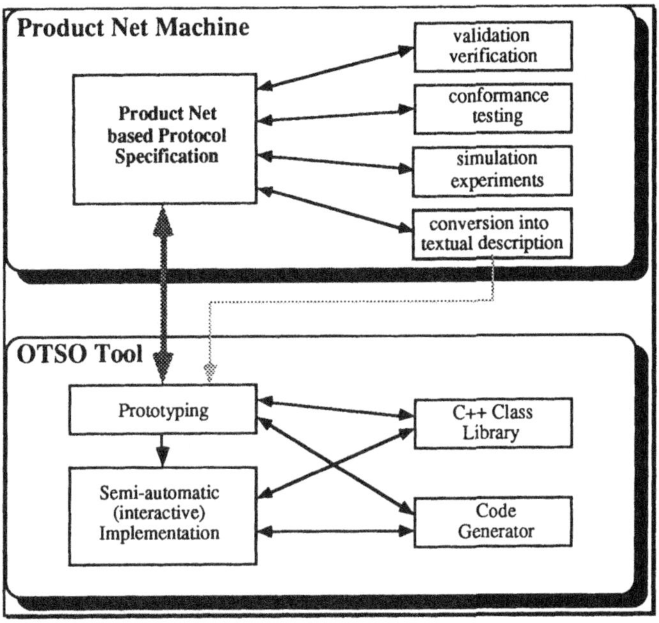

Figure 3.1: Protocol Design

The protocols are subdivided into a number of small functional units, each of it provides a dedicated functionality. This step of protocol development is already determined in the specification phase where higher level Petri Nets are used to uncover and describe protocol inherent parallelism.

The step from specification to implementation is an iterative process based on the OTSO Tool [KoRe92]. OTSO is an environment for development, interactive testing, simulation, and execution - including distributed runtime system management. OTSO is based on the principles of object-oriented programming and is built on top of C++. It consists of the OTSO C++ class library and a code generator. OTSO supports concurrency, asynchronous communication, synchronization, and transparent distribution of objects. Automatic generation of interactive user interfaces, support for finite state machines, and multiple concurrent service instances are offered as well. OTSO provides support for simulation and efficiency optimization. System configurations as well as scheduling, queueing and communication methods can be customized using the class library and virtual time. Most of the services are provided by automatically generated interface objects called agents. OTSO hides the physical implementation (such as operating system processes, threads, virtual tasks) of concurrent objects, allowing the fine-tuning of the system in a late development phase.

3.1 Formal Specification

We utilize the idea of dividing the program code into small modules which can be executed in parallel. Our approach is to combine formal specification techniques with a highly efficient concurrent execution of protocol software. Since Petri Nets are well suited to describe and uncover protocol inherent parallelism we have developed a programming language based on this concept. The Petri Net type we are using is based on product nets [BOP89] which were developed especially for specification and analysis of protocols. We have introduced some modifications resulting in the so-called restricted product nets. A detailed description of petri nets and product nets is beyond the scope of this paper. The interested reader refer to eg. [Rupp91].

Roughly spoken, product nets use places consisting of typed tokens, erase and inhibitor arcs, combined with arc labels and transition inscriptions (called predicates). Once having specified a protocol (as described in the following section) the step to implementation is straightforward. Since the granularity of Petri Net based implementations can be manipulated easily, we use the product net machine of the GMD (Gesellschaft für Mathematik und Datenverarbeitung) to monitor concurrently executable protocol functions. The results are illustrated in the next sections.

3.2 Simulation Results

The product net machine consists of a graphical editor supporting the design of product nets, and a parser which checks the syntactical correctness of the descriptions. Concerning a dynamical analysis of the specification, reachability graphs can be computed. Additionally, a simulation generates paths through the reachability graph. The analysis of large reachability graphs is supported by features like:

- search for markings with specified properties,
- generate a path from the initial marking to a given marking,
- computte the set of all token appearing in the markings of a given place,
- etc.

The following simulation results were achieved by using the product net machine. Since a complete analysis of *XTP* is beyond the scope of this paper we will focus on the connection setup in *XTP*. The results can be used to calculate the optimal number of processors necessary to achieve a given performance of a software implementation. The number of concurrently executable actions during a multicast setup is depicted in figure 3.2.

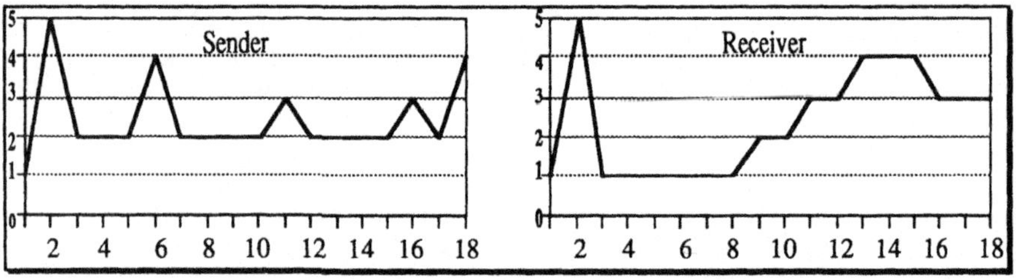

Figure 3.2: Number of concurrently executable actions during the connection setup

To obtain more general information about the number of concurrently executable actions several types of connection scenarios have been simulated, differing in the respective settings of SREQ (Status requested), DREQ (Delivery status requested) and NOCHECK (Disable checksum function). Furthermore, it has been distinguished between:

- a successful packet transmission: no packet loss occurred and the packet is errorfree.
- an erroneous packet transmission: loss of FIRST packet, duplicated FIRST packet, loss of one receiver´s CNTL packet, checksum error.

The number of concurrently executable tasks at the sender differs from the respective numbers at the receivers. Fig 3.3 illustrates the numbers of concurrently executed actions during connection setup, based on about 200 simulations runs.

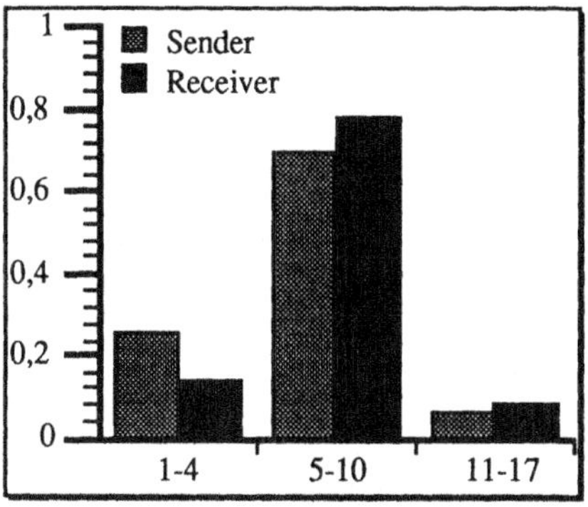

Figure 3.3: Cumulative frequencies of specific intervals

4. CONCLUSISION

Within this paper we have demonstrated a possibility to deal with the inadequacy of communication systems to support various application demands in an adequate manner. Depending on the traffic situation an suitable protocol stack is configured in run-time based on the requested QOS of the applications. The problem with existing flexible architectures is that they strive to provide a coarse-grained modularity of the configurable modules in order to reduce the communication and computation overhead. There are various prototype systems, which have nothing or little common and therefore are difficult to be compared or even integrated. The reason for that is the higher-level of the offered abstraction (vertical growth).

The solution we propose is to provide a facility with varying granularity of the configurable modules in run-time. To allow this, we introduce a new configuration dimension: we change the implementation strategy in run-time and by this also the behaviour of the configuration facility.

To uncover the protocol modules which could be used independently in a flexible protocol architecture we use our Petri-Net approach. It provides us with knowledge about what is feasible when trying to parallelize protocol architectures.

REFERENCES

[BOP89] H.J. Burkhardt, P. Ochsenschläger, R. Prinoth: *Product Nets - A Formal Description Technique for Cooperating Systems*, GMD-Studien No. 165, Gesellschaft für Mathematik und Datenverarbeitung, September 1989

[CiOf91] I. Cidon, Y. Ofek: *Metaring: A Full-Duplex Ring with Fairness and Spatial Reuse*, Technical Report, T.J. Watson Research Center, Yorktown Heights, 1991

[Clar85] D.D. Clark: *The Structuring of Systems using Upcalls*, Proc. 10th ACM Symposium on Operating System Principles, pp. 171-180, Washington, Dec. 1985

[CRMA91] H.R. van As et al.: *CRMA II: A Gbit/s MAC Protocol for Ring and Bus Networks with Immediate Access Capability*, Proceedings EFOC/LAN'91, pp. 262-277, London, 1991

[DPH92] P. Druschel, L. Peterson, N. Hutchinson: *Beyond MicroKernel Design: Decoupling Modularity and protection in Lipto*, Proc. of the 12th. IEEE Conf. on Distr. Comp. Sys., 9-12 June, Yokohama, Japan, 1992

[DQDB90] *DQDB Metropolitan Area Network P.802.6_/D12*, IEEE, Feb. 7, 1990

[FDDI89] *FDDI*, International Standard ISO 9314-1, 2, 3, 1989

[FDDI90] P. Martini, T. Meuser: *Service Integration in FDDI*, Proceedings of the 15th Conference on Local Computer Networks, Minneapolis, Minnesota, Sept. 30 - Oct. 3, 1990

[Ferr90] D. Ferrari, *Client Requirements for Real-Time Communication Services*, RFC 1193, Nov. 1990

[Haas91] Z. Haas: *A Protocol Structure for High-Speed Communication over Broadband ISDN*, IEEE Network Magazine, pp. 64-70, January 1991

[Hein92] B. Heinrichs: *XTP Specification and Parallel Implementation*, Proc. International Workshop on Advanced Communications and Applications for High Speed Networks, pp. 77-84, Munich, March 16-19, 1992

[Hein93] B. Heinrichs: *Modular Specification and Implementation of High-Speed Transfer Systems*, GI/ITG Workshop "Distributed Multi-media Systems", Stuttgart, 18.-19.02.1993

[Heli91] Perihelion Software Limited: *The Helios Parallel Operating System*, Prentice Hall International (UK) Ltd., Hertfordshire, Great Britain, 1991

[HeRu92] B. Heinrichs, M. Rupprecht: *XTP - Efficient Parallel Software Implementation based on a Petri Net Specification Technique* IFIP TC6 Workshop on Broadband Communications Estoril Portugal 20/22 January 1992

[HIPP92] *HIPPI-FP High Performance Parallel Interface - Framing Protocol*, Preliminary Draft Proposed American National Standard for Information Systems, February 14, 1992

[HuPe91] N.C. Hutchinson, L.L. Peterson: *The x-Kernel: An Architecture for Implementing Network Protocols*, IEEE Transactions on Software Engineering, vol. 17, no. 1, 1/1991

[MaPe90] S.W. O'Malley, L.L. Peterson: *A Highly Layered Architecture for High-Speed Networks*, Proc. 2nd International Workshop on High-Speed Networks, Palo Alto, November 27-29, 1990

[KoRe92] J. Koivisto, J. Reilly et al.: *OTSO User's Guide*, Technical Research Centre of Finland, Telecommunications Laboratory, 1992

[LeBl92] H. Leopold, G. Blair et al.: *Distributed Multimedia Communication System Requirements*, Esprit Project 5341, Document Number OSI95/Deliverable ELIN-1/P/V4, 1992

[PoSc91] T. F. La Porta, M. Schwartz: *Architectures, Features, and Implementation of High-Speed Transport Protocols*, IEEE Network Magazine, pp. 14-22, Mai 1991

[ReTi93] W. Reinhardt, M. Tierney: *Requirements Analysis for the Protocol Selector*' RACE II Project EuroBridge, Deliverable 26, 1993

[Rupp91] M. Rupprecht: *Implementierung und parallele Verarbeitung von Kommunikationssoftware*, PhD-thesis, Aachen, University of Technology, 1991

[SBS92] D. C. Schmidt, D. F. Box, T. Suda: *ADAPTIVE: A Flexible and Adaptive Transport System Architecture to Support Multimedia Applications on High-Speed Networks*, University of California, Irvine, Techn. Report 92-46, 1992

[Stil92] B. Stiller: *PROCOM: A Manager for an Efficient Transport System*, Proc. IEEE Workshop on the Architecture for Implementing of High Performance Communication Subsystems, February 1992

[Stai92] R. Stainov: *A Dynamic Configuration Facility For Multimedia Communications*, Aachen University of Technology, Department of Computer Scienece, Techn Rep. 92-25, 1992

[Tsch91] Ch. Tschudin: *Flexible Protocol Stacks*: Proc. of SIGCOMM 91, Computer Communication Review, Vol. 21, No. 4, Sept 1991

[Zitt91] M. Zitterbart: *High-Speed Transport Components*, IEEE Network Magazine, pp. 54-63, January 1991

Integrated Broadband Communication Networks
and Services (C-18) / V.B. Iversen (Editor)
Elsevier Science B.V (North-Holland)

An Experimental Architecture Supporting Multimedia Applications on Broadband Networks

G. Girardi[a], D. Ercole[a], M. Ullio[a], R. Melen[b], A. Albanese[c]

[a]Centro Studi e Laboratori Telecomunicazioni, via G. Reiss Romoli 274, 10148 Torino, Italy.

[b]Politecnico di Milano, Dipartimento di Elettronica, via Ponzio 34/5, 20133 Milano, Italy.

[c]Bell Communication Research, 445 South Street, Morristown, NJ, 07960, USA.

Abstract

This work describes a part of a two years Bellcore/CSELT collaborative research agreement on broadband internetworking services, to investigate service requirements for supporting multimedia applications on heterogeneous broadband networks and specifically explore the interworking of ATM and IEEE 802.6 based networks. The effort will be on the experimentation on a real environment of a multimedia application that can stress the networks, to serve as a testbed for multimedia services research.

This paper presents a multimedia application chosen for the experiment, a reference architecture under investigation and the structure of the service management software.

Keyword Codes: C.2.1; C.2.2; C.2.4
Keywords: Network Architecture and Design, Network Protocols, Distributed Systems

1. Introduction

The evolution of video technologies, personal computers and workstations, reducing costs, is boosting the diffusion of more complex customer equipments, and is offering new opportunities to the multimedia market.

Software developers are taking full advantage of this trend, releasing multimedia applications on different platforms. Although many products for stand-alone interactive applications have become now available, the main interest is in distributed multimedia applications to solve educational, medical and interpersonal communication problems: these applications require the transport of large amounts of data related to different media (i.e. with different quality of service requirements) over a wide area.

The Broadband Integrated Services Digital Network (B-ISDN) should provide the users with a large bandwidth and low delay connectivity over a geographical area. Therefore a key issue today is broadband multimedia service definition to support new multimedia applications.

Regarding broadband networks, a great effort has been done in standardization bodies to define architectures, services and protocols; the work is still in progress and a unifying scenario is emerging. Two solutions have been proposed: Metropolitan Area Networks (MAN) based on IEEE 802.6 technology [7] and Broadband Switching Systems (BSS) based on Asynchronous Transfer Mode (ATM) [5]; the different time schedules for the release of these technologies and their different characteristics, which make them suitable for different exploitations in an operating scenario, suggest the investigation of the interoperability requirements [4].

In this scenario, as part of a two years Bellcore/CSELT collaborative research agreement on broadband internetworking services, it seems very attractive to investigate two main items:

- service requirements for supporting multimedia applications on heterogeneous broadband networks;
- internetworking of ATM and IEEE 802.6 based networks.

The following part of this paper presents a multimedia application chosen for the experiment, a reference architecture under investigation and a structure for the service management software.

2. The Multimedia Application

The scope of Bellcore/CSELT collaboration is to investigate how a broadband network can support multimedia applications. The focus of the activity is on: service definition and management; network requirements in support of multimedia applications; interworking problems between heterogeneous networks.

An application suitable for this activity should possibly become part of everyday work of a significant number of users and stress the experimental network to give useful indications on users requirements and on hidden weak points in service provisioning and management. Several applications, developed either in Bellcore or in CSELT, have been considered: the possibility investigated at present is to experiment the Bellcore Cruiser; if a commercial multimedia application will become available in the meantime, it will be considered as an attractive alternative.

Cruiser is an application designed to support informal communication between members of a company that work in different buildings or on different floors of the same building. The use of a broadband wide-area network can improve this system allowing communications between different cities with little degradation of the end-to-end performances. Cruiser improve the informal communications creating what is called a "virtual workplace": with the multimedia terminal every user can "walk" around the offices and "meet" randomly other people during the workday.

The main idea is to have a terminal that is always turned on and a multimedia application that allows the users to form groups. The application running on every multimedia terminal calls every other member of the group with a round robin strategy during the day: no answer to the incoming call is necessary to set up the connection.

This application requires a high fidelity audio quality, to be used for personal communications, such that the voice of speakers can be easily recognized by the other party. The requirements for the video are not so strict, because the video communication involve slowly motion pictures, so a low quality video may be sufficient to recognize the counterpart.

The topic problem to solve in order to support this application in a wide area digital network is to minimize the coding and transport delay: to use this application to enhance personal communications minimum delay is required to ensure that the conversation could be as easy and nice as in a "live" situation. As a consequence, even if the quality of the video could be low, it seems that the use of a large amount of bandwidth is necessary to reduce delay and to assure the end users the same performances they are used to in a local area. For these reasons it looks like only broadband networks are suitable to extend this application over a wide geographical area.

3. The Experimental Environment.

In B-ISDN based on ATM technology, point-to-point virtual connection capabilities can be used to provide a number of connection oriented services. Additional network elements, Connectionless Servers, allow the provisioning of connectionless services like SMDS.

Broadband networks based on MAN technology were developed for the provisioning of SMDS; however Connection Oriented Isochronous service can be easily supported with this technology and even if a complete standardization activity has not been considered, this service is provided in some field trials.

SMDS based on MAN technology has been the first broadband service to reach the commercial phase in a number of countries. At the same time a number of trial on connection oriented services in networks based on ATM technology has been announced. The target architecture requires the integration of the two technologies in a unique broadband network.

In Figure 1 is depicted the architecture of the heterogeneous network that will provide SMDS service during phase 3 (a complete description of SMDS phase 3 can be found in Bellcore documents [2]). It must be reminded here that the Switching System (SS) may have either a centralized or distributed technology [3], i.e. it may use MAN technology; on the contrary the B-ISDN SS (BSS) uses the ATM technology.

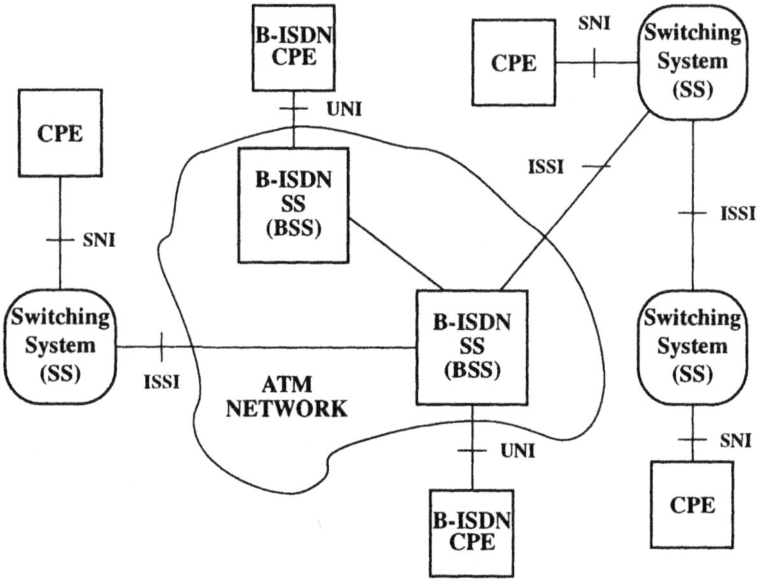

ATM	**Asynchronous Transfer Mode**
B-ISDN	**Broadband Integrated Services Digital Network**
BSS	**B-ISDN Switching System**
CPE	**Customer Premises Equipment**
ISSI	**Inter Switching System Interface**
SNI	**Subscriber Network Interface**
SS	**Switching System**
UNI	**User Network Interface**

Figure 1 - Broadband architecture for SMDS phase 3

The provisioning of multimedia services on this heterogeneous environment has not been planned. However it is possible with the redefinition of Subscriber Network Interface (SNI), and of Inter Switching System Interface (ISSI), to support also connection-oriented traffic (we will refer to those most general interfaces as SNI* and ISSI*) and the implementation of isochronous capabilities in the MAN networks.

Figure 2 represents the architecture subset under study within the Bellcore/CSELT collaboration: the MAN is a particular case of distributed SS, realized with IEEE 802.6 technology; the ATM MAN Interface (AMI), is a proprietary interface, i.e. it is a particular instance of ISSI* suitable for the case under discussion. Some solutions adopted in this interface may be different from solutions described in the definition of the ISSI interface.

Figure 3 depicts a possible configuration of the experimental environment. The

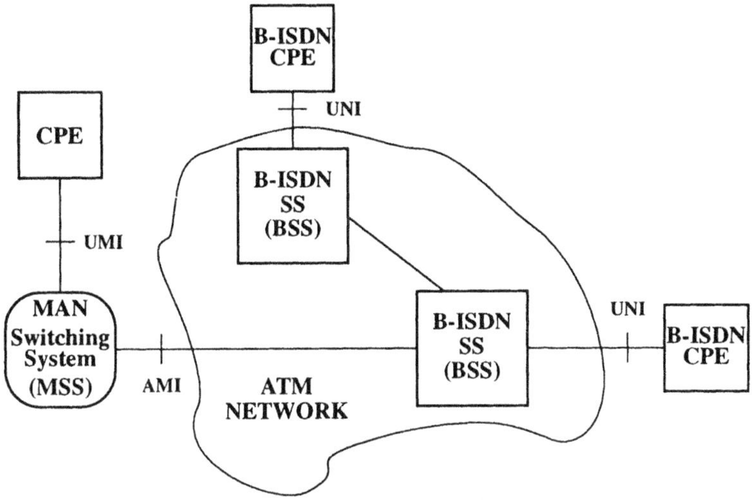

AMI	ATM MAN Interface
ATM	Asynchronous Transfer Mode
B-ISDN	Broadband Integrated Services Digital Network
BSS	B-ISDN Switching System
CPE	Customer Premises Equipment
ISSI	Inter Switching System Interface
MSS	MAN Switching System
UMI	User MAN Interface
UNI	User Network Interface

Figure 2 - Broadband architecture for multimedia services

Customer Premises Equipments (CPEs) are multimedia environments that include multimedia terminals, two analog networks and a Local Area Network (LAN).

A multimedia terminal consists of a workstation, a monitor, a video camera, an audio microphone and a speaker: this terminal allows the user to manage three media, audio, video and data. The two analog networks are used for audio and video interconnection of many users within the same Customer Premises Network (CPN). A switch controller accepts connections requests from the users, processes them, notifying the called party of the incoming call, and sets up the connections.

Signalling message are exchanged between a user and the switch controller using the Local Area Network. In addition to that LAN is used for data transfer and resource management. The analog networks are connected to the broadband networks with a Multimedia Interface Module (MIM): a commercial video codec and an audio codec transform the analog signals in digital signals, which are fed into the MIM. A LAN board connects the MIM to the LAN used for signalling and data transfer. The function of the MIM is to accept data flows from three separated input and multiplexing them onto the broadband interface, that may be UNI or UMI. Moreover the MIM must translate the signalling of the private domain in the customer-to-network signalling of

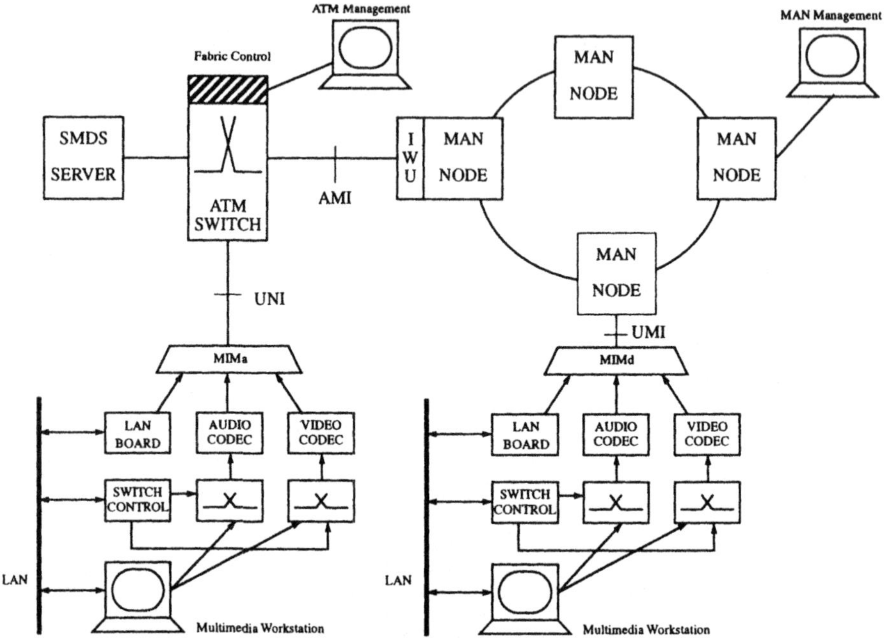

Figure 3 - Broadband network architecture for the experiment

the broadband network.

Concerning the access protocols to the broadband networks, different MIMs can be defined: a MIMa, based on the ATM protocol according to the definition of UNI, and a MIMd, based on IEEE 802.6 protocol and suitable for the UMI interface.

The general architecture of the UMI has not been studied in standard bodies and also signalling aspects of the UNI have not been completely defined: for the purpose of this experiment, some hypotheses should be done. This argument is a subject to further study and will be the content of a future paper.

The ATM network, in the experimental environment, consists of an ATM switching fabric, a control and management unit of the switching fabric, a Switched Multi-megabit Data Service (SMDS) server. To support a multimedia service in a broadband network, a possible solution is to use SMDS and broadband bearer services, like connection-oriented non-isochronous and connection oriented isochronous, as a platform on which more sophisticated teleservices can be defined.

The SMDS server solves the routing and multi-cast problems, and performs other functions like address validation, address screening and Closed User Group management. It is connected to the ATM switching fabric and can be reached from an UNI using some specific VCIs assigned to the customer at subscription time by the SMDS service administrator.

The SMDS server cannot be seen from a UMI: if an SMDS data unit sent by a customer connected to a UMI is not local to the MAN, the MAN/ATM InterWorking Unit (IWU) sends it to the SMDS server, i.e. in the MAN the presence of the server is hidden to everybody but the IWU.

The IWU is a part of a MAN node and it is connected to the ATM switch through the AMI. The structure of the IWU supporting connectionless and isochronous connection-oriented basic services has been described in [8]. The AMI is still under study; it will be based on the ATM protocol and two physical layers will be allowed: SONET STS-3c and SDH STM-1.

The IWU must solve many problems: the first one is the mapping of isochronous MAN connections onto asynchronous ATM connections; the second is the translation of MAN protocols into ATM protocols and vice-versa. A third problem is the exploration of how the native MAN management commands can be translated into native ATM signalling functions, and vice-versa, and evolve to a common signalling and Customer Network Management (CNM) for MAN and ATM.

The signalling protocol must support also multimedia services, i.e. it must support multimedia, multiparty calls: it is beyond the scope of the experimentation to define such a signalling protocol, while either protocol being defined in Bellcore [9] or CSELT [6] could be used for the experiment.

To support the connection-oriented services a specific unit must be defined in the MAN environment that can play an analogous role to the call processing unit of the ATM node: we will refer to this unit as signalling Termination and Bandwidth Manager; this unit may be realized either in a distributed or in a centralized structure. It examines connection requests coming from the users and checks for free bandwidth and for responses of the called parties. If the operation succeeds, it notifies to the calling users the virtual channels reserved for the connections.

4. The Software Architecture

This section describes the architecture of the control and management software, that will allow the experimental testbed described in the previous section to offer multimedia broadband services with dynamic resource allocation. This software architecture must be both sophisticated and highly flexible, to allow fast provisioning of new services and customer management functions.

In the next years, Customer Premises Equipments (CPEs) and Customer Premises Networks (CPNs) are expected to share with the Switching System (and other network entities) the task of service control and management. By allowing the customer to do his own management, the Operating Companies will be able to satisfy sophisticated customer needs, while simplifying and speeding up their network management functions.

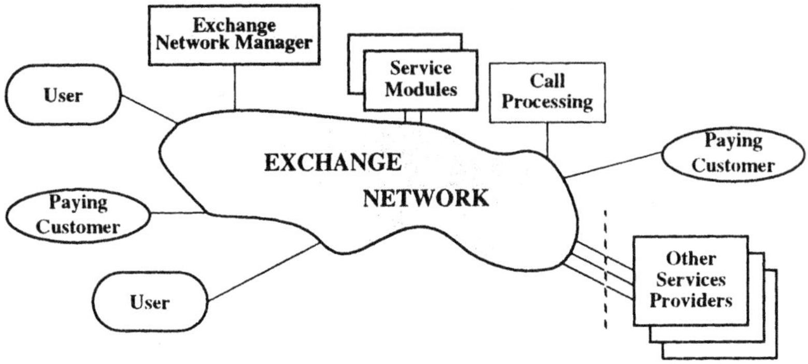

Figure 4 - Administrative Domains in a Telecommunication System

In Figure 4 the administrative domains involved in telecommunication services provisioning are shown:

- the User selects the access network and requests communication and information services using an access network;
- the Paying Customer represents the paying entity for the services provided to users; subscribes to network services, selects service providers and sets service constraints to users;
- the Exchange Network is made up of transmission and switching facilities, and provides basic transport services; it can be built around many different networks technologies;
- the Service Provider uses the basic exchange network services for offering services to the users within the constraints imposed by the paying customer;
- the Exchange Network Manager directly manages the exchange network, translating generic user service requests into network-dependent requests, within the constraints imposed by contracts with paying customers and service providers. The operating company can be at the same time a service provider, for the services it is able (or allowed) to offer. A number of different Service Modules are connected to the Exchange Network to provide these services. The Exchange Network Manager programs these modules according to customer requests.

In Figure 5 two networks, connected through an InterWorking Unit (IWU), constitute the exchange network: one network is based on MAN technology, the other on ATM technology. In each one a Network Management Entity offers users and service providers the transport services they need using network basic connectivity. Users, customers and service providers can be connected to either access network; the operating company is responsible to provide the communication path between domain portions residing on different access networks. The IWU is under the control of an InterWorking Unit Management Entity.

The three Management Entities of this picture are a distributed realization of the Exchange Network Manager of Figure 4. The MAN Management Entity, as the ATM Management Entity, can require services to the other network management entity. The IWU Management Entity, in this example, is under the MAN Management Entity control.

Every network management entity can be located everywhere in the exchange network. The exchange network itself provides the physical links between management entities and their related network elements.

A software architecture named B-SOS (Broadband Services Operating System) is being explored at Bellcore, to demonstrate service and network management concepts [1]. Each network domain may be seen as an administrator (sometimes a group of administrators) with a generic template, called Service Machine (SM), responding to client requests by allocating native resources obtained by other SMs. The template provides a generic Management Information Base (MIB) and the support for communication with the clients and the resources of the SM. The core functions of the

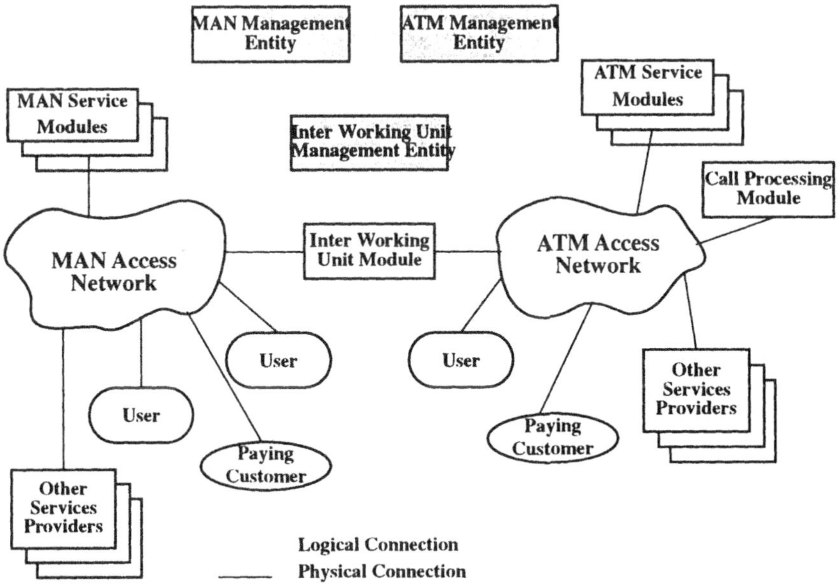

Figure 5 - Management Entities in a telecommunication system in which a MAN Access Network and an ATM Access Network are connected trough an Inter Working Unit

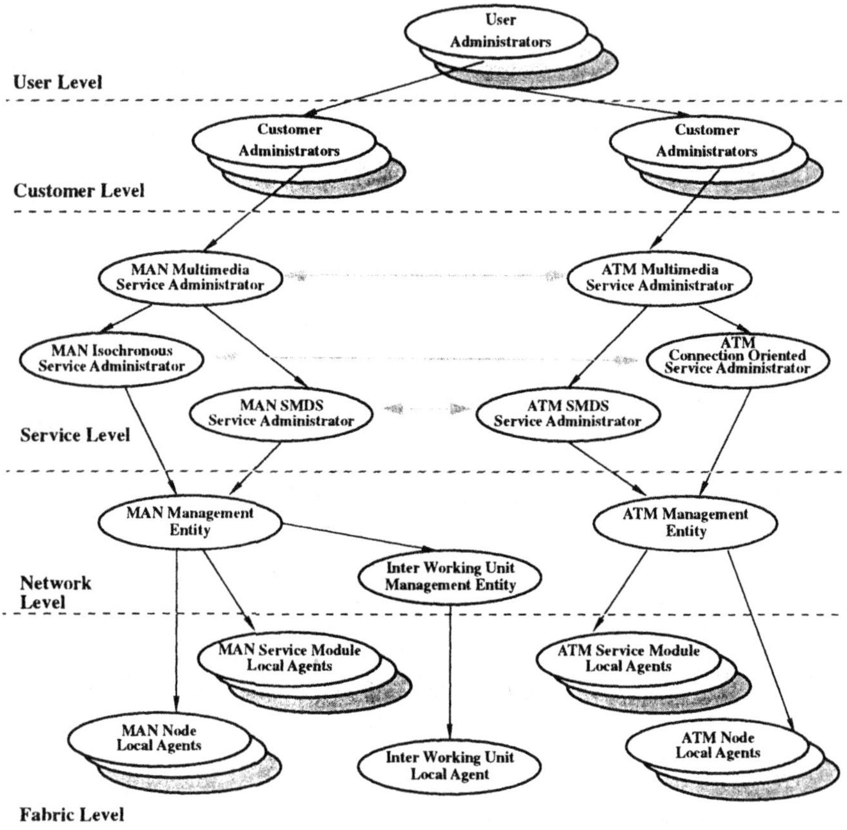

Figure 6 - Experimental Management Architecture

SM are peculiar to every instance and depend on the strategy that the management entity implements.

The experimental management architecture is depicted in Figure 6. CPE or CPN connected to the MAN or to the ATM network are managed be User Administrators. A User Administrator can require services to its Customer Administrator (CA) using the exchange network infrastructure.

The Customer Administrator subscribes different Service Administrator and performs service management functions for the CPE or CPN. A "directory" service gives informations about available services to the CAs. Every Customer Administrator may have users connected either to the MAN network, or to the ATM network. It can also subscribe to every service provider in the exchange network domain or connected to the exchange network.

At service level, in the experimental management architecture, two Multi-Media Service Administrators (MMSAs), one for each network, guarantee the maximum independence regarding the means by which the two networks provide multimedia services to their customers. The service management and service access primitives will be the same, but the way they are implemented may depend on the underlying services offered by the networks and on the operating company strategy.

To provide a multimedia communication between a user connected to the MAN network and a user connected to the ATM network, the two MMSAs are peer entities that provide services to each other.

To set-up a multimedia call to an ATM customer, a MAN customer must issue a service access request to his MMSA, specifying the called party and the characteristics of the call. The MAN MMSA will issue a service access request to the ATM MMSA, on behalf of his customer: in this case the ATM MMSA is a server of the MAN MMSA. An equivalent procedure is followed to set-up a multimedia call to a MAN customer originated on the ATM network.

The MMSA makes use of the underlying basic services, on the MAN side SMDS and Connection Oriented Isochronous service, on ATM side Connection Oriented service and SMDS. The customer can also subscribe to these services, although this option is not shown in Figure 6. Even if the Isochronous Service on MAN is different from a Constant Bit Rate Connection Oriented service on the ATM network, a mapping can be defined so that a connection oriented communication is possible between two users connected to the two different networks.

Under the service layer, there is the network layer with the native transport capabilities: the ATM virtual path and virtual channel capabilities, managed by the ATM Management Entity, and MAN bandwidth allocation capabilities, performed by the Head Of Bus (HOB) of the IEEE 802.6 dual bus and managed by the MAN Management Entity. The IWU Management Entity (IME) is responsible for resources (e.g. bandwidth an buffers) allocation within the IWU and preforms management functions such as setting the VCI translation tables for the MAN-ATM interworking. The IME is under the control of the MAN Management Entity which can decide the resource allocation strategy.

At the Fabric Control level, Local Agents (LAs) manage single network elements: in the networks there is one LA for every node and one for every service module (e.g. SMDS Connectionless Server and Signalling Termination and Bandwidth Manager). In the InterWorking Unit there is another LA.

5. Conclusions

As a part of a two-year Bellcore/CSELT collaboration, a joint effort has been made

to define an architecture that could be used as a framework in the realization of the experimental prototypes for multimedia service provisioning on a heterogeneous broadband network.
This paper has summarized the scope of the project and the major results on general network architecture and on management and control software architecture for the provisioning of multimedia services on the network. The detailed testbed infrastructure, together with evaluation of the activity results will be the subject of future works.

References

[1] A.Albanese, T.N.Devetzis, M.Maszczak, L.Ambrosoli, M.Marini, "A Network and Services Integrated Management System Prototype", COMPEURO '91, May 13-16 1991, Bologna, Italy.

[2] Bellcore, "Inter-Switching System Interface Generic Requirements in Support of SMDS Service", Bellcore Technical Advisory TA-TSV-001059, Issue 1, December 1990.

[3] Bellcore TR-TSY-000772, "Generic System Requirements in Support of Switched Multimegabit Data Service," May 1991.

[4] A. Biocca, G. Freschi, A. Forcina and R. Melen, "Architectural issued in the interoperability between MANs and the ATM network," in Proc. Int. Switching Symposium (ISS '90), Stockolm, Sweden, vol. II, pp. 23-28.

[5] CCITT Recommendation I.361 (1990): "B-ISDN ATM Layer Specification".

[6] E. Contini, L. Piovano, M. Ravera, "Signalling Protocols for Multimedia Applications: Prototype for an ATM Test-bed", in 2nd CNR Seminar on "Broadband Communication Networks and Services", Roma, October 29-30 1991.

[7] IEEE 802.6 Standard (1990), "Distributed Queue Dual Bus (DQDB) Subnetwork of a Metropolitan Area Network (MAN)".

[8] A. Ippoliti, A. Albanese, P. Coppo, D. Ercole, M. Ullio, "Interworking of Bearer Services on IEEE 802.6 and ATM Networks.", in Proc. Tenth Annual EFOC/LAN '92 Conference, Paris, France, June 24-26, 1992.

[9] S. Minzer, "A signalling protocol for complex multimedia services," to be published on IEEE Journal on Selected Area on Communications.

Integrated Broadband Communication Networks
and Services (C-18) / V.B. Iversen (Editor)
Elsevier Science B.V (North-Holland)

The poor gain from Statistical Multiplexing in the homogeneous and the heterogeneous case

Theo A. Smit

PTT Research, Dept. of Applied Mathematics and Signal Processing
P.O. Box 421, 2260 AK Leidschendam, The Netherlands
Telephone: +3170 3323023, Telefax: +3170 3326477, E-mail: T.A.Smit@research.ptt.nl

Abstract

The success of using Statistical Multiplexing for traffic admission control to an ATM-link strongly depends on the source characteristics. In this paper the direct relation between the statistical multiplexing gain and the source parameter values is shown for the homogeneous case: multiplexing sources with the same traffic characteristics. For the heterogeneous case the gain is smaller and we found that this is mainly caused by differences in peakrate rather than differences in burstiness.

Keyword Codes: B.4.m; C.2.m; G.3
Keywords: ATM, Connection Acceptance Control, Statistical Multiplexing

1. INTRODUCTION

The possibility of achieving a high resource utilization by applying Statistical Multiplexing is mentioned as one of the prominent reasons for developing ATM. Statistical Multiplexing has been studied by many researchers, but opinions still are very different.

This paper studies the influence of source characteristics on the resource utilization gain, in case that Statistical Multiplexing is used instead of Peakrate Allocation. Based on previous work, we will describe the burst level model without buffer and justify the use of it to calculate the Statistical Multiplexing Gain. This model was used to determine for each type of sources the effect of using Statistical Multiplexing in the homogeneous case, i.e. when only one type of sources is multiplexed.

Also very interesting is the heterogeneous case: multiplexing sources with different traffic characteristics. We investigated this situation and found that the gain suffers from the heterogeneity. This phenomenon was further studied and turned out to be allmost totally caused by differences in a single source parameter: the peakrate.

The outline of this paper is as follows. Section 2 describes the model, the input traffic characterization, the multiplexer performance measures and the analysis.

In section 3 a justification of this model is given. A rule for dimensioning the buffer is proposed and it is shown that the burst level model without buffer is appropriate for calculating the acceptable input traffic. Section 4 defines the Statistical Multiplexing Gain and elaborates numerical results in the homogeneous and the heterogeneous case, from which several conclusions are drawn. Finally section 5 summarizes the results.

2. BURST LEVEL MODEL WITHOUT BUFFER

This section describes a burst level model of the multiplexer. On burst level traffic is modelled in terms of information flows with certain rates rather than cells handled individually. The number of sources (connections) and their stochastic behaviour are assumed to be constant. First the input traffic is modelled, secondly the performance is characterized. Finally the analysis is given.

2.1 Basic Model Description

Figure 1 shows the model which is studied here. Sources put ATM-cells into a common buffer. The transmission link extracts cells from the buffer at a constant rate. If the sources offer more cells than the link can transmit, the buffer gets filled and cells may be lost due to buffer overflow. At connection establishment control functions which restrict the input traffic are needed to guarantee a certain connection quality. These functions are known as Connection Acceptance Control (CAC). It is important to investigate the input traffic limitations in order to get the maximum capacity out of the resources on one hand and to provide a certain quality of connections on the other.

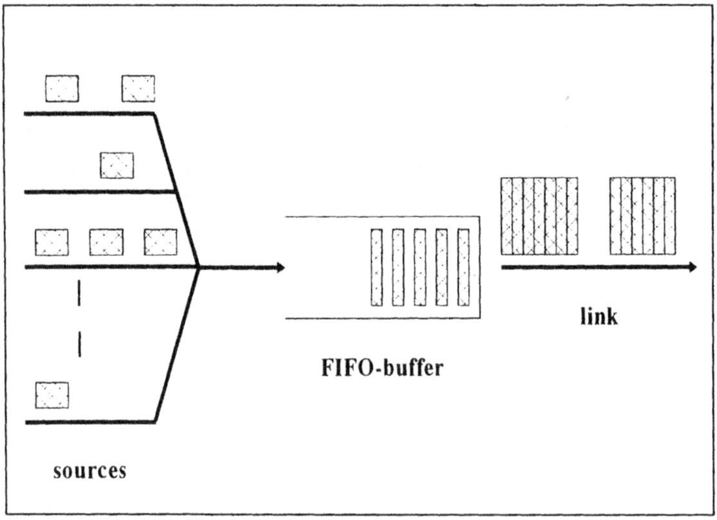

Figure 1: Model of Multiplexing in ATM

In this paper, a burst level model, or "fluid flow"-model, without buffer is studied. At burst level the process of connections changing their activity level is studied rather than the arrival process of individual cells.

2.2. Input Traffic Characterization

Modelling the input traffic we make the following assumptions:

1. The traffic is generated by ON/OFF-sources. These sources alternate between periods of silence and periods in which a traffic flow at a constant rate (the peakrate) is offered.
2. The sources are stochastically independent.
3. There is a finite number of different types of sources.

The input traffic is fully described by the following parameters:

- k number of source types
- i index to indicate the source type, $1 \leq i \leq k$
- r_i peakrate to linkrate ratio of sources of type i
- a_i activity of sources of type i, which means the long term fraction of time the source is in the ON-state
- n_i number of sources of type i

2.3 Performance Characterization

Two performance measures are studied here. The "overflow probability" is the probability the offered traffic exceeds the transmission capacity at an arbitrary instant. This measure can be interpreted as the long term fraction of time congestion occurs. The "loss probability" is the probability an arbitrary cell will be lost. This performance measure can also be interpreted as the long term fraction of traffic which will be lost.

Other performance measures which may describe the quality of the connection on other topics, e.g. the cell delay, cell delay jitter and consecutive cell loss, are not studied here.

2.4. Analysis

The state of the sources at an arbitrary instant is symbolized by the stochastic variable X_i, the number of sources of type i in the ON-state ($0 \leq X_i \leq n_i$). For short, vector notation is used:

$$\underline{X}^T = (X_1, X_2, \ldots X_k)$$

This notation will also be used for the parameters r_i and realizations x_i of X_i.
Given a realization $\underline{x}$ of $\underline{X}$, the total offered traffic $q(\underline{x})$ is computed by

$$q(\underline{x}) = \sum_{i=1}^{t} r_i x_i = \underline{r}^T \underline{x}$$

Using independency (assumption 2. in section 2.2), the probability that x_i of the n_i sources of type i are active is binomially distributed:

$$\Pr(X_i=x_i) = \begin{pmatrix} n_i \\ x_i \end{pmatrix} a_i^{x_i}(1-a_i)^{n_i-x_i}$$

Again using independency, the probability that the system is in state $\underline{x}$ at an arbitrary instant is given by

$$\Pr(\underline{X}=\underline{x}) = \prod_{i=1}^{k} \Pr(X_i=x_i)$$

The overflow probability is the sum of the probabilities on states at which the offered traffic exceeds the transmission capacity:

$$\Pr(\text{overflow}) = \sum_{\underline{x}|q(\underline{x})>1} \Pr(\underline{X}=\underline{x})$$

The loss probability is the ratio of the average lost traffic to the average offered traffic per unit of time:

$$\Pr(\text{loss}) = \frac{\sum_{\underline{x}|q(\underline{x})>1} \left(q(\underline{x})-1\right) \Pr(\underline{X}=\underline{x})}{\sum_{i=1}^{k} n_i r_i a_i}$$

This completes a method for calculating the cell loss probability and the overflow probability for the heterogeneous case from a set of input traffic parameters and the linkrate.

In the homogeneous case, multiplexing sources with activity level a and peak to link ratio r, the maximal load is given by

$$\rho(r,a) = \underset{n\in N}{\text{MAX}} \left\{ n \cdot r \cdot a \mid \Pr(\text{overflow}) \leq \alpha \right\}$$

Binary search on the number of sources n is used to calculate the acceptable traffic given a certain accepted overflow probability α. In the heterogeneous case, binary search was used to calculate the number of sources of a certain type which can be added to the traffic offered by a constant number of other sources.

3. JUSTIFICATION

This section explains why the burst level model without buffer, as described and analyzed in the previous section, is appropriate to calculate the overflow and cell loss probability. The relation between the congestion and the buffersize is explained, resulting in a proposal strategy for dimensioning the buffer.

As already commonly known [1-4] the relation between buffersize and congestion exists out of two distinct parts. See figure 2. These references show that congestion can be seen as the total of two causes:

1. On one hand congestion is caused in periods at which too many sources are in the active state. The buffer overflows by the offered bitrate exceeding the linkrate. The length of such a period of congestion is of the same order as the burst durations. We will call this type of congestion "burst level congestion".

2. On the other hand congestion is caused by too many sources offering a cell simultaneously. Congestion of this type only occurs when the buffer is so small that it cannot contain one cell per source. This type of congestion is incidental and is called "cell level congestion".

As shown in figure 2 the cell level congestion dominates at small buffersizes. Without any buffering the congestion is of the same order as the load. At small buffersizes the congestion can be approximated by the M/D/1/S-queue. At larger buffersizes the cell level congestion becomes negligible so the burst level congestion dominates at higher buffersizes. By the

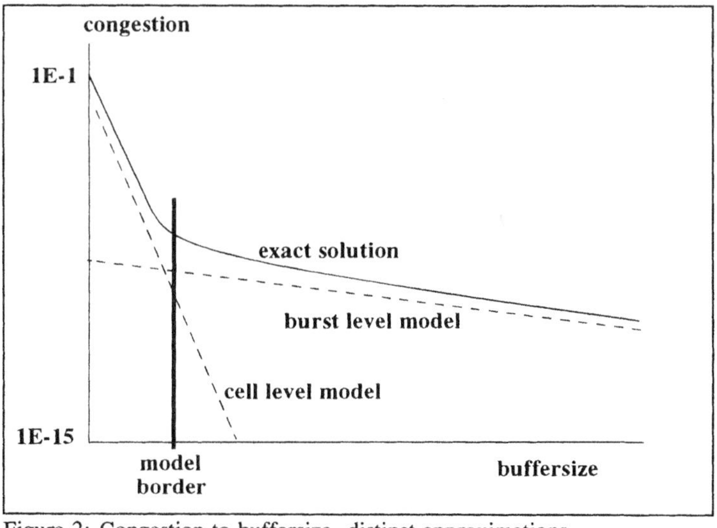

Figure 2: Congestion to buffersize, distinct approximations

absence of the cell level congestion at large buffers, the burst level model is appropriate to approximate the congestion here.

Also commonly known (e.g. [2]) is the fact that burst level congestion decreases very slowly at increasing buffersize. Large buffers in the order of several burstvolumes are needed before any significant load increase is reached. This makes burst buffering not worthwhile. Moreover a buffersize of several burstvolumes causes unacceptable delays. From this follows that the buffer is only useful to avoid cell level congestion. The buffer should be dimensioned such that the cell level congestion is smaller than a significant portion (e.g. 1%) of the acceptable congestion.

Assuming a buffer which suppresses exactly the cell level congestion only burst level congestion remains. This validates the burst level model without buffer (described in section 2.) for computing the congestion and the acceptable load.

4. NUMERICAL RESULTS

This section presents some numerical results obtained from the burst level model without buffer (See section 2). In particular the relation between traffic parameters and the statistical multiplexing gain is shown and the difference between homogeneous and heterogeneous traffic is investigated.

4.1. The Statistical Multiplexing Gain in the homogeneous case

To determine the suitability of a source type for Statistical Multiplexing it is interesting to know which gain can be reached in comparison with Peakrate Allocation. At Peakrate Allocation each source occupies a portion of the available bandwidth equal to its peakrate.

Figure 3 shows the acceptable load in the homogeneous case for sources with a peakrate to linkrate ratio r varying from 0.003 to 1.0 (plotted horizontally) and an activity level a varying from 0.01 to 1.0 (plotted vertically). Each point (r,a) in the plain represents an ON/OFF-source type. The thin lines connect points with an equal acceptable load at Statistical Multiplexing with an overflow probability of 10^{-9}. Two areas can be distinguished. In the area where the curves are horizontal the source activity equals the acceptable load. In this area there is no statistical multiplexing gain. In the vertical area the acceptable load exceeds the source activity, so in comparison with Peakrate Allocation more sources can be accepted. In this area Statistical Multiplexing is beneficial.

We define the Statistical Multiplexing Gain (SMG) of a source type (r,a) by

$$\mathrm{SMG}(r,a) = \frac{\rho(r,a)}{a} - 1$$

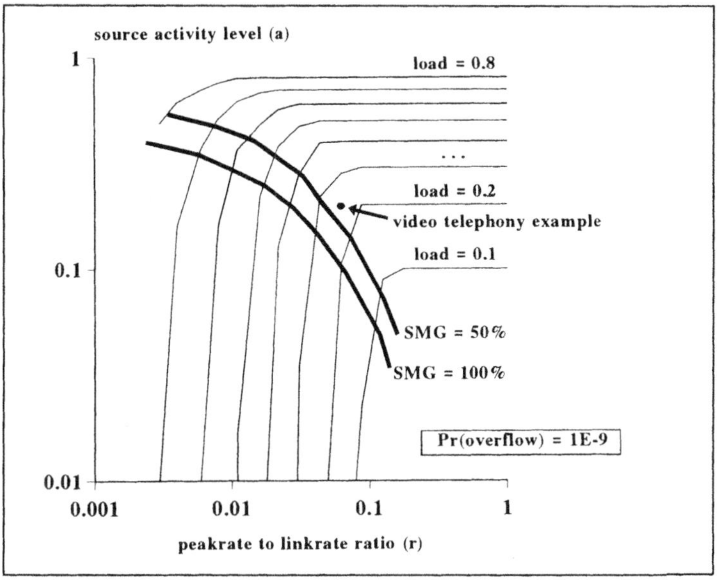

Figure 3: The Statistical Multiplexing Gain

From figure 3 it can be easily deduced which gain can be reached for a specific source type. E.g. for video telephony with peakrate 10 Mbit/s and burstiness 5 (see [5]), the peak to link ratio r on a 155 Mbit/s transmission link is 0.067 and the activity level a equals 0.2. As can be seen from Figure 3 the acceptable load $\rho(r,a)$ is about 0.25. Substitution of these values in the formula above leads to a gain of 0.25 (coincidental the same value as the acceptable load). This number shows the poor gain of statistical multiplexing for this application.

The fat lines in Figure 3 connect the points with equal SMG. Two iso-gain lines, of 50% and 100%, are shown.

A global indication of the resource savings can be computed from the Statistical Multiplexing Gain. We define the Maximum Resource Savings (MRS) as the portion of the transmission resources that can be saved by using Statistical Multiplexing instead of Peakrate Allocation. Suppose we have n sources with parameters r and a. The total offered traffic is the result of the multiplication of n, r and a. Each virtual path (VP) accepts at Statistical Multiplexing load $\rho(r,a)$ and at Peakrate Allocation load a. The MRS is the ratio of the number of VP's saved by SM divided by the number of VP's needed at Peakrate Allocation:

$$\mathrm{MRS} = \frac{nr - \dfrac{nra}{\rho(r,a)}}{nr} = 1 - \frac{a}{\rho(r,a)} = 1 - \frac{1}{\mathrm{SMG} + 1}$$

In reality resource savings are lower due to rounding off values. In our video telephony example with a SMG of 25%, the MRS is 20%.

Some warnings should be mentioned here. First, figure 3 presents values for the homogeneous case. As pointed out in the next section mixing different source types causes a higher congestion and consequently a lower acceptable load. This phenomenon "consumes" a part of the statistical multiplexing gain.
Second, the source parameters must be policed. Some distinction arises between real source parameter values and the parameter values to accept by the policing function. Since connection acceptance control must reckon with "worst case traffic" it should be based on the accepted parameter values rather than the real values. So some statistical multiplexing gain is lost here.

At first glance few present-day applications are suitable for statistical multiplexing. Telephony, facsimile and file transfer have not enough variability (source activity near to 1.0) to reach a statistical multiplexing gain. The high bitrates of Video Retrieval and HDTV cause high peakrate to linkrate ratios at which also no statistical multiplexing gain can be obtained. Maybe LAN-LAN-interconnection and low quality video conferencing are suitable applications. Further investigations are needed to point this out.

4.2. The heterogeneous case
Figures 4 to 6 present admissible traffic streams which are a combination of two different source types. In these figures the solid lines connect combinations of loads with the same overflow probability (10^{-8}). The area below this line is the admissible region. The dotted lines represent the "homogeneous approximation". This approximation determines the admissible region by drawing a straight line through the two points on the axes which represent the admissible load in the homogeneous case. Another interpretation is that each source takes a portion of the available bandwidth namely the linkrate divided by the acceptable number of sources in the homogeneous case.

In figure 4 the admissible region of two types of sources with different rates as well as different activity levels is shown. The load from the first traffic type, with peak to link ratio $r=0.1$ and activity level $a=0.02$, is plotted horizontally. In the homogeneous case 54 sources of this type are acceptable without exceeding the overflow level of 10^{-8}. The load from the second type, with parameters 0.02 and 0.5 respectively, is plotted vertically. In the homogeneous case, 59 sources of this type are acceptable. Overflow probabilities at the homogeneous approximation are plotted along the dotted line.

This example shows a large deviation between the homogeneous approximation and the exact solution. The exact solution curve (solid) bends far below the homogeneous approximation line (dotted).

Furthermore overflow probabilities of heterogeneous traffic mixes accepted by the homogeneous approximation may become a factor 10^2 higher than the accepted overflow probability. The homogeneous approximation is broadly overestimating the acceptable traffic for heterogeneous traffic mixes and is therefore not usable in practice.

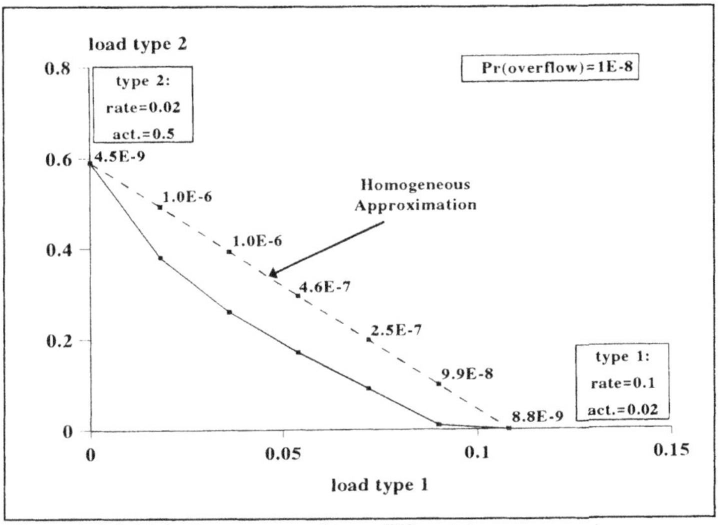

Figure 4: Multiplexing sources of two different types

In all our investigations it appeared that the deviation between the homogeneous approximation and the exact solution is mainly caused by differences in peakrate. Appearantly sources with the same peakrate supplement eachother well, even if they have different activity levels. Figures 5 and 6 illustrate this phenomenon. Figure 5 shows an example of two source types having the same activity level but different peakrates. In this case there is a substantial difference between the acceptable load and the homogeneous approximation. In figure 6 all sources have same peakrates but the types differ in activity level. Now there is not much difference between the acceptable load and the homogeneous approximation.

5. CONCLUSIONS

A simple model for evaluating Statistical Multiplexing is presented here. We studied the homogeneous and the heterogeneous case: multiplexing sources with same respectively different traffic characteristics.

For the homogeneous case it appeared that the burst level model without buffer is appropriate to determine the statistical multiplexing gain. Buffers are only suitable to avoid cell level congestion. Calculations show that only a few applications, namely LAN-LAN-inteconnection and low quality videoconferencing, reach a worthwhile gain. Other applications have a poor gain because of high peakrate to linkrate ratio (HDTV) or low variability (Telephony, facsimile and file transfer).

In the heterogeneous case, the statistical multiplexing gain is smaller than could be expected from linear interpolation of the gains in the homogeneous case. It appears that this is caused by differences in peakrate, since multiplexing sources with same peakrates but different activity levels have allmost the same statistical multiplexing gain as in the homogeneous case.

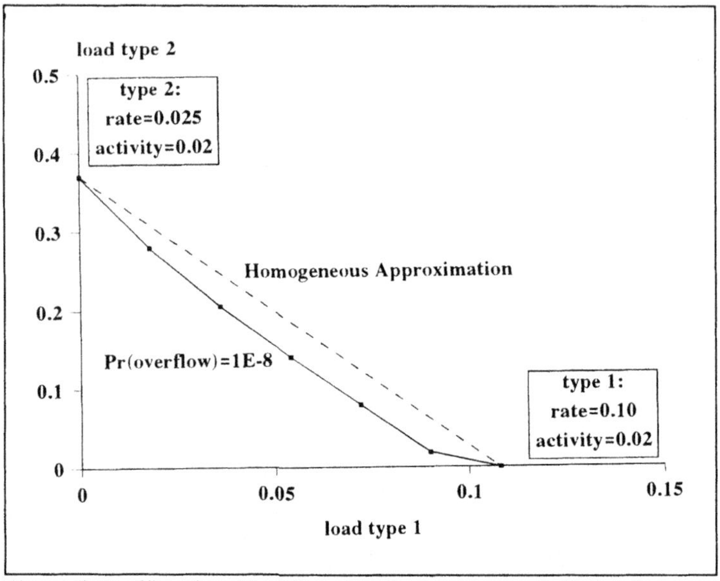

Figure 5: Traffic mix with different peakrates, same activity levels

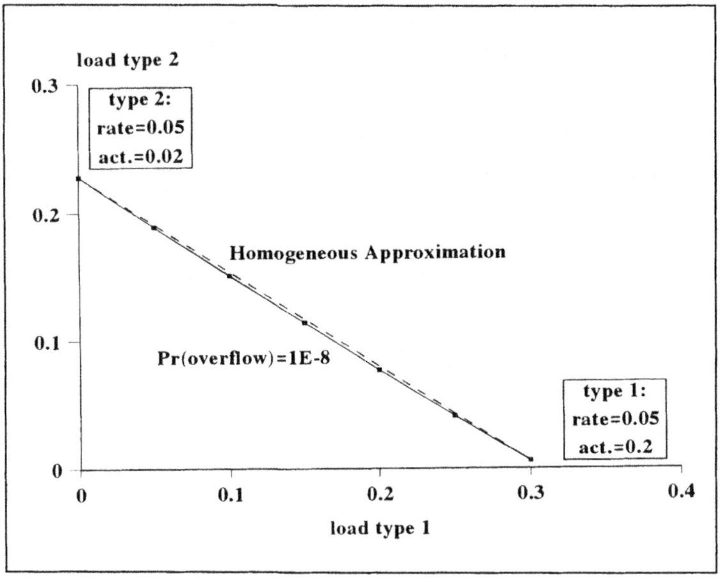

Figure 6: Traffic mix with different activity levels, same peakrates

REFERENCES

1. Hans Kröner, Statistical Multiplexing of Sporadic Sources - Exact and Approximate Performance Analysis, 13th International Teletraffic Congress, Copenhagen, June 1991, pp 847-852

2. Nikolaos M. Mitrou, Kimon P. Kontovassilis, Sophia G. Nalpadidou, Emmanuel N. Protonotarios, Bursty Traffic in ATM Networks, An Engineering Approach for CAC & Traffic Shaping, Race 1022 Workshop: ATM Network Planning and Evolution, London, April 1991

3. Carsten Rasmussen, Jens Sørensen, Kenn S. Kvols, Søren B. Jacobsen, Source Independent Call Acceptance Procedures in ATM Networks, IEEE Journal on Selected Areas in Communications, April 1991, Vol. 9, No. 3, pp 351-358

4. James W. Roberts, Variable-Bit-Rate Traffic Control in B-ISDN, IEEE Communications Magazine, September 1991, Vol. 29, No. 9, pp 50-56

5. G. Gallassi, G. Rigolio, L. Verri, Resource management and dimensioning in ATM networks, IEEE Network Magazine, May 1990, pp 8-17

Integrated Broadband Communication Networks
and Services (C-18) / V.B. Iversen (Editor)
Elsevier Science B.V (North-Holland)

An Effective-Rate Enforcement Algorithm for ATM Traffic and its Hardware Implementation

N.M. Mitrou and A.K. Koukos*

Abstract

A method is presented for shaping ATM traffic streams in order to enforce them to exhibit desired Effective Rates (ER) when statistically multiplexed with similar streams. Some background material with the definition and the basic properties of the ER is first reviewed. Then, the shaping algorithm, consisting in controlling the cell emission rate and the volume distribution of the generated bursts, is described. Finally, an architecture for implementing the shaping algorithm is outlined and its realisation on a VLSI chip is presented.

1. INTRODUCTION

Using the ATM as the ubiquitous transfer mode in the Broadband Communication Networks (IBCN) [1] allows a flexible bit rate allocation to connections of different services, with divergent characteristics and QoS demands. The main objective in such a network (as in any other telecom network) is simply maintaining the required QoS for each connection. The great challenge, however, for the traffic analyst and engineer is how to fulfil this objective with the minimum expenditure of network resources or, equivalently, with the maximum throughput.

In the case of bursty traffic sources, efficient utilisation can be achieved by allowing the sum of the peak rates of all multiplexed connections on a link to temporarily exceed the output link rate. Using buffers of appropriate sizes to hold the overflow traffic can aid in maintaining a cell loss probability lower than an affordable limit, one of the two quality indices of the transport service (the other being related with the delay of the cells). The maximum achievable multiplexing gain (defined as the increase in the admissible load, compared to the peak-rate-allocation case) depends on the traffic characteristics, the desired QoS and the network resource dimensions (buffer space and output link speed). The main problem is, given the network, how to control the traffic appropriately in order to fulfil the above two-fold objective, i.e. maximising the multiplexing gain while maintaining the desired QoS for every connection.

A critical step towards devising efficient control procedures is the definition and usage of appropriate traffic models. A small set of measurable and controllable parameters that can adequately describe the traffic streams entering the network (also

* The authors are with the National Technical University of Athens, Electrical and Computer Eng. Dept., Computer Science Division, 9 Heroon Polytechneiou, Zographou 157-73. E-mail addresses: mitrou@ektor.ntua.gr and antonis@ektor.ntua.gr

referred to as *traffic descriptor* [2]) is thus required. The trade-off, however, between simplicity and accuracy is not always that easy to solve. Under certain assumptions, a single parameter, termed *Effective Rate (ER)* or *Effective Bandwidth* [3], [4], [5], [6] can be used as a simple and efficient traffic descriptor. In [6] a formal definition of the ER of streams is given, its main properties are investigated and efficient procedures for its approximate calculation are presented.

The purpose of this contribution is to propose an Effective-Rate enforcement scheme and an architecture for its hardware implementation. It is realised through a peak-rate control mechanism which is adaptive to the traffic characteristics. Off-line computed data in the form $c=f'(V,T,R_e)$, (where c is the controlled peak rate, V the mean volume of the emitted bursts, T the mean interburst time and R_e the desired effective rate) are stored in memory for an anticipated range of parameter values. By getting measurements of the parameters V and T of the produced stream, the value of c is adjusted according to the stored data in order to achieve the agreed-upon R_e. Moreover, the volume distribution of the emitted bursts is enforced to exhibit a hypo-exponential tail, in order to fulfil the basic modelling assumptions.

The paper is organised as follows: For the sake of completeness, section 2 gives in short the definition and some basic properties of the ER (for more details the reader is referred to [6]). The ER enforcement algorithm is described and roughly validated in section 3. Section 4 presents an architecture for a hardware implementation of the algorithm, while section 5 outlines an actual single-chip implementation. Finally, section 6 draws some concluding remarks. Due to the space limitations, a detailed validation of the algorithm has been omitted here and the reader is referred to [9].

2. DEFINITION AND PROPERTIES OF THE ER

Consider N independent and statistically identical traffic streams of the simple, exponential ON/OFF type with a parameter vector (V,r,c), feeding a multiplexer with a buffer space V_b and a service rate C (equivalent parameter vectors can be obtained from (V,r,c) through substituting any one of the r or c by their ratio $B=c/r$, the so defined *burstiness* of the stream).
As far as $rN<C<cN$, there is always a finite probability P_b, $0<P_b<1$, of cell loss (as seen by each individual stream) which is uniquely determined by the fluid flow analysis [7]. As shown in [10], this probability is a uniformly continuous function of all the parameters involved, including C, for any V_b0. By using general probabilistic arguments it is easy to deduce that this function is decreasing with C, tending to zero, as C tends to cN, and to 1 as C tends to rN. Hence, by fixing the cell loss probability P_b we can always find a unique value of $C=C_N$ associated with the specified QoS point (V_b,P_b). This association defines the sequence C_N of points on the C-axis corresponding to N=2,3,... traffic streams which can at maximum be multiplexed without violating the QoS point. Each of the capacities C_N can be found by iteratively applying the analysis procedure, until the difference of the actual cell loss probability from the specified P_b becomes less than a predefined small value.

Definition : The Effective Rate, R_e of a simple, exponential ON/OFF traffic stream with a parameter vector (V,B,c), for a given cell loss probability P_b in a multiplexer with a buffer space V_b and an output link rate C is determined by a suitable interpolation between the points with co-ordinates (C_N, C_N/N), where N=2,3,... and C_N is the output capacity required for N multiplexed streams, as determined above.

Remark 1 : Obviously, the ER of a traffic stream depends not only on the characteristics of the stream itself but, also, on the multiplexing environment (i.e. the output rate C and the buffer space, V_b) as well as on the desired QoS (here determined by the cell loss probability, P_b).

Remark 2: As a result of the fluid flow analysis [7], it is the ratio V/V_b that affects the multiplexer's performance and, consequently, the ER of a stream, and not the absolute values of V.

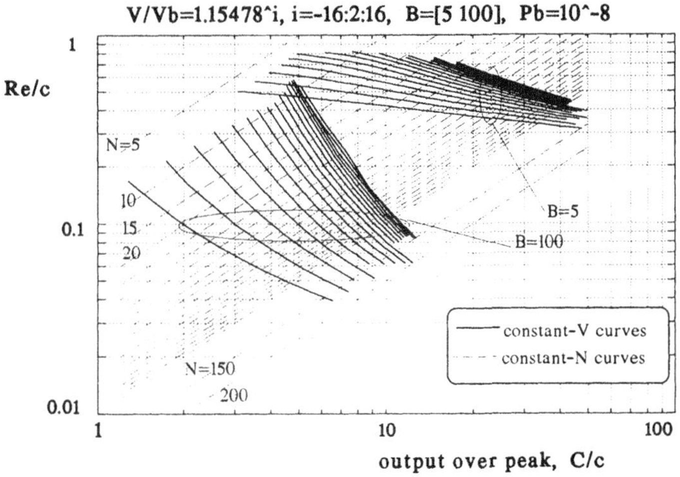

Figure 1: Relationship between R_e, V, B, C/c: variation of R_e/c with C/c

Figure 1 shows the R_e-curves (normalised over the peak rate) derived by a linear interpolation between the points with co-ordinates (C_N,C_N/N) in a log-log scale, for various values of V/V_b in the range (0.1,10), and two different values of B (a low one, equal to 5, and a high one, equal to 100). For each N, the corresponding C_N was derived by applying successively the fluid-flow analysis procedure described in [7].

Remark 3: The very smooth variation of the curves R_e/c in Figure 1 suggests that the linear interpolation in a log-log scale is a suitable one for the Definition above.

Remark 4: The curves R_e/c shown in Figure 1 represent also the inverse of the multiplexing gain, since $R_e/c = C_N/c/N$.

Properties of the ER :

Property 1: The ER of a stream is always between the peak and the mean rate.

This property follows directly from the fact that as C varies in the range (rN, cN) the corresponding probability of buffer overflow (for any length) varies continuously between one and zero).

Property 2 : The ER increases with V/V_b, provided that the other parameters are kept constant. The slope of this increase is large for small values of V/V_b (<1) and diminishes for large values of V/V_b (Figure 1).

Property 3 : The ER increases with c, provided that the other parameters are kept constant (see Figure 2 in the next section).

Property 4 : For "sufficiently large" C, the ER of a stream decreases as the output capacity C increases. The slope of the corresponding curves is larger for larger values of B, as it can be seen from Figure 1. For $B=1$ the slope is zero, i.e. the ER of constant-bit-rate streams is constant (equal to the constant rate) for any C.

Properties 2 to 4 are deduced only experimentally from figures, like Figure 1 above, although there are strong evidences for their general validity.

As a final comment, it must be stressed that the definition of the ER given above is fairly general and not restricted to the ON/OFF case. The only prerequisite for its applicability is the existence of a continuous, decreasing function of C giving the overflow probability P_b and the availability of an appropriate analysis tool to determine the points $(C_N, C_N/N)$, for specific values of P_b.

3. EFFECTIVE RATE ENFORCEMENT THROUGH CELL RATE CONTROL

By applying the ER definition, given in the previous section, nomographs of the function

$$R_e = f(V, c, T), \quad (1)$$

like those of Figure 1, can be produced, for any given multiplexer dimensions, C, V_b and cell loss probability, P_b. Relationship (1) is referred to as the *effective rate analysis function.*
Using the fluid-flow analysis tool, relation (1) can also be evaluated in the inverse form

$$c = f'(V, T, R_e) \quad (2)$$

giving the peak rate at which the bursts must be emitted in order to maintain (enforce) the Effective Rate of the produced stream constant. For any parameter values of interest, tried in practice, the inversion of (1) in the form of (2) was successful. (2) is referred to as the *(peak rate) shaping function.* Figure 2 shows typical curves of the shaping function. For implementation purposes it is more appropriate to use the intercell distance, $1/c$, instead of using the peak rate, c and the mean interburst distance, *T*, instead of *r*. It is also suggested the usage of a logarithmic grid for both independent variables, *V* and *T*. Figure 3 gives an example of the variation of $1/c$ with *T*, for various values of *V* and a certain R_e

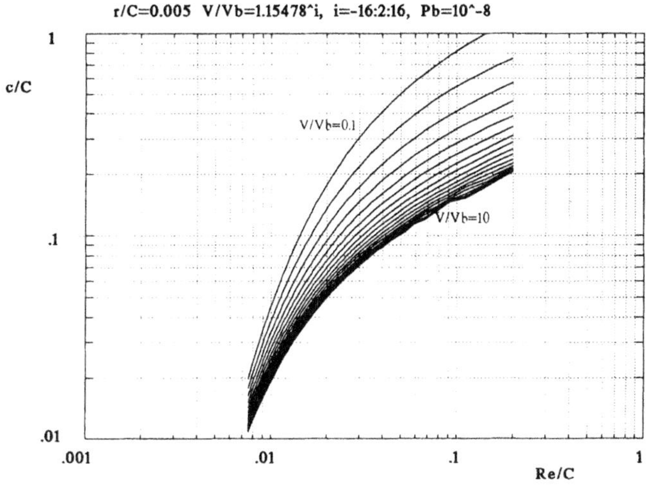

Figure 2 : Peak Rate versus effective rate curves

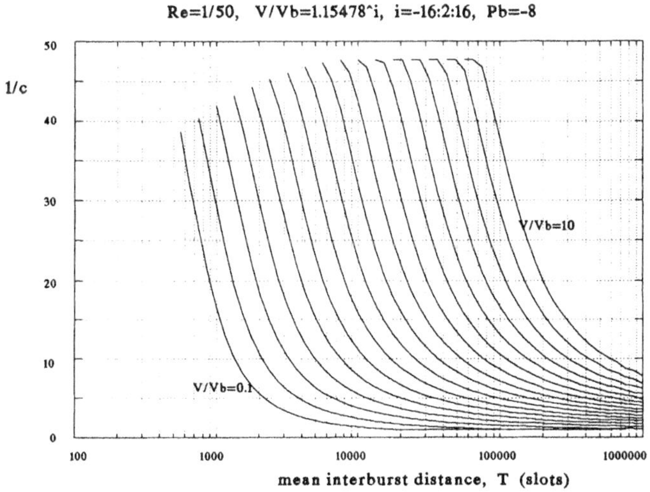

Figure 3: curves of constant Re (=1/50), in the V,T,1/c space.

4.1. The Effective Rate enforcement algorithm

In principle, any parameter that affects the ER of a stream could be used for shaping. For example, in the case of ON/OFF streams the mean burst volume and/or the mean interburst distance and/or the peak rate within bursts could be controlled appropriately with the aim of keeping the ER of the produced stream (as given by (1)) constant. Since, however, it is difficult to efficiently control statistical averages, we choose to control the peak rate for that purpose. The key idea here is the implementation of a shaping function (2) in terms of results derived off-line (like those of figures 2, 3). This data (giving either c, or $1/c$) is tabulated and retrieved quickly for any combination of parameter values, which are estimated through on-line measurements. Figure 4 shows the environment of such a shaping unit as well as the main interfacing signals.

The proposed shaping process is as follows :

Step 1. Get measurements of the burst volumes and the interburst intervals of the stream at the output of the local buffer over an estimation window.

Step 2. Using the measured samples, assess the mean burst volume and the mean interburst time.

Step 3. Choose a suitable value for the peak rate at which the next burst is emitted, in order to regulate the effective rate of the produced stream close to the agreed value.

Step 4. Segment the bursts that exceed in volume a certain (dynamically adjusted) upper bound and enforce a silence period between the segments.

Specific implementations of steps 2 and 4 are presented in the following two subsections, while a hardware-oriented algorithm is given in [9].

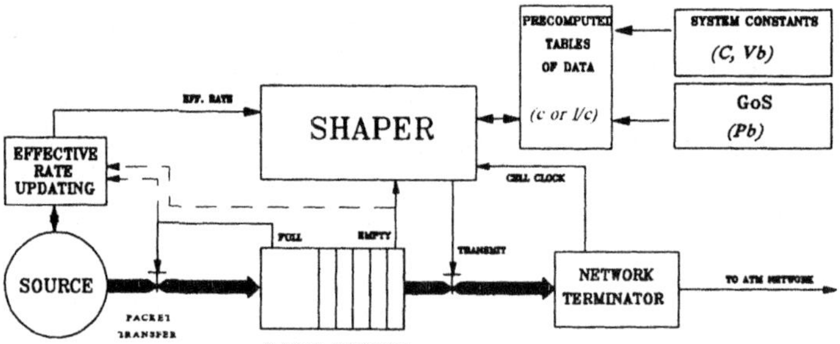

Figure 4: Environment of an ER enforcement device

4.2. Parameter estimation

In order to apply the shaping function (2), we need estimates of the V and T parameters (step 2 of the shaping algorithm). With reference to Figure 4, these parameters concern the traffic stream generated at the output of the local buffer, and not the packet stream generated by the traffic source. Packets may be emitted back-to-back, forming bursts of larger volumes. Burst delineation is therefore performed by means of the EMPTY signal of the local buffer.

Estimates of the V and T can be obtained by a simple, finite-window, averaging estimator, e.g.

$$\left.\begin{aligned} \overline{V} &= \frac{V_1 + V_2 + \ldots + V_w}{w} \\ \overline{T} &= \frac{T_1 + T_2 + \ldots + T_w}{w} \end{aligned}\right\} \tag{3}$$

where V_i, T_i, i=1,2,..., w, are samples taken within a time window of size w. The latter is expressed in number of bursts and not in absolute time values. Thus the window is floating over an integer number of ON/OFF cycles, being self-adaptive to changes of the traffic statistics. The window moves smoothly with time, by including a new sample at a time in the place of its oldest sample (this is different from the jumping window estimation, where adjacent windows do not overlap).

It is well known from Statistics that the estimators (3) are unbiased (for any w) and consistent (as $w \rightarrow \infty$), provided that the samples are independent and the traffic source is stationary. However, in practice there is no guarantee about the stationarity of the source and, hence, a small window size is preferable in order to follow possible changes of the source statistics. This becomes more important from the fact that the rate shaping is applied with a time lag relatively to the measured V and T samples (the V_i and T_i of past cycles are used to determine the cell rate of forthcoming bursts). This lag, which is proportional to the window size, must be kept as small as possible. Simulation results have shown that a window size of 8 samples gives a good balance between the confidence to the estimates (for exponentially distributed samples the coefficient of variation of the estimates in (3) is inversely proportional to $\sqrt{w}$) and the need for keeping the window size small.

4.3. Long burst segmentation and silence enforcement

One of the basic modelling assumptions used in the fluid-flow analysis is that the underlying chain is Markovian. In our case this means that the ON and OFF periods of the multiplexed streams are exponentially distributed. However, although we can realistically assume an exponential distribution for the interburst distance (i.e. the full ON-OFF cycle duration) and, approximately, the same for the OFF period (provided that the source is a burtsy one), we can hardly assume the same for the burst volumes. Even if it holds true for the packets generated by the source, the outcoming bursts may follow a different distribution (hyper-exponential) due to the phenomenon of consecutive packet merging, which occurs during overload periods.

Experiments have shown that the distribution of the burst volume plays a very significant role in the multiplexer's performance. Specifically, a hyper-exponential distribution usually gives rise to a worse performance than the exponential one with the same mean (see Figure 6 in the performance results subsection). This is so because a hyper-exponential distribution is composed of parallel exponential stages and the components with the smaller (in magnitude) eigenvalues, which correspond to longer bursts, prevail in the tail (see example in Figure 5). An analytical approach, however, is not that easy to be carried through in this case. Some sophisticated models that can handle specific non exponential distributions,[8], are not suitable for on-line implementations (especially in hardware) due to their computational complexities. The method that is suggested here consists in establishing an upper bound for the volume of bursts and enforcing a silence period whenever this bound is exceeded. Both, the burst volume bound and the enforced silence duration are determined dynamically in terms of the current estimates, as in the following.

Consider a stationary source which generates bursts according to a certain (hyper exponential) distribution with a mean V_0. Consider also the mean volume estimator (3) at the output of the shaper, which we assume to give a volume estimate equal to $\overline{V}'$ at the beginning of an ON period.Whenever the burst currently emitted exceeds in volume a certain multiple of the current estimate, i.e. $V > k\overline{V}'$, it is segmented into as many segments of size $k\overline{V}'$as necessary and a remnant, which is less than or equal to $k\overline{V}'$. Between the segments a silence period is enforced equal to the corresponding mean silence estimate. It must be stressed that the estimator applies on the segmented-burst streams, not on the original one, so that $\overline{V}'$ normally changes from segment to segment as the estimation window moves forward. The segment size, however, is not updated during segmentation. k is referred to as the *segmentation factor*.

<u>*Assertion 1*</u> (See [9] for a proof): For any $V_e > 2V_0$ there can be found a segmentation factor $k>1$ such that the burst volume of the stream after segmentation (characterised by the random variable v' with mean V_o') exhibits a hypo-exponential tail distribution (above V_e), i.e.

$$\Pr\{v' > V\} < e^{-\frac{V}{V_o'}}, \quad \forall\, V > V_e \tag{4}$$

A sufficient condition for k is given by:

$$k \le n/(\sqrt{\frac{2n}{w}} + 1) \quad with \quad \equiv \frac{V_e}{V_o} \tag{5}$$

The practical significance of (5) is that it gives a very simple figure for controlling the burst volume tail distribution of the shaped stream. Obviously, for a certain n (which can be chosen empirically, e.g. in the range 5-20), the largest allowable k must be used in order to minimise the delay caused by segmentation and silence enforcement.

Figure 5 is a demonstration example of the suggested segmentation method. The original burst stream is a hyper-exponential one, composed of two parallel stages with rates 1/20 and 1/95, respectively. The second stage is significant for the tail of the distribution, which is located considerably above the exponential one with the same mean (equal to 50, in this case). There are also shown in Figure 5 the complementary error functions , 0.5*erfc(V/k), for various k-values, which constitute upper bounds for the corresponding distributions of the bursts after segmentation, as shown in [9]. The simulation results for the case of k=5, which are also shown in Figure 5, confirm the analysis.

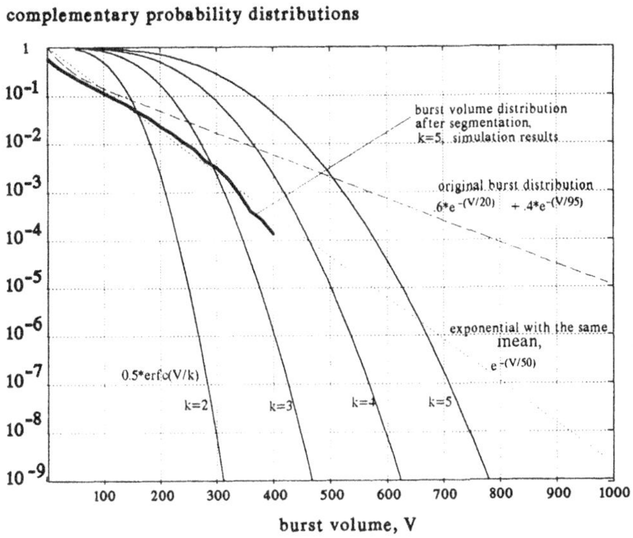

Figure 5 : Hyperexponential (two parallel stages), exponential (with the same mean), complementary error functions and distribution of the segmented bursts

4.4. Performance simulation results

Simulation experiments have been carried out which validated the proposed ER enforcement algorithm. In each experiment, a number of independent traffic streams were generated, shaped according to the rate control and burst segmentation procedure described above and multiplexed together on the same output link. With the enforced effective rates summed up to unity, the resulted *CPDF(x)* (assuming an infinite buffer configuration) was found to pass close by the specified QoS point (V_b, P_b), as in the examples of Figure 6. For very low values of P_b, the *CPDF(x)* gives approximately the cell loss probability in a finite buffer of capacity x.

Figure 6(a) has been derived with exponential sources without activating the segmentation procedure. Two cases are shown, one with 67 streams, shaped with a common R_e, equal to 0.015 (=1/67), and the other with two different R_e values (33 streams , shaped with R_e=0.015, and 17 streams, shaped with R_e=0.03). For both cases the driving sources were identical, with V=50 and r=0.005. The performance simulation curves, as we can see from Figure 6(a), lie close to the theoretically expected one for the 67 identical streams, calculated by the fluid flow analysis for a constant peak rate equal to 0.116.

Figure 6(b) shows respective results derived by using hyper-exponential sources with and without burst segmentation. Each source is Erlang type, composed of two exponential stages in parallel, with volume constants V_1=20 (with a probability 0.6) and V_2=95 (with a probability 0.4), respectively. The curves of Figure 6(b) correspond to 33 multiplexed sources, shaped with R_e=1/33, with a segmentation factor $k=\infty$ (no segmentation) and k=5. We can see that the former is far from satisfying the specified QoS. For comparison purposes, the curve corresponding to the same settings with exponential sources is also shown. The burst volume distributions for the three cases are shown in Figure 5.

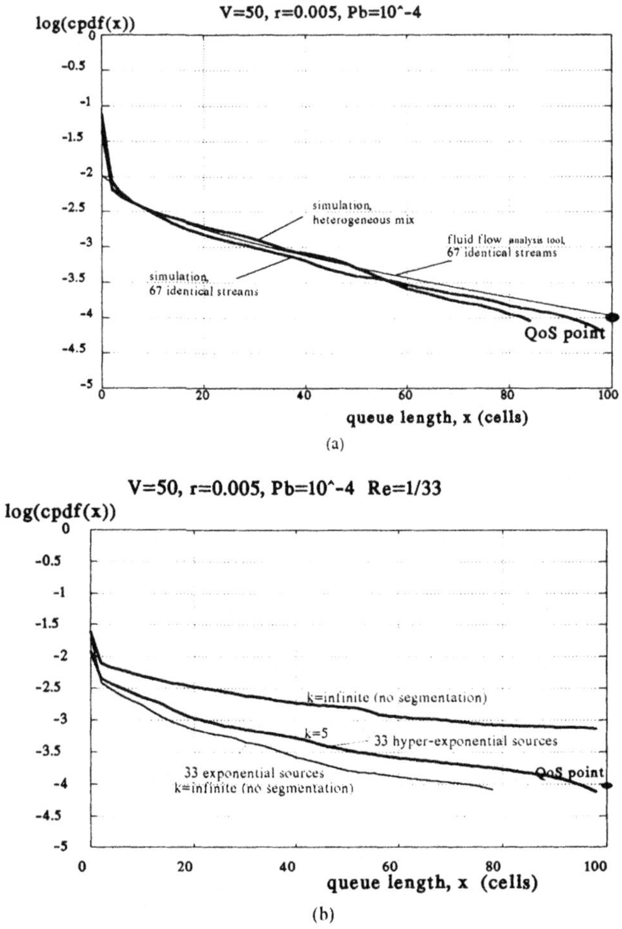

Figure 6: Simulated performance of the shaping algorithm
(a) exponential (b) hyper-exponential sources

5. ARCHITECTURE OF THE SHAPING UNIT

5.1. General description

Figure 7 shows the architecture of the shaper in a block diagram form, consisting of six main blocks. The BURST-VOLUME ESTIMATOR and the INTERBURST-INTERVAL ESTIMATOR blocks get samples of the burst volume and the interburst time distance, correspondingly, and realise the finite-window estimators (3), as explained in the next subsection. Having obtained estimates of the parameters V and T, the intercell-distance TABLE is accessed through the TABLE ACCESS CONTROL unit and the appropriate value of $1/c$ is retrieved. Since the table contains values of $1/c$ on a discrete grid of points in the R_e-V-T space, either an interpolation process is performed by the INTERPOLATOR or a conservative value of $1/c$ is retrieved from the table by accessing the nearest tabulated point with the smallest R_e, the largest V and the smallest T. Here we follow the second approach. The intercell distance is used by the TRANSMIT CLOCK GENERATOR unit to trigger the transmission of cells from the local buffer periodically. The BURST SEGMENTATOR block is responsible for comparing the segment currently transmitted with an upper bound (V-bound, calculated by the product of the current volume estimate times the segmentation factor), masking transmission as soon as the former exceeds the latter and re-enabling transmission after the expiration of an appropriate silence interval. Details of the architecture of the main blocks are given in the next subsections.

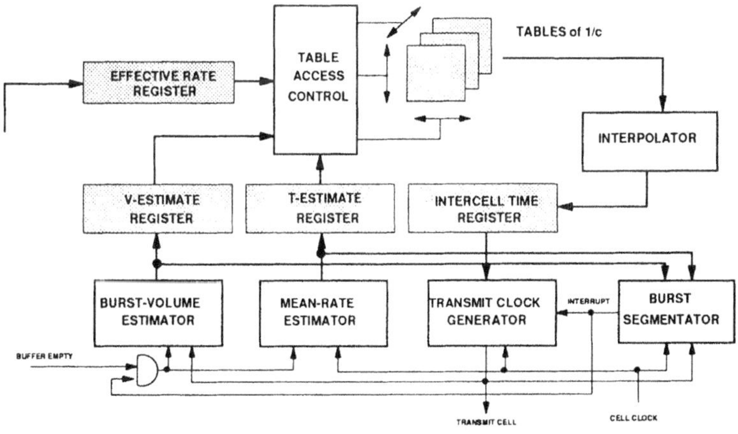

Figure 7: Block diagram of the shaping device

5.2. The Burst-Volume and the Interburst-Interval Estimators

The purpose of these two units is to monitor the ON/OFF cycles of the produced stream at the output of the local buffer (by sensing the EMPTY signal, see Figure 4), to get measurements of the number of cells (V_i) and the duration (T_i) of each ON/OFF cycle (i) and to calculate estimates of the mean values V and T, respectively, using relation (3). A generic structure of both units is shown in figure 8.

A bank of registers is used to hold the samples of the current window, the start of which (oldest sample) is being addressed by the *sample index*. The new sample is measured by a counter, which is reset at the beginning of each ON/OFF cycle and driven by the appropriate signal. Another register is used to hold the sum of the samples, which gives a multiple of the sought estimate. At the end of the ON/OFF cycle the counter contains a valid new sample, which updates the estimate and replaces the oldest one in the register bank. This is performed in two steps, as shown in Figure 8. At the first step the oldest sample is subtracted from the maintained sum of samples while, at the second step, the new sample is added to the sum and also transferred into the oldest sample's place in the register bank. At the beginning of the next ON/OFF cycle the counter is reset and the sample index advances by one position, thus implementing the window movement.

5.3. The Table-Access-Control unit

This unit gets the V and T-estimates together with the agreed effective rate value, R_e, and returns a value for the intercell distance, $1/c$. First, each of the two estimates and the R_e value is positioned on the corresponding axis by means of successive comparisons with the grid elements (V__grid, T__grid, and ER__grid respectively). Thus the point having the closest lower R_e, the closest upper V, and the closest lower T, is determined in the R_e- V- T space (point C_1 in figure 9(a)) and the corresponding $1/c$ value is retrieved. This is a conservative approach, leading to a lower rate than actually needed. If a higher accuracy is required, an interpolation process could be implemented between the points A,B,C,D on each of the planes, and then between the two planes. Should however, a dense grid be implemented, the complexity introduced by such an interpolation process hardly is justified. In figure 9(b) a piece of code is given for the table access algorithm implementing the conservative approach (with no interpolation).

5.4. Selection of the grid

The selection of a suitable grid of points in the R_e-V-T space is very critical, since it will determine the amount of storage required to hold the $1/c$ values. It must be emphasised that each specific multiplexing set up (i.e. certain V_b and C) combined with a specific QoS (i.e. certain cell loss probability, P_b) is associated with its own individual table of $1/c$ values and any change to the above parameters would require a new table. Normally, what may change in a certain network configuration are the link rates and the required cell loss probabilities associated with Virtual Paths, when a dynamic bandwidth allocation scheme is implemented. Such cases should be treated with care in order to avoid inconsistencies between the stored tables and the actual multiplexing set up.

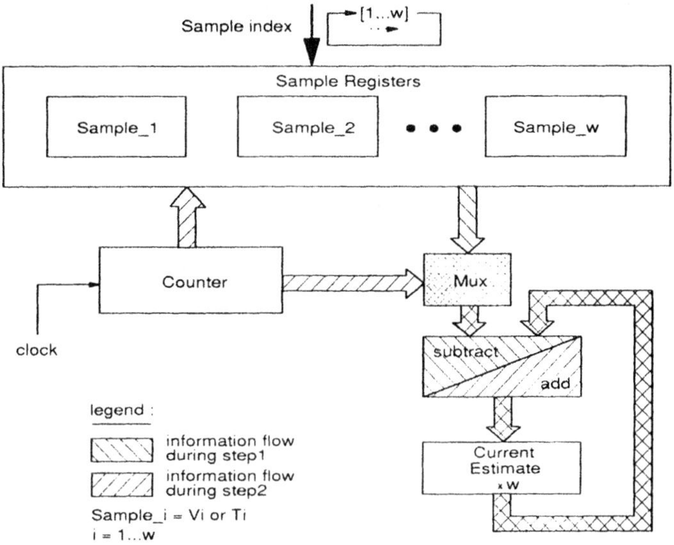

Figure 8 : Block diagram of the *V*,*T*-estimators

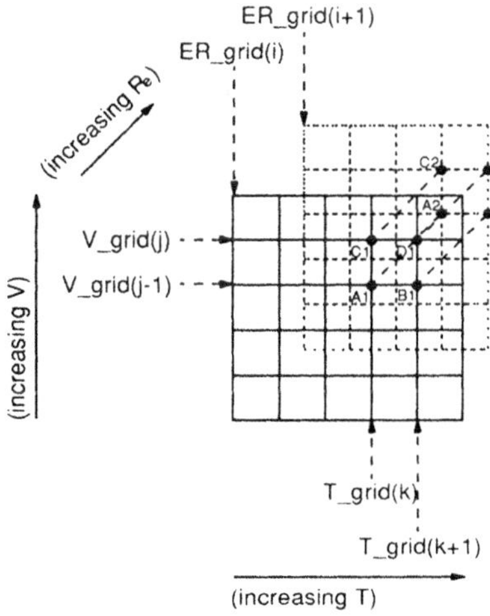

(a)

Code for the table access algorithm

```
i=1, j=1, k=1 ;
while (Re> ER_grid(i)), next i ;
if i>1, let i=i-1
while (V_estimate > V_grid(j)), next j ;
while (T_estimate < T_grid(k)), next k ;
if k > 1, let k=k-1 ;
let c_inverse = TABLE(i,j,k) ;
```

(b)

Figure 9 : Table access mechanism
(a) grid configuration (b) access algorithm

One approach would be to store many tables corresponding to different configurations and/or QoS values. In such a case any saving of memory is welcome.

As it can be observed from Figure 3, the usage of a logarithmic scale for the parameters *V* and *T* leads to a wide range coverage and a smooth variation of the $1/c$ curves. Furthermore, not all of the combinations (R_e, V, T) within the selected range are valid, since it is

$$R_e > r = \frac{V}{T} \qquad or \qquad T > \frac{V}{R_e} \tag{6}$$

Therefore, for any value of *V*, we use a *T*__grid satisfying (6). For storage economy reasons, we do not use a separate *T*__grid for each *V* value; instead, we use a common one extending all over the *T* range and choose the appropriate starting point for each *V*. Specifically, with each *V*= *V*__grid(j) we associate a pointer, *p(j)*, pointing to the first valid *T*__grid element satisfying (6) (Figure 10). With such an arrangement, the last code line of the access algorithm in Figure 7(b) should be modified as :

let c__inverse=TABLE(i,j,k-p(j)+1)

A 50% saving of memory has been obtained by this method in specific realisations.

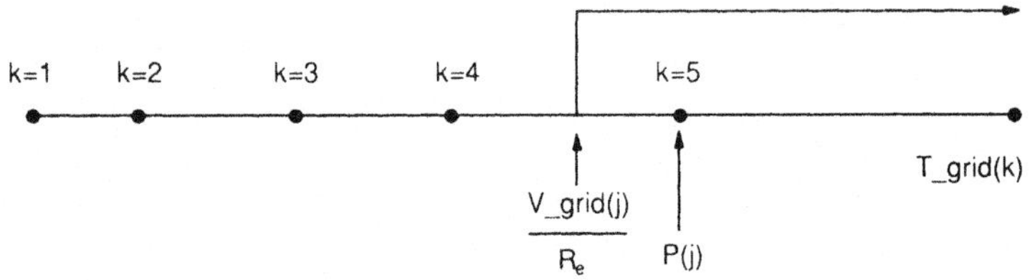

Figure 10 : T__grid and the starting pointer, *p(j)*

6. VLSI IMPLEMENTATION

The Traffic Shaper, described in the previous section, has been designed as a VLSI chip using the ECPD15 CMOS technology of ES2. The standard-cell design tool SOLO 1400 has been used for that purpose, resulting in a silicon area of approximately 20 square millimetres. The main features of the realised design are the following:

- Byte-parallel output with a transmission clock up to 20 Mbytes/sec
- Programmable window size up to 8
- V sample registers of 10 bits (V max $\approx$ 1000 *cells*)
- T sample registers of 19 bits (T max $\approx$ 500000 *slots*)
- Interface to access a 16x128x128 (R_e-V-T) table of $1/c$ values. Each $1/c$ value is stored in a fixed point format 8.2 (8 bits for the integer and 2 bits for the fractional part).
- Ability to realise cell rates corresponding to non-integer intercell distances with a decimal precision equal to 0.25.

The last feature, although not that important for low cell rates, may become quite significant for large rates.

7. CONCLUSION

The effective rate of an ATM traffic stream in a multiplexer depends on the statistical properties of the stream as well as on the multiplexer's dimensions (buffer space and link rate) and the desired QoS (cell loss probability). ON/OFF exponential streams can be described by three parameters, namely the mean burst volume, V, the mean interburst distance, T, and the cell rate during bursts, c. By fixing the dimensions of the multiplexer and the QoS, it is possible to derive a shaping function $c=f'(V,T,R_e)$ through applying the fluid flow method (or any appropriate analysis tool) on a discrete grid of points in the (V,T,R_e) space. This function gives the cell rate at which the bursts of the stream should be emitted in order to maintain a certain effective rate, R_e.

Based on tables of values of the shaping function it is possible to control the effective rate of a stream at network entrance points. A buffer is used to decouple the cell rates of incoming and outgoing bursts. The effective rate enforcement method proposed here consists in getting measurements at the output of the decoupling buffer, estimating the parameters V and T and determining the suitable cell rate by accessing the corresponding entry of the stored tables. The next burst is then emitted at the so-determined cell rate. A long-burst-segmentation process is also performed which guarantees a hypo-exponential distribution (in the tail) for the emitted bursts. Theoretical bounds for such a segmentation have been presented.

REFERENCES

1. CCITT Recommendation I.150 on "B-ISDN Asynchronous Transfer Mode, Functional Characteristics", *Temporary Document 65*, Geneva 1992.

2. CCITT Recommendation I.371, "Traffic Control and Congestion Control in B-ISDN", *Temporary Document 64*, Geneva 1992.

3. K. Mase and S. Shioda, "Real-Time Network Management for ATM Networks," in J.W. Cohen and C.D. Pack, editors, *Queueing*, .

4. T. Murase, H. Suzuki, S. Sato and T. Takeuchi, "A Call Admission Control Scheme for ATM Networks Using a Simple Quality Estimate," *IEEE JSAC*, Vol. 9, No 9, pp. 1461-1470, Dec. 1991.

5. R. Guerin, H. Ahmadi and M. Naghshineh, "Equivalent Capacity and its Application to Bandwidth Allocation in High-Speed Networks," *IEEE JSAC*, Vol. 9, No. 7, pp. 968-981, Sept. 1991.

6. N.M. Mitrou and K.P. Kontovasilis, "Using Effective Rates for Traffic Description and Control in ATM Networks," submitted for presentation in the *5th International*

Conference on Data Communication Systems and Their Performance, Oct. 25-28, 1993, Research Triangle Park, North Carolina.

7. D. Anick, D. Mitra and M.M. Sondhi, "Stochastic Theory of a Data-handling System with Multiple Sources," *The Bell System Technical Journal*, Vol. 61, No. 8, pp. 1871-1894, Ocober 1982.

8. I. Norros, J. Roberts, A. Simonian and J. Virtamo, "The Superposition of Variable Bit Rate Sources in an ATM Multiplexer," *IEEE J-SAC* Vol. 9, No 3, pp. 378-387, April 1991.

9. N.M Mitrou, A.K. Koukos and S. Perissakis, "Analysis and VLSI Implementation of an Effective-Rate Control Algorithm for ATM Streams," submitted for publication in the *Computer Networks and ISDN Systems*, 19993.

10. K.P. Kontovasilis and N.M. Mitrou, "Bursty Traffic Modeling and Efficient Analysis Algorithms via Fluid-Flow Models for ATM IBCN," to appear in the *Annals of Operations Research*, special issue on Methodologies for Performance Analysis of High Speed Networks.

11. K.P. Kontovasilis and N.M. Mitrou, "Bursty Traffic Description via Effective Rates and Associated CAC and Traffic Enforcement Algorithms," RACE document NTUA_123_0044_CC_CC, June 1992.

Integrated Broadband Communication Networks
and Services (C-18) / V.B. Iversen (Editor)
Elsevier Science B.V (North-Holland)

Effective bandwidths with different quality of service requirements

N. G. Bean*

University of Cambridge, Statistical Laboratory, 16 Mill Lane, Cambridge, U.K.

Abstract

The concept of an *effective bandwidth* for controlling ATM networks with multiple source classes is quite well understood, for example see [6, 5, 7]. However, there are two main gaps in the literature. Firstly, the concept of an effective bandwidth has mainly been investigated under the cell loss quality of service (QoS) measure describing the proportion of time that cell loss occurs, which we call the *resource based QoS measure.* We feel that it is essential to extend the concept to the QoS measure describing the proportion of cells that are lost, which we call the *stream based QoS measure.* Secondly, there has been no investigation of effective bandwidths when the different source classes require radically different qualities of service. This investigation is important because certain types of call will require almost error-free transmission (such as HDTV or real-time data transfer) and other calls will be willing to receive a much lower QoS for a reduced price (such as telephony). The following work applies to any stationary source model with a finite number of possible bandwidth requirements and the bufferless resource model of Hui [6].

Keyword Codes: C.2.0
Keywords: Computer-Communication Networks, General

1. INTRODUCTION TO EFFECTIVE BANDWIDTHS

The concept of an effective bandwidth received general acceptance after the companion papers of Kelly [7] and Gibbens and Hunt [5]. They showed, for a variety of systems, that the required QoS is guaranteed if $\boldsymbol{n} = (n_1, n_2, \ldots, n_J) \in \mathbf{Z}_+^J$ satisfies,

$$\sum_{j=1}^{J} \alpha_j n_j \leq C, \tag{1}$$

where n_j represents the number of calls in progress from source class j. C is interpreted as the capacity of the resource, and α_j as the *effective bandwidth* at the resource of a source of class j. The effective bandwidth may depend on the characteristics of the sources and the resource, but is *independent* of the number of calls in progress.

The basic idea behind the concept of an effective bandwidth is that the acceptance region is approximately linearly constrained. The approximate linearity of the acceptance region for an ATM resource was first noted by Hui [6].

*Supported by The George Murray Scholarship and The Cambridge Commonwealth Trust

The state space of a circuit switched network with a link of capacity C and each call of type j requiring α_j circuits is also given by Eq. (1). Therefore, an ATM network can be thought of as operating *as if* it were a circuit switched network with these requirements. This implies that there is potential for numerous insights from the mature literature of circuit switched networks (for example see Kelly [8]) to be applicable to the developing broadband networks. This explains the strong research interest in the concept of an effective bandwidth.

Most papers in this field use buffered resource models. We use the bufferless model of Hui [6] with the intention of controlling cell loss by controlling bandwidth saturation. This model has the advantage of being able to ignore the correlations over time and naturally directs the research at the burst level as buffers will be too small to manage burst level saturation. However, the proposal of the Stop-and-Go service discipline at switching nodes and the associated framing concept [14] also supports the use of bufferless models at the cell level. The framing concept has arisen to provide bounded end-to-end delay and controlled delay jitter for delay sensitive traffic. The concept is that any cell that can not be transmitted in the current frame should be discarded, rather than buffered for later frames as this can cause more and more cells to be delayed. In other words, it is often better to lose a few cells than cause a significant delay to a large number of cells. Under this type of service discipline the cell loss probability does not depend on the accumulation of cells due to earlier behaviour and so the bufferless resource model of Hui [6] is applicable. In this paper we are motivated by the burst level and use parameters that are suitable for that level but are aware that the techniques may also be suitable at the cell level.

In Section 2 we show that the concept of an effective bandwidth is applicable under the *stream based QoS measure*. In Section 3 we investigate the issues involved in accepting calls with radically different QoS requirements at the same resource. We propose four techniques, designed to ensure that each source class receives its required QoS, and show that one simple technique is able to offer all the efficiency gains of the best possible techniques with almost none of the overheads. In Section 4 we present a numerical investigation and show that effective bandwidths are applicable under these techniques. In Section 5 we give our conclusions about this research.

2. STREAM BASED QUALITY OF SERVICE MEASURE

In this section we develop the concept of an effective bandwidth for the simple model of a bufferless resource due to Hui [6]. However, we consider the *stream based QoS measure*, rather than the *resource based QoS measure* used in most papers on effective bandwidths. We define the source models and the model of the resource and then elaborate on the two different QoS measures. Finally, we investigate the effective bandwidth concept.

There are J source classes labelled $j = 1, 2, \ldots, J$. Each source is stationary and is modelled as an independent integer-valued non-negative simple random variable: that is, a random variable with only a finite number of possible outcomes. A source of type j has a peak bandwidth requirement of $a_j \in \mathbf{Z}_+$ and a mean bandwidth requirement of $a_j x_j$. The a_j could be in units of bits per second or any other measure. The resource has capacity (or bandwidth) C and let n_j describe the number of calls in progress of type j. We are interested in the acceptance region, $\mathcal{A}$: the largest subset of $\mathbf{Z}_+^J$ such that if $\boldsymbol{n} = (n_1, n_2, \ldots, n_J) \in \mathcal{A}$ then the necessary QoS requirements are obeyed.

Let the random variable X_{ji} describe the bandwidth requirement of source i of type j, where $i = 1, 2, \ldots, n_j$ and $j = 1, 2, \ldots, J$, and the X_{ji} are independent of each other. The total bandwidth required by all $\boldsymbol{n}$ sources is then given by the integer-valued random variable,

$$S(\boldsymbol{n}) = \sum_{j=1}^{J} \sum_{i=1}^{n_j} X_{ji}.$$

We use the Chernoff Approximation [1] to approximate the tail probabilities of $S(\boldsymbol{n})$ and so consider approximations to the acceptance region. Define $M_j(s)$ to be the logarithmic moment generating function of the random variables X_{ji}, and so

$$M_j(s) = \log \mathsf{E}\left[\exp(sX_{ji})\right], \tag{2}$$

where log represents the natural logarithm and E the expectation operation. If $\sum_{j=1}^{J} n_j a_j > C$ and $\sum_{j=1}^{J} n_j a_j x_j < C$, the Chernoff approximation states that,

$$\mathsf{P}(S(\boldsymbol{n}) \geq C) \approx \exp\left(\inf_{s \geq 0}\left[\sum_{j=1}^{J} n_j M_j(s) - Cs\right]\right). \tag{3}$$

For finite $\boldsymbol{n}$ this is an upper bound, and it is exact in the limit as $n_j \to \infty$ for all $j = 1, 2, \ldots, J$, with C increasing in proportion.

In [6, 5, 7] the authors use as their QoS measure the proportion of time that the resource is saturated, or that cells are lost. We call this the *resource based QoS measure* because it is looking from the viewpoint of the resource at the probability of saturation or cell loss. This is also referred to as P_{sat} in [13]. We contrast this with the proportion of cells that are lost. We call this the *stream based QoS measure* because it is looking from the viewpoint of the stream, or each source, at the probability that an individual cell will be lost. This is referred to as P_{loss} in [13], where an initial investigation into the concept of an effective bandwidth under this QoS measure is undertaken. The intention is to find a very simple expression for the effective bandwidth involving just the mean and peak bandwidth requirements, or the mean and variance of the bandwidth requirement.

The *stream based QoS measure* is likely to be more useful in practice, since it will be important to know more about the cell loss process than just how often loss occurs. In fact, the stream based measure is also likely to be inadequate as many types of traffic will find more refined measures, such as the probability of consecutive cell loss, very important.

The *resource based QoS measure* has been favoured by previous authors for good reason; it only requires knowledge of the the tail probability of saturation, rather than anything more detailed, and so is readily available from the Chernoff approximation. In contrast, the *stream based QoS measure* involves the proportion of cells that are lost and so requires knowledge of how many cells are offered and how many cells are lost. The stream based measure is clearly less analytically tractable and less numerically attractive than the resource based measure. However, it is necessary to address this issue before the effective bandwidth concept can be considered for use in a real network.

The expected number of offered cells is given by,

$$\mathsf{E}\left[S(\boldsymbol{n})\right] = \sum_{j=1}^{J} n_j \mathsf{E}\left[X_{j1}\right] = \sum_{j=1}^{J} n_j a_j x_j,$$

and the expected number of lost cells is given by,

$$\mathsf{E}\left[(S(\boldsymbol{n}) - C)^+\right] = \sum_{k=C}^{L} (k - C)\mathsf{P}(S(\boldsymbol{n}) = k),$$

where $L = \sum_{j=1}^{J} n_j a_j$, but will usually be omitted from the notation. Since $S(\boldsymbol{n})$ is a positive integer-valued random variable, this can be simplified to

$$\mathsf{E}\left[(S(\boldsymbol{n}) - C)^+\right] = \sum_{k>C} \mathsf{P}(S(\boldsymbol{n}) \geq k). \tag{4}$$

The Chernoff Approximation, Eq. (3), can be used to approximate this by,

$$\mathsf{E}\left[(S(\boldsymbol{n}) - C)^+\right] \approx \sum_{k>C} \exp\left(\inf_{s \geq 0}\left[\sum_{j=1}^{J} n_j M_j(s) - ks\right]\right). \tag{5}$$

Let A represent the QoS required by each source class. Under the *resource based QoS measure* the requirement becomes,

$$P_{\text{sat}} \triangleq \mathsf{P}(S(\boldsymbol{n}) \geq C) \leq A.$$

Under the *stream based QoS measure* the requirement becomes,

$$P_{\text{loss}} \triangleq \frac{\mathsf{E}\left[(S(\boldsymbol{n}) - C)^+\right]}{\mathsf{E}\left[S(\boldsymbol{n})\right]} \leq A,$$

or equivalently,

$$\mathsf{E}\left[(S(\boldsymbol{n}) - C)^+\right] \leq A \sum_{j=1}^{J} n_j a_j x_j.$$

$Q(\boldsymbol{n}) = A \sum_{j=1}^{J} n_j a_j x_j$ is the necessary QoS requirement under the stream based QoS measure and we define the acceptance region induced by the Chernoff Approximation as,

$$\mathcal{A} = \left\{\boldsymbol{n} \in \mathsf{R}_+^J : \sum_{k>C} \exp\left(\inf_{s \geq 0}\left[\sum_{j=1}^{J} n_j M_j(s) - ks\right]\right) \leq Q(\boldsymbol{n})\right\}. \tag{6}$$

In Figure 1 we numerically investigate the accuracy of the Chernoff Approximation and the linearity of the acceptance region, $\mathcal{A}$. The QoS parameter, A, is chosen to be 10^{-9} and we restrict the source models to two of the On-Off type so that we can make exact calculations. As a simple model of silence suppressed voice we use the On-Off model with $a_2 = 1$ (peak rate of 64 kbits/s) and $x_2 = 0.4$ [2]; as an attempt at modelling a video source we use an On-Off model with $a_1 = 200$ (a peak rate of 12.8 Mbits/s) and $x_1 = 0.3$. We choose the capacity, C, to be 10000 (or 640 Mbits/s).

In Figure 1 we show the exact acceptance region, and the acceptance region found using the Chernoff Approximation, $\mathcal{A}$. Note that the boundaries of both acceptance regions are highly linear, but in fact the complements of the acceptance regions are convex sets. Also, the Chernoff Approximation is very accurate at determining the shape of the acceptance region but is slightly

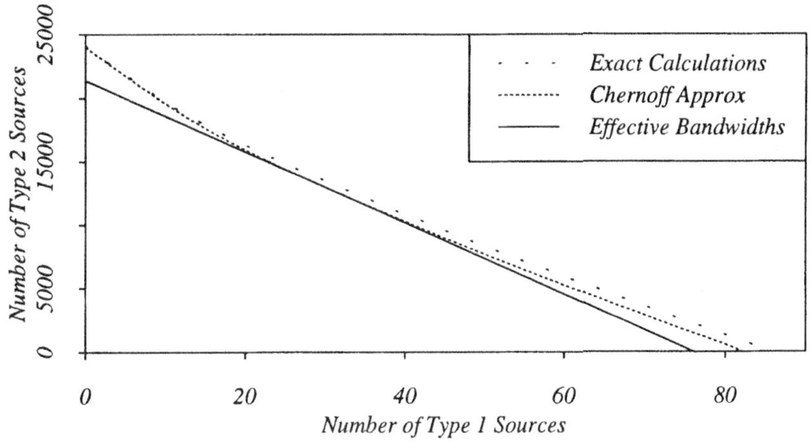

Figure 1. Acceptance Regions for the Stream based QoS measure

conservative. It is the near linearity of the boundary that is most important for the practical application of effective bandwidths as this implies that there exists a linearly constrained region that is a good, and conservative, approximation to $\mathcal{A}$.

What is the purpose of observing that the complement of $\mathcal{A}$ is a convex set? This implies that the tangent plane to $\mathcal{A}$ taken through a given point $\boldsymbol{n}^*$ on the boundary of $\mathcal{A}$ necessarily defines the best possible *linearly* constrained acceptance region through the point $\boldsymbol{n}^*$ that is entirely contained within $\mathcal{A}$. This linearly constrained acceptance region defines the effective bandwidth of each source exactly as in Eq. (1). It would perhaps make sense to choose $\boldsymbol{n}^*$ according to the expected traffic mix. Let $\boldsymbol{s}^* = [s_k^*]_{k>C} \in \mathbb{R}_+^{L-C}$ achieve the required infima in Eq. (6) with $\boldsymbol{n}$ replaced by $\boldsymbol{n}^*$. In other words, for all $k > C$,

$$\left[\sum_{j=1}^{J} n_j^* M_j(s_k^*) - k s_k^*\right] = \inf_{s \geq 0} \left[\sum_{j=1}^{J} n_j^* M_j(s) - ks\right]. \tag{7}$$

The co-ordinates of the normal to the tangent through $\boldsymbol{n}^*$ are given, for each $i = 1, 2, \ldots, J$, by,

$$D_i(\boldsymbol{n}^*) = \sum_{k>C} M_i(s_k^*) \exp\left[\sum_{j=1}^{J} n_j^* M_j(s_k^*) - k s_k^*\right] - \frac{\partial Q(\boldsymbol{n}^*)}{\partial n_i}, \tag{8}$$

and so the equation of the tangent plane can easily be found.

In Figure 1 we show the effective bandwidth approximation deduced from the Chernoff Approximation with $\boldsymbol{n}^* = (30, 12966)$. Note that it only depends weakly on the choice of $\boldsymbol{n}^*$ along the boundary of $\mathcal{A}$.

The effective bandwidth approximation can be improved by sharpening the estimate of a single point on the boundary, $\boldsymbol{n}^*$, through the use of large deviations theory [6, 3] or the saddlepoint approximation [4, 12] to more accurately approximate the exact acceptance boundary. The co-ordinates of the normal, Eq. (8), can then be used to find the equation of the tangent plane through $\boldsymbol{n}^*$.

3. DIFFERENT QUALITY OF SERVICE REQUIREMENTS

This section contributes to the field of connection acceptance control and bandwidth allocation for ATM networks with multiple source classes, where different source types require different qualities of service. It is well known that QoS parameters in the range of 10^{-9} are typical of ATM networks. However, for certain sources such as voice calls, this is far more strict than required. That is, regardless of other QoS issues, different sources will be willing to accept different cell loss probabilities. Accordingly, considerable efficiency gains may be available if different levels of QoS are offered. In the remainder of this section we investigate this issue and suggest a suitable scheme for achieving these efficiency gains. As in the previous section we consider the model of a bufferless resource due to Hui [6] and so concentrate on the burst and call levels, but recall our earlier comments on possible application at the cell level through the use of the Stop-and-Go service discipline [14].

Kraimeche and Schwartz [9, 10] and Ngo and Lee [11] consider the allocation of resources at the call or burst level. They consider various schemes with some delayable traffic and where wideband calls can be transformed (through either bandwidth compression or a tradeoff between transmission rate and time) into narrowband calls. In [9], Kraimeche and Schwartz also consider a moving boundary pre-emptive scheme.

In comparison, we consider real-time traffic and so assume that calls can not be queued or delayed but instead are lost. We determine rules for call acceptance and call level bandwidth allocation with the intention of controlling the QoS. The *stream based QoS measure* is used, as in the previous section. For simplicity we shall only consider the problem for a single network resource and one source type within each QoS class.

There are two obvious ways to allocate the bandwidth and achieve the correct QoS measures. Firstly, partition the bandwidth between the QoS classes and dimension the partitions to guarantee the QoS measures. This partition is maintained at all costs and we call the technique *no sharing* (NS). Secondly, partition the bandwidth between the QoS classes but allow the calls to use spare bandwidth in other partitions at the risk of being pre-empted. This technique is called *complete sharing* (CS) and achieves a considerable efficiency gain since there is greater scope for multiplexing. Also, the calls with low QoS requirements can often use the bandwidth kept in reserve for the other classes. Nevertheless, the cost of this efficiency is much greater complexity since the bandwidth partition is not calculated per source class but relies on a calculation involving all the traffic. In order to discuss the efficiency gains, we consider the reference model of offering only one QoS class and providing all traffic with the strictest QoS required, as in Section 2. This is known as the *one QoS* technique (1QoS) and is clearly the simplest technique, but not necessarily the least efficient at managing the bandwidth.

We consider the case where there are two different QoS classes. The analysis of only two QoS classes is supported by the fact that there is only one bit allocated to cell loss priorities in the header of a B-ISDN ATM cell. That is, the CCITT has decided against offering many priority options and have instead chosen a simple priority scheme. We support their choice as we feel that simplicity has to be the primary target with bandwidth efficiency as an important but secondary consideration. We show in this section that offering two classes will provide an opportunity for considerable efficiency gains with no appreciable complexity overheads. To offer more classes is a case of diminishing return; less gain will be achieved and more complexity will be required for every additional QoS class. We refer to the QoS classes as class 1, with a strict requirement of very low cell loss probability, $A_1 \sim 10^{-9}$, and class 2, with a less strict requirement of $A_2 \sim 10^{-4}$.

Considering the orders of magnitude difference in QoS requirements, there is little to be lost by giving absolute priority to one QoS class. This means that one QoS class has complete protection, in that they have access to all the bandwidth, and the other class receives no protection and just uses any spare bandwidth. We refer to this technique as *complete priority* (CP). We are particularly concerned with the comparison between complete sharing and complete priority and conclude that there is so little to be lost by complete priority that the additional effort required to partition the bandwidth for complete sharing is not warranted.

As in the previous section, we assume that the random variable describing the bandwidth requirement of a source of type j has logarithmic moment generating function $M_j(s)$ defined in Eq. (2). Accordingly, a source of type j has a peak bandwidth requirement of a_j, a mean bandwidth requirement of $a_j x_j$ and the resource has capacity (or bandwidth) C. For simplicity alone, we assume that there is only one source class per QoS class.

3.1. One QoS

The *one QoS* technique is the same as the technique proposed and analysed in Section 2. Only one QoS class is offered and so all calls must receive the strictest QoS required. The boundary of the acceptance region is defined as: for each value of n_1 choose the largest integer n_2 such that,

$$\mathsf{E}\left[(S(n_1, n_2) - C)^+\right] \leq Q(n_1, n_2),$$

where, $Q(n_1, n_2) = (n_1 a_1 x_1 + n_2 a_2 x_2) A_1$. This guarantees that the proportion of cells lost will be at most A_1 and hence much less than A_2.

3.2. No Sharing

In the *no sharing* technique the bandwidth is partitioned between the two classes. We use the term "partition" in the strict mathematical sense, namely that the two portions are disjoint and exhaustive. If the traffic from class j overflows partition j, then that traffic is lost.

If there are n_1 calls of type 1 in progress, the bandwidth allocated to the calls of type 1 must be at least C_1, where C_1 is the unique positive solution to,

$$\mathsf{E}\left[(S(n_1) - C_1)^+\right] = n_1 a_1 x_1 A_1,$$

and $\mathsf{E}\left[(S(n_1) - C_1)^+\right]$ is defined by Eq. (4) or the approximate version Eq. (5) with $J = 1$.

Therefore, bandwidth $C_2 = C - C_1$ is allocated to calls of type 2 and so the resource can support at most n_2 calls of type 2, where n_2 is the largest integer such that,

$$\mathsf{E}\left[(S(n_2) - C_2)^+\right] \leq n_2 a_2 x_2 A_2.$$

This defines both the acceptance region and the bandwidth partitions required to operate the resource and guarantee the required QoS.

It is not clear at this point which of these two techniques will be the least efficient at managing the bandwidth. The one QoS technique has the advantage of using the entire bandwidth for multiplexing, but does not make use of the different QoS parameters. Whereas, the no sharing technique does make use of the different QoS levels but, in doing so, partitions the bandwidth and loses some of the multiplexing gain.

3.3. Complete Sharing and Complete Priority

Complete sharing involves the entire bandwidth being shared on a demand basis by all the traffic sources. However, when the resource is congested the bandwidth is partitioned to ensure that the correct QoS is received by each QoS class. In contrast, *complete priority* is defined as calls of type 1 have access to all the bandwidth and calls of type 2 use any spare bandwidth at the risk of being pre-empted. The benefits from these techniques are obvious: the resource not only benefits from increased multiplexing gain due to sharing the whole capacity, but also from the low priority calls using the spare bandwidth kept in reserve for the high priority traffic.

The necessary QoS requirement for (n_1, n_2) sources, $Q(n_1, n_2)$, is given by,

$$Q(n_1, n_2) = \begin{cases} n_1 a_1 x_1 A_1 + n_2 a_2 x_2 A_2, & \text{complete sharing;} \\ n_2 a_2 x_2 A_2, & \text{complete priority,} \end{cases}$$

where these are expected to be very similar because $A_1 \lll A_2$.

The expected number of lost cells is given by Eq. (4) or the approximate version Eq. (5). Therefore, assuming there are n_1 calls of type 1, the resource can successfully transmit at most n_2 calls of type 2, where n_2 is the largest positive integer such that,

$$\mathsf{E}\left[(S(n_1, n_2) - C)^+\right] \le Q(n_1, n_2).$$

Note that for complete sharing, cells may be lost from type 1 and type 2 calls. However, for complete priority, all the lost cells will be from calls of type 2; unless n_1 is sufficiently large that $n_1 a_1 > C$. In this case cells may be lost from both streams and as long as the upper bound on n_1, $\mathsf{E}\left[(S(n_1) - C)^+\right] \le n_1 a_1 x_1 A_1$, is obeyed both QoS requirements will be met. The same upper bound is also required for the complete sharing technique.

Determining the correct bandwidth partition for the complete sharing technique is a difficult task. If the traffic models are any more complicated than the simple "On-Off" model then determining the partition becomes nearly impossible.

4. NUMERICAL INVESTIGATION

Having developed these techniques, we investigate their performance. For the purpose of this investigation we choose the two QoS parameters to be $A_1 = 10^{-9}$ and $A_2 = 10^{-4}$. All other parameters are the same as in Figure 1.

There are two very striking observations to be made about Figure 2. The acceptance boundary for the complete sharing and complete priority techniques are indistinguishable (hence plotted only once) and very nearly linear. On closer examination it can be seen that the complements of the acceptance regions are convex sets and the non-linearity is greatest in the region where there are very few calls of type 1. This is because there are so few calls of type 1 that essentially no multiplexing can occur. There is still multiplexing between the two types of traffic, but very little within the calls of type 1. This explains why the non-linearity is far more noticeable in the no sharing case because there is no multiplexing gain to be achieved through the intermixing of traffic types. Note that the Chernoff Approximation exhibits this feature and determines the shape of the curve very accurately, but is conservative.

We note, for this example, that the one QoS technique is generally superior to the no sharing technique. This can be explained by the fact that the multiplexing gain achieved through

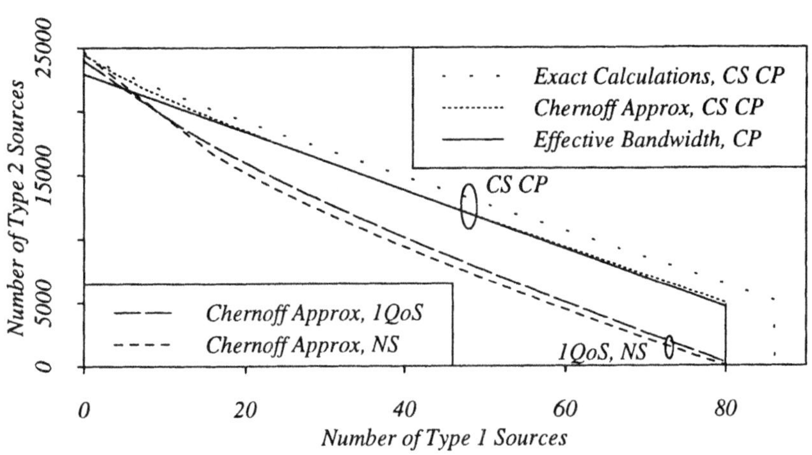

Figure 2. Acceptance Regions.

intermixing the traffic is greater than the gain achieved by offering the lower QoS class and keeping the traffic apart.

In Figure 2 we also show the effective bandwidth approximation to the complete priority technique. It is highly accurate and depends only weakly on the choice of $(n_1^*, n_2^*) = (30, 16064)$. Figure 3 shows the exact cell loss probabilities achieved by *complete priority* when this effective bandwidth approximation is used. For both classes, there seems to be a strong attraction for the cell loss probability to lie one order of magnitude below the intended value. This could be reduced by determining a choice of (n_1^*, n_2^*) that is nearer the exact acceptance boundary.

We also investigate the less likely, and diametrically opposed, scenario of the high bandwidth calls requiring less protection from cell loss than the low bandwidth calls. That is, we use the same parameters as for the previous figures except the calls of type 1 are now the low bandwidth calls, with $a_1 = 1, x_1 = 0.4$, and $A_1 = 10^{-9}$, and the calls of type 2 are the high bandwidth calls, with $a_2 = 200, x_2 = 0.3$, and $A_2 = 10^{-4}$. Similar features can be observed in Figure 4 as in Figure 2; again the effective bandwidth approximation to the complete priority technique is highly accurate and only depends weakly on (n_1^*, n_2^*). However, unlike Figure 2, the one QoS technique is generally far inferior to the no sharing technique. This is because the one QoS technique does not take advantage of allowing the high bandwidth calls a significant proportion of loss. This contrasts with Figure 2 because, in that case, the low bandwidth low priority calls multiplex so efficiently that very little advantage can be gained through offering the lower QoS class.

To complete the numerical work we present an example for which the complement of the acceptance region (for both complete sharing and complete priority) is not a convex set but in fact the acceptance region itself is a convex set. We again use just two calls, both modelled as On-Off sources, but with identical parameters; that is, $a_1 = a_2 = 1$ and $x_1 = x_2 = 0.4$. However, we make the calls of type 1 higher priority than the calls of type 2 through setting $A_1 = 10^{-9}$ and $A_2 = 10^{-4}$. This could be used in practice to offer a 'silver service'. In Figure 5

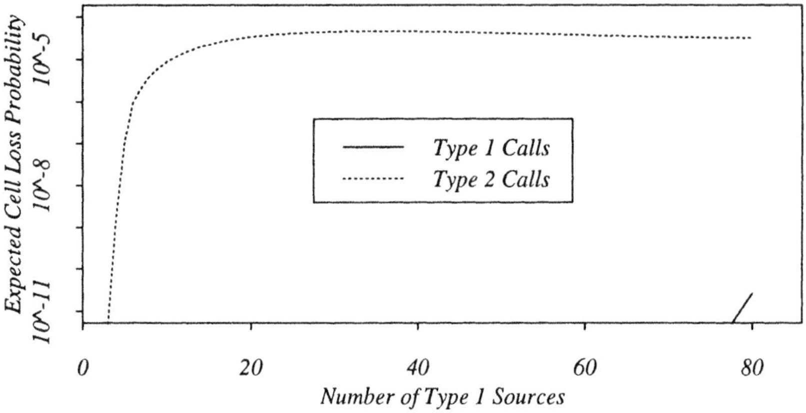

Figure 3. Exact Cell Loss Probabilities for Complete Priority.

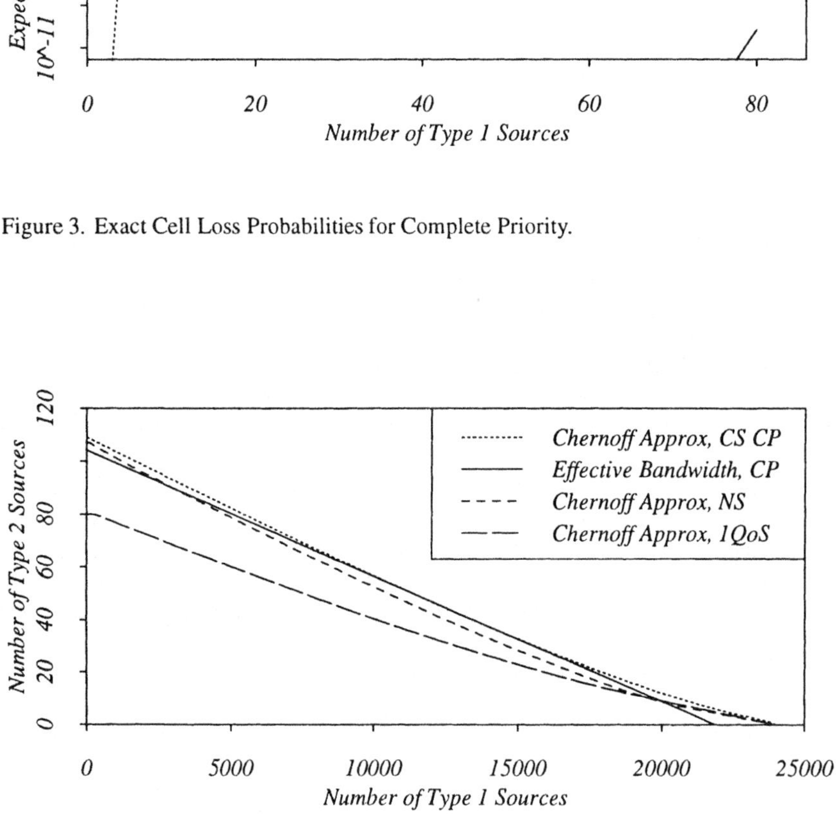

Figure 4. Acceptance Regions for the Alternative Parameter set.

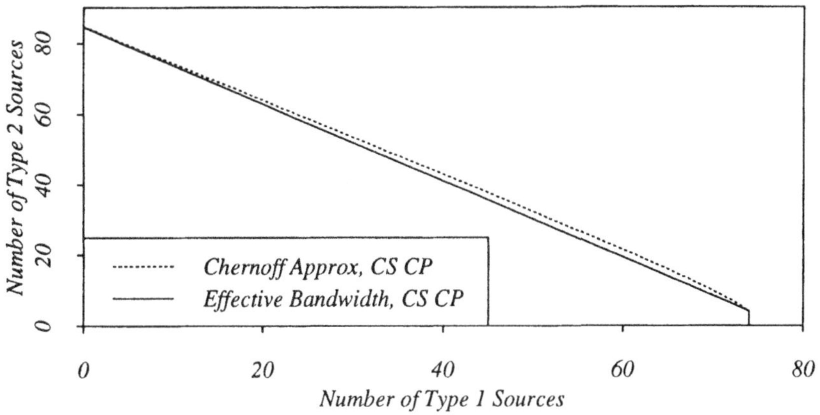

Figure 5. An Acceptance Region with a Concave Boundary.

it can be seen that $\mathcal{A}$ is itself a convex set; but again it has a highly linear boundary and so we can associate effective bandwidths by choosing the best possible linearly constrained, and conservative, acceptance region. This is achieved by joining the endpoints of the curve, ignoring the closing of the acceptance region due to the upper bound on n_1.

5. CONCLUSIONS

In this paper, we have addressed the problem of connection acceptance control for the situation of multiple call classes. We specifically addressed two situations, firstly where all calls have the same QoS requirement and secondly where different source classes have radically different QoS requirements. For the first problem we confirmed that the effective bandwidths concept could also be applied when the *stream based QoS measure* is used instead of the *resource based QoS measure*. It is widely agreed that the notion of an effective bandwidth could be enormously helpful in the management of ATM networks, as they make available the well-developed literature of circuit switched networks, for example see [8].

In Section 3 we proposed four possible techniques for offering different QoS parameters to different source classes. We considered the situation where there are two different QoS classes on offer because the technique needs to be simple and to offer more than two classes will only add severely to the complexity at each resource. We investigated all four techniques and found, as expected, that two are quite inefficient (one QoS and no sharing) and that two offer indistinguishable performance (complete sharing and complete priority). Since the complete priority technique is so much simpler than the complete sharing technique we propose this technique for future use. The complete priority technique is simple to describe and implement. In order to aid implementation we have shown that the effective bandwidth concept is suitable for this technique. The effective bandwidth approximation is highly accurate and conservative. We have also shown, through Figure 5, that care must be taken in determining the effective bandwidths as the acceptance region may or may not be a convex set.

It is clear that the complete priority technique is not restricted to the situation of just two call classes. All the results can also be developed for the situation of two QoS classes, each of which consists of many source classes. In this situation the effective bandwidth concept can again be used. In conclusion, the complete priority technique offers quite considerable efficiency gains over the one QoS and no sharing techniques with few, if any, complexity overheads. The efficiency advantage for complete sharing over complete priority is so negligible that it does not seem sufficient to warrant the significant additional complexity of the bandwidth partition.

Acknowledgements

I take great pleasure in thanking Frank Kelly for his assistance and infectious enthusiasm.

REFERENCES

1. P. Billingsley. *Probability and Measure. Wiley Series in Probability and Mathematical Statistics.*, Wiley Interscience, New York, Second Edition, 1986.
2. P. T. Brady. A Model for Generating On-Off Speech Patterns in Two-way Conversations. *Bell. Syst. Tech. J.*, 48:2445–2472, Sept. 1969.
3. J. A. Bucklew. *Large Deviation Techniques in Decision, Simulation and Estimation. Wiley Series in Probability and Mathematical Statistics*, Wiley Interscience, New York, 1990.
4. H. E. Daniels. Saddlepoint Approximations in Statistics. *Ann. Math. Statist.*, 25:631–650, 1954.
5. R. J. Gibbens and P. J. Hunt. Effective Bandwidths for the Multi-type UAS Channel. *Queueing Systems*, 9:17–28, Sept. 1991.
6. J. Y. Hui. Resource Allocation for Broadband Networks. *IEEE J.S.A.C.*, SAC-6(9):1598–1608, Dec. 1988.
7. F. P. Kelly. Effective Bandwidths at Multi-Class Queues. *Queueing Systems*, 9:5–16, Sept. 1991.
8. F. P. Kelly. Loss Networks. *Ann. Appl. Prob.*, 1:319–378, 1991.
9. B. Kraimeche and M. Schwartz. Analysis of Traffic Access Control Strategies in Integrated Service Networks. *IEEE Trans. Commun.*, COM-33(10):1085–1093, Oct. 1985.
10. B. Kraimeche and M. Schwartz. A Channel Access Structure for Wideband ISDN. *IEEE J.S.A.C.*, SAC-5(8):1327–1335, Oct. 1987.
11. B. Ngo and H. Lee. Queueing Analyses of Traffic Access Control Strategies with Preemptive and Nonpreemptive Disciplines in Wideband Integrated Networks. *IEEE J.S.A.C.*, SAC-9(7):1093–1109, Sept. 1991.
12. Z. Ren and J. S. Meditch. A Two-Layer Congestion Control Protocol of Broadband ISDN. In *IEEE Infocom '92*, pages 7A.4.1 – 7A.4.9, Florence, Italy, 1992.
13. J. Roberts (Editor). *Performance Evaluation and Design of Multiservice Networks. Information Technologies and Sciences*, Cost 224, Commision of the European Communities, Luxembourg, 1992.
14. L. Trajković and S. J. Golestani. Congestion Control for Multimedia Services. *IEEE Network Magazine*, 6(5):20–26, Sept. 1992.

Integrated Broadband Communication Networks
and Services (C-18) / V.B. Iversen (Editor)
Elsevier Science B.V (North-Holland)

Equivalence of policing mechanisms

A.M. Møller

Network Planning Department, Jydsk Telefon, Sletvej 30, DK-8310 Tranbjerg-J, Denmark.

Telephone: + 45 89 45 45 45
Telefax: + 45 86 29 09 05
E-mail: moller_amm@jt.dk

Keyword Codes:

Keywords:

1. Introduction

Managing an ATM network may require control of sources. Two proposed policing mechanisms are the leaky bucket and the moving window. This paper contains a performance result which makes it possible to compare the two mechanisms. It is shown that a leaky bucket corresponds to an infinite number of moving window mechanisms in parallel. The result is used in the following discussion about which mechanisms are suited for peak cell rate policing.

The paper is organised as follow. Section 2 contains a basic description of the policing mechanisms, in section 3 the proposition is stated and proved. The fourth section is on peak rate policing, and the fifth section concludes the paper.

2. Definitions and basic descriptions

The system to be analysed consists of ATM cells arriving at a policing mechanism. One time unit equals the time it takes a cell to pass a measurement point on the incoming link. The time is slotted according to this unit with cells always arriving at the beginning of a time slot. More than one cell may arrive in a single time slot and, in that case, the policing mechanism treats them one by one, possibly accepting some of them and rejecting the rest. Accepted cells are also denoted legal cells, and, similarly, rejected cells will sometimes be referred to as illegal cells.

By $LB(d,B)$, I denote a leaky bucket algorithm with increment factor 1, decrement factor d, and bucket size B. This means that an arriving cell will increment the buffer level (the bucket) by 1 unit if the cell is accepted, i.e. if the buffer level does not exceed B units. Moreover, for each time slot, the buffer level is decremented by d units, if possible.

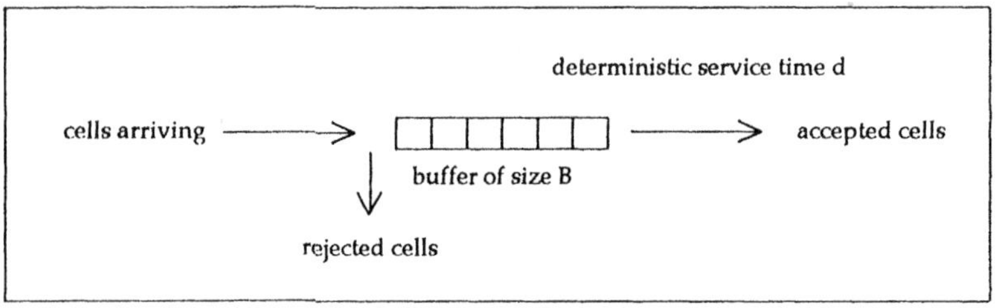

Figure 1. The leaky bucket $LB(d,B)$ - a queuing system

The average output from $LB(d,B)$ is at most $B/t+d$ within a time period of length t.

A moving window algorithm denoted $MW(k,w)$ has a window size w in which it accepts k cells. In each time slot, the mechanism divides the arriving cells into legal and illegal cells, and the latter ones are rejected. The maximum average load for a cell stream policed by $MW(k,w)$ is k/w.

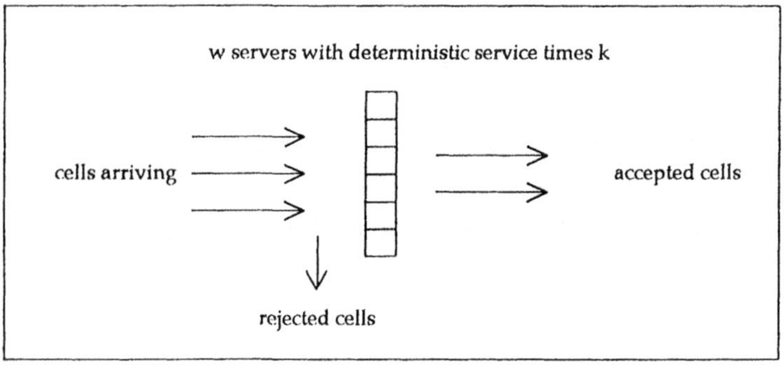

Figure 2. The moving window $MW(k,w)$ - a rejection system

Putting policing functions in parallel means that an entering cell is checked simultaneously in all policing functions, and the cell is accepted only if it is accepted by all of them.

A more rigorous description of the policing mechanisms is found e.g. in [COST224 1991].

3. The main result

Proposition

Policing by $LB(d,B)$ is the same as policing by the moving windows MW_k in parallel for $k = 0,1,2,\ldots$ where

$$MW_k = MW(k + \lfloor B \rfloor, w_k) \text{ and } w_k = \lceil 1 + k - (b - \lfloor B \rfloor)) / d \rceil.$$

Note

$\lfloor x \rfloor$ is the lower integer part of x, i.e. the biggest integer smaller than or equal to x.

$\lceil x \rceil$ is the upper integer part of x, i.e. the smallest integer bigger than or equal to x.

Example 1

A leaky bucket can be replaced by a single moving window in degenerate cases ($B < 1$) or if $B = 1$ and $d = 1/D$ with D integer. In the latter case, $LB(1/D,B)$ corresponds to $MW(B,D)$.

Proof of Proposition

If the first cell of a source is accepted by the $LB(d,B)$, it means that $B \geq 1$. Then all MW_k's also accept at least one cell in the corresponding windows w_k. On the other hand, if $LB(d,B)$ does not accept the first cell, it will not accept any cell, therefore $B < 1$. In that case, the moving window MW_0 also rejects all cells.

Thus I may assume that the two mechanisms of policing have accepted the same cells of the source until the present time slot with some cell arrivals. I will show that the action taken by $LB(d,B)$ is also taken by the MW_k's in parallel.

In a period of successive time slots, the following simple equivalence is true regarding the contents and the flow of the buffer (the bucket) of the leaky bucket $LB(d,B)$

$$\text{initial content} + (\text{no of cells accepted}) - \text{flow} = \text{final content}. \tag{1}$$

Now assume that m of the present cell arrivals are accepted by $LB(d,B)$, i.e. the contents of the buffer are less than or equal to $B - m$ at time of arrival. The maximum flow from the bucket in the preceding $w_k - 1$ time slots is $(w_k - 1)d$, and the initial contents in that time span are at least 0. Thus (1) gives that

$$(\text{no of cells accepted}) \leq (B - m) + (w_k - 1)d < k + \lfloor B \rfloor - (m - 1)$$

where it is used that $\lceil x \rceil < x+1$. Since the number of accepted cells in the preceding $w_k - 1$ time slots is kept below $k + \lfloor B \rfloor - (m-1)$ it means that at most $k + \lfloor B \rfloor - m$ cells are accepted, and it may be concluded that MW_k accepts at least the m cells.

Secondly, assume that m out of the present n cell arrivals are rejected by $LB(d,B)$. If the time of arrival is τ then at time τ the bucket contains more than $B-(n-m)$ units. Let

$$T' = \max\{t \le \tau : \text{bucket at start of time slot } t \text{ is empty}\}$$

i.e., for exactly $\tau - T'$ time slots, the bucket has been non-empty. Now define k to be the smallest integer with $w_k \ge \tau - T' + 1$ so that $w_{k-1} \le \tau - T'$. The window comprising time slots $\tau - w_k + 1 \ldots \tau + 1$ of size w_k time slots includes the $\tau - T'$ time slots where the bucket has been non-empty. Thus it is known that the flow from the bucket in that time span has been $(\tau - T')d \ge w_{k-1}d$ as a minimum. The initial content of the bucket may be put to 0 since it matches the flow of the bucket before it is known to be empty. Therefore

$$(\text{no of cells accepted}) > B-(n-m)+w_{k-1}d \ge k-(n-m)+\lfloor B \rfloor$$

so at least $k-(n-m)+\lfloor B \rfloor+1$ cells are already accepted in w_k and MW_k does not accept the m cells.

4. Peak cell rate policing

A source with a declared peak cell rate of $1/T$ has an illegal cell whenever it comes closer than T time slots to the preceding legal cell. Policing functions which police the peak cell rate in this strict sense is easily set up both with a moving window mechanism and with a leaky bucket mechanism. In most cases, the customer will not be satisfied with a strict peak cell rate policing at the UNI because he does not have fully control of the source entering the policing function. Moreover, at the NNI the source is almost surely disturbed having passed multiplexers in a network. Therefore, the customer requires acceptance of some cell delay variation.

In [CCITT 1992], a method of controlling the cell clumping is suggested which is based on a leaky bucket. H. Hemmer in [Hemmer 1992] compares this method with some selected window flow control mechanisms and finds that the window mechanisms are too generous. The parameter choices are as in the following Example 2 with Δ as the maximum cell delay variation allowed.

Example 2

The leaky bucket $LB(1/T, 1+\Delta/T)$ corresponds to the moving windows $MW(k, Tk-\Delta)$ put in parallel for $k = 1,2,\ldots$, with $\Delta < T$.

The conclusion drawn from Example 2 is that with regard to peak rate policing, window mechanisms like the moving window, the jumping window and the stepping window would fail. Either they would discard legal cells if they are chosen stricter than the ones in the example, or they would not catch all illegal cells, just like H. Hemmer observes.

The effect of invoking different moving windows and moving windows in parallel is illustrated, choosing the policing mechanisms of Example 2. The cell stream is supposed to be of type CBR and cells are randomly and independently delayed in an $M/D/1/m$ queue, i.e. a queue with Poisson arrivals, deterministic service times, 1 server and m queuing places.

Parameter choices are: cell inter-arrival time $T = 20$ time slots, maximum allowed CDV $\Delta = 3$ time slots, load in delay queue $\rho = 0.8$, service time $D = 1$, number of queuing places $m = 9$. In this specific example, only one cell is rejected in between accepted cells, thus the cell loss probability, CLP, is found by

$$\begin{aligned} \text{CLP} &= \sum_{x=1}^{\infty} \frac{1}{x+1} \text{P}(x \text{ cells in sequence}) \\ &= \frac{1}{2} - \sum_{x=2}^{\infty} \frac{1}{x(x+1)} \text{P}(\geq x \text{ cells in sequence}). \end{aligned} \quad (2)$$

By brute force, the probability is calculated of having more than x cells in sequence accepted in between rejections of two cells. This is the same as the inverse distribution function of the waiting time until a cell is rejected. The distribution is shown in figure 3 for different cases: invoking $LB(1/T, 1+\Delta/T)$, and invoking combinations of $MW(k, Tk-\Delta)$.

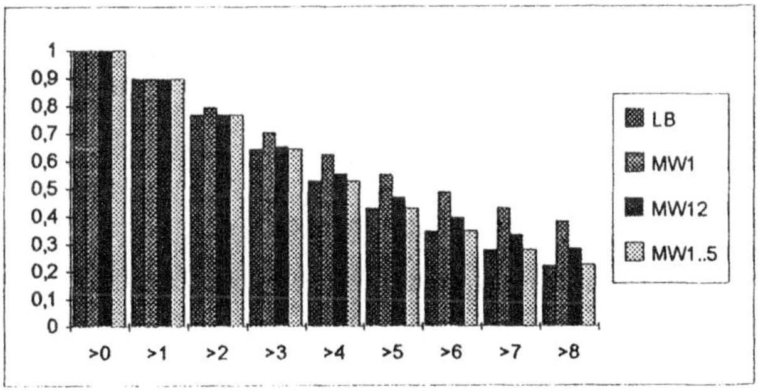

Figure 3. The inverse waiting time distribution of rejecting cells.

The effects of 4 different policing mechanisms are shown: the leaky bucket, the moving window MW_1, the moving windows MW_1 and MW_2 in parallel, and the five moving windows $MW_1 \ldots MW_5$ in parallel. Invoking more moving windows is clearly seen to approximate the leaky bucket.

Figure 4 shows the increasing tolerance of accepting longer series of cells when increasing the window size. The distribution shown in front is the one for the MW_6 and at the back it is the MW_1. The cell loss probability is calculated from (2) - a slight overestimation is expected due to summing only finitely many terms.

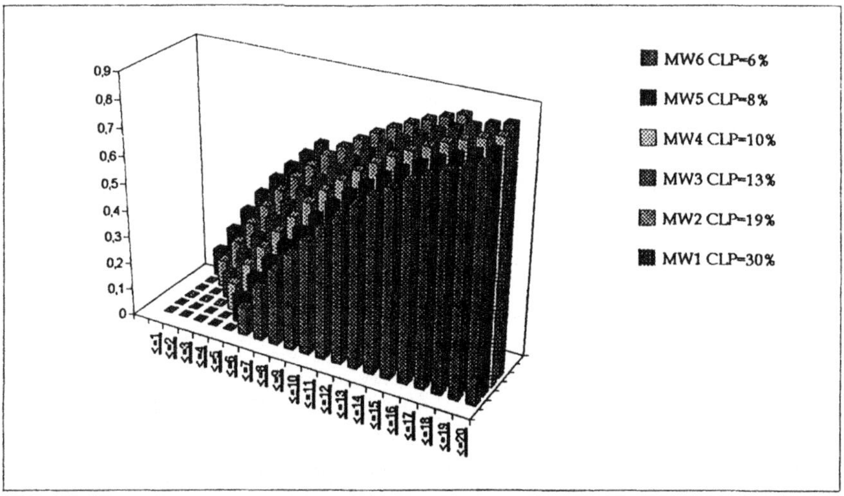

Figure 4. The waiting time distribution of rejecting cells.

Apart from leaky buckets and the variety of window flow mechanisms it has been suggested by V.B. Iversen and A.B. Nielsen in [Iversen, Nielsen 1992] to police by means of a so-called jumping leaky bucket. It is briefly described as a jumping window mechanism using a leaky bucket to control the overall load. The advantage of the jumping leaky bucket over the ordinary leaky bucket is that the updating takes place only once every w time slots when w is the window size.

The jumping leaky bucket can be defined not to reject legal cells but, depending on the window size, it is more generous than the leaky bucket because the misplaced cells might be closer than allowed to the neighbouring cells. Nevertheless, it is a valuable alternative to the leaky bucket since it requires less computational effort.

5. Conclusion

In this paper, an equivalence between a leaky bucket and moving windows in parallel has been given. The relationship gives insight into the different performances of the policing mechanisms. Using the result, it is found that window mechanisms are not suited for peak cell rate policing as this is defined in CCITT draft recommendation I.35B. Nevertheless, the combination of a leaky bucket and a jumping window mechanism in the jumping leaky bucket may be an alternative and more cost efficient solution.

References

[CCITT 1992] CCITT SG XVIII Draft Recommendation I.35B. Geneva, June 1992.

[COST224 1991] COST 224 Performance evaluation and design of multiservice networks. Final report, October 1991. Edited by J.W. Roberts.

[Hemmer 1992] H. Hemmer: Overvåkning av ATM-celler utsatt for variabel forsinkelse. NTS 10, Århus, August 1992. (In Norwegian.)

[Iversen, Nielsen 1992] V.B. Iversen, A.B. Nielsen: Traffic Management in ATM Networks Based on a Counting Mechanism. IFIP Workshop on Broadband Communications. Estoril, January 20-22,1992.

Integrated Broadband Communication Networks
and Services (C-18) / V.B. Iversen (Editor)
Elsevier Science B.V (North-Holland)

A Flow Management Architecture for B-ISDN *

Finn Arve Aagesen

Department of Computer Systems and Telematics, The University of Trondheim,
The Norwegian Institute of Technology, 7034 Trondheim, Norway.

Abstract

Flow management is the set of functions related to the control of the traffic flow during information transfer. This paper is proposing a total flow management scheme considering both the applications, the transport system and the ATM system. The basic principle has been to adapt the flow management to the pattern of the *applications*. This is possible by using transport system flow management mechanisms that reflect the application flow parameters and by making important flow parameters *visible* for the ATM system. The flow management is related to various time resolution regions. The *transport* system flow management comprises: rate-control and ack-based flow-control. The *ATM system* flow management comprises: source spacing, rate-control/policing and ack-based flow-control. Spacing schemes based on *stretching* are proposed for the various application classes.

Keyword Codes: C.2.1; C.2.2; C.2.3
Keywords: Computer-Communcation Networks, Network Architecture and Design; Network Protocols; Network Operations

1. INTRODUCTION

A vital part of the future high capacity networks seems to be B-ISDN based on ATM. B-ISDN will carry traffic from a broad set of traffic sources, and one important objective is a high degree of statistical multiplexing. *Flow management* is here defined as the set of functions related to the control of the traffic flow during the information transfer. Flow management is necessary because of limited resources, but is also a tool for reaching the statistical multiplexing objective.

The total system is divided into *three* components: the applications, the transport system and the ATM system. The applications comprise the ISO/OSI layer 5-7 functions, and the transport system the layer 3-4 functions. This paper considers a flow management concept related to this *total* context. A spacing scheme based on *stretching* is an important element within this concept. The flow patterns from the end-systems are spaced before the transfer via the ATM system, but this spacing shall *not disturb* the aspects of the flow pattern that are of significance as seen from the applications.

Section 2 gives a short overview of various flow management mechanisms. Section 3 gives a short discussion of architectual trends, while Section 4 presents a flow paradigm

*Part of the work reported in this paper was done under contracts with Norwegian Telecom, Research Department (NT-RD) as a part of the project TRASBI (Traffic studies for B-ISDN Networks).

to be used in the following discussion. Section 5 and 6 give proposed solutions for the transport system and ATM system flow management, respectively. Section 7 is related to the choice of *stretching* intervals, while Section 8 is discussing the validity of spacing.

2. FLOW MANAGEMENT MECHANISMS

Flow management mechanisms for high capacity networks have been a discussion topic for several years. The traditional *ack-based* stop-and-wait and sliding-window flow-control mechanisms are in the general case insufficient. Important *additional* techniques are: source spacing, peer-to-peer rate-control and policing based on rate-control.

Source spacing comprises active changes done with the information flow for the spreading of bursts of information over non-active periods. *Stretching* is the uniform spreading of the information within *stretching intervals.*

Peer-to-peer *rate-control* was introduced by transport system protocol designers as an alternative to the ack-based mechanisms. Information sending is based on a contracted *maximum* information quantity W during a defined *reference* time interval τ.

Policing is an ATM-related concept considered by many authors. Policing is based on an agreed *contract* between the ATM system user and the ATM system. Such a contract can either be related to traffic flow *parameters* of the ATM system user process or *variables* resulting from this process. A variable-based contract can be based on the amount of information during defined reference time intervals. In [Aage91] a variable-based policing scheme denoted as the EPPS (Excess Process Policing Scheme) was presented. EPPS is based on a contracted *reference rate* λ and a contracted *maximum excess* count L(t) between the *real count* N(t) and the *contracted count* λt after an elapsed *life-time* t of an excess process with period *duration* Tp. This policing function can be described by the following relationship:

$$N(t) - \lambda t \leq L(t); \quad 0 \leq t \leq Tp \tag{1}$$

Peer-to-peer rate-control and policing based on rate-control are conceptually analogous in the sense that both are based on an agreed information quantity during a reference time interval. Policing based on rate-control is in this paper denoted as rate-control-based policing and is abbreviated as *rate-control/policing.*

3. SOME ARCHITECTUAL ISSUES

The basic idea of the ISO/OSI model is to put different functions on various layers and to standardize the peer-to-peer protocol relationship. However, since the early 80-ies, criticism has been made against the way of using this basic idea. This criticism is related to solutions with extensive use of *multiplexing* and *flow-control* on various layers and a transmission-system-oriented definition of QOS (Quality Of Service) offered by the various layers. The principle has been the enhancement of the transmission service *upwards* and *not* the satisfaction of the application requirements *downwards.*

The first one that argued against this tendency was McQuillan [Mcqu80] and Saltzer [Salt81]. This discussion has once again blown up, now related to high capacity networks

([Tenn89], [Clar90] and [Feld90a]). It goes along the following three axes: i): layered co-operation versus layered encapsulation, ii): multiplexing on one layer versus multiplexing on many layers and iii): application-requirement-oriented versus transmission-system-offering-oriented QOS parameters. The solution proposed in Section 6 is based on a layered model. This solution is further characterized by layered cooperation, multiplexing within the ATM system and application-requirement-oriented QOS.

4. A FLOW PARADIGM

4.1. Classification of the applications

The classification of the applications, as illustrated in Figure 1, is inspired by the CCITT ATM adaption layer service classification [Boud92], but is slightly different.

Classification criteria	FBR	VBR	AAM	ALM
Timing relation between source and destination :	Synchronous		Asynchronous	
Bit-rate mode :	Constant	Variable		
Association mode :	Associations			Letters

Figure 1. Classification of the applications.

The FBR (fixed bit-rate) and VBR (variable bit-rate) classes comprise fixed and variable bit-rate synchronous applications, respectively. The classes AAM and ALM comprise asynchronous applications. Class AAM (asynchronous associated mode) is further split into the *transaction* class AAM_t and the *block-transfer stream* class AAM_s. Class ALM (asynchronous letter mode) is representing *letter-oriented* applications.

The transaction class AAM_t contains both *interactive* and *distributed system* traffic. Distributed system traffic comprises traffic between application programs as well as operating system traffic. The characteristics of interactive traffic and distributed system traffic are different because of the different "think time" for human users and programs.

The interactive traffic can further be split into interactive *graphics* and interactive *text/data*. These sub-classes are different with respect to the amount of bits per interaction. Let $AAM_{t,g}$ be the interactive graphics class, $AAM_{t,t}$ the interactive text/data class and $AAM_{t,d}$ the distributed system class.

4.2. Time resolution aspects

A high capacity network will have activities in various time-resolution regions. The *extreme*, *burst*, *frame* and *average* time resolution regions are defined. See Figure 2. The corresponding processes describe the activities in the related time resolution regions. Within the context of ATM, the *extreme* process is related to the flow pattern of individual ATM cells. The *burst process* is some time-related grouping of extreme processes, and the average process is the flow pattern related to a predefined *average interval* in the order of seconds or higher.

The concept *frame* is a generalization of the concept video-frame, which is typically between 25 and 40 msec. A frame is a *significant application-related* quantity of information. The frame-time is the time for the "presentation" of this information. For image-related sources, the frame is one terminal screen. For sources with a "fixed" period, i.e. FBR, VBR and AAM_s, the frame-time is *the period* of the generated flow pattern. For the classes $AAM_{t,d}$ and ALM, the frame has no direct intuitive interpretation.

In addition to these defined regions, the *message*, *packet* and *stretching interval* process regions are used. Message is used synonymous with service data unit (SDU) and packet with protocol data unit (PDU).

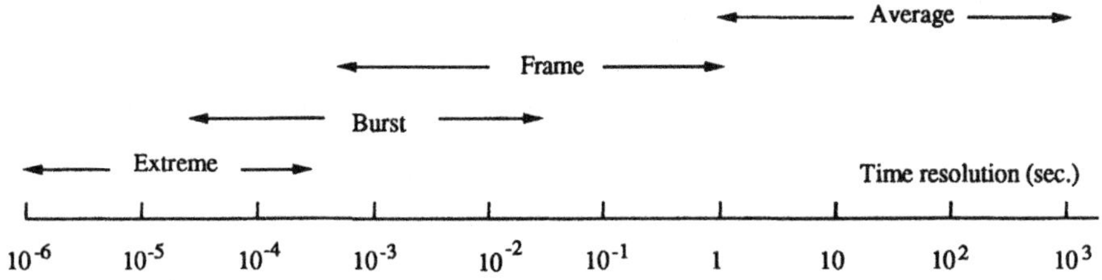

Figure 2. Time resolution regions.

4.3. Flow and performance requirement parameters

In the context of a protocol layer, the flow management parameters can be classified as *performance requirement* parameters related to individual information units (Q-parameters) and *traffic flow* parameters related to the overall structure of the flow (F-parameters). These are just denoted as *performance* and *flow* parameters and are further defined as:

F_S	: Service flow parameters	Q_S	: Service performance parameters
F_P	: Protocol flow parameters	Q_P	: Protocol performance parameters
F_L	: Layer flow parameters		

The *service* parameters are the parameters related to the service, the *protocol* parameters are the parameters related to the peer-to-peer-interaction and the *layer* parameters are the total set of parameters "visible" within a layer. See Figure 3. An *additional* index (N) indicates the (N)-layer. In the following sections, the indexes "T" and "A" are used for indicating the transport system and the ATM system, respectively.

Let Z be a general stochastic quantity. The operators: E(Z), σ(Z), Min(Z), Max(Z), c(Z), b(Z) and ϑ(Z) are the mean, the standard deviation, the minimum, the maximum, the coefficient of variation (σ(Z)/E(Z)), the burstiness (Max(Z)/E(Z)) and the jitter ((Max(Z)-Min(Z))/E(Z)). Such parameters will be the specific members of the Q- and F-parameter sets. The indexes {e,b,f,a,p,m,s} indicate the parameters for the extreme, burst, frame, average, packet, message and stretching interval processes, respectively.

The *Q-parameters* will comprise quantities such as delay (D) and error rate (P_E) related to specific information units. Both E(D), c(D), Max(D) and ϑ(D) are of importance. For a usable definition of error-rate, the time-resolution concepts are needed. This because losses caused from buffer shortage can be high in short overload intervals. So the loss-rate quantities $E(P_E)$ and $\{Max(P_{E,i}); i=b,f,a\}$ are of importance.

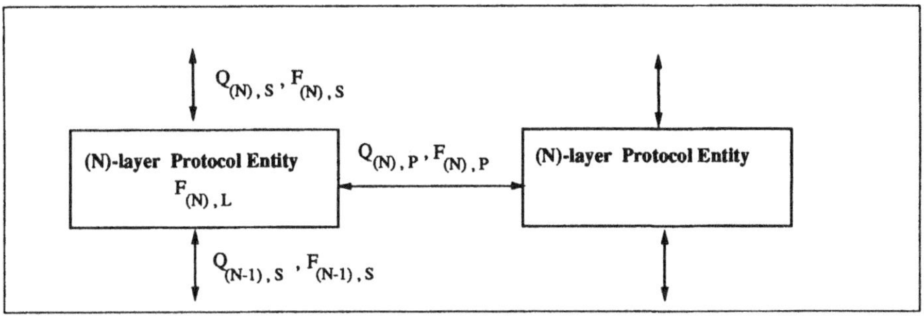

Figure 3. The F- and Q-parameters related to a layer.

The *F-parameters* do also need the time resolution concepts. The F-parameters can comprise quantities such as:

τ_f	: the frame-time
τ_a	: the average interval
$\{t_q; q = e,b,f,a,p,m,s\}$	: the elapsed duration time for process q
$\{X_q; q = e,b,f,p,m\}$	: the length *in bits* of an information unit for process q
$\{Tp_q; q = e,b,f,a,p,m,s\}$	: the period duration for process q
$\{Ta_q; q = e,b,f,a,p,m\}$	: the activity duration for process q
$\{\Lambda_q; q = e,b,f,a,p,m,s\}$	: the throughput in *bits/sec* during process q

The importance and the use of these various parameters depend on the specific layer and also on the specific context, i.e. service or protocol. The *throughput* Λ_q is an important resulting quantity. Both $E(\Lambda)$ as well as $b(\Lambda_q)$ and $c(\Lambda_q)$ for the various regions q are of importance. In addition to the above listed quantities, the layer flow parameters F_L can comprise quantities such as *state*-variables, *context*-variables and *timers* related to the protocol layer peer-to-peer-operation.

The F_S-parameters can differently from the Q_S-parameters to some extent be "readable" from the F_L- and F_P-parameters. However, the relationship between the service flow parameters and the protocol flow parameters is dependent on the specific service and protocol. By definition : $F_L = F_P \cup F_S$, i.e. the service and the protocol flow parameters are both subsets of the *layer* flow parameters. Thus, the service flow parameters $F_{(N),S}$ can be visible for layer (N-1) through $F_{(N-1),S}$, which can be based on the layer parameters $F_{(N),L}$.

5. TRANSPORT SYSTEM FLOW MANAGEMENT

A large set of new transport protocols and combined transport/network layer protocols are proposed for use in high capacity networks. The following gives a non-exhaustive list: Delta-t, NETBLT (The Network Block Transfer Protocol), CMP (Check Mode Protocol), NADIR, VMTP (The Versatile Message Transfer Protocol), XTP (The Express Transfer Protocol), RTTP (The Real-Time Transport Protocol), TPR (Transport Protocol for Real-time Services) and TP++. General overviews are given in [Doer90], [Dabb89] and [Hehm89]. Details on VMTP can be found in ([Cher86], [Nord89]), NETBLT in [Clar87], XTP in ([Prot90], [Ches89]), RTTP in [Dabb89], TPR in [Anas90] and TP++ in [Bier92]. The *interesting* features within the context of flow management is that:

- there are various protocols related to various *application classes* like FBR, VBR, AAM_t and AAM_s and
- the flow management is in most cases based on a combination of *rate-control* and *ack-based* flow-control.

TPR is an example protocol for the FBR application class, XTP for VBR, VMTP for AAM_t and NETBLT for AAM_s. Rate-control is used by CPM, NADIR, NETBLT, RTTP, VMTP and XTP. Because of weak performance requirements, any transport system protocol can be used for ALM. The most likely choice, however, is a transaction-oriented or a block-transfer-stream-oriented transport system protocol. In the following discussion of flow management, ALM can therefore be considered as either being analogous to AAM_t or AAM_s.

5.1. Some example rate-control mechanisms

In *XTP*, the *volume* is controlled by end-to-end flow-control and the *intensity* by rate-control. The rate-control operates in *two* time-resolution regions by using the PDU-parameters *Rate* and *Burst*. Rate is the maximum number of bytes per second and Burst is the maximum number of bytes in a burst of packets. The *protocol layer* operation is based on the quantities *Credit* and *Rtimer*. Credit is the remaining number of bytes in the burst transferred, and Rtimer is a timer controlling the intervals between bursts.

XTP implementations are not expected to implement Rtimer with resolution greater than the expression 1/(10*K), where K is the number of Mega-instructions available for networking tasks. So K=1 gives a resolution of 100 msec and K=100 gives a resolution of 1 msec. This covers the processes from the upper part of the burst region.

The *NETBLT* rate-control does only operate in *one* time-resolution region. A reference interval *Interval*, the packet length and *Rate* (the maximum number of packets per *Interval*) are agreed.

In *VMTP* a different mechanism is used. The maximum burst denoted as a message is 16 Kbytes. This message is segmented into packets of minimum length 512 bytes data plus a 64 bytes header. The *message* flow is controlled by an end-to-end stop-and-wait protocol, while the stream of *packets* is controlled by rate-control. This rate-control is based on the quantity *Intergap*, which is the minimum time between packets.

5.2. A transport system flow management solution

In the general discussion of multiplexing and flow-control by Feldmeier [Feld90a], a combination of rate-control and ack-based flow-control is recommended. It is felt here that a chosen solution shall have the potential for i): adaption to the real-time flow pattern of the applications, ii): prevention of information bulking and iii): cooperation with the ATM system.

Rate-control is such a mechanism, and is considered as the appropriate mechanism for the *non-interactive* application classes FBR, VBR and AAM_s. The rate-control of the message and packet flow should be done in the frame and average time resolution regions. For the *interactive* applications, the VMTP solution, with stop-and-wait flow-control on concecutive transactions and rate-control on the packet flow within each transaction, seems to be optimal. The total scheme is illustrated in Table 1.

Table 1
Transport system flow management scheme

Application Class	Message Flow	Packet Flow
FBR,VBR,AAM_s	Rate-control	Rate-control
AAM_t	Ack-based flow-control	Rate-control

5.3. Transport system layer flow parameters $F_{T,L}$

Transport system flow parameters are given in Table 2. For VBR applications a protocol with flow mechanisms like XTP is assumed used, and for AAM_t a protocol like VMTP. NETBLT is assumed used for AAM_s. For XTP we have $\tau_a = 1$ sec, Burst $= \text{Max}(X_m)$, Rate $= \text{Max}(\Lambda_a)$, Rtimer $= \tau_f$ and Credit:=Burst at the start of every new Rtimer interval. For VMTP we have X_p = constant and Intergap=Min(Tp_p). For NETBLT the flow parameters will be Rate=X_f/X_p and Interval=τ_f.

Table 2
Flow parameters $F_{T,L}$

Application Class	Message and Packet Process Parameters	Frame and Average Process Parameters
FBR	$X_m = X_p =$ constant, $Tp_m = Tp_p =$ constant	
VBR	$\text{Max}(X_m)$, X_p, Tp_m, Tp_p	τ_f, τ_a, $\text{Max}(X_f)$, $\text{Max}(X_a)$
AAM_t	$\text{Max}(X_m)$, X_p, $\text{Min}(Tp_p)$	
AAM_s	$X_m =$ constant, $Tp_m \approx$ constant, $X_p =$ constant, $Tp_p \approx$ constant	τ_f, $X_f \approx$ constant

6. ATM SYSTEM FLOW MANAGEMENT

6.1. A cooperative scheme based on framing visibility

The ATM system has a defined sublayering [Boud92]. Independent of this sublayering, the ATM system must contain a call acceptance and contract negotiation function (*CACF*) and an ATM system flow management function (*AFMF*). The ATM system is planned to offer *four* service classes. Class A is the circuit emulation fixed bit-rate service class,

class B is the variable bit-rate synchronous service class, class C is the connection-oriented asynchronous service class and class D is the connection-less asynchronous service class.

The ATM-classes A and B are analogous to the defined application classes FBR and VBR. There is obviously not a one-to-one relationship between the application classes {AAM, ALM} and the ATM classes {C, D}. There is rather not a one-to-one relationship between the application classes {VBR, FBR} and the ATM-clases {A, B}. Using an ATM service with zero jitter for the classes FBR and VBR may not be optimal with respect to delay because the transport system also has to provide zero jitter.

Let now the ATM system flow and performance parameters be defined by: $F_{A,S} \subseteq F_{T,L}$ and $Q_{A,S} = Q_{T,P}$ as illustrated in Figure 4. $F_{A,S}$ is a necessary subset of the layer flow parameters $F_{T,L}$, so that the ATM system can "see" the pattern of the application flow. The *layer* flow parameters $F_{T,L}$ can give frame-related information such as frame *duration*, frame *start* and also dynamic frame *information length*. This is in the following denoted as *framing visibility*. Dynamic frame-related information can be transferred to the ATM-system as an information preamble.

With such a scheme, the transport service class will be known independent of the ATM service class chosen. The ATM system flow management can then comprise 1): spacing of the process $F_{T,P} \subseteq F_{A,S}$ based on the *framing* information from $F_{T,L}$ and the constraints $Q_{A,S} = Q_{T,P}$, 2): rate-control/policing of the resulting spaced process $F_{A,P}$ and 3): flow-control for the the flow of messages for the transaction class AAM_t.

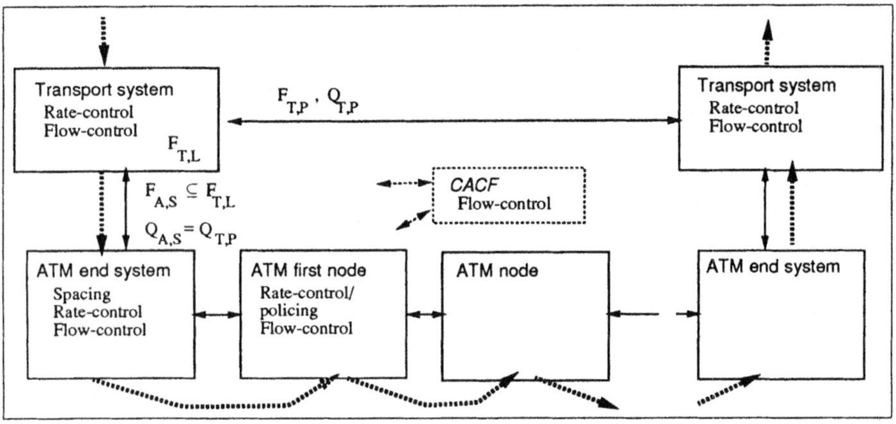

Figure 4. The ATM system flow management.

6.2. Spacing

Let Δ be the stretching interval (in sec). Proposed stretching intervals for the various applications are summarized in Table 4 in Section 7. For applications with stochastic patterns, the seemingly "optimum" stretching interval is in the region 1-40 msec.

6.3. Rate-control/policing and flow-control.

A policing relationship was described by the relation (1) in Section 2. In the general case, one such relationship is defined for *each* of the time resolution regions {q ; q=e,b,f,a}. However, with stretching intervals as proposed, this scheme is simplified. There will be no need for rate-control/policing in the burst and frame regions. Proposed flow management mechanisms for the *extreme* and *average* process regions are given in Table 3.

Table 3
The ATM system rate-control/policing and flow-control.

Application class	Extreme process	Average process
FBR	Rate-control/policing	(No flow management)
VBR	Rate-control/policing	Rate-control/policing
AAM_t	Rate-control/policing	Ack-based flow-control
AAM_s	Rate-control/policing	Rate-control/policing

The policing in the *extreme* region is the policing of a *regular* cell process. By using the quantities defined in Section 2 and 4, the *extreme* and *average* process policing/rate-control schemes can be described as follows:

$$N_e(t_e) - \lambda_s\, b_s(\Lambda_s)\, t_e \leq 1; \quad 0 \leq t_e \leq Tp_e \tag{2}$$

$$N_a(t_a) - \lambda_a\, t_a \leq L_a(t_a); \quad 0 \leq t_a \leq \tau_a \tag{3}$$

For AAM_t, ack-based stop-and-wait flow-control is used on concecutive messages. The network can act *back-pressure* on each message. But how shall the *message* flow be managed within the ATM system? *Should* there be a contract related to the intended rate of messages? And *should* there be some flow management procedure within the ATM network for each message? Seen from a traffical point of view, the answers to both of these questions are *yes*. However, research and experimentation are still needed to see if this is feasible.

7. A DISCUSSION OF STRETCHING INTERVALS

This section is discussing stretching intervals Δ for various application classes. The discussion is related to the layer flow parameters $F_{T,L}$ defined in Section 5.3. A transmission bit-rate of 155.52 Mbits/sec is assumed. The cell-size is 53 bytes of which 48 bytes are used for data. Let "m" be the number of ATM cells per stretched information unit, "n" the number of ATM cells per stretching interval and "ϕ" the relationship n/m . The quantity ϕ is the potential burstiness reduction.

The cell and burst processes are denoted as *intra-frame* processes and the frame and average processes as *inter-frame* processes. It is the inter-frame processes that have significance for the applications. A tempting choice is to make Δ as large as possible, *while*: i): ϕ is still increasing, ii): the stretching does not influence the inter-frame processes, iii): the delay is satisfactory and iv:) the additional buffer space needed is acceptable.

The additional delay within the ATM end-system will under most of the cases be much less than Δ. By utilizing the *framing visibility* described in Section 6.1, dynamic frame-related information can make pipelining possible. The discussion is given in the following subsections. The conclusions are summarized in Table 4 below. The table does also indicate throuhgput regions as well as burstiness and coefficient of variation.

Table 4
Proposed stretching interval Δ.

Application class	$E(\Lambda)$	$b(\Lambda_f)$	$c(\Lambda_f)$	Δ	Δ
FBR video	1-50 Mbits/sec	1	0	Tp_p	50-500 μs
FBR voice	64Kbits/sec	1	0	Tp_p	1-40 ms
VBR video	1-30 Mbits/sec	1.5-10	0.1-0.5	τ_f	1-40 ms
$AAM_{t,t}$	$\approx$0-64 Kbits/sec	1-∞	-	$\geq$Intergap	1-40 ms
$AAM_{t,g}$	$\approx$0-30 Mbits/sec	1-∞	-	$\geq$Intergap	1-40 ms
$AAM_{t,d}$	$\approx$0-10 Mbits/sec	1-∞	-	$\geq$Intergap	1-40 ms
AAM_s	$\approx$0-10 Mbits/sec	$\approx$1	$\approx$0	Interval	1-40 ms

7.1. Fixed bit-rate applications (class FBR)

FBR applications are primarily video and voice. The following three cases are from [Anas90]:

Case 1 : An X-ray ultrasonic application. Each frame has 512 lines with 512 pixels a 8 bit. τ_f=40 msec. One line per TPDU gives Tp_p= 78μsec. $E(\Lambda)$=52.4 Mbits/sec.

Case 2 : A nuclear application. Each frame has 128 lines with 128 pixels a 8 bit. τ_f=62.5 msec. One line per TPDU gives Tp_p= 488μsec. $E(\Lambda)$=2.09 Mbits/sec.

Case 3 : Voice with $E(\Lambda)$=64 Kbits/sec and τ_f=125μsec. 192 voice samples a 8 bit per TPDU gives Tp_p=24 msec.

Let the TPDU be used as the stretching information unit. For Case 1, we have m=11 and ϕ=2.7, for Case 2: m=2.6 and ϕ=66 and for Case 3: m=4 and ϕ=2201. For all these cases the minimum value choice is $\Delta=Tp_p$. The multiplex gain will not increase with larger Δ. This minimum choice also gives the minimum delay and jitter.

7.2. Variable bit-rate applications (class VBR)

The **Cases 4-7** are from [Aage91]. These are based on the same material as published in [Verb88] and [Verb89]. All data are based on a codec with τ_f=40 msec. A tempting choice is to use a stretching interval $\Delta=\tau_f$. Most VBR applications will know the whole frame content before sending, so any additional delay in the application part of the system does not need to be introduced. If a frame-related TSDU is assumed, i.e. $Tp_m=\tau_f$=40 msec, we get for the four cases m=456,440,1783,3080 and $\phi\approx$32,33,8,5. The multiplex gain will increase with increasing Δ because of significant burstiness reduction. Stretching intervals larger than τ_f can be used. This will give an additional end-to-end delay without any expected significant increasing effect on the multiplex gain. The frame content in consecutive frames is highly correlated. The benefit from mixing independent frame-stretched sources is expected to be higher than from using longer stretching intervals for the individual sources.

Table 5
Some VBR-codec statistics. (Intensities in Mbits/sec).

	Case 4 (Video-phone)	Case 5 (Video-conference)	Case 6 (TV-serie)	Case 7 (TV-sport)
$E(\Lambda)$	4.38	4.23	17.12	29.57
$b(\Lambda_f)$	6.27	6.50	1.87	1.34
$c(\Lambda_f)$	0.44	0.36	0.11	0.11

7.3. Transaction and block-transfer stream applications (class AAM).

For the *transaction* class AAM_t, consider an example **Case 8** given in [Cher89] for a 4.46 Mbits/sec transfer based on VMTP. This case covers the sub-classes $AAM_{t,d}$ and $AAM_{t,t}$. The maximum message is 16 Kbytes and the minimum packet is 512 bytes plus a 64 bytes header. In the example this minimum packet length is used. Let the packet be used as the stretching information unit. This gives m=12, Intergap=0.91 msec and ϕ=29. Using a stretching interval Δ=Intergap seems to be a reasonable choice. Larger Δ can be used. However, with appropriate *transport system* rate-control within the *frame* region, a stretching interval Δ=Intergap can be as good as using $\Delta=Ta_m$. Using larger intervals than Ta_m is possible. However, this will need significant additional buffering capacity.

If VMTP is to be used for the *graphics* sub-class $AAM_{t,g}$, some parameters must be increased, and the flow-control mechanism should be used differently. The screen content should be used as the end-to-end flow-control-related quantity, rather than the 16 Kbytes VMTP message. **Case 9, 10** and **11** are defined with screen sizes 1 Mbits, 10 Mbits and 30 Mbits, respectively. Without any local subnetwork involved, the 16 Kbytes message can be sent in one VMTP packet. If FDDI is involved, this will limit the TPDU size to 4500 bytes. Assuming a VMTP packet length equal to 4000 bytes, a screen will for these cases comprise 30, 300 and 900 TPDUs. Let the TPDU also in this case be used as stretching information unit. We then have m≈85. Assuming the total transfer time equal to 1 sec, gives Tp_p(=Intergap) equal to 33 msec, 3.3 msec and 1.1 msec, and ϕ equal to 142, 14 and 5. This gives Intergap values in a feasible region.

Framing visibility can also be utilized in this case. By letting the ATM system know the whole TSDU length, the ATM system can decide to use *larger* spacing intervals than Intergap, depending on the traffic situation, the remaining length of the TSDU and the available buffer space for the ATM spacing process.

For the *block-transfer stream* class AAM_s, a **Case 12** is considered [Clar87]. This case comprises the transfer of large blocks of data by using NETBLT. The minimum packet size is 128 bytes, however, "efficient" packet size is likely to be in the area 500-2000 bytes. Example rate-control parameters for PCs are Interval= 20 - 40 msec, and Rate=5 packets/Interval. Using Interval=40 msec, this represents a throughput in the region 128 Kbits/sec - 2 Mbits/sec. A natural choice is Δ=Interval. This gives m=13-208 and ϕ=1128-70.

7.4. Letter applications (class ALM)

This is an important class. Letters can represent large amounts of information (data, voice, still graphics and video) and can be a frequent traffic type. The spacing depends on the protocol used. It is referred to the discussion of the spacing for AAM_t and AAM_s.

8. ON THE SPACING VALIDITY

A spacing solution has a *trade-off* related to increased throughput, reduced cell-loss, increased policing simplicity, increased protocol architecture complexity, increased end-system complexity, increased delay and increased need for buffers in the end-systems. There is still a lot of work to be done before the optimality of the proposed flow management solution is understood. Models are needed that gives *throughput*, *cell-loss* and *delay* for the non-spaced as well as for the spaced solutions.

An end-system-model has been studied by Klovning [Klov92] for the simultaneous spacing of several VBR applications. The model gives *delay* and the need for *buffers* as well as indications of the spacing algorithm complexity.

Some simple node-model considerations have been done in [Aage92]. The performance measures are *cell-loss*, *buffer need* and *throughput*. For FBR traffic, there should not be any *traffic loss*. However, it also seems as the *interference loss*, i.e. loss caused by the interference between the cell-pattern on the input channels bound for the same output channel, can be avoided. Simulations and calculations for VBR traffic gave a significant potential increase in transmission capacity utilization by using stretching. However, work remains related to the analysis of the important transaction graphics class $AAM_{t,g}$ and also to the mix of the various application classes.

9. CONCLUSIONS

Flow management has been considered within the context of a total system consisting of the applications, the transport system and the ATM system. A total flow management scheme has been proposed, where the transport system *layer flow* parameters are used as the basis for the ATM system service flow definition. *Framing visibility* means that important application-related flow parameters are visible for the ATM system.

The transport system flow management is assumed to be based on the VMTP solution for transaction-oriented traffic and the XTP solution for non-transaction-oriented traffic. For *transaction*-oriented transport system protocols, stop-and-wait flow-control should be used on the consecutive transactions, and rate-control should be used on the packet flow related to one transaction. For *non-transaction* transport system protocols, rate-control should be used on the message and packet flow within the frame and average time resolution regions.

The proposed ATM system flow-management scheme comprises source spacing based on *stretching*, *rate-control/policing* and *stop-and-wait flow-control*. Stretching intervals have been proposed for the various application classes. On one hand the stretching makes the policing simpler. The *extreme* process will get the characteristics of the *stretching* interval process, so that policing of the extreme and the average processes in many cases are satisfactory. On the other hand, the burstiness will decrease, so that a *high* increase of the transmission capacity utilization can be expected.

Acknowledgements

I would like to thank the TRASBI-project group for fruitful and stimulating discussions. In addition special thanks to B.E. Helvik, DELAB and K. Moldeklev and E. Klovning, NT-RD for constructive comments to the basic content of this paper.

REFERENCES

[Aage91] F.A. Aagesen and B.E. Helvik, On Excess Process Policing in B-ISDN Networks, 13th Int. Teletraffic Congress (ITC 13), Copenhagen, June 1991.

[Aage92] F.A. Aagesen, On Source Spacing in B-ISDN, SINTEF report STF 40 A92140, October 1992.

[Anas90] G. Anastasi a.o, TPR: A Transport Protocol for Real-Time Services in a FDDI Environment, IFIP WG 6.1/6.4 International Workshop on Protocols for High Speed Networks 1990.

[Appl91] J. Appleton, Traffic Shaping in Asynchronous Transfer Mode Networks, 8th UK Teletraffic Symposium, April 1991.

[Bier92] E.W.Biersack a.o, Gigabit Networking Research at Bellcore, IEEE Network, Vol.6, No.2, March 1992.

[Boud92] Jean-Yves Le Boudec, The Asynchronous Transfer Mode: A Tutorial, Computer Networks and ISDN Systems 24, 1992.

[Boye92] P.E. Boyer a.o., Spacing Cells Protects and Enhances Utilization of ATM Network Links, IEEE Network, Vol.6 No.5, September 1992.

[Butc89] B. Butcher, A Flexible Transport Service in the Berkom Broadband Environment, IFIP WG 6.1/6.4 International Workshop on Protocols for High Speed Networks, Zurich, May 1989.

[Cher86] D.R. Cheriton, VMTP: A Transport Protocol for the Next Generation of Communication Systems, ACM Proc. Sigcomm 86.

[Cher89] D.R. Cheriton, VMTP as the Transport Layer for High-Performance Distributed Systems, IEEE Communication Magazine, June 1989.

[Ches89] G. Chesson, XTP/PE Design Considerations, IFIP WG 6.1/6.4 International Workshop on Protocols for High Speed Networks, Zurich, May 1989.

[Clar87] D.D. Clarc, M.L. Lambert and L. Zhang: NETBLT: A Bulk Data Transfer Protocol, Network Working Group RFC 998.

[Clar90] D.D. Clarc, Architectual Issues for a New Generation of Protocols, Proceedings of Sigcomm 90, September 1990.

[Dabb89] W.S. Dabbous, RTTP: Real-time Transport Protocol, IFIP WG 6.1/6.4 International Workshop on Protocols for High Speed Networks, Zurich, May 89.

[Doer90] W.A. Doeringer a.o., A Survey of Light-Weight Transport Protocols for High-Speed Networks, IEEE Trans. on Comm., Vol.38, No.11, November 1990.

[Feld90a] D.C. Feldmeier, Multiplexing Issues in Communication System Design, Proceedings of the Sigcomm 90, September 1990.

[Feld90b] D.C. Feldmeier and E.W. Biersack, Comparison of Error High Bandwidth-Delay Product Networks, IFIP WG 6.1/6.4 International Workshop on Protocols for High Speed Networks, Palo Alto, November 1990.

[Hehm89] D.B. Hehman a.o., High-Speed Transport Systems for Multi-Media Applications, IFIP WG 6.1/6.4 International Workshop on Protocols for High Speed Networks, Zurich, May 1989.

[Hemm89] H. Hemmer a.o., Policing Functions in ATM networks, 8th Nordic Teletraffic Seminar, Helsinki, August 1989.

[Ingv91] T. Ingvaldsen, Traffic Studies of an ATM PABX, MSc.-thesis, IDT/NTH, Trondheim, January 1991.

[Jain92] R.Jain, Myth about Congestion Management in High-Speed Networks, Internetworking: Research and Experience, Vol.3, 1992.

[Klov92] E. Klovning, Cooperative Flow Management Protocol Architecture, M.Sc.-thesis, IDT/NTH, Trondheim, December 1992.

[Lea92] C.T Lea, What Should Be the Goal of ATM, IEEE Network, Sept. 1992.

[Mcqu80] J.M. McQuillan, Local Network Technology and the Lessons of History, Computer Networks, no 4, 1980.

[Nord89] E.Nordmark and D.R. Cheriton, Experiences from VMTP: How to Achieve Low Response Time, IFIP WG 6.1/6.4 International Workshop on Protocols for High Speed Networks, Zurich, May 1989.

[Prot90] Protcol Engine Inc., XTP Protocol Definition Rev.3.5, September 1990.

[Salt81] J.H. Saltzer a.o., End-to-end Arguments in System Design, Proceedings of Second International Conference on Distributed Systems, April 1981.

[Tenn89] D.L. Tennenhouse, Layered Multiplexing Considered Harmful?, IFIP WG 6.1/6.4 International Workshop on Protocols for High Speed Networks, Zurich, May 1989.

[Verb88] W. Verbiest, L. Pinnoo and A. Boeten, The Impact of the ATM concept on Video Coding, IEEE Journ. on Selected Areas in Comm., Vol.6, No. 9, 1988.

[Verb89] W. Verbiest, and L. Pinnoo, A Variable Bit Rate Video Codec for Asynchronous Transfer Mode Networks, IEEE Journal on Selected Areas in Communications, Vol. 7, No. 5, 1989.

Integrated Broadband Communication Networks
and Services (C-18) / V.B. Iversen (Editor)
Elsevier Science B.V (North-Holland)

TRAFFIC MEASUREMENTS AND TEST EQUIPMENT IN BATMAN

by

Lars Dittmann (*) and Kenn Kvols (**)

(*)
CBT/EMI
Technical Univ. of Denmark
Building 348
DK-2800 Lyngby
Tel.: +45 42881444
Fax.: +45 45931634
EMAIL : ld@emi.dth.dk

(**)
NUN
KTAS
Nørregade 21
DK-1199 Copenhagen K
Tel.: +45 33996644
Fax.: +45 33143356
EMAIL : kk@ktas.dk

1. INTRODUCTION

In these years the first experimental networks based on the asynchronous transfer mode (ATM) are presented and first generation ATM products are soon available. ATM uses fixed size packets, called cells, for the information transport. The cells consist of 48 octets user information and a 5 octets header field. ATM makes full integration of services possible, as cells are dynamically allocated to the users. Furthermore ATM gains better utilisation of the network resources by use of statistical multiplexing.

The asynchronous nature of ATM and the use of statistical multiplexing implies that the information flows directed towards a given link might occasionally exceed its capacity, which makes it necessary to introduce multiplex buffers in the switches. Therefore, cells passing an ATM switch will experience variable delays, and they get lost with a non-zero probability. To characterise these phenomena new QOS parameters have been introduced: Cell loss ratio, cell delay and cell delay variation [1]. Which of these parameters that are relevant depends on the type of service that is being considered. Constant bit rate (CBR) services such as voice require low delay but they are not that sensitive to cell losses. On contrary the variable bit rate (VBR) services such as file transfer require a very low cell loss ratio but they are not sensitive to delay and variation in delay.

ATM uses a call admission control (CAC) function at each call set-up [2][3]. Based on negotiated bandwidth parameters and QOS requirements this function decides whether the call can be accepted or not. In order to prevent the CAC from being too optimistic a policing function, also referred to as an usage parameter control (UPC) function, is applied. It supervises the traffic of each connection, and if the negotiated bandwidth parameters are violated, the cells causing the violation are discarded or marked with a violation tag and dropped later at a switch node if congestion occurs. Deliberate attempts to exceed the negotiated bandwidth parameters and malfunctioning terminals are the main reasons for employing a policing function. To characterise the actions taken by the policing function the QOS parameters cell discard ratio and violation tag ratio can be used.

A critical issue when introducing ATM networks is the problem of measuring and testing[4]. The QOS parameters cell loss ratio, cell delay and cell delay variation are some of the ATM characteristics that need to be measured and tested. Important is also the performance testing of ATM specific functions like the call admission control and the usage parameter control.

When testing the relevant QOS parameters it is necessary to generate a variety of traffic types, including CBR and VBR traffic, using an ATM test generator (ATG). The generator generates two kinds of traffic: background traffic and test traffic. The background traffic is generated on one or more ATM connections in order to stress the system under test (SUT). Background cells have well defined channel identifiers (VCIs/VPIs) but they have no requirements with respect to the payload information. The test traffic is generated on some specific ATM connections and time stamping and sequence numbering is written in the payload of the test cells. At the receiving side an ATM traffic analyser (ATA) filters out the test cells and writes an additional time stamp into the payload. Finally the ATA processes the payload information in order to calculate the relevant performance data.

When testing the CAC function it is necessary to get information about e.g. the aggregate cell loss ratio at each outgoing multiplex. This information may be calculated if the number of arrived cells and the number of lost cells is measured at each outgoing multiplex during specific time periods. Testing of the UPC function may be performed by measuring the number of discarded cells and the number of violation tagged cells per connection over pre-defined periods of time.

The aim of this document is to describe the traffic measurements and the test abilities in the Danish research and experimental ATM network BATMAN. Firstly, section 2 gives a brief description of the BATMAN network. The following section describes the measurements performed at each outgoing multiplex of the switches and in each UPC function. Finally, section 4 describes the test equipment developed within the BATMAN project.

2. THE BATMAN NETWORK

The BATMAN network is an ATM network with 3 switch nodes: One in Aarhus at the research laboratories of Jutland Telephone, another in the centre of Copenhagen at KTAS and the final one at the Technical University of Denmark (TUD), in Lyngby (about 15 km north of Copenhagen). Each of the three switch nodes is connected by two bi-directional transmission links to both its neighbour nodes - see figure 1. The network forms a platform that allows the different partners to carry out experiments with a real ATM network.

The BATMAN project is divided into several phases. In phase I the switch nodes are interconnected with 2 Mbit/s PCM transmission lines. Although the transmission speed is low this phase makes it possible to gain experience with different topics of the ATM concept, such as usage parameter control, call admission control, cell delay variation and cell loss behaviour. Furthermore it makes it possible to make use of the test equipment developed in BATMAN at an early stage. In phase II the transmission speed is upgraded to 155 Mbit/s. This upgrade is planned to be made concurrently with the introduction of SDH in the Danish backbone network by the end of 1993.

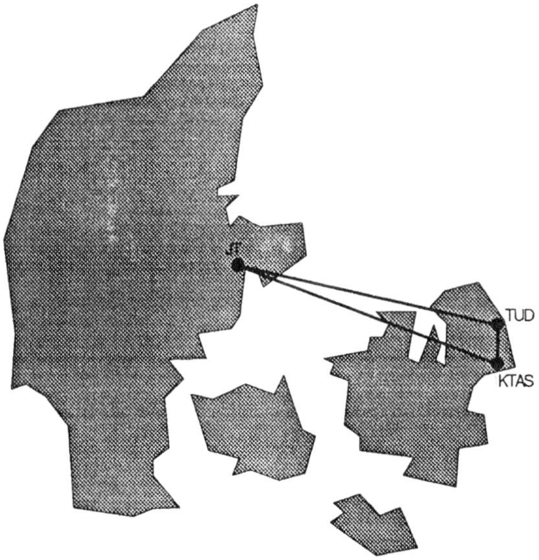

Figure 1
The triangular BATMAN network

The BATMAN switch is based on an existing PDH switch, where the main part of the hardware has been replaced, while the software has been re-used. In phase I it is a 16 by 16 output buffered switch with software controlled buffer sizes at the outlets, and in phase II it is a self-routing ring switch. At each inlet the individual ATM connections are controlled by three leaky buckets in parallel, i.e. up to three different traffic parameters may be controlled. The policing function is not only present at the UNI but also at the NNI, where a passive supervision can be used to gain information about the changes in characteristics of the traffic on a given ATM connection.

In order to load the network with different types of traffic patterns and to allow measurements of a number of key parameters an ATM traffic generator (ATG) and an ATM traffic analyser (ATA) have been developed.

The ATG can generate a number of connections with different statistical behaviour by a superposition of one or more mini-sources. The mini-source is a so-called on/off source i.e. it alters between an on state where cells are generated with a fixed rate and an off state where no cells are generated. By using a proper mix of mini-sources a large number of different VBR connections can be emulated. CBR connections can be generated by keeping a mini source in the on-state all the time.

In the payload of the cells generated by the ATG time stamping and sequence numbering may be applied. This information together with information provided by some ATM local analysers (ALAs) placed within the switch nodes can be used by the ATA to calculate delay, delay variation and cell loss at each of the nodes, as well as the corresponding end-to-end figures.

3. MEASUREMENTS IN THE SWITCH NODES

The measurement parameters described in this section are realised in phase I of BATMAN. Similar parameters will be realised in phase II.

Three sets of data are measured in the switch nodes:

1) Loss data
2) Delay data
3) Policing data

The loss data can be used to calculate the overall cell loss ratio and the cell loss ratio during congestion at each outgoing link, and the delay data can be used to determine the occupancy distribution at each multiplex buffer. The policing data gives information about cell discard and violation tagging occurring at the usage parameter control.

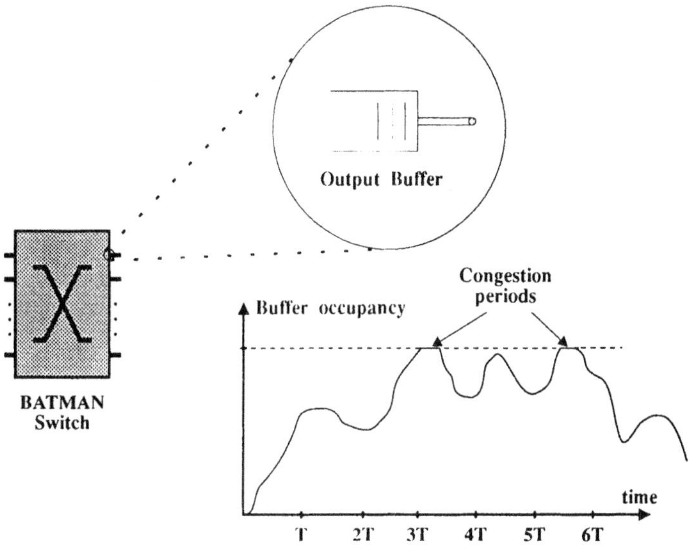

Figure 2
Switch node with outgoing multiplex buffers

The measurement data is transmitted to an external PC and stored in a memory storage, ready to be used for performance evaluation.

The multiplex buffer size and a priority threshold are programmable. High priority cells get lost if they arrive to a full buffer whereas low priority cells get lost if the buffer level exceeds the priority threshold when they arrive. The buffer size and the priority threshold are identical for all multiplex buffers in the network.

DELAY DATA

For each multiplex buffer the buffer level is read with a sample period T, and send to the PC as a 1 byte data message. The minimum sample period depends on the data speed and the number of multiplex buffers under observation.

LOSS DATA

At each multiplex buffer the cell loss ratio is measured, i.e. the ratio between the number of lost cells and the number of arrived cells during a long period of time. In consecutive intervals of duration kT the number of cell arrivals and the number of lost cells are counted, and this information is send to the PC as an 8 byte data message.

Another loss parameter measured at each multiplex buffer is the cell loss ratio during congestion, i.e. the ratio between the number of lost cells and the number of arrived cells during periods of congestion. A congestion period starts when a cell is lost and ends when n consecutive cells are not lost. For each such period the number of cell arrivals and the number of lost cells are counted, and this information is sent to the PC as a 4 byte data message.

POLICING DATA

At each entrance to the network there is an usage parameter control based on the leaky bucket algorithm that takes action whenever a connection violates its agreed parameters. Violating cells are either deleted or violation tagged. When the UPC takes action, it informs the PC about the type of action and about the connection identifier.

The UPC function has three leaky buckets in parallel. Each leaky bucket is characterised by a decrement and an increment factor, two threshold values, C_1 and C_2, and a state parameter PS. The UPC may operate in four different states:

PS:	Action taken:
00	No action is taken
01	Violation tagging if C_1 is exceeded
10	Cell deletion if C_1 is exceeded
11	Violation tagging if C_1 is exceeded and cell deletion if C_2 is exceeded

The measurements in the switch nodes are of interest from a network management point of view and when testing the CAC and the UPC functions.

Both the switch data and the UPC data can be used to calculate management information about the network status on the ATM level. This information can e.g. be used to generate alarms if the cell loss ratio at a given outgoing link rises to an unacceptable level or if an UPC discards a too large fraction of an users cells.

As the CAC uses a QOS estimate for each link on the route between the originating and terminating point of an ATM connection to decide whether a request for a new connection can be accepted or not, the switch data can be used to test the CAC function. It is possible to check whether the QOS estimation algorithm is reliable or not, and the accuracy of the estimates can be tested.

4. TEST GENERATOR AND ANALYSER

The tool-set developed for the evaluation of the integrated BATMAN network consists of 3 different parts :

1) ATM Traffic Generator (ATG)
2) ATM Traffic Analyser (ATA)
3) ATM Local Analyser (ALA)

The traffic generator and the traffic analyser is used to carry out end-to-end evaluations, while the local analyser is used to make measurements on a single network component. However, all information (global or local) is finally processed and presented to the operator by the ATA.

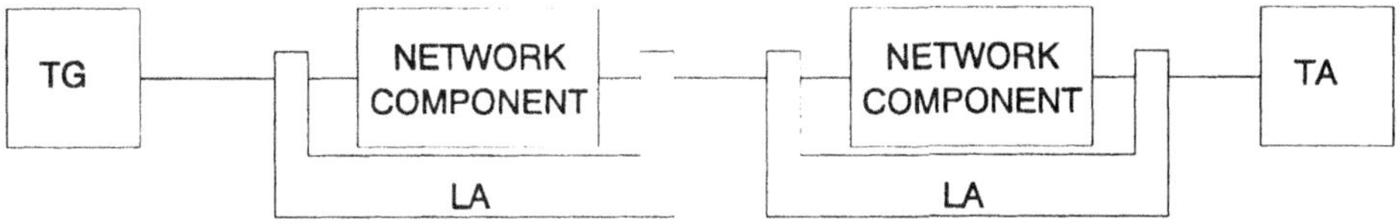

Figure 3
Measurement configuration with traffic generator, traffic analyser and a number of local analysers.

The information required to carry out the specified evaluations is stored within the datafield of the cell and the cells used for these measurement are to be considered as pure probe cells and cannot be used to carry any "user data".

While all cells emitted by the ATG can contain global time stamp and sequence number, only cells with a specific VPI are used for the local evaluation. This allows to mix probe cells with real user data without disturbing the contents of the "real cells".

If only end-to-end parameters are measured the use of local analysers are not required.

ATM TRAFFIC GENERATOR

The main objective for the traffic generator is to load a network (or just a single network component) with a basic load which enables the operator to carry out measurement in an environment with a stochastic behaviour.

The basic module in the traffic generator is a so-called on/off source (here denoted a mini-source), that alternates between a state where the source is inactive and a state where cells are emitted with a constant rate. The time the source remains in either on or off state follows a geometric probability distribution. By a special parameter setting it is possible to force the source to remain active and thereby create a CBR source.

The final traffic flow from the ATG is created by superimposing a number of these on/off or mini-sources, as illustrated in figure 4.

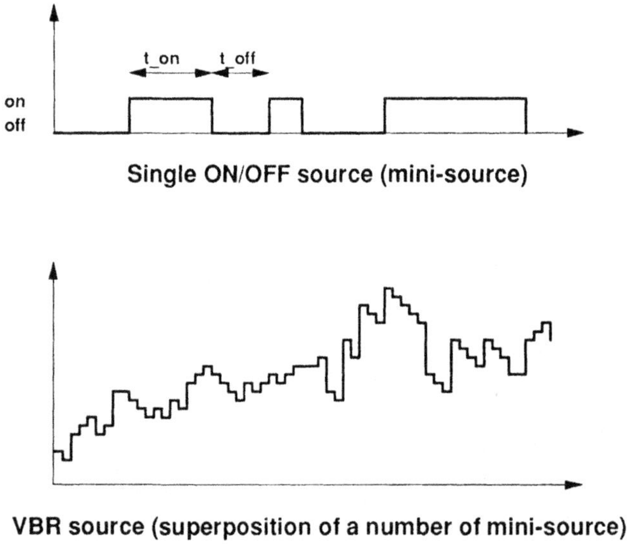

Figure 4.
Source characteristics of basic source and VBR source.

The mini-sources are grouped in groups of 4. These 4 mini sources are using the same parameters, which means : the average time in on-state, the average time in off-state, the transmission rate in on-state, VPI and VCI. Further it is possible to control the number of sources within a group that is active.

One or more of these groups of 4 sources can used to emulate a VBR source.

Dependent on the configuration of the traffic generator, it will be possible to implement between 200 and 1600 mini sources (or 50 to 400 VBR sources).

To every VBR source is related a 53 byte memory, used to store the header and data segment of the cells. Alternatively the data segment can be initialised with a global time stamp and sequence number to be used by the ATA, as shown in figure 5.

The traffic generator is based on a pseudo random number generator with a cycle time larger than 1 month (based on 155 Mbit/s transmission rate).

The ATG is implemented on a PCB to be located in a 386 PC-AT. The setting of parameters for the sources and the contents of the header and data segment is made using a Microsoft Windows based interface.

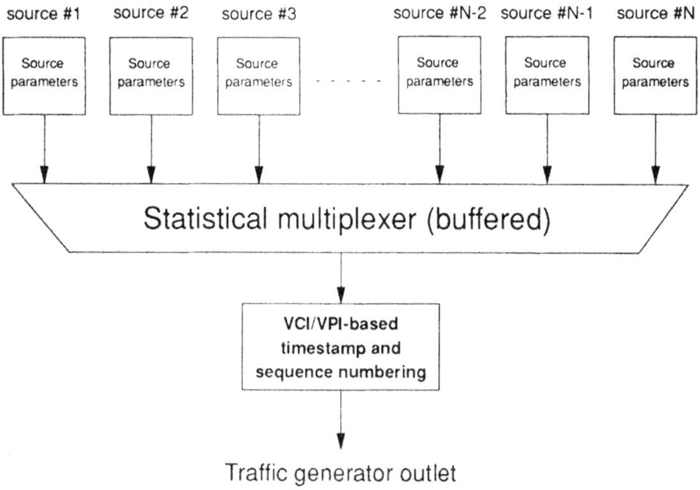

Figure 5
Configuration of traffic generator

ATM TRAFFIC ANALYSER

The ATA is used to calculate the global delay (GD) and global cell loss (GC). At the entrance to the ATA a filter is selecting the probe cells that are to be used in the evaluation. After this the GD and GC are calculated and written into the datafield of the cell. Finally the information is stored in a memory storage, ready to be used for performance evaluation. The amount of information stored is selectable. The ATA can either store the entire probe cells (with or without header) or filter out information only related to a cell loss, see figure 6.

To carry out the GD measurement it is required that both the ATG and the ATA are located within the same PC.

All the statistical processing and presentation of results are based on software in the PC running under Microsoft Windows.

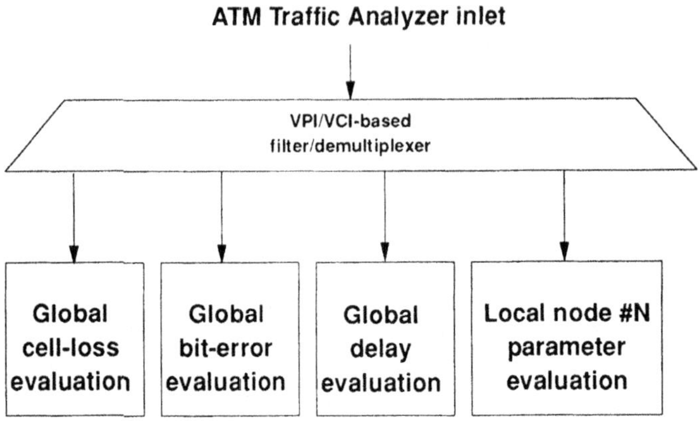

Figure 6.
Configuration of Traffic Analyzer.

ATM LOCAL ANALYSER

The objective of the local analyser is to measure the local delay and cell loss. At the entrance to the network component (that has been selected for measurement) the test cells are given a local time stamp and sequence number. At the output this information is used to determine whether a cell loss has occurred and to calculate the cell delay. A pointer is used to indicate where this local information shall be stored within the datafield (Dependent on the number of network component that the cell has passed prior to this one).

As all the information measured within the Local Analyser is stored within the test cell payload, this information can be used by the ATA to derive detailed information about the global measurements, i.e. in which nodes the delay and cell loss have occurred.

CONFORMANCE VS. PERFORMANCE TESTING

An important part of the establishment of a communication test network is the verification and evaluation process. In the case of ATM, that differs significantly compared to the existing POT network, it is important to divide this process into 2 distinct parts:

1) test of the basic functional behaviours (conformance test - debugging)
2) performance evaluation and verification of integrated functional interaction.

The first type of test is mainly carried out during the integration phase of the network and relates to test a fully predictable result. During these tests it will be verified that e.g. the switch can switch cells from one inlet to another outlet (without cell loss and bit errors during a short period) and that the police function can delete cells.

The second type of test will be performed when the initial test has shown that the individual functional modules behave as expected. The main objective of these tests is to gain a better conceptual understanding of ATM in areas where the behaviour is difficult to predict by using simulations or analysis. As the result from these tests are more or less self standing (without the possibility to be theoretically verified) it is important that the initial test has gained confidence in the basic functional behaviour.

The tools developed within the BATMAN project for evaluation have mainly focused on the second type of test which is considered the most difficult to carry out. Although these tools as well could have been used for the initial test, more simple and easy to use tools where selected (simplified versions of the final tool set).

NETWORK INTERPRETATION OF TEST SET-UP.

One of the objectives with the BATMAN test network is to allow the users with interest in traffic theory to carry out tests, that to some extend, can be used in comparison with analytical predictions of the behaviour of an ATM network. However, before these tests can be carried out it is important to clarify the differences between the physical implementation and the (often) simplified network configuration used in the analysis.

Considering a B-ISDN network configuration it is possible to use the ATG for different applications. In addition to a configuration where the ATG is used as a single terminal (and a large number of ATG are used to carry out the test scenario), the test generator can as well be considered as a B-NT2 (i.e. the traffic originating from a number of terminals within the users premises). Due to the (statistical) multiplexing of a number of sources and finite capacity of the output from the traffic generator (155 Mbit/s) a certain amount of cell delay variations (compared to the source model) must be foreseen even before the cells leave the generator. This is an important difference compared to common theoretical evaluations of queuing systems, where cells from different sources can arrive concurrently to the queue.

5. SUMMARY

The BATMAN testbed is considered an experimental network, which allows the users to carry out a variety of different measurements and tests. Performance data giving information about overall cell loss and delay is measured in the switch nodes and transmitted to an external PC. Background traffic and test traffic with different statistical behaviour is generated by superposition of a number of on/off sources. The test traffic is analyzed with respect to cell loss, cell delay and cell delay variation.

The first performance tests that will be carried out will be used to investigate the following topics:

* Violation tagging
* Cell delay and cell delay variation
* Cell loss and consecutive cell loss
* Call admission control
* Usage parameter control (leaky bucket with violation tagging)

The impact of tagging on the network performance will be clarified. Cell delay variation (CDV) as defined by CCITT SG XVIII will be carefully tested. The burst character of cell losses, or the consecutive cell loss phenomena, will be evaluated by the parameter: cell loss rate during congestion. And the efficiency of different CAC proposals will be investigated.

ACKNOWLEDGEMENT

The authors will like to thank the project partners who have carried out the actual implementation for detailed information about the different functionalities.

REFERENCES

[1] CCITT Recommendation I.35B, "B-ISDN ATM layer cell transfer performance".

[2] C. Rasmussen, J.H.Sørensen, K.S. Kvols, S.B. Jacobsen, "Source-Independent Cal Acceptance Procedures in ATM Networks", Journal on Selected Areas in Communications vol 9, no. 3, April 1991.

[3] CCITT Recommendation I.371, "Traffic control and congestion control in B-ISDN".

[4] R. Grünenfelder and R. Brander, "Testing will assure the quality of service in an ATM environment", *Telephony*, pp40-43, October 5, 1992.

Integrated Broadband Communication Networks
and Services (C-18) / V.B. Iversen (Editor)
Elsevier Science B.V (North-Holland)

The Synthesized Traffic Generator; Objectives, design and capabilities

Bjarne E. Helvik[a], Ole Melteig[b] and Leif Morland[a]

[a] SINTEF DELAB, N-7034 Trondheim, Norway
[b] Norwegian Telecom Research, N-2007 Kjeller, Norway

Abstract

To validate the performance of an ATM based B-ISDN, extensive traffic measurements are necessary. The network performance measures obtained, e.g. cell loss, delay and jitter, will depend heavily on the load put on the system. Hence, the validity of the measurements is conditioned by the realism of the load put on the system during the measurement. This paper describes an ATM traffic generator, the STG (Synthesized Traffic Generator), which is designed to generate a traffic load as realistic as possible. The modelling framework on which the STG is based, is outlined. The source type models includes the behaviour of physical sources and the formation of ATM cell streams done by the terminal equipment. Load scenarios defines time varying traffic mixes towards various destinations. An STG can generate the multiplexed traffic stemming from up to 2048 independent sources. A nearly infinite number of independent and reproducible cell load sequences, with the same statistical properties, may be generated for each scenario. The work is as a part of the RACE project PARASOL (R 1083).

1 INTRODUCTION

New generations of high capacity networks and switching systems based on the ATM (Asynchronous Transfer Mode) principle are on their way both in the public and private domain. The capacity (bandwidth) on demand idea underlying ATM poses a wide range of challenges with respect to traffic engineering. The characteristics of the load put on the systems are partly unknown, new call handling and switching techniques are introduced and very strict requirements are put on grade of service parameters like cell loss frequencies and cell delays.

Unfortunately, our possibility to perform a thorough validation of the traffic carrying capabilities of these new systems is restricted. A sufficiently detailed mathematical analysis is impossible and simulations are too demanding. Hence, measurements and experimentation are our prime means to gain confidence in this new and yet not fully proven transfer technology.

Carrying out these measurements, the traffic load put on the system is the most important constituent with respect to the results obtained. The best approach is of course to put controlled, real traffic on the system. This, however, is not feasible. Real traffic can not be kept sufficiently stable and be varied with respect to load and mix to obtained controlled and accurate results. A realistic loading of a system requires an array of independent generators, each producing the traffic stemming from a large number of independent sources. In addition to established sources/ services, it should be possible to load ATM systems with

- traffic from foreseen not yet available services, and
- traffic corresponding to various choices with respect to service specifications and terminal equipment design,

to test effect of these on the network performance and the QoS before decisions are made.

The above triggered our work on traffic generation which so far has resulted in the Synthesized Traffic Generator, denoted the STG during the rest of the paper. Its properties can be outlined as follows:

- The traffic load stems from a large number of independent sources (maximum 2048) each with its unique virtual circuit (VC), payload, etc.
- The models of the sources are physically interpretable, reflects both the information transfer caused by the application/service, and the protocols and equipment of the end systems. The models are adaptable to a wide range of source types, e.g. speech, variable bitrate coded video, various kinds of data traffic, and new and yet unknown services/equipments.
- The generated traffic reproduces the short and long term properties of real traffic, i.e. the range from microseconds to hours.
- Generation scenarios with a mix of traffic from many different types of sources with different and time-varying traffic interests.
- A nearly infinite number of independent and reproducible generations of each scenario.
- A compact and efficient hardware implementation.

Most of the work reported here is carried out within the RACE project PARASOL (R 1083). PARASOL was established to define measurement and validation methods for ATM systems and to develop prototype equipment. At the end of the project, beginning of 1993, a complete modular measurement system was available. This system is briefly described in Section 3.1. For a more detailed presentation of PARASOL see [9] and [3].

The next section presents our source modelling approach and the traffic model on which the generator are based. Section 3 presents the STG design and how it interacts with the rest of the PARASOL measurement system. The two final sections outline the capabilities of the generator and its intended use. The paper is concluded in Section 6 with the current status and a sketch of our future plans.

2 SOURCE MODELLING

2.1 Modelling approach

In the context of traffic generation, we can divide between the following approaches:

Memory based generation; where a prerecorded or predefined cell sequence is reproduced during generation. Due to its determinism and simplicity, this approach is suited for initial functional testing. It is, however, unsuitable for validation of traffic handling capabilities since limited cell sequences will be available with a reasonable memory.

Generation according to a class of stochastic processes; for instance renewal processes or (multiple) on-off sources. The parameters of the stochastic process are chosen to fit some of the statistics of a real source. This may be regarded as a black box approach. In this case it is possible to generate long independent sequences, but it still remains open how well the real traffic is represented.

Physically based source models; which also are based on stochastic processes, but in this approach the detailed behaviour of the source is attempted reproduced. It may be regarded as a white box approach were the inner workings of the physical source and end system are modelled.

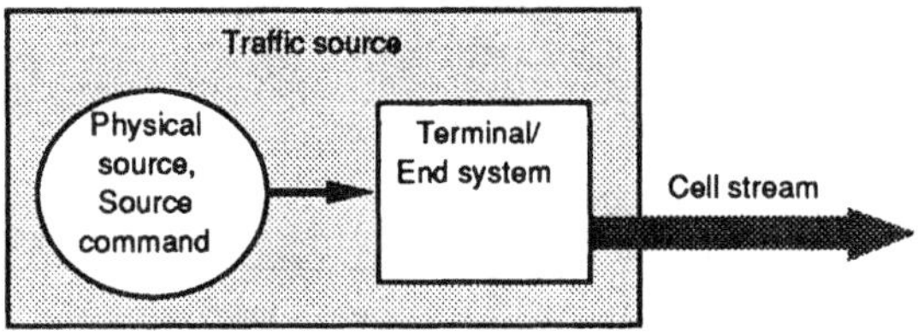

Figure 1 Basic elements of a source type.

As evident from the introduction, the last of these approaches is advocated. Before presenting the source type model itself, it is useful to consider the general traffic generation process. Regard the sketch in Figure 1, showing the basic elements which we consider in the modelling process.

- The *physical source* itself. This may for instance be speech from a man, a still picture, a video, a text, etc. In addition, there is a command function associated by the physical source, i.e. when and with whom to communicate.
- The *terminal* and/or *end system* which the physical source uses to communicate. For instance, the sampling of voice, removal of "silent samples", assembling of "active" voice samples into ATM cells and transmission of these. Note that the terminal - end system may include signal processing, passage through a number of interfering protocol layers and even transmission over intermediate non ATM networks.
- The *cell stream* resulting from the above processes.

While the black box approach is to derive a model which has some of the statistical characteristics of a sample cell stream, see for instance [5], the white box approach is to base the model on knowledge about the traffic source itself. That is,

- the information content and the dynamics of the physical source,
- the transformation and compression of the original bit stream performed by the terminal equipment. For instance by video codecs, signal processing and compressing.
- the packetizing, splitting, spacing etc. performed by various protocol layers, e.g. the transport layer and the AAL (ATM Adaptation Layer).

The Composite Source Model, to be presented in a later subsection, is primarily aimed at the white box approach. The model may, however, also be adapted to produce a cell stream having a required set of statistics. Work is going on to establish a varied set of traffic sources within the composite source modelling framework [6].

2.2 Activities modelled

The basic element in the composite source model is the source type definitions which models the behaviour of various kinds of ATM sources. Before describing the modelling of source types and how models are defined by a number of sources of the various types, the activities of ATM sources should be regarded. These are often split into four (or three) levels corresponding to their time scale. See Figure 2 for an illustration.

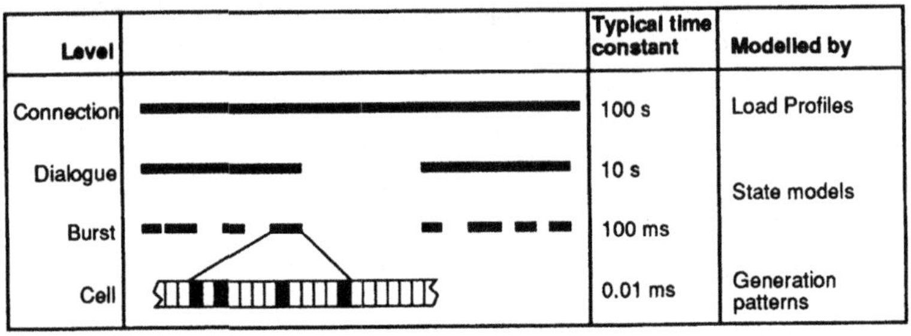

Figure 2 Depiction of ATM source activity and corresponding model levels.

Connection level, describing connects and disconnects of the virtual circuit. Typical time constants are in the order of 100 s. This level is related to the command function associated by the physical source. When to communicate (the time when a call is established) is in the STG modelled by the load profile. Where to communicate (call destination) is modelled by the cell header (VCI, VPI, destination). These characteristics are defined in the scenario set-up, see Section 2.4, which also includes a traffic mix towards various destinations, which may vary with the time.

Dialogue level, describing the alternations between the A and B party in a connection. Typical time constants are in the order of 10 s. Since the STG generates simplex traffic, only one side may be included in the model.

Burst level, describing the behaviour of the active party. The activities on this level are primarily governed by the physical source. Time constants are in the order of 100 ms. The dialogue- and burst levels are modelled by a finite state machine.

Cell level, describing the actual cell generation. This level is primarily governed by the terminal and/or end system. Time constants are in the order of 0.01 ms. The cell level is encompassed in the Composite Source Model by a repetitive cell pattern associated with each state.

2.3 The composite source model

As indicated in the section above, the activities of a source type are modelled by state diagrams and generation patterns. A more thorough presentation and discussion of the model is found in [7]. Note that we first model source types and in the final model define a number individual and independent sources with a behaviour defined by each of the source types.

State models

As mentioned, the dialogue- and burst levels of a source type are modelled by a finite state machine (state model). Its behaviour is that of a finite state continuous time Markov process [2]. In its simplest form, each activity is represented by a single state. However, the sojourn time of an activity can not always be properly modelled by a negative exponential distribution. Furthermore, it can not always be modelled as independent of the previous state/activity. Hence, such activities are modelled by groups of states, which may have Coxian, Erlang, Hyper-exponential or general phase type distributed sojourn times [12]. Dependencies may be introduced by letting the entry probability of a group be dependent on the previous state. See Figure 3 for an example. In the figure the "large task" have a non exponential sojourn time when it starts after the source has been idle. Entered from above, it has a negative exponential sojourn time distri-

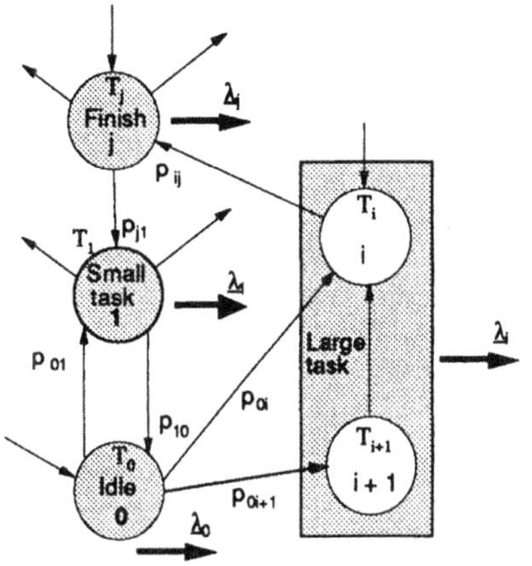

Figure 3 Generalized example of a source type model.

bution. Formally, the activity on the burst and dialogue level of source type k is modelled by a Markov process with transition matrix $\Omega^{(k)}$. Since it is more descriptive we have used the transition probabilities $p_{ij}^{(k)}$ and the state sojourn times $T_i^{(k)}$ in the model definitions, i.e.

$$\Omega^{(k)} = \{\omega_{ij}^{(k)}\} \qquad \omega_{ij}^{(k)} = \frac{p_{ij}^{(k)}}{T_i^{(k)}} \tag{1}$$

Patterns

Cell generation patterns models the cell stream resulting from the information transfer associated with each state. This stream is formed by the terminal and/or end system. Each state has a pattern λ associated with it, see Figure 3. The pattern is a binary vector where 1 corresponds to the generation of a cell, otherwise 0. As a trade off between modelling power and implementation we let this pattern be of constant length L and cells are generated cyclically according to this pattern as long as the source is in the corresponding state. Figure 5 gives an illustration of two patterns.

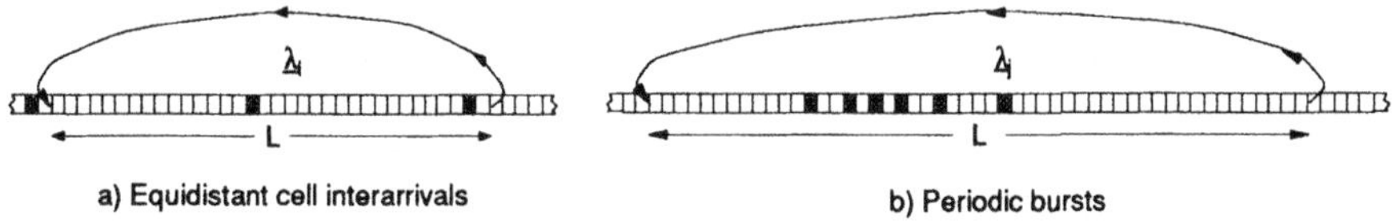

Figure 4 Examples of cell generation patterns.

Constant length patterns give a reasonable model of the behaviour of the terminal and end system since this usually has a repetitive behaviour. In the generator we have two kinds of patterns:

- ordinary, with a length $1024 \le L_B \le 4096$ with a default value 2048 which roughly corresponds to a repetition frequency of 64 khz on a 155 Mbit/s link. Most source-types and states may be modelled with ordinary patterns.

- long patterns, with a length $n \cdot L_B$ $n = 2, 3, \ldots, 32$. These are used to model slower cycles, for instance video frames.

The properties of this model is studied in [8]. The source type in the Composite Source Model may also be reformulated as a DMAP process, introduced by Blondia, [1].

Composition

So far, it is dealt with the source type. In the rest of this subsection, it is outlined how many sources of different types are handled. Let say we have K different source types with $m^{(1)}, \ldots, m^{(k)}, \ldots, m^{(K)}$ sources of the different types. Of the k'th type, let $m_i^{(k)}(t)$ denote the number of sources in (source type specific) state i at time t. Since the composition of Markov processes also is a Markov process, the set of $\{m_i^{(k)}(t)\}_{\forall(k, i)}$ constitutes the global state of the part of model defined by state machines. The selection of the next global event, state change and time when it occurs, is governed by the probabilities shown in Table 1.

Table 1 Parameters governing the next transition of the composite model.

The time until next state change is negative exponentially distributed with parameter	$\sum_k \sum_i \frac{m_i^{(k)}(t)}{T_i^{(k)}}$	(2)
The next type k and state i combination to be affected by a state change is given by the probability	$\frac{m_i^{(k)}(t)}{\sum_k \sum_i m_i^{(k)}(t)}$	(3)
The specific source among these which is changing state is given by the probability	$\frac{1}{m_i^{(k)}(t)}$	(4)
The next state j of this source is given by the probability	p_{ij}	(5)

The cell pattern transmitted from the composition of $m^{(1)}, \ldots, m^{(k)}, \ldots, m^{(K)}$ sources is denoted the periodic cell transmission pattern (PCTP). It is, with two modifications, the sum of the patterns of all the sources added together. When a source is added the identity (e.g. VCI) of the source is included. The modifications are:

- The start of the pattern is aligned with the position in the PCTP which determines the cell generated when the source entered its current state.
- If a cell from a pattern is attempted added into an already occupied position in the PCTP, it is shifted ahead to the first vacant position in the PCTP.

This procedure is illustrated in Figure 5. For a formal description, it is referred to [7]. If the PCTP at time t is regarded as a vector $\underline{S}(t)$, cells are transmitted cyclically from this pattern, i.e. at time t a cell from source $S_{t \bmod L}(t)$ is transmitted.

2.4 Scenarios

The sources of the various types sending traffic toward the destinations in the network is defined in a scenario. A setup is exemplified in Table 2. Most of the entries are self explaining. Foreground sources are sources on which it is performed end-to-end measurements by the PARASOL system. The destination tag is an indicator derived from the VCI/VPI in the cell header which highlights a terminal- or a point within the target system, the traffic is routed to.

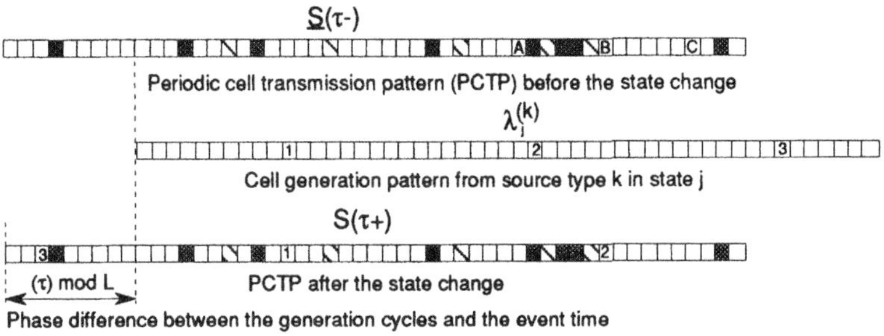

Figure 5 Example of the periodic cell transmission pattern (PCTP) immediately before and after a source changes state. Cells from the actual source are denoted A, B, C, before and 1, 2, 3 after.

Table 2 Example of a scenario setup table.

Source Id	For-ground	Source Type	Destin. Tag	Cell Header	Cell Payload	Load Profile
001-100		Phone	A	`Range of Names´	`Name´	Jumpy
200	Y	HDTV	A	`Name´	`Name´	Const.
300-350		File Transfer	B	`Name´	`Range of Names´	Step

The load profiles needs a further explanation. The load profiles models the connection levels as indicated in Figure 2. During generation it defines the relative number out of the total number of sources which are active at the various times. By active is meant that they are generating cells. See Figure 6 for an example on how six profiles governs the traffic load during a measurement.

In Figure 7 all the elements introduced above which constitute a scenario are illustrated. The "other parameters" are measurement period, random number seeds, initialisation, which generator to run it on, etc.

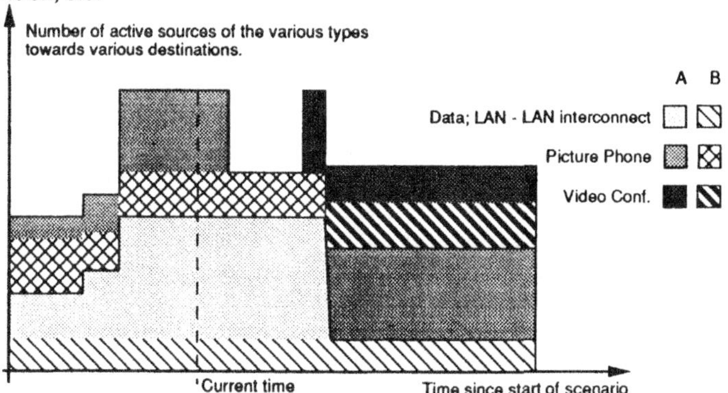

Figure 6 Example of scenario load profiles in terms of number of sources.

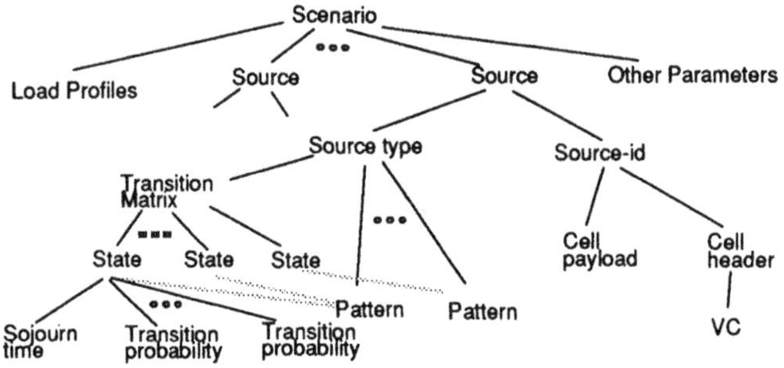

Figure 7 Elements contained in the traffic model constituting a scenario.

3 DESIGN

3.1 The PARASOL measurement system

The measurement system consists of three main parts as shown in Figure 8. The *generator* which generates load including test cells towards the system under test (SUT), the *analyser* which analyses the test cells from the SUT and the control workstation which is used to set up the instrument, and for presentation the results. These three parts can in principle reside in the

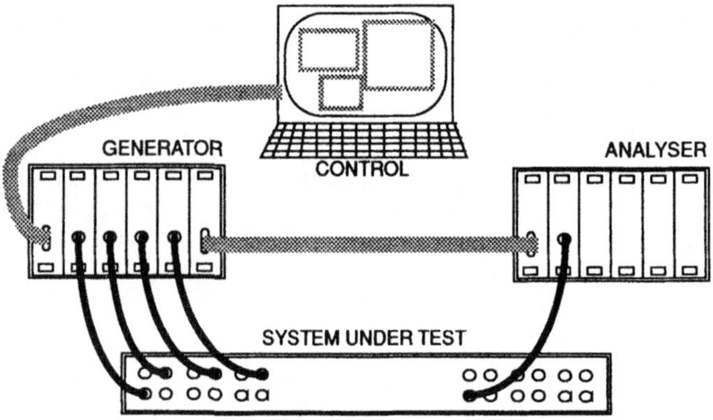

Figure 8 The parasol measurement system in a bulk load configuration.

same physical cabinet, or they can be distributed over a wide area, connected by an ethernet. The current implementation requires the three parts to be in the same room. The test tool is based on modules, as shown in Figure 9. Each module performs a specific task. This makes it a flexible tool, which is easily reconfigured to perform a wide range of measurements. The modules are listed in Table 3 and examples of system configurations are shown in Table 4.

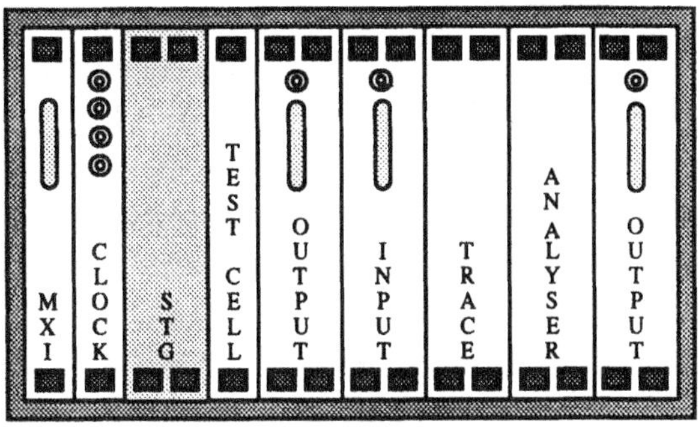

Figure 9 The modular system.

Table 3 Modules of the PARASOL measurement system.

	Module	Function
TCM	TIMING CONTROL	Controls the clocking of the system
IM	INPUT	Input an ATM/SDH cell stream into the system
OM	OUTPUT	Outputs an ATM/SDH cell stream from the system.
MTG	MEMORY TRAFFIC GENERATOR	Produces cell based on an sequence downloaded to the module
STG	SYNTHEZISED TRAFFIC GENERATOR	Produces realistic traffic in real time.
TGM	TEST CELL GENERATOR	Adds/Replaces cell with test cells on the outgoing cell stream
IRM	IMPAIRMENT RECORDER	Records cell losses/delay in live traffic.
IGM	IMPAIRMENT GENERATOR	Produces cell loss/delay to live traffic, based on a fixed or a downloaded pattern.
TM	TRACE	Traces the cellstream (parts or complete)
AM	ANALYSER	Analyses the cell stream, and sends the results to the controller/MMI

3.2 The STG subsystem

STGs are used when a realistic traffic from the test tool is required. The traffic can either be foreground on which measurements are performed, background, or both. Other modules which are required in this generator configuration are the TCM, the TGM and the OM. The software and hardware elements that constitutes the STG sub-system are shown in Figure 10. The software which includes the man machine interface (MMI) consisting of several graphical editors for the STG, the Traffic Handler with the STG database performing the necessary checking and compilation of source types and scenarios, and the STG specific drivers, are dealt with at the end of the section.

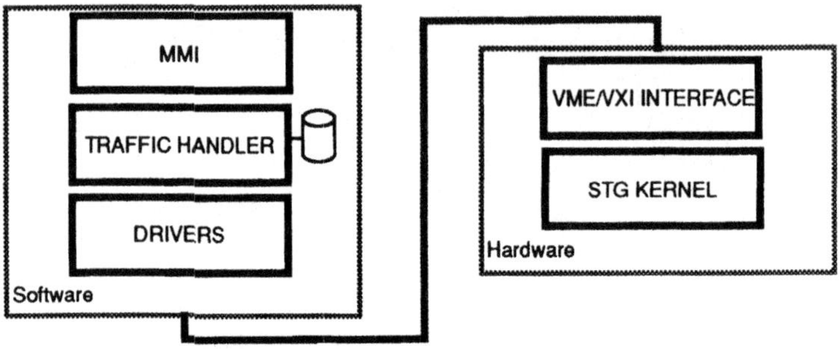

Figure 10 STG sub-system block diagram.

Table 4 Examples of system configurations.

Measurement configurations	TCM	IM	OM	MTM	STG	TGM	IRM	IGM	TM	AM
A complete test instrument in one box	X	X	X		X	X				X
A bulk generator	X		3X		3X					
A specialized analyser	X	X							2X	2X
Adding test cells to live traffic	X	X	X			X				
Impairment to live traffic	X	X	X					X		
Non Measurement configurations										
A multiplexer	X	4X	X							
A demultiplexer	X	X	4X							

Hardware

The hardware is made as a two slot C-size VXI module. It consists of 12 FPGAs (field programmable gate arrays), each in the range 10-20 000 gates and 10 Mbit of RAM. One of the slots serves as the VME/VXI interface, which handles the communication with the rest of the instrument. The other slot is the STG kernel and handles the composite source model. The functionality is arranged as shown in Figure 11.

In the block to the left, the next time, and old/new state are found for the next state change, cf. (2) and (3) of Table 1. These are transferred to the next block, and the block is ready for the next operation. The FIND SOURCE block finds the source for the state change, cf. (4) and (5), and finds the pattern associated with the new state. These data are sent down to the UPDATE PATTERN block together with the time. The FIND SOURCE block also handles the activations/ deactivations sent by the software, and the long patterns. Long patterns are handled as a set of normal patterns resulting in a "state change" for each new part of the long pattern.

The UPDATE PATTERN block updates up to 8 state changes in parallel, at the time requested. The old patterns are removed from the PCTP (periodic cell transmission pattern) and the new ones added, see Figure 5. For each 2.7 microsecond a source_id is read out from the PCTP and

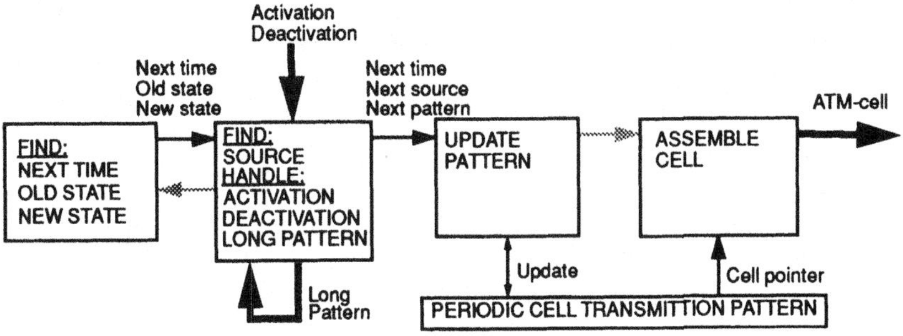

Figure 11 The STG hardware block diagram.

sent to the ASSEMBLE CELL block. This block uses the source_id from the PCTP as a pointer to a cell with header and payload, and sends this cell out to the next module in the test tool. A cell may be transformed into a test cell (foreground traffic) before it ends up in the output module (OM).

The two blocks to the left, which handle the stochastic parts of the model, use four Tausworthe pseudo-random generators with primitive trinomials between 25-34 bit [14], [15]. The sequence lengths of each random generator are mutually prime. The generators are clocked as many time as the used length between each time a number is read out, to ensure no correlation between the numbers.

The activation and deactivation are used to emulate the connection level. The software sends activation and deactivation messages to the hardware, at the time these should take place according to the load profiles. There will be one profile for each source type to a specific destination in the network, see Table 2 for an example. In a profile the number of sources of a source type to a destination will vary according to the graph. When the level changes, the software selects which sources to activate/deactivate. Since the properties of all source of the same type to the same destination are identical, an arbitrary of these may be chosen. The software handles all profiles for all source types to all destination simultaneously, cf. Figure 6. When a profile changes, the activation/deactivation which corresponds to change will be sent to the hardware. The changes are effectuated in the hardware as soon as possible, which are maximum 8 activations or deactivations per 0.1 ms and on the average 7 at nominal load.

Software

The software is object oriented and programmed i C and C++. The total code length is in the about 200 000 lines. The software is build according to a server / client model, and all communications between the modules passes through a *postmaster*. This makes it easy to make changes in the internal communication.

The *traffic handler* is the core of the STG software and handles the database for all the definitions and libraries. It also checks the parameters and calculates values for the user, like average and maximum load, average rate of state and pattern changes. Graphs like the expected load as a function of time and type/destination are also calculated and presented to the user. The hardware is initialized through the *instrument manager / agent*, by calculating the necessary parameters to set the STG in a steady state, and then downloading the data to the hardware.

The *Man Machine Interface* runs under X-windows / MOTIF. It includes a number of editors. Source types, patterns, profiles and scenarios are named and available from a library.

- In the source types editor the sources are defined as a state diagram with a "bubble" for each state and the sojourn time and a pattern_id inside. Arrows between the states hold the (non null) next state probabilities, see Figure 3.
- The patterns are defined graphically as a bitmap on the lowest level. It is possible to zoom out to see the complete pattern, and then the bits are represented as bar-graph with both colour and height as an indication of the cell rate. The bitmap looks similar to the cell level in Figure 5.
- Load profiles for activation / deactivation are defined in an editor as a graph like those indicated in Table 2. For each measurement, a *scenario* is defined. In this editor the number of sources, source type, destination, cell header / payload and load profile are put together, see Table 2.

4 GENERATOR CAPABILITIES

Number of different sources

One module is capable of generating traffic from up to 2048 sources. Assuming that telephony would occupy all the link capacity and the source bandwidth requirement for the user will be equivalent to that in the existing PCM system, a 100% utilisation for this purpose would constitute some 2175 sources. Taking into account that at least some of the sources will require higher bandwidth than 64 Kbit/s, and that even 80% utilisation of the capacity is an optimistic estimate, it is believed that 2048 sources more than covers the need for describing the link traffic. For extreme requirement more sources can be provided by merging traffic from two or more adjacent STGs in a mainframe.

Source type complexity in a scenario

A source type characterizes sources with identical statistical behaviour, see Section 2.3. The STG can handle 128 states, which may be used for ten source types each comprising 12-13 states. This was regarded as sufficient. The user is not restricted in the utilization of the available states, but may define anything between the extremes a single source type of 128 states or 128 types each having one state.

Pattern lengths and long patterns

The traffic stemming from one source in an active state is contained in a pattern, cf. Figure 4. As default the length of this pattern is 2048 cell slots. By putting one cell into the pattern of this length the lowest bandwidth of the default pattern length is achieved and given by $155 \times 10^6 \text{bit/s} \cdot 1/2048 \cdot 48/53 = 69 \times 10^3 \text{bit/s}$ as the link bandwidth is 155 Mbit/s and the payload to cell length ratio is 48/53.

If one anticipates that a narrow band constant bitrate service (CBRS), like conventional PCM telephony above, should constitute a significant part of the traffic in the ATM network it becomes important to generate this type exactly. This can be accomplished by setting the pattern to the length of 2193 cell slots. The pattern length may be varied between 1024 and 2048 cells

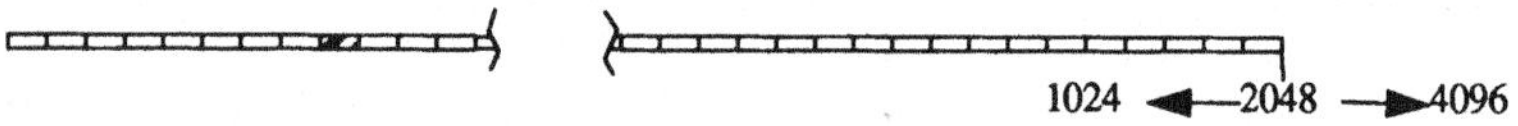

Figure 12 Range of the lengths of basic patterns in the STG.

Some source types may have specific cell transmission sequences exceeding the maximum pattern length, for instance a variable bitrate (VBR) video codec and transfer system exhibiting a frame periodic behaviour. The STG is provided with long patterns to allow the user to model such events. A long pattern is simply a concatenation of between 2 and 32 ordinary patterns thus covering a cell sequence lasting for up to 0.179 s. The cells from long patterns are transmitted to the link as long as the source stays in the corresponding state.

Sojourn times and activity rate

The ratio between maximum and minimum mean sojourn times of the states of source types in a specific scenario is limited by the number of digits in various variables in the hardware arithmetic of the STG. It will also depend on the number of sources in the traffic specified by the user. A scale factor ranging from 0 to 15 allows the range of the sojourn time to be altered as shown in Table 5. The time between events at burst and dialogue level has a granularity of 2.7 μs and can be up to 3.5 hours irrespective of the scenario size.

Table 5 Ranges of the mean sojourn times of the states of source types dependent on scale factor [x] and number of sources in a scenario.

Number of sources	Lowest range [0]		Highest range [15]	
	min.	max.	min.	max.
1	2.7 μs	22 s	88 ms	205 hours
2000	2.7 μs	11 ms	88 ms	6 min.

Random number generation

For verification purposes, it is mandatory to load the system under test with non periodic and non repetitive traffic. The variates used in the STG, cf. Table 1, are generated by feedback shift registers [14]. The periods for these are 2^{33}-1, 2^{34}-1, 2^{25}-1 and 2^{31}-1, which are relative prime numbers. The number of shifts between draws are 33, 34, 25 and 31 respectively and are not divisible the cycle length. Thus the number of state changes (variate sets) before the sets of numbers from the four generators will be repeated equals the product of the individual generator periods. If the nominal value of one state change per 0.1 ms is observed, the STG will produce non repetitive traffic for 33×10^{24} years. The traffic will then be repeated only if all states contain exactly the same number of sources as when the STG was started, which is most unlikely.

By reading the content of the shift registers at the end of a measurement block, and use these as generator seeds for the next experiment, the user has a guarantee of independent measurement blocks. Note also that the STG may be initialized to a global state drawn from its steady state distribution at the beginning of each experiment.

It should also be mentioned that the STG may serve as a deterministic pattern generator with a period equal to that of the maximum long pattern, see the subsection above, and one state. Thereby the STG also offers the features of debugging tools like memory based traffic generators.

Load extremes

By specifying patterns where all cells are assigned, a 100% load is obtainable. This is of interest as a debugging feature. For traffic experiments, the maximum load is defined by the user, and may exceed 100% and cause internal cell losses. The superposition of the traffic from the individual sources in a scenario, Section 2.3, may be regarded as a multiplexer with a limited buffer size. Cell losses, internal to the generator, has no effect on measurement accuracy, as sequence numbering and time stamping etc. is done the subsequent test cell module.

The lowest periodic traffic the user can specify is about three cells per second. This is attained by the use of maximum long patterns and a one state source. Non periodic traffic has no practical lower bound – one cell per fortnight without activation and deactivation.

Activation and deactivation

In the hardware an activation or a deactivation of a source represent a workload similar to a state change. The instantaneous capacity of any one of these operations is 8 per 0.1 ms. Requests exceeding the instantaneous value will be buffered and the generation of traffic will deviate slightly from what the user has specified.

When selecting an operating system several aspects had to be taken into consideration. The decision of using Interactive UNIX in the PARASOL system, left us with no upper bound for the transfer of commands from software to hardware. This, however, does not seem to cause any major problem as activations are loaded block-wise to a FIFO in the hardware with the accuracy of a few hundred millisecond. A change from 0 to 2048 active sources, or from zero to full load, can be undertaken within a couple of seconds. This is beyond the expected rate of load change at the connection level.

Lab experienced people envisaged the need for "knobs" to adjust the load on the fly. The software is designed for a later implementation of this feature.

5 INTENDED USE OF THE GENERATOR

The STG is designed to generate traffic for measurement of ATM-related performance parameters. The most important are cell loss rate, delay and delay variation, all closely related to load characteristic. There are several areas where such measurements are of interest; Validation/verification, statistical gain enhancement, and for the network operators, also procurement.

The first large scale application of the STG is in the broadband laboratory facilities of the Norwegian Telecom [13],[4]. This laboratory will also have switches especially designed for traffic measurements [11].

Validation and verification of equipment

The verification of the ATM concept has already been going for some time struggling with the limitations of mathematical analysis and computer simulation mentioned in the introduction. The figures for parameters like cell loss ratio in question are 10^{-9} to 10^{-11}. To cover ten cell losses, in the case of the largest cell loss rate, the expected measurement period is 52 hours if the foreground traffic is a 64 kbit/s application, or more realistically 1.63 hours on a 2 Mbit/s application. It appears therefore quite clear that memory based traffic generator with capacity to produce traffic with period in the order of seconds are insufficient as validation tool.

The elaboration of source types which describe existing traffic and the more uncertain forecast of new applications taking advantage of the new features of ATM is a vast area to cover. Norwegian Telecom has ongoing activities to survey the demand for bandwidth requirements and to assess the characteristics of WAN and intra LAN traffic which is regarded as early users of ATM network. These assessments will give input for the modelling of traffic for B-ISDN laboratory experiments. More than one switching architecture should be investigated for performance comparison.

Parallel to this work, network operators will have to prepare and carry out experiments with synthetic traffic under observation of network performance and quality of service parameters to arrive at a valid requirement specification.

Statistical gain enhancement.
The STG will be a useful tool in the estimation of effective bandwidth of the various applications as they interface to ATM. Effective bandwidth is the term for the smallest network resource that has to be reserved for an individual virtual connection with given traffic parameters.

Procurement
When purchasing exchanges or other network components based on concepts different from those rated through lab experiment it may be necessary to proceed in two steps:
- Invitation to tender for providing a test device where traffic measurements can be performed.
- Invitation to tender for the procurement of the equipment approved in the first step.

Background for controlled QoS experiments.
The traffic from an STG is reproducible, both exact and with respect to statistical properties. This is an important feature when the objective is to compare the sensitivity to ATM-related network imposed impacts on the bit/cell stream of different implementations of a service. The STG therefore provides with the facility to establish bench mark tests for e.g. codecs for sound and video. Such an experiment will, among others, be undertaken in the RACE project BIPED.

6 CONCLUDING REMARKS

The STG is nearly finished. The hardware is up and running. A thorough testing of its statistical properties and handling of boundary conditions remains. The software for the STG is in a fairly good condition, except the man machine interface which still is immature. Work is going on towards defining a library of source types. The first application of the STG in the BIPED (RACE 1056) demonstration [10], where an array of STGs will provide load according to the specification of the BIPED project.

An important lesson learned from the STG development is that when a stochastic machine is implemented in hardware, you end up with a "stochastic hardware", which was very difficult to test and debug. This is caused by the enormous number of conditions which may arise, some of them occurring very rarely. Due to a structured design, and ACTEL´s internal chip probing, the testing and debugging was still manageable.

The Norwegian Telecom will use a number of STGs for bulk load generation in their broadband laboratory facilities and prototype STGs are made available for interested partners in PARASOL. It is likely that commercial versions will be made. A next generation of the generator with increased functionality, e.g. signalling, is under consideration.

Acknowledgements

The development of the STG has been a joint effort from a number of persons and companies. Steinar Brevik (NTR), Jens C. Michelsen (KTAS), Lars Nedergaard (KTAS) and Per Skovgaard (KTAS) have done a great job in the design specification of the generator and in the hardware development. The software for defining and handling source types, scenarios, etc. and for setting up the generator is developed by Sanjay Morjaria (QMW) and Hilmar Strand (DELAB). The man machine interface is developed by Serge Zloto and others at IXplus. The authors would like to thank all of the above for their enthusiasm, hard work and the fruitful cooperation.

References

[1] U. Briem, T. H. Theimer, and H. Kröner. A general discrete-time queueing model: Analysis and applications. In *Tele and Data Traffic; Proc. from the ITC 13*, pages 13 – 19. Elsevier Science Publishers B. V., June 19 – 26 1991.

[2] D. R. Cox and H. D. Miller. *The Theory of Stochastic Processes*. Chapman and Hall, 1965.

[3] Gregan Crawford and Wolfgang Grupp. Network performance measurements in ATM systems. *Telecommunications*, pages 22–32, August 1991.

[4] Kjell Gjære. OCTOPUS, an ATM switch for traffic studies. Draft Report R-10/93, Norwegian Telecom Research, 1993. In Norwegian.

[5] R. Grunenfelder, J. Cosmas, S. Manthorpe, and A. Odima-Okafor. Measurement and ARMA model of video sources in ATM environment. In *Tele and Data Traffic; Proc. from the ITC 13*, pages 981 – 985. Elsevier Science Publishers B. V., 1991.

[6] Poul E. Heegaard and Bjarne E. Helvik. Modelling of some ATM traffic sources in the composite source framework. Report under preparation, SINTEF DELAB, 1993.

[7] Bjarne E. Helvik. A model for real-time generation of ATM traffic from a large number of sources. In *The 9'th Nordic Teletraffic Seminar*. Norwegian Telecom, Research Department, August 21 - 23 1990. Also available as ELAB-RUNIT report STF40 A90162.

[8] Bjarne E. Helvik, Per Hokstad, and Norvald Stol. Correlation in ATM traffic streams – Some results. In C. D. Pack, editor, *Queueing Performance and Control in ATM*, pages 25–32. Elsevier Science Publishers B. V., June 19 –26 1991.

[9] Ole Melteig. Introduction to the PARASOL project. In *The 9'th Nordic Teletraffic Seminar*. Norwegian Telecom, Research Department, August 21 - 23 1990.

[10] T. Migotti, V. Teacher, and N. Bjorkman. Performance evaluation of broadband connections and services under varying traffic loads. *Electronics and Communication Engineering Journal*, 4(4):243 – 251, August 1992.

[11] Trond Myrstad. Test-switch for broadband ISDN. *Telektronikk*, 88(3):63 – 65, 1992. In Norwegian.

[12] Marcel F. Neuts. *Matrix-Geometric Solutions in Stochastic Models*. John Hopkins University Press, 1981.

[13] Terje Ormhaug. Plans for a broadband laboratory at Norwegian Telecom Research. Internal document F91/u/304, -305, Norwegian Telecom Research, March 1991. In Norwegian.

[14] R. C. Tausworthe. Random number generated by linear recurrence modulo two. *Math. Comp.*, 19:201 – 209, 1965.

[15] Neal Zierler and John Brillart. On primitive trinomials (Mod 2). *Information and Control*, 13:541 – 554, 1968.

Integrated Broadband Communication Networks
and Services (C-18) / V.B. Iversen (Editor)
Elsevier Science B.V (North-Holland)

Measurement Techniques and Equipment for B-ISDN

Reto Grunenfelder

Alcatel STR, Friesenbergstr. 75, 8055 Zurich, Switzerland
grunenfelder@alcatel.ch

Abstract

This contribution addresses the problem of measuring and testing in the *Broadband Integrated Service Digital Network* (B-ISDN). The paper outlines important aspects and basic facts of B-ISDN such as the particularities of the *Asynchronous Transfer Mode* (ATM) layer and the *ATM Adaptation Layers* (AALs). This builds the basis for deducing the most relevant requirements for broadband testing and measuring as shown in section 3. Finally, the paper presents a framework and techniques for the development and evaluation of relevant broadband test equipment such as generators and analyzers.

Keyword Codes: B.4.1; B.4.5; C.2.m
Keywords: Data Communications Devices; Reliability, Testing and Fault-Tolerance; Miscellaneous

1. Introduction

The International Telegraph and Telephone Consultative Committee (CCITT) has chosen the ATM as the switching and multiplexing technique for the future B-ISDN. In ATM the information flow is organized in fixed size packets called cells each consisting of 48 octets information field and 5 octets header field. ATM allocates cells dynamically on demand and is based on statistical multiplexing. This is the key for fully guaranteeing integration of services of any type and bit rate.

At the end of 1993, first ATM products will be on the market. The realization of these products requires very sophisticated methods and technologies of design. To gain experience with ATM, administrations such as, France Telecom [1], Deutsche Bundespost Telekom [2] and the Belgian Broadband Associators start B-ISDN pilot projects. Finally, these projects will serve as a basis for the smooth evolution from today's existing network infrastructure to the B-ISDN.

In a B-ISDN switch, internal delays may occur because several cells may simultaneously be directed to the same trunk (output contention). Due to this effect, the delay between end users of the B-ISDN becomes variable (delay jitter) and cells get lost with a very small probability (cell loss ratio). Transfer delay time, delay jitter and cell loss ratio are the most important *Quality of Service* (QoS) parameters. Others are cell error and mis-insertion rate.

2. Requirements for Broadband Tests

2.1 Background

Basically, a high quality of telecommunications equipment can be guaranteed while performing *conformance, functionality* and *performance* tests. This is very important to meet the service specific QoS requirements.

When starting a test, it is first necessary to know whether the *System Under Test* (SUT) meets the formal requirements (conformance) of the referenced standard. This means, e.g., that the transport of ATM cells over the Synchronous Digital Hierarchy (SDH) works correctly. SDH guarantees error-free transmission. The critical points in B-ISDN testing are the ATM layer and the AALs.

The functionality test validates that the SUT can perform the basic cell-oriented functions such as the ATM layer specific cell routing through a switch. It may act at the Virtual Path (VP) level, the Virtual Channel (VC) level or both. Furthermore, testing of functionalities includes the checking of AAL specific functions such as the mapping of particular types of traffic onto the underlying ATM layer. The measurement of cell error ratio and mis-insertion rate are part of the functionality test, too.

For performance testing, generators (artificial sources) load the SUT with ATM traffic and analyzers receive and interpret the measurement data to provide useful information about the performance of the SUT. Figure 1 schematically shows the B-ISDN measurement set-up.

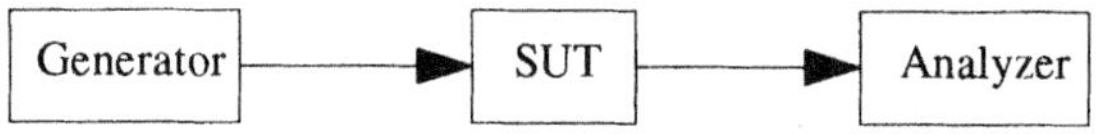

Figure 1. Measurement set-up

The profile should be as realistic as possible. ATM traffic is a heterogeneous mix of quite different services. Generally, in a B-ISDN the traffic can be split into four service classes depending on the requirements for constant or variable bit rates and timing relationships, namely

class A: circuit emulation
class B: real time variable bit rate
class C: data, Connection Oriented (CO)
class D: data, ConnectionLess (CL).

The provision of these four service classes is guaranteed with the AALs defined by CCITT [3].

A successful platform for a B-ISDN test system requires accurate generation of traffic profiles for each of the service classes including mixes and sophisticated tools which are able to analyze all the important service specific parameters. To highlight the impact of the service classes and the AALs on the requirements of the test system, it is adequate to briefly outline the main features of the AALs in section 2.2.

2.2 Background on AALs

To simplify the realization of the AAL functions and to avoid protocol proliferation, maximum commonality between the protocol elements of the AAL protocols has been worked out. The "data-oriented AALs" such as 3/4 and 5 are divided into two sublayers a

Segmentation And Reassembly sublayer (SAR) and a Convergence Sublayer (CS) which may itself be composed of a Common Part of CS (CPCS) and a Service Specific CS (SSCS). The SAR processes the data units (e.g. octets, bit streams, variable length frames) so that they fit into cells (segmentation) and conversely reconstructs the original data units from information contained in cells (reassembly). Figure 2 summarizes the basic features of the AALs.

AAL Definitions	AAL1	AAL2	AAL3/4	AAL5
Timing Relation	required		not required	
Bit Rate	constant	variable		
Connection Oriented	yes		yes/no	yes

Figure 2. ATM Adaptation Layers

AAL1 provides Constant Bit Rate (CBR) services such as video and circuit emulation. This AAL is relatively simple as there are no requirements for error detection and recovery. Transfer of timing information over the call is the major function in order to ensure synchronization of data.

AAL2 provides support for Variable Bit Rate (VBR) traffic which requires maintenance of timing information across the call. Much work still needs to be done on AAL2.

Although AAL3 and AAL4 were separated until recently, AAL3/4 is now a single entity. It provides support for VBR traffic which does not require timing information. This AAL supports Connection Oriented (CO) and ConnectionLess (CL) data traffic. It supports a number of services including Cyclic Redundancy Check (CRC) used to detect bit errors in the SAR Protocol Data Unit (SAR-PDU). It has a length of 44 octets only.

AAL5 is a recent development and is designed to be a "simple" AAL to support VBR data traffic with no timing relationship. AAL5 does not offer enhanced services as AAL3/4 and consequently does not require any octets of the cell payload for protocol information. AAL5 uses 48 octets payload and CRC is performed per CS-PDU. AAL5 is seen as the primary AAL that will be used to provide LAN interconnection over ATM networks.

3. Testing

The very first introduction of ATM is a Virtual Path (VP) cross connect infrastructure, giving access to CL data switches offering services such as the Connectionless Broadband Data Service (CBDS) [1] and the Switched Multi-megabit Data Service (SMDS). CO services, e.g. voice, can be offered on the same infrastructure. The installation of pilot networks (1994), providing this infrastructure, requires delivery of well-behaving components. This necessitates that measurement techniques and equipment actually have to be able to perform tests on the "component level", e.g. validation of the specification of the switches. In the near future (1994) equipment must be provided for testing all the necessary aspects of these networks and broadband services (CO and CL). The successful test platform evolves with the B-ISDN. The success of ATM strongly depends on the ability to test both the equipment and the services [4].

3.1 Testing ATM Components

The test of ATM components such as switches and multiplexers necessitates powerful traffic generators and analyzers at the User Network Interface (UNI) and the Network Network Interface (NNI). Tests have to be performed on the ATM layer and on the AALs. A platform has to be provided that allows testing of all the imaginable functionalities and combinations of them. Alcatel STR solved this problem by offering several possibilities for completely editing cell header, cell payload and providing adequate analysis tools [5]. This allows accurate testing of all the functionalities on the ATM layer and the AALs.

While measuring maximum throughput and stability of the SUT, e.g. switch or multiplexer, it is loaded with continuously increasing traffic (*load test*). This gives comprehensive performance information allowing the verification of the initially imposed system specification such as Cell Loss Ratio (CLR). While dimensioning switches the objective was to achieve a cell loss probability due to buffer overflow in the range of 10^{-10} [6]. This kind of test gives a relationship between throughput and CLR. Under the given traffic conditions the imposed CLR finally determines the maximum throughput of the switch. An accurate estimate of the CLR is only obtained if performing long-term measurements. Figure 3 illustrates the 95% confidence interval for a CLR of 10^{-10} versus the number of submitted cells n and the standard deviation $\sigma(n)$ of the losses.

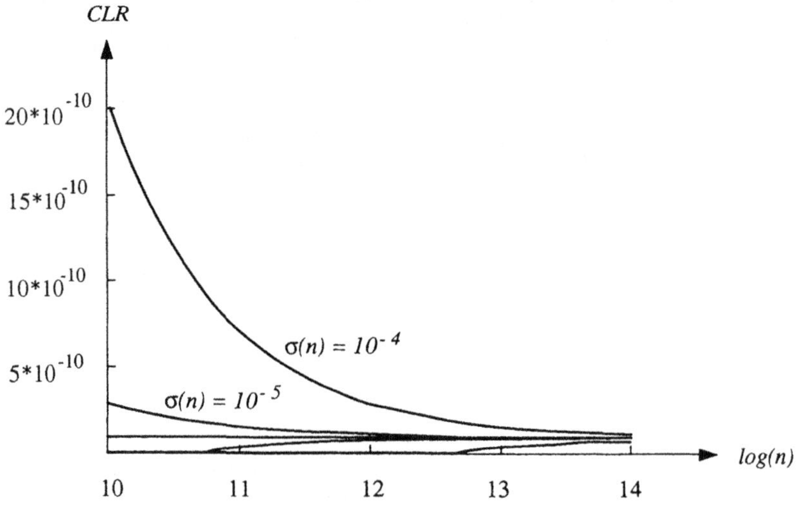

Figure 3. 95% confidence interval of CLR 10^{-10} versus number of submitted cells n

Practice has shown that the confidence interval given in (1) generally has close coverage to the desired confidence level.

$$\left[max\left\{0, CLR(n) - t_{n-1,1-\alpha}\ \frac{\sigma(n)}{\sqrt{n}}\right\}\ ,\ CLR(n) + t_{n-1,1-\alpha}\ \frac{\sigma(n)}{\sqrt{n}} \right] \qquad (1)$$

$t_{n-1,1-\alpha}$ is the upper $1-\alpha/2$ critical point for the t distribution with $n-1$ degrees of freedom. As $n >> 40$ it is adequate to approximate n by infinite and taking into account a 95% confidence interval yields $t_{\infty,0.95} = 1.960$.

If estimating a CLR of 10^{-10} with 95% confidence it is necessary to consider a sample size n in the range of 10^{11} to 10^{12}. The time used for such a measurement is inversely proportional to the load (with n constant). Figure 4 shows the relationship while assuming 155.520 Mbit/s links able to transport an ATM gross rate of 149.76 Mbit/s.

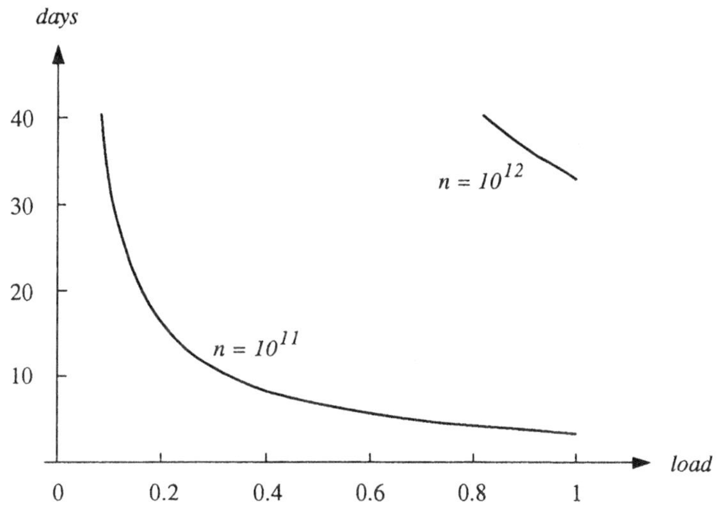

Figure 4. Measurement time for sample sizes $n = 10^{11}$ and 10^{12} respectively versus load

Obviously, the measurement time drastically increases while reducing the load. The closer to 1 Erlang the faster the measurement. To reduce measurement time, it is recommended to test the SUT under heavy load and then to estimate the CLR at the required load by some extrapolation methods. The application of some results known in extreme value theory [7] seems to be promising. The shape of such an extrapolation can easily be illustrated with the CLR of the M/D/1/s queue. The Poisson assumption is justified as long as the total load consists of a huge number of different connections with small load [8]. The CLR can be expressed [9] by the following approximation formula (2). It approximates the correct result with a relative error of about 0.3 ‰. Figure 5 shows a logarithmic plot of the CLR (formula 2) versus load. This closed form expression of the CLR gives relevant indications of accurate extrapolations.

$$CLR \approx \frac{(1-\rho)^2}{\frac{\alpha(\rho)}{2+\rho}\, e^{(2/3\alpha(\rho)s)} - \rho(1-\rho)} \qquad (2)$$

$$\textit{with} \;\; \alpha(\rho) = 1 - \rho - 2\ln(\rho)$$

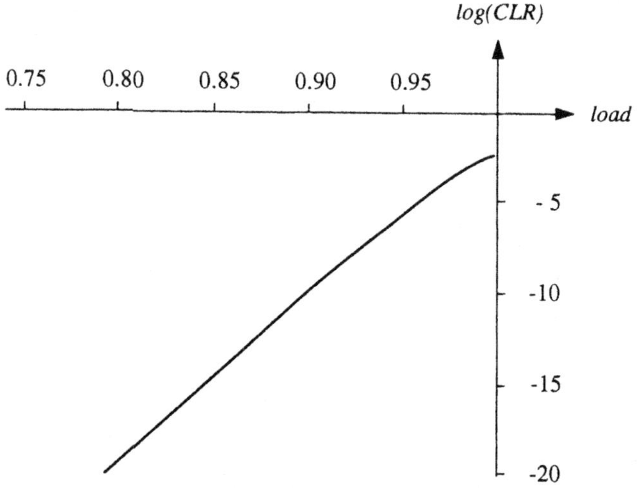

Figure 5. CLR versus load of the M/D/1/100 queue

Performance tests require loading of the whole switch. Obviously, it is too expensive to connect a generator to each inlet of the SUT. Hence, the measurement set-up has to be evaluated such that a small number of generators are enough for accurately loading the whole SUT. The measurement data, e.g. CLR and cell transfer delays of the SUT, strongly depend on the traffic profile. Alcatel STR provides with its whole test equipment product line, currently the ATM Traffic Generator Analyzer (ATGA) [5], the possibility of defining versatile traffic profiles of any type.

From the network and service performance point of view, correct operation of Usage Parameter Control (UPC) at the UNI and Network Parameter Control (NPC) [10] at the NNI are strictly necessary. The problem of testing UPC and NPC, which are implemented in a switch, does not restrict itself to the control of the peak cell rate, but also to the Cell Delay Variation (CDV) [11] and the control mechanism's reaction time. This is the time the control mechanism needs to detect that a non-conforming cell. To test UPC and NPC, generators must cover a wide range of CDVs for both single and mixed CBR sources. The analyzer evaluates the reaction time of the control mechanism while comparing the statistical properties of the submitted and the received cell stream.

3.2 Terminal Equipment

In most test cases ATM traffic generators replace the equivalent terminal as defined in [10]. Such a terminal generally produces a mix of load of different traffic classes. Generators should be able to produce both pre-defined traffic as well as traffic conforming with measured data. Generators are exclusively used at the UNI and the NNI. For a CBR traffic it is therefore necessary that e.g. the peak cell rate and the CDV specified at these interfaces are correctly produced. For VBR traffic, at the moment CL traffic, it can be derived from measurements on LANs that short term and long-term correlation have to be taken into account [12]. This is a consequence of the findings in [12] that are particularly important because their data is extremely accurate and covers a whole year. A striking feature of LAN traffic, hence also for CL data traffic in ATM, was its burstiness at every time scale, indeed from milliseconds to months. In fact no time scale was found where the process would appear stationary, because the random variation between months was already mixed with the historical increase of load.

Only a very flexible implementation of the generator can guarantee the production of such complex traffic profiles. Each generator is able to generate more than 2048 different addresses. Analyzers have to be able to fully measure such "real" equivalent terminals and generators. It is relevant that the analyzer can distinguish compliant or non-compliant cells with respect to control mechanisms monitoring each particular source. Furthermore, analyzers check whether the AALs correctly map the data into the ATM layer.

It is very important that received data are compliant with the respective AAL. This is particularly true for circuit emulation (AAL1). It guarantees delay equalization, cell loss concealment and end-to-end synchronization while using the playout buffer in the receiving terminal. It is the task of the analyzer to measure and process these data in order to indicate compliance on the receiving side.

3.3 Testing ATM Adaptation Layers

Firstly, the AAL test checks the segmentation and the subsequent reassembly process. This kind of test shall be performed on an end-to-end or local (between service and ATM interface) basis. Secondly, AAL testing includes the check of the Convergence Sublayer (CS). For AAL3/4 and AAL5 a full test of both the Common Part of CS (CPCS) and the Service Specific CS (SSCS) is necessary. The CPCS protocol for AAL3/4 is directly derived from the specification of 802.6. The proper mapping of CS-PDUs into SAR-PDUs is relevant when considering e.g. the interworking between DQDB and ATM.

For AAL3/4 error detection is performed per cell payload and for AAL5 per CS-PDU. This obviously shows the necessity for test procedures on end-to-end and local bases. A further example of the necessity of end-to-end and local test is the interworking between DQDB and ATM. Such an interworking may base on the streaming mode, the messaging mode or both simultaneously. The streaming mode requires end-to-end AAL testing, since there is no reassembly on the CS-PDU level within the network. If using messaging mode, tests can be performed on a local basis. CL interworking of data requires also the ability to test a ConnectionLess Server (CLS).

A CLS performs a lot of functionalities such as address validation, address screening, access class enforcement, group address features, monitoring of the maximum number of interleaved data units, sequence preservation per user packet, volume counting for charging and routing of data units through the ATM network. In order to maintain compatibility with the ISDN numbering plan, addresses are coded in the CCITT E164 format. Address checking can be taken over from ISDN testing. Monitoring and routing require loading of the CLS with a multitude of different CL traffic streams. The characteristics of such streams and the requirements for their generation were shortly outlined in section 3.2. The "quality", e.g. congestion of the CLS and retransmission of PDUs, of a CLS strongly depends on the accuracy of the access class enforcement. The test of this function is mainly similar to the load test, described in section 3.1.

3.4 Measuring Network Performance

When testing ATM networks, the delay measurement becomes more complicated than for components since the transmitting and the receiving site have to be synchronized with an accuracy of microseconds, e.g. via a satellite connection. This allows measurement of the absolute delay time of cells and indication of delay quantiles. For measuring the delay variation (jitter), synchronization is not needed. It can easily be performed by comparing received and transmitted time profile.

For CO traffic such as voice, the delay measurement shows whether the connection is compliant with respect to delay jitter. This is very important for guaranteeing QoS. Normally, such measurements are conducted over well-defined reference connections. The cell stream of a particular service is called Traffic Under Test (TUT). The Background Traffic (BT) consists of all kinds of traffic passing the network except the TUT. The BT is a mix of different kinds of traffic of all the service classes. BT models take into account mixes of CBR and VBR sources; the latter mainly consist of data traffic with non-negligible short-term and long-term correlation. The structure of the BT has an important impact on the characteristics of the TUT. Methods for measuring and evaluating the statistics BT and TUT were presented in [13]. The generation of complex enough BT requires adequate models and hardware. The first and most relevant step towards BT generation is given with Alcatel 8643 ATGA [5].

Apart from delay quantiles, retransmission of Protocol Data Units (PDUs) of CL services is a very important parameter. The loss of one cell causes retransmission of the whole PDU. A PDU may consist of up to about 200 ATM cells. In a CL service environment cell loss may mainly occur due to the ATM layer's control mechanism UPC [14]. The supplementary load due to retransmission of PDUs can drastically increase the network load. At the same time retransmission increases delay. Obviously, it is very important that analyzers can detect the retransmission rate in order to give an accurate estimate of the delay quantile. In a further step such measurements may provide some helpful data for network regulation.

4. Conclusions

The ultimate success of ATM and broadband services will depend on the ability to test and measure both equipment and services. The successful platform for broadband testing and the B-ISDN evolve in line. Methods and equipment of broadband testing have to be derived from the broadband specific functions, namely ATM layer and AALs.

Testing consists of three phases, namely conformance, functionality and performance testing. Performance testing, especially the measurement of the CLR of an ATM component, is very time-consuming. The improvement of the CLR measurement by introducing strict confidence, e.g. 95% confidence interval, requires extremely long measurement sequences. The paper illustrates a method which allows a considerable reduction of the measurement time. Basically, the method is based on the trade-off between CLR and load. Accurate extrapolation techniques allow the CLR estimation based on heavy load measurements.

Generally, it is very important to point out that generators are able to provide accurate mixes of traffic profiles of all service classes. Short- and long-term correlation of CL traffic is very important since studies have shown that correlation has a very important influence on the QoS.

5. References

[1] C. Rozmaryn, "Telecommunications: Towards the Year 2000", *Proc. XIV Int. Switching Symposium, Yokohama,* First Day Program, October 1992

[2] G. Wachholz and K-D. Grohs, "Concept of Deutsche Bundespost Telekom for a Broadband-ISDN Pilot Network", *Proc. XIV Int. Switching Symposium, Yokohama,* paper A6.1, October 1992

[3] CCITT Recommendation I.363, "B-ISDN ATM Adaptation Layer (AAL) Specification"

[4] R. Grunenfelder and R. Brander, "Testing Will Assure the Quality of Service in an ATM Environment", *Telephony*, pp. 40-43, October 5, 1992

[5] "Alcatel 8643 ATM Traffic Generator Analyzer ATGA", Edition 2, February 1993

[6] M. A. Henrion et al., "Switching Network Architecture for ATM Based Broad- band Communications", *Proc. XIII Int. Switching Symposium, Stockholm,* paper A7.1, May 27 - June 1, 1990

[7] V. Dijk et al., "Extrapolating ATM-Simulation Results Using Extreme Value Theory", pp. 97-104, *Queueing, Performance and Control in ATM*, Elsevier, 1991

[8] E. Cinlar, "Superposition of Point Processes" in *Stochastic Point Processes: Statistical Analysis, Theory and Applications*, pp. 549-606, Wiley, 1972

[9] M. Luoni, "ATM Traffic Characterization with Applications to Connection Acceptance Control", *PhD thesis, EPFL,* Lausanne Switzerland, 1991

[10] CCITT Recommendation I.371, "Traffic Control and Congestion Control in B-ISDN"

[11] CCITT Recommendation I.35B, "B-ISDN ATM Layer Cell Transfer Performance"

[12] H. J. Fowler and W. E. Leland, "Local Area Network Traffic Characteristics, with Implication for Broadband Network Congestion Management", *IEEE J. Sel. Areas in Comm.*, pp. 1139-1149, 1991

[13] R. Grunenfelder et al, "Measurement of ATM Traffic on the Cell, Burst and Activity Level by Traffic Sampling", *Proc. IEEE Infocom'92*, pp. 479-486, Florence, May 6-8, 1992

[14] S. Rao et al., "Interworking of DQDB MAN and ATM based B-ISDN", *Proc. Interworking' 92*, Bern, paper 7.5, November 1992

Integrated Broadband Communication Networks
and Services (C-18) / V.B. Iversen (Editor)
Elsevier Science B.V (North-Holland)

B-ISDN Signalling and Intelligent Networks; how two worlds can come together

P.F.C. Blankers

Royal PTT Nederland NV, PTT Research
P.O. Box 421, 2260 AK Leidschendam, The Netherlands

ABSTRACT

At the moment, the development of future signalling systems in the Broadband ISDN is an important topic in international standardization. In parallel, standards and recommendations for Intelligent Networks are being developed. The world of IN and the world of B-ISDN signalling seem to be rather separated. The functional models of both IN and B-ISDN signalling describe control functions for offering telecommunication services. Because the models partly describe similar functions and because advantages of the IN concept (e.g. fast introduction of new services, modularity, centralization of control) are also applicable to the B-ISDN, it is very important to clarify the relationship between them, and even to integrate the two models. A first obstacle in the comparison of both models is the difference in terminology. Furthermore, several directions in the development of the two concepts can be detected, that may be conflicting. It is important to detect any discrepancy in an as early stage as possible, in particular with regard to the division of Call Control functions (B-ISDN) and Service Control functions (IN). Conflicting developments may lead to severe problems in the future, when IN concepts and B-ISDN signalling concepts are integrated in one network.

Keyword codes: C.2.1; H.1
Keywords: Network Architectures and Design; Models and Principles

1. INTRODUCTION

International standardization bodies put a lot of effort in the development of signalling systems for the Broadband ISDN (B-ISDN). In parallel to this, standards and recommendations for Intelligent Networks (IN) are being developed. There seems to be a gap between the world of IN and the world of B-ISDN signalling. Co-ordination between both developments does not take place. This is a rather undesirable situation, knowing that both concepts describe functions, systems and protocols for offering telecommunication services.

Functional modelling is the basis for the development of systems and protocols. Furthermore, a functional model provides insight in the distribution and allocation of functions in the telecommunication network. For this reason, the functional model is taken as a starting point for the comparison of the developments of B-ISDN signalling on the one hand, and Intelligent Networks on the other hand. Both functional models will be presented in the following chapters. Because both models have a strong relationship with the functional model of the basic call in the Narrowband ISDN (N-ISDN), this will be presented firstly.

2. FUNCTIONAL MODEL FOR N-ISDN SIGNALLING

For the development of protocols in the N-ISDN, the three stages methodology has been applied [1]. Based on a description of the user's perception of a telecommunication service (stage 1), a functional model is being developed (stage 2). A functional model is a description of the functions in the network, that are needed to offer a particular service [2,3]. The stage 2 description is the basis for the specification of signalling procedures, message and information elements (stage 3 description).

The functional model of the N-ISDN basic call is shown in figure 1. As can be seen in this figure, a functional model shows Functional Entities (FEs) and relationships. An FE (symbolized in the figure as a circle) is a group of interrelated functions, that can be allocated to a particular physical place. A relationship (represented by a line between two FEs) describes the information that is exchanged between FEs, when a particular service is being offered. This exchange of information is modelled in information flow diagrams. The information flow diagrams are not shown in figure 1, but they are a very essential part of the stage 2 description.

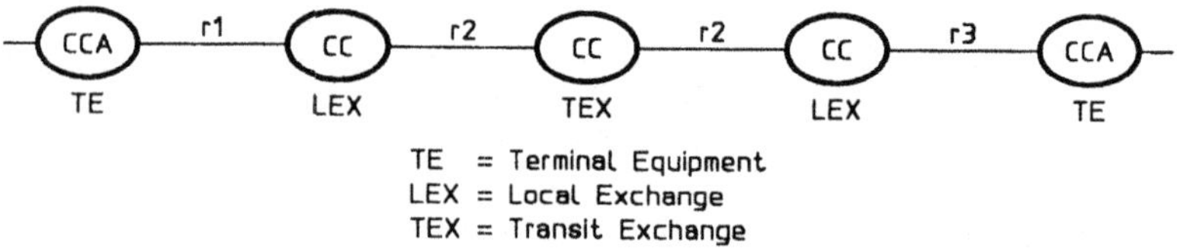

Figure 1. The basic call model for N-ISDN.

In figure 1 two different FEs have been identified: CCA (Call Control Agent) and CC (Call Control). The responsibility of CCA is to formulate and control service requests, and to interact with the (human) user. The functions of CC are mainly setting up and releasing a connection (end-to-end transmission channel) in the N-ISDN. The CCA could be allocated to a terminal, the CCs to switching nodes in the public network: Local Exchange (LEX) and Transit Exchange (TEX).

Based on the N-ISDN basic call model, models for supplementary services (i.e. services in addition to a basic call, e.g. call forwarding) in the N-ISDN have been developed as well. Since the stage 2 description is service dependent, for each service a new model has to be generated.

3. INTELLIGENT NETWORKS

3.1. Concept of Intelligent Networks

The user's demand for new supplementary services evolves faster than the development of new telecommunication devices and networks. Therefore it is important that such devices can be adapted to new user requirements without having to redesign the complete network. In order to cope with this problem, the concept of Intelligent Networks (IN) is being developed. The IN describes an architecture to offer supplementary services. This architecture is network and service independent. Its main objective is to allow a rapid and easy introduction of new or modified supplementary services. The IN is characterized by a modular service description by means of Service Independent Building Blocks (SIBs) and by a centralization of control functions for supplementary services [4].

3.2. Intelligent Network model

In order to model the Intelligent Network, the Intelligent Network Conceptual Model (INCM) has been designed [4]. This model consists of four planes, each one corresponding to a different level of abstraction of both services and network. One of those planes is the Distributed Functional Plane (DFP), which corresponds with a functional model in the terminology of the CCITT three stages methodology. Figure 2 shows the FEs that represent the control functions of the IN-DFP model [5]. Several FEs can be distinguished:

SCF:	Service Control Function
SSF:	Service Switching Function
SRF:	Specialized Resource Function
SDF:	Service Data Function
CCF:	Connection Control Function
CCAF:	Connection Control Agent Function

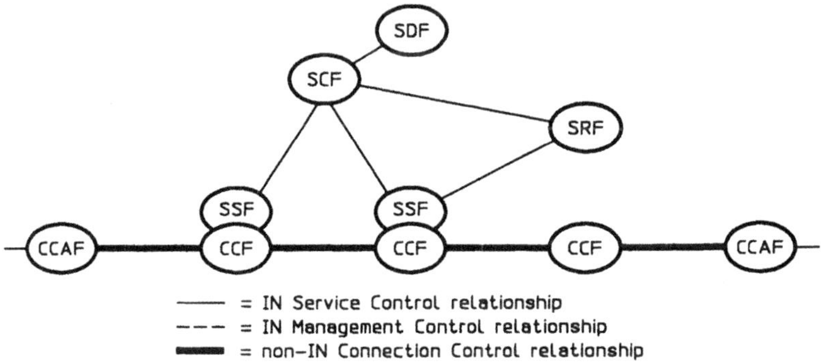

Figure 2. The Distributed Functional Plane model of IN

The five FEs at the bottom (CCF and CCAF) represent the basic call process, the main function being the control of an (end-to-end) connection. This part of the model is very similar to the basic call model of the N-ISDN, as shown in figure 1. The other FEs are IN specific, and their functionality can be explained as follows. In case a supplementary service is activated by an end-user the SSF will detect a 'trigger' event and will provide the means to the SCF to take over the control of the call. During execution of the IN service the SCF has the possibility to query databases, residing in the SDF. The SCF can invoke special functions (e.g. announcements, voiceboxes) in the SRF. In this way, each SIB can be modelled as a set of predefined transactions, or information flows, between the Functional Entities. After having carried out a certain set of SIB's, the control of the service is given back to the basic call process.

The SCF is capable of providing a large number of functions, that are needed to control a wide range of services. This makes the IN-DFP model service independent. This is a contrast with the N-ISDN functional models (for basic call and for supplementary services), which are developed completely according to the CCITT stage 2 description, and thus are by definition service dependent.

3.3. Evolution of IN

For the stepwise evolution of Intelligent Networks, several 'Capability Sets' will be defined. CCITT has nearly completed the recommendations for IN Capability Set 1 (CS1). Capabilities of CS1 are intended to support so-called 'Type A' services: services that are 'single ended' and have a 'single point of control' [4]. This means that services apply to only one party in a call, and at one moment in time, only one SCF is controlling a certain aspect of the call, residing in one SSF. Examples of CS1 services are call forwarding and credit card calling. The target set of CS2 services has not yet been defined, but CS2 will in addition to 'Type A' services offer also 'Type B' services, such as CCBS (Call Completion to Busy Subscriber) and conference calling.

The IN-DFP model for CS1 has been finished and standardized [5]. For CS2 changes can be expected on the area of service management (influenced by the Telecommunication Management Network, TMN) and new relationships may be defined, such as an SCF-SCF relationship and an SDF-SDF relationship. The IN-DFP model is supposed to be network independent and it will also be applicable to a future network such as the B-ISDN. However, the specific concepts of B-ISDN signalling have not yet been taken into account in the evolution of the IN model.

4. B-ISDN TARGET SIGNALLING

4.1. New requirements for signalling

Signalling systems in the B-ISDN will be different from existing signalling systems, such as DSS1 and ISUP/C7 [6,7]. Requirements for B-ISDN signalling systems can be divided in four categories:

1. Requirements that are set by the Asynchronous Transfer Mode (ATM). These requirements will have an impact on the bearers in the User Plane, thus demanding changes in the higher layers of the signalling system (e.g. other parameters are required in a setup message).

2. Requirements that are set by ATM, for the transfer of signalling information. If signalling messages will be transferred by ATM cells, the lower layers of the signalling systems will have to be adapted (e.g. the development of the ATM Adaptation Layer for Signalling (SAAL).
3. Requirements by new telecommunication services in the future B-ISDN, such as the expected rapid growth of multimedia and multiparty services. These requirements will mainly have an effect on the upper layers of the signalling system.
4. Requirements with regard to evolution from and compatibility with existing systems.

In the development of signalling systems for the B-ISDN, category 3 plays the most important role at the moment. For the comparison between B-ISDN signalling and IN concepts, this category of requirements is the most important. Requirements of category 3 have led to the concept of the separation between Call Control and Connection Control, which will have a large impact on signalling systems in the 'target' B-ISDN.

4.2. Concept of target B-ISDN signalling

In present signalling systems of the Narrowband ISDN, Call Control and Connection Control are integrated in one monolithic system. One (bidirectional) connection is being established as a part of the call setup procedure. A call is therefore supported by exactly one connection. In future, this may no longer be the case. Future multiparty and multimedia services require a different understanding of the terms 'call' and 'connection'. A call can be regarded as an instance of a telecommunication service, while a connection refers to an end-to-end transmission channels. A call may be supported by more than one connection. A so-called multiconnection call is very important for the support of multimedia services [8,9].

In order to offer future services in an efficient and flexible way, functions for Call Control (CC) will be separated from functions for Connection Control (also called Bearer Control, BC). Call Control can be described as the network functions that control the requested telecommunication service. Bearer Control is responsible for setting up, maintaining and releasing connections (transmission channels) in the User Plane. It includes switching and routing functions [8,9].

B-ISDN signalling will be introduced in three releases. Since release 1 is a short-term solution, offering a limited set of services, the specifications will not be based on a separation between CC and BC. The functional model for B-ISDN signalling release 1 is very similar to the functional model of the N-ISDN (figure 1). Release 2 and 3 will offer more advanced services, such as multiparty and multimedia services. Specifications of signalling protocols are being developed, based on a separation between CC and BC [10]. For reasons of simplicity, the word 'target' is used to address releases 2 and 3.

4.3. Model of target B-ISDN signalling

Figure 3 shows the functional model for a basic call in the target B-ISDN. The model contains two levels of control: Call Control (CC) and Bearer Control (BC). Some nodes in the network are equipped with CC and BC functions, while other nodes only have BC functions. CC functions are needed only at the access sides of a network (in the Local Exchange). The following FEs have been identified in the B-ISDN basic call model:

CCA: Call Control Agent

CC: Call Control
BC: Bearer Control

The basic call in the B-ISDN will also include look-ahead procedures (e.g. for service negotiation before resource allocation), multiparty calls and multiconnection calls. The model in figure 3 therefore not only describes a call with a single point-to-point connection (as the figure would imply), but be interpreted as a basis for other, more complex calls. For example, multiparty and multiconnection calls can be modelled by using multiple instances of particular FEs.

It should be noted that the term 'Call Control' in the B-ISDN model is different from the term 'Call Control' as it is used in the N-ISDN model (figure 1). 'Call Control' in the N-ISDN corresponds to the combination of 'Call Control and 'Bearer Control' in the B-ISDN. To be more exact, the 'Call Control' functions (N-ISDN terminology) in the N-ISDN model are mainly 'Bearer Control' functions (B-ISDN terminology) and for a very little part 'Call Control' functions (B-ISDN terminology).

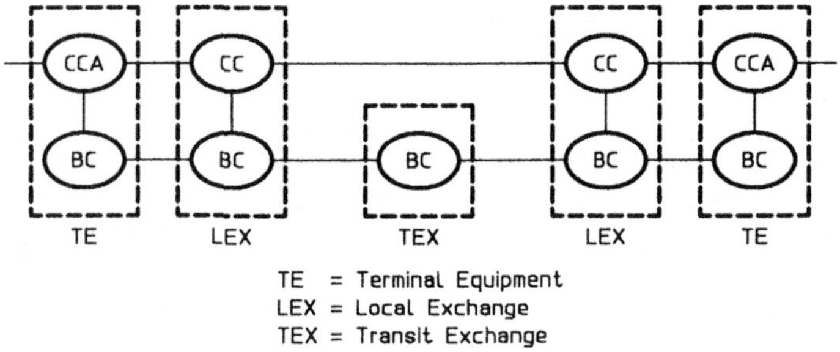

Figure 3. The functional model for target B-ISDN signalling, based on CC/BC separation.

4.4. Evolution of the model

The functional model that is shown in figure 3 can be regarded as stable [1]. Information flow diagrams are being developed for calls with point-to-point connections and calls with point-to-multipoint connections. These information flow diagrams are expected to be finalized in 1993. There are no concrete ideas of how the model will evolve in order to describe other advanced (supplementary) services. There is a tendency to extend the set of CC functions. On the other hand, it is required that functional models of supplementary services in the N-ISDN (non IN based) can be coupled to the B-ISDN basic call model, but these requirements have not yet been taken into account. Concepts of the IN have not yet been considered either.

5. COMPARISON BETWEEN IN AND B-ISDN FUNCTIONAL MODEL

5.1. Terminology

Before we can start comparing IN and target B-ISDN, it should be noted that the terminology of both concepts differs very much and may lead to a lot of confusion.

On the one hand, the IN concept makes a distinction between 'Connection Control' and 'Service Control'. On the other hand, in the target B-ISDN a need has been identified to distinguish 'Call Control' from 'Bearer Control'. To make it even more confusing, 'Connection Control' in IN is also referred to as 'Call Control', while 'Bearer Control' in the B-ISDN is often used as a synonym for 'Connection Control'. The term 'Connection Control' definitely has a different meaning in both worlds.

It is clear however, that the split of functions in the IN is at a different level than the split of functions in the B-ISDN. This is illustrated in figure 4. When IN concepts and B-ISDN concepts are combined as they are at present, three levels of control can be distinguished: Service Control, Call Control and Bearer Control.

IN terminology	**B-ISDN signalling**	**Proposal**
SERVICE CONTROL		*SERVICE CONTROL*
----------------------		----------------------
	CALL CONTROL	*CALL CONTROL*
CONNECTION CONTROL		
or CALL CONTROL	----------------------	----------------------
	CONNECTION CONTROL	
	or BEARER CONTROL	*BEARER CONTROL*

Figure 4. Terminology and split of functions in IN, B-ISDN signalling and the combination of the two existing concepts.

5.2. An integrated model for IN/B-ISDN signalling

The functional models that have been presented in figure 2 and 3 look rather different. However, it is very important to realize that these models have something in common: both models describe control functions for offering telecommunication services. In spite of this, the models have been developed completely independently. The exact relationship between both models is very unclear. Because the models partly describe similar functions and because advantages of the IN concept (e.g. fast introduction of new services, modularity, centralization of control) are also applicable to the B-ISDN, it is very important to clarify the relationship between them, and even to integrate the two models.

The IN-DFP model can be seen as an extension of the already existing N-ISDN basic call model, trying to keep the N-ISDN basic call model (and procedures) as much as possible unchanged. The B-ISDN basic call model is a redesign of the existing N-ISDN basic call model (interpreting the basic call in a wider sense, namely including multiparty and multiconnection calls). Thus, integration of the two models would possibly look as the model shown in figure 5. Figure 5 is the result of replacing the basic call part in the IN-DFP model (figure 2) by the basic call model of B-ISDN (figure 3). As can be seen in the figure, a relationship will be defined between the SCF and the CC level, in order to invoke the SCF at the detection of a trigger event. Also there is a relationship between

SCF and the BC level, in order to control special actions with regard to bearer setup and release.

The model in figure 5 is a first attempt to combine the functions of IN and B-ISDN signalling in one figure. Considering figure 5, the following issues are important starting points for the integration of IN concepts and B-ISDN signalling concepts:

- The functions at BC level are very similar to the functions of the N-ISDN basic call model. The relationships between SCF and BC/SSF would therefore not be completely redesigned, but can be based on existing relationships in the IN model.
- During the development of IN CS1 some difficult problems were caused by the fact that the basic call had already been defined. The relationship between the SCF and basic call FEs had to be defined keeping the basic call process as much as possible untouched. These problems can be avoided when developing the relationship between the SCF and CC/SSF, because CC is still in a very early stage of development. When CC FEs will be specified, a possible relationship with future IN FEs should be taken into account.

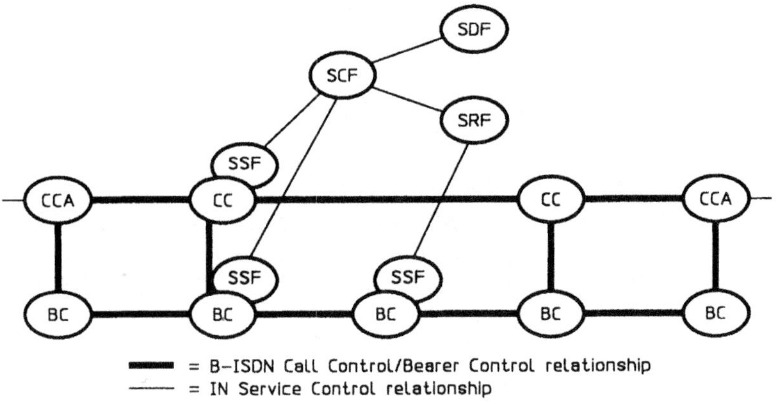

Figure 5. A possible model for the integration of IN and B-ISDN signalling concepts.

5.3. Discrepancy in allocation of functions

As has been mentioned before, the IN-DFP model can be seen as an extension of the already existing N-ISDN basic call model. One of the principles of IN in this respect is, that functions for advanced services are specified *outside* the basic call process. On the other hand, the B-ISDN functional model is a redesign of the existing N-ISDN basic call model. The main intent here is to specify more functions for advanced services *within* the basic call process. For example, multiparty and multiconnection calls will be regarded as 'basic calls'. In B-ISDN signalling, functions that are not related to the basic control of 'connections' (transmission channels in the User Plane) are allocated to the Call Control level.

These different points of departure may lead to conflicting developments. The question arises, whether there is a possible overlap in functions that are allocated to CC (in B-ISDN signalling) and to SCF (in IN). Unless there is a clear understanding of which functions are activated at certain service requests, an overlap of functions should be avoided as much as possible. Overlap would lead to duplication of functions and overhead in control systems. The functional model is the basis for the development of protocols, so any kind of overlap could lead to serious interworking problems between protocols and between networks. In other words: discrepancies in the allocation of functions should as much as possible be avoided.

At the moment, there is no essential discrepancy between functions of the SCF and functions of the CC level. However, the B-ISDN functional model is still in an early stage and developments continue. There is a tendency to allocate more and more functions to CC FEs, e.g. user authentication/authorization. On the other hand, IN Capability Set 2 will include some capabilities that are also offered by the target B-ISDN basic call procedures: the control of multiparty calls and multimedia calls. In addition, new relationships may be defined: between SCF and SCF (see figure 6), and between SDF and SDF. These relationships allow the support of a look-ahead procedure, similar to the one modelled by CC entities in the B-ISDN. The level of the split of functions between Call Control and Service Control may therefore not be as obvious as it was shown in figure 4.

More co-operation should be established between IN experts and B-ISDN signalling experts on an international level. The division of functions between Call Control FEs and the SCF is a very important issue. If the functions of Call Control FEs and Service Control FEs are not unambigeously allocated, it may result in serious interworking problems between systems and networks in the future.

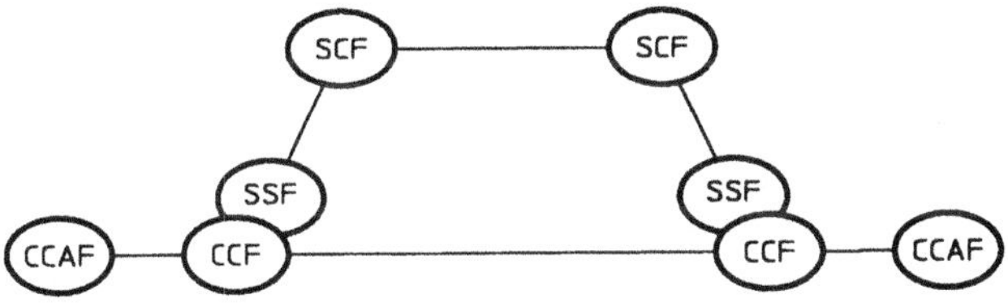

Figure 6. The SCF-SCF relationship for IN CS2: leading to a possible discrepancy between CC functions in the B-ISDN and SCF functions in IN ?

6. CONCLUSION

In this contribution an attempt has been made to make a comparison between two concepts which have been developed completely separately up till now. The IN concept and the B-ISDN concept are different, but not completely independent of each other. Several conclusions can be drawn:

- Differences between IN-terminology and B-ISDN terminology may be leading to a lot of confusion. A proposal for coherent terminology has been made.
- The separation between Call Control and Bearer Control is one of the main characteristics of target B-ISDN signalling systems. So far, it has not yet been considered, how this concept can be combined with the assets of IN structured networks. On the other hand, IN concepts will offer advantageous for the B-ISDN

network as well as for other telecommunication networks. IN concepts have not yet been considered in the development of B-ISDN signalling systems.

- If no special action is taken, the development of the IN and the B-ISDN signalling systems will each go their own way. This may be leading to various solutions to similar problems, and may even result in serious interworking problems between systems and networks in the future. This expected growing overlap of capabilities is shown in figure 7. There is an urgent need for more co-ordination and co-operation between experts of both areas. Now that IN Capability Set 2 and B-ISDN signalling release 2 are still in an early stage of development, a close co-operation is possible and meaningful. The functional description (stage 2) offers a good way to detect any discrepancy and to define the future direction for attunement between developments in IN and B-ISDN signalling.
- A proposal has been made for the development of a functional model, that integrates the functional models of IN and B-ISDN signalling. Several study items can be detected that need special attention, when developing an integrated model:
    - the impact on the IN FEs, when IN concepts are considered in a redesigned basic call model,
    - the division of functions between Call Control and Service Control,
    - the development of a relationship between a CC FE and an SCF.

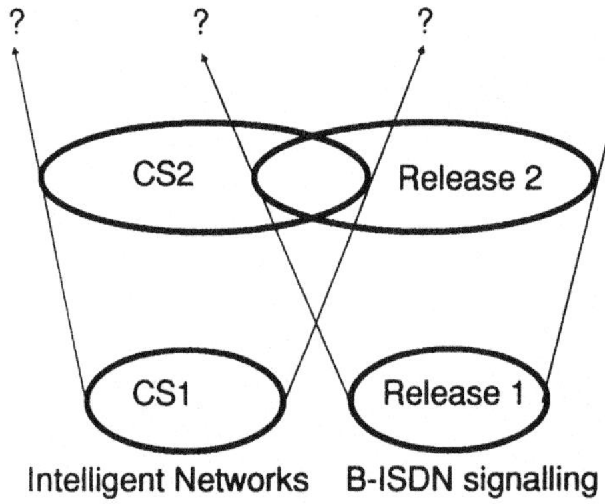

Figure 7. Likely ways of development of B-ISDN signalling and IN

Acknowledgement

The author would like to thank the members of project groups 'Signalling in Broadband Networks' and 'Long Term IN' at the laboratories of PTT Research for their help and fruitful co-operation.

References

1 CCITT Rec I.130 (Blue Book), Method for the characterization of telecommunication services supported by an ISDN and network capabilities of an ISDN, Geneva, 1989.

2 CCITT Rec Q.65 (Blue Book), Stage 2 of the method for the characterization of services supported by an ISDN, Geneva, 1989.

3 CCITT Rec Q.71 (White Book), ISDN 64 kb/s circuit mode switched bearer service, Geneva, 1992.

4 CCITT WP XI/4, Draft Recommendation Q.1211, Introduction to Intelligent Network Capability Set 1, Geneva, September 1992.

5 CCITT WP XI/4, Draft Recommendation Q.1214, Intelligent Network Distributed Functional Plane Architecture for Capability Set 1 (CS1), Geneva, September 1992.

6 CCITT SG XI, Recommendation Q.930-Q.931 (Blue Book), ISDN User-Network Interface Layer 3 specification for basic call control, Geneva, 1989.

7 CCITT SG XI, Recommendation Q.721-Q.766 (Blue Book), Specifications of signalling system no 7, section 5: Integrated Services Digital Network User Part (ISUP), Geneva, 1989.

8 CCITT WP XI/4 - Baseline document for target signalling requirements, editor M. Testa, Geneva, September 1992.

9 Blankers, P.F.C. and Keskin, K., The separation between Call Control and Connection Control, Proceedings of the Workshop on Broadband Communications, IFIP TC6, Estoril, January 1992.

10 CCITT WP XI/6, BAP Baseline document, Editor A. Paglialunga, Geneva, September 1992.

Integrated Broadband Communication Networks
and Services (C-18) / V.B. Iversen (Editor)
Elsevier Science B.V (North-Holland)

Compatibility aspects of B-ISDN signalling

Jack Tuyt

Royal PTT Nederland NV, PTT Research P.O. Box 421, 2260 AK Leidschendam, The Netherlands

ABSTRACT

B-ISDN imposes a number of extra requirements on signalling. Almost the entire N-ISDN signalling stack has to be replaced by other protocols to support these new signalling requirements. Signalling for B-ISDN will be introduced in three releases. These three releases must be compatible. For each release the signalling stacks will be modified. The major changes for Release 1 will take place in the layers responsible for transporting and routing the signalling information. The modifications for release 2 and 3 only have an impact on the signalling protocols. Separation of call control and connection control will be one of these modifications, but also the adoption of the OSI Application Layer Structure. These fundamental differences between Release 1 and releases 2 and 3 introduce compatibility problems. A number of solutions have been suggested by the involved standardization groups. None of them has been accepted yet. A decision should be taken before the end of 1993 since at this date the CCITT recommendations for the Release 1 signalling protocols will be finished.

Keyword Codes: C.2.1; C.2.2

Keywords: Network Architecture and Design; Network Protocols

1. INTRODUCTION

Signalling is needed to setup, maintain and release bearers between users. For **B**roadband-**ISDN** (B-ISDN), not only the characteristics of bearers will change -e.g. flexible bandwidth-, but also the number of bearers allowed between two users and the maximum number of users involved in one call. These new features impose a large number of modifications on the existing **N**arrowband-**ISDN** (N-ISDN) signalling protocol stacks. In order to have a smooth evolution to the target B-ISDN signalling, the revaluation of the signalling stacks will take place in three steps. It is essential that these three releases are compatible. As will be shown in this paper, a number of fundamental problems have to be solved in order to meet this compatibility requirement.

2. WHERE DO WE START?

The N-ISDN signalling stacks will be used as a starting point for the development of the B-ISDN signalling stacks. Therefore a short overview of the existing signalling stacks will be given. In Figure 1, it is shown that two different types of signalling stacks are used. One at the **U**ser **N**etwork **I**nterface (UNI) -e.g. **D**igital **S**ubscriber **S**ignalling **S**ystem No. **1** (DSS1)- and the other at the **N**ode **N**ode **I**nterface (NNI) -e.g. **C**ommon **C**hannel **S**ignalling **S**ystem No. **7** (CCSS 7)- . Although the protocols used in both stacks are all different, a large similarity between them exist.

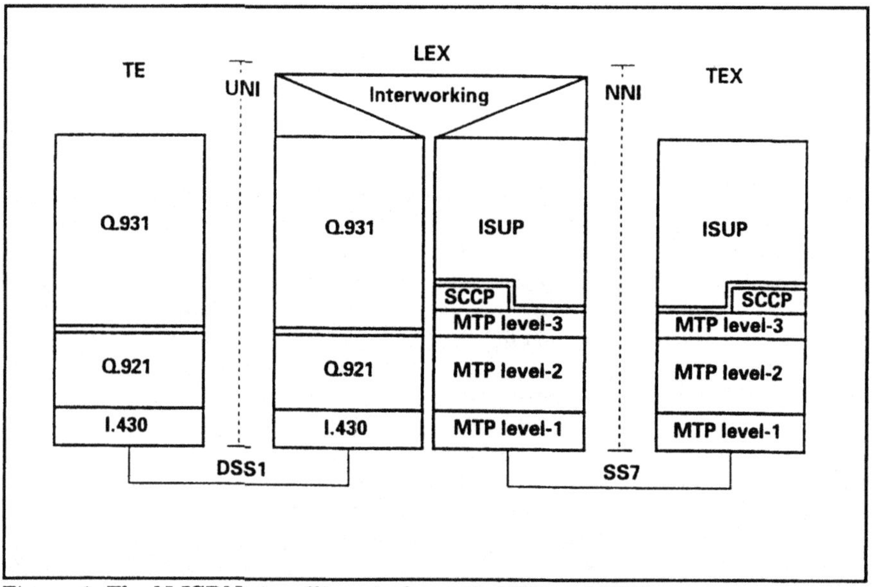

Figure 1: The N-ISDN signalling stacks at theUNI and NNI.

Q.931 and ISUP: These layers are the actual signalling protocols. They exchange messages necessary to setup, maintain and release bearers between users. In N-ISDN , the basic signalling service is only able to establish **one** bidirectional bearer of **64 kbits/sec** between **two** users. [4],[5]

SCCP and MTP 3: These layers are responsible for routing the messages provided by the signalling protocols to the correct locations. [7],[8]

Q.921 and MTP 2: These layers are responsible for an error free transport of messages on the link level. [6],[7]

I.430 and MTP 1: These layers are responsible for a correct transmission of bits. [7]

3. B-ISDN SIGNALLING REQUIREMENTS

Signalling is needed to setup, maintain and release bearers between users. In N-ISDN , the basic signalling service is only able to establishe **one** bidirectional bearer of **64 kbits/sec** between **two** users. In B-ISDN this basic signalling service will be extended. Not only the characteristics of the bearers will be enhanced, but also the number of bearers and users involved in one call will be extended.

3.1. New bearer characteristics

In B-ISDN, the bandwidth of the bearers to be established, will not be limited to 64 kbits/sec, but ranges from 1 kbits/sec to 150 Mbit/sec. Two ways of bandwidth allocation will be offered.

Peak allocation: The bandwidth needed for a given bearer is specified. The specified value indicates the maximum-bandwidth=minimum-bandwidth=average-bandwidth. For a number of applications the used bandwidth varies. For these kind of applications, this way of allocation is less attractive.

Variable allocation: Not only the maximum bandwidth is specified but also the average bandwidth and the burstiness. This way of allocation is attractive for applications that need a variable bitrate.

3.2. New services

In B-ISDN, the basic call services of N-ISDN will be enhanced. The following enhancements have an impact on signalling:

Multi-party: The number of users involved in one call will not be limited to two anymore.

Multi-connection: The number of connections between two users will not be limited to one anymore.

Service negotiation: In B-ISDN, the establishment of the associations between the users involved and the eststablishment of bearers between these users, can be separated. This means that the users involved, may exchange information before bearers are established. This mechanism allows different routes for call and bearer control information, which may be advantageous for mobile applications. Furthermore it may prevent the establishment of bearers to an incompatible or busy subscriber.

Distribution: Allowing the distribution of information to many subscribers. This service could be used for broadcasting radio and television programs.

3.3. Three releases to target solution

The realization of all B-ISDN signalling requirements will have a large impact on the existing N-ISDN signalling stacks. In order to stimulate the progres and to force an early solution, it has been decided in CCITT to introduce B-ISDN signalling in 3 releases. The most important new signalling requirements for each release are summarized in Table 1.

Table 1: New signalling requirements for each release.

	Release1 (1993)	Release2 (1995)	Release3 (1997)
Peak allocation of bandwidth	YES	YES	YES
Variable allocation of bandwidth		YES	YES
Multi-party service		YES	YES
Multi-connection service		YES	YES
Service negotiation		YES	YES
Distributive services			YES

4. B-ISDN SIGNALLING RELEASES.

B-ISDN imposes a number of extra requirements on signalling. Almost the entire N-ISDN signalling stack has to be replaced by other protocols to support these new signalling requirements. Signalling for B-ISDN will be introduced in three releases. Each release will be discussed.

4.1 Release 1.

The only new signalling requirement for Release 1 is related to the bandwidth of bearers which is not restricted to 64 kbits/sec anymore ([1]). As will be shown, this new signalling requirement not only has an impact on the contents of the signalling information but also on the transport of this information. A comparison between Figure 2 (depicting the Release 1 signalling stacks), and Figure 1 (depicting the N-ISDN signalling stack), shows that except for the **S**ignalling **C**onnection **C**ontrol **P**art (SCCP) and the **M**essage **T**ransfer **P**art-**3** (MTP-3) all protocols have been replaced.

The transport of signalling information

For the transport of B-ISDN information, the **A**synchronous **T**ransfer **M**ode (ATM) is selected. B-ISDN information is not restricted to user information but also applies to signalling information. The use of ATM has an impact on the physical layers which must be able (amongst other things) to support higher bandwidth. Above the ATM layers, the **S**ignalling **ATM** **A**daptation **L**ayer is used. This S-AAL offers a service equivalent to that of Q.921 at the UNI and MTP-2 at the NNI. The procedures which are used to offer that service are however different.

The contents of signalling information

Because the bandwidth of the bearers is not restricted to 64 kbits/sec anymore, extra information must be exchanged related to the required characteristics of the bearers. This has only a very limited impact on the signalling messages and signalling procedures. Therefore it has been decided to base the Release 1 signalling protocol, on the N-ISDN signalling protocols -e.g. Q.931 and **ISDN** **U**ser **P**art (ISUP)-.

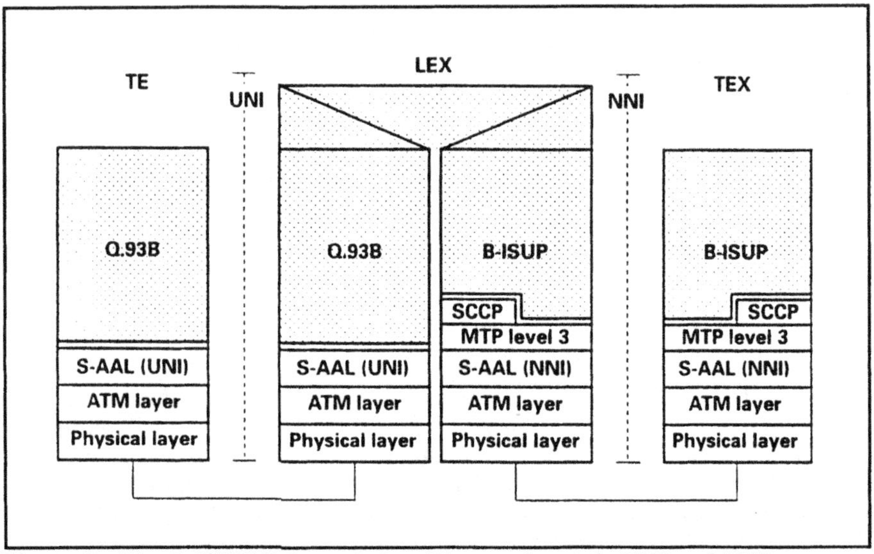

Figure 2: The B-ISDN Release 1 signalling stacks at theUNI and NNI.

4.2 Release 2,3.

Most of the new signalling requirements imposed by B-ISDN, are supported by Release 2 ([3]). The extra Release 2,3 signalling requirements do not have an impact on the layers involved in transporting and routing signalling information -e.g. Physical layer, ATM, S-AALs MTP-3, SCCP- . Therefore, these Release 1 layers will be reused in Release 2,3. The impact of the Release 2 signalling requirements on the signalling protocols however, is very large. First a separation of call control and connection control is required, second a more flexible protocol structure has to be adopted. Because of these differences, the Release 2 and 3 protocols -e.g. **B**roadband **A**pplication **P**art (BAP)- will not be a simple extension of the Release 1 signalling protocols.

4.2.1 Separation of call control and connection control.

For B-ISDN, new services will be introduced for calls with many users and many bearer connections. In order to reduce the possibility of establishing bearers between incompatible or unavailable users, a negotiation procedure may be invoked before the establishment of the bearers. Also other procedures exist who are not related to bearer control. In order to make a clear distinction between these different types of procedures a separation between connection control (bearer related) and call control (non bearer related) has been introduced ([3],[12]).

4.2.2 A different structure for Release 2 and 3 signalling protocols.

Besides separated protocols for Connection Control and Call Control also other protocols are expected such as Supplementary Services or Release 3 upgrades. Most requirements for these protocols and their mutal relations are not known yet. Therefore a structure has been choosen for the BAP which offers the flexibility of adding new protocols and new relations between them. This structure is based on the **OSI** **A**pllication **L**ayer **S**tructure (OSI ALS) ([2],[9],[10]). A simplified drawing of this Structure is given in Figure 3. The OSI ALS has the folowing advantages:

- Before a dialogue between two entities is established, the context of the dialogue is negotiated. The context of the dialogue, called the application context, identifies the set of protocols needed for that dialogue and the coordination functions needed between these protocols. The coordination functions are represented by the **S**ingle **A**ssociation **C**ontrol **F**unction (SACF) block in Figure 3. The protocols needed are represented by the **A**pplication **S**ervice **E**lement (ASE) blocks.

- One of these ASEs is mandatory independently of the application context. This one is called **A**ssociation **C**ontrol **S**ervice **E**lement (ACSE) . This ACSE supports all procedures necessary to establish a dialogue and to negotiate the application context. A good alternative for this ACSE is the **D**ialog **Ha**ndler (DHA) as used in **T**ransaction **C**apabilities **A**pplication **P**art (TCAP) [11]. This DHA will be used in BAP (see Figure 3). The **T**ransaction **S**ub**l**ayer (TSL) which is also part of TCAP will be used as an interface between BAP and the lower layers -e.g. SCCP/MTP-3/S-AAL-.

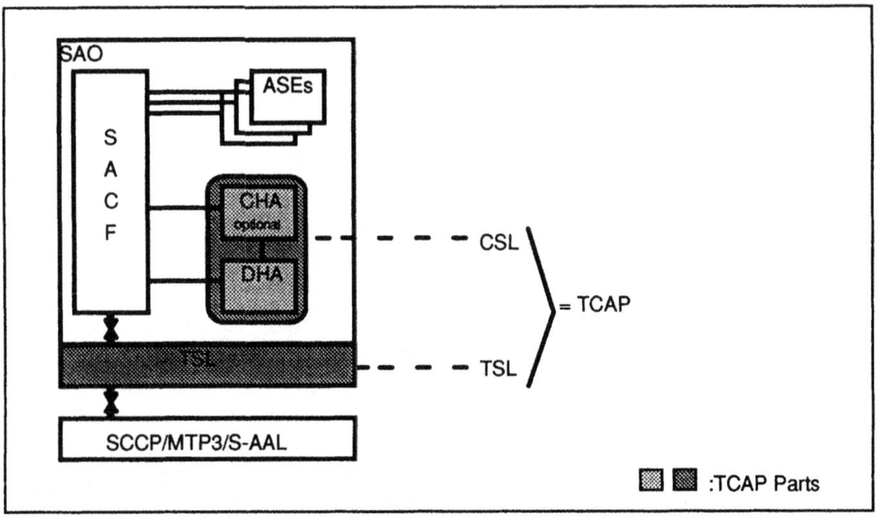

Figure 3: Protocol structure as used for Release 2 and 3 BAP protocols.

5. COMPATIBILITY ASPECTS

Two fundamental differences exist between Release 1 and releases 2,3 signalling protocols. One is the separation of call control and connection control the other is the use of the OSI ALS for Releases 2 and 3. These differences introduce compatibility problems. It will be shown that these problems can be solved.

5.1. Impact of separation of call control and connection control on compatibility.

Compatibility between two protocols, one having separation of call and connection control and the other having combined control, is difficult to realize. A number of solutions are being studied. Two promising solutions are:

1. Compatibility with Release 1 is solved by special ASEs.
2. The Bearer control ASE of Release 2,3 is also able to support the Release 1 signalling protocol.

5.1.1 Use of special ASEs

In this solution, compatibility with Release 1 is solved by including special ASEs in Releases 2 and 3. These special ASEs , Q.93B ASE and **B**roadband-**ISUP** (B-ISUP) ASE, support the Release 1 signalling protocols. This is shown for the NNI in Figure 4.

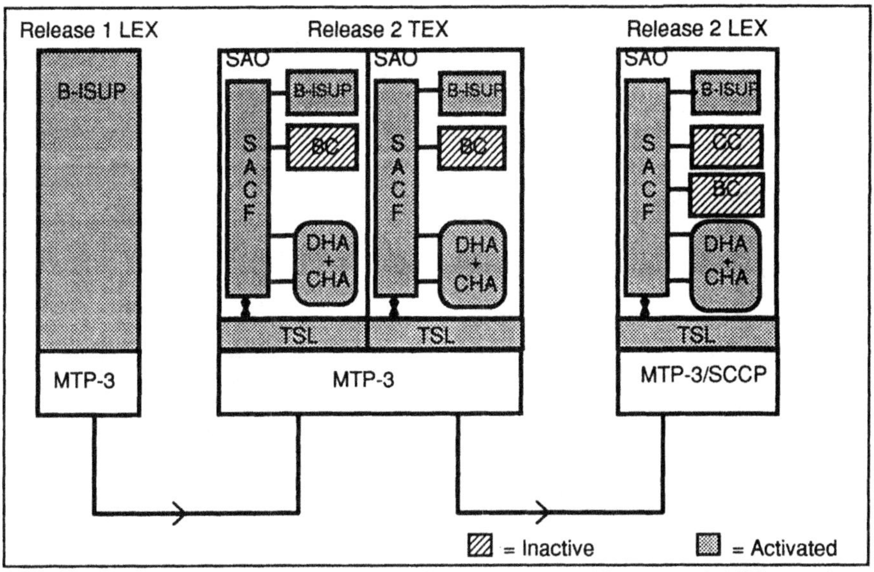

Figure 4: Release 1 signalling protocols are supported by special ASEs used only for this purpose.

If a Release 1 signalling message is received at a Release 2 exchange, this special ASE will interpret this message. It should be noted that the Release 2 **B**earer **C**ontrol (BC) ASEs and the **C**all **C**ontrol (CC) ASEs will not be activated in this case. This solution has the following advantages/disadvantages:

Advantages:

- It does not constrain the Release 2 protocols by any design rules imposed by Release 1, resulting in flexibility in designing/modifying Release 2 and 3 ASEs.
- The Release 1 ASEs (Q.93B or B-ISUP) can be omitted in environments where Release 1 protocols are not needed. This implies that the disadvantages will disappear eventually.

Disadvantages:

- A large commonality is expected between the procedures, messages and parameters of B-ISUP and BC ASE.

5.1.2 BC ASEs supporting Release 1 signalling protocols

In this solution compatibility between Release 2,3 and Release 1 is solved by designing the Bearer Control ASEs of Release 2,3 in such a way that they are able to support also Release 1 signalling protocols. These special ASEs are called BC/Q.93B ASE or BC/B-ISUP ASE. This is shown for the NNI in Figure 5.

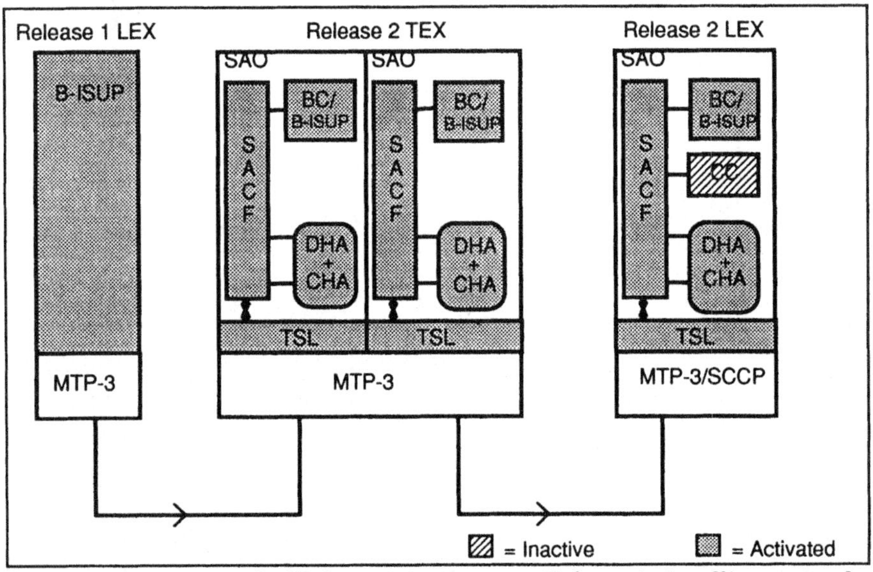

Figure 5 Bearer control of Release 2,3 to support Release 1 signalling protocols.

If a Release 1 signalling message is received at a Release 2 exchange, this special ASE will interpret this message. It should be noted that the CC ASEs (only available at a Release 2 LEX) will not be activated in this case. This solution should only be selected if most procedures, messages and parameters are common to BC and B-ISUP. It is expected that this overlap exist because the ISUP protocols on which Release 1 signalling is based are designed to establish a bearer between two LEXs. This solution has the following advantages/disadvantages:

Advantages:

- It does not constrain the Release 2,3 call control ASEs by any design rules imposed by Release 1, resulting in flexibility in designing/modifying these ASEs.

Disadvantages:

- It does constrain the Release 2,3 bearer control ASEs by design rules imposed by Release 1, resulting in inflexibility in designing/modifying these ASEs.

5.2. Impact of using different structures.

In contrast to Release 1 signalling protocols, the Release 2 and 3 signalling protocols are based on the OSI Application Layer Structure. As a consequence, the Release 2 and 3 signalling protocols use a standardized procedure to start/release and negotiate the context of a dialogue. These procedures are located at the TSL and Component Handler (CHA) modules (see Figure 4 and 5). These procedures are not supported by the Release 1 signalling protocols. This means that a Release 2 call request is not recognized by the Release 1 signalling protocol and that an alternative context can not be negotiated. On the other hand, although the Release 1 signalling protocol is supported by the Release 2 and 3 signalling protocols, a Release 1 request will not be accepted since the Release 1 signalling protocol does not use the CHA and TSL service needed for starting a dialogue.

One way of solving this problem is to introduce the Release 2,3 mechanism to start/release and negotiate the context of a dialogue, also in Release 1. This proposal has not been accepted yet, since it implies extra complexity for the Release 1 signalling protocols which is redundant in an environment restricted to Release 1 equipment. Alternatives to overcome this problem are being studied. A decision should be taken before the end of 1993 since at this date the CCITT recommendations for the Release 1 signalling protocols will be finished.

6. CONCLUSIONS

The B-ISDN signalling requirements will be introduced in three releases. All three releases must be compatible. It has been shown that fundamental differences exist between Release 1 and the releases 2 and 3. The compromises that are needed to meet this compatibility requirement may result in inefficient solutions for one of these releases. The aim must be the target solution which should not be constraint by the intermediate ones Two solutions have been presented in this paper that meet this requirement.

Several standardization groups are involved in developing these signalling protocols. A close co-operation between these groups is needed to reach the final goal -e.g. powerfull and efficient target signalling protocols-.

References.

1. CCITT SWP XI/6-1 B-ISDN Release 1 acces protocol., Geneva, September 1992.
2. CCITT WP XI/6, BAP Baseline document, Editor A. Paglialunga, Geneva, September 1992.
3. CCITT WP XI/4 - Baseline document for target signalling requirements, editor M. Testa, Geneva, September 1992.
4. CCITT SG XI, Recommendation Q.930-Q.931 (Blue Book), ISDN User-Network Interface Layer 3 specification for basic call control, Geneva, 1989.
5. CCITT SG XI, Recommendation Q.721-Q.766 (Blue Book), Specifications of signalling system no 7, section 5: Integrated Services Digital Network User Part (ISUP), Geneva, 1989.
6. CCITT SG XI, Recommendation Q.920-Q.921 (Blue Book), Digital subscriber signalling system No. 1 (DSS 1), data link layer, Geneva 1989.
7. CCITT SG XI, Recommendation Q.702-Q.704 (Blue Book), Message transfer part (MTP), Geneva 1989.
8. CCITT SG XI, Recommendation Q.711-Q.714 (Blue Book), Signalling connection control part (SCCP), Geneva 1989.
9. CCITT XI/4 'Architecture Framework for the Development of OSI Based Signalling and OAM Protocols' Geneva, September 1991.
10. ISO International standard ISO 9545 'Information Processing System, Open Systems Interconnection, Application Layer Structure', Geneva 1989.
11. CCITT SWP XI/2-4 Q.771 - Q.775 , TCAP, Geneva September 1992.
12. Blankers, P.F.C. and Keskin, K., The separation between Call Control and Connection Control, Proceedings of the Workshop on Broadband Communications, Estoril, January 1992.

Integrated Broadband Communication Networks
and Services (C-18) / V.B. Iversen (Editor)
Elsevier Science B.V (North-Holland)

TIMER HANDLING IN HIGH-PERFORMANCE TRANSPORT SYSTEMS

Bernd Heinrichs, Kai Jakobs

Aachen University of Technology
Informatik IV; Department of Computer Science
Ahornstr. 55, 52074 Aachen

Tel.: +241/80-21410; FAX: +241/80-21429
e-mail: heibe@informatik.rwth-aachen.de, jakobs@informatik.rwth-aachen.de

Abstract

Timer handling has a strong impact on various transport layer mechanisms. For instance to achieve an efficient provision of multicast and congestion control timer management is crucial. In this paper we describe and evaluate performance of the Slow Start / Tri-S flow control mechanism. The mechanism is source controlled and the window size adjustment is based on round trip times (RTTs). One main objective of Slow Start / Tri-S is the reduction of window oscillation and the provision of low delay compared to TCP/IP Slow Start. The second part of the paper depicts the error control scheme of XTP in multicast mode. Inefficiencies like redundant retransmissions are illustrated and the influence of respective timers on the performance is explained. To overcome the problems we introduce some enhancements.

Keyword codes: C.2.0; C.2.1; C.2.2
Keywords: Computer Communication Networks, General; Network Architecture and Design; Network Protocols

1. Motivation

In the area of high performance networking the subject of congestion control is of increasing importance. Congestion is defined as a network state, in which an increase of the offered load does not raise the throughput. The congestion problem mainly results from a non existing or poor interaction of error and flow control mechanisms on transport layer with routing, buffering and protocol conversion functionality on the network layer. The development of high-speed networks and fast media access protocols results in a growing discrepancy between high (Gbit-range) and small (kbit-range) bandwidths in a global heterogeneous network. Gateways and routers easily get overloaded when interconnecting such different networks. We call a network **collapsed**, when a small increase of the offered load causes the throughput to decrease drastically. In one documented case the throughput decreased by the factor 10^3 [1]. Other consequences are longer or even unacceptable transfer delays, packet loss and the break down of connections. Without external manipulation through load regulating algorithms the throughput cannot be increased anymore, and the network remains in an undesired, but stable condition.

Work described in this paper has been partly funded by the RACE II project EUROBRIDGE. The EUROBRIDGE Consortium consists of the following members: Ericsson EUROLAB, D; RWTH Aachen, D; Teles, D; Detecon, D; Broadcom, IRL; CTS, IRL; CET, P; Intracom, GR; VTT, FIN; Siemens-Nixdorf, IRL

Various approaches to reduce the probability of a congestion collapse are known [2,3]. Depending on the application and the network, the strategies' suitability varies. The mechanisms are divided into algorithms with ***implicit*** and ***explicit*** report of overload. Implicit algorithms work locally in the endsystems, using expiring acknowledgement timers as reporting mechanism [1,4]. Since conventional implicit strategies do not become active until the network is congested, they are also called ***reactive*** techniques. Explicit algorithms report the load to the endsystems by setting flags in crossing packets [5] or by generating and sending control packets [6]. Explicit mechanisms start to work before the network becomes congested, that means they have a ***preventive*** nature. Although the explicit report mechanisms seem to have advantages over implicit mechanisms they cannot be used on global heterogeneous networks. Feedback signals have to traverse various networks before reaching the sender which probably does not understand the signal from the other network. Therefore, in heterogeneous environments implicit schemes are useful. How to achieve congestion avoidance with implicit schemes is one of our main research topics. The problem is how to provide the sender with sufficient information without generation of feedback signals in intermediate nodes.

The limitations of the *Slow Start* congestion control policy become obvious when observing the high throughput oscillation of this algorithm, the periodic packet loss and the difficulty to adjust timers correctly in order to achieve an efficient implicit congestion control. In most cases each connection loses a packet during the same congestion period; that means all of the connections are decreasing their congestion windows at the same time. Since the original window adjustment algorithm used in Slow Start resets the window to one, the sum of the window sizes after congestion may be too small to fully utilize the bandwidth of the line when the network has relatively large pipes (especially on gigabit links). An oscillation of such high magnitude in window size leads to drastic round trip time and queue length fluctuation. The periodic packet loss at each peak of the fluctuation has severe impact on the network performance. The pipeline of the packet stream is already broken when the timeout is detected and a full window worth of packets has to be retransmitted. If only one connection loses a packet during a congestion period, the total window size would not vary so abruptly. At least in situations where only a small number of connections has led to congestion, use of a multiplicative window decrease algorithm would be advantageous since it reduces the congestion window only by half [7].

The following chapter starts with a brief overview of existing congestion control mechanisms and depicts their limitations. Chapter 2.1 describes the influence of timers on the efficiency of implicit algorithms. Chapter 2.2 introduces Slow Start / Tri-S. Chapter 2.3 describes our simulation model and chapter 2.4 ends with the illustration of first performance results. In chapter 3 we investigate the error recovery problem of semi reliable XTP multicast. We have figured out various drawbacks in the XTP algorithm, which mainly result from an incorrect setting of the timers controlling the error control flow. We describe the XTP heuristics for multicast provision, show some inefficiencies and give propositions how to enhance the performance of XTP multicast. Chapter 4 ends with a summary and an outlook on further work.

2. Congestion

As mentioned in chapter 1, throughput oscillation is a major problem when using Slow Start. A possible solution to this problem is the use of *Slow Start / Tri-S* [4]. This mechanism has the aim to reduce the oscillation characteristics of Slow Start. The problem of an appropriate timer adjustment remains the same. The acknowledgement timer is based on round-trip time estimations and the respective algorithms for the timeout calculation. Reducing the probability of false timeout alarms will improve the efficiency of congestion control mechanisms.

Implicit as well as explicit transport layer based congestion control schemes are only keyed to the receiver's resource capabilities. Overflow of buffers or overloading of resources in the intermediate nodes are not covered. In the Internet world the only control mechanism is provided by ICMP (Internet Control Message Protocol) which generates source quenches to inform the sender to reduce its load. The sender increases its load as long as no such packet arrives. The gateways can only discard packets to control the traffic, when buffers are out of space.

Service strategies of relay systems for the minimization of waiting times do not influence directly the data volume introduced into the network, but the relation of the connections and their specific handling. The service strategies determine the degree of fairness of the network, because they control the succession, in which the incoming packets are served.

All of the mentioned strategies operate on layers 3 and 4 according to the ISO/OSI RM. Additionally an efficient load management requires the use of functions from higher and lower layers. An OSI layer 2 protocol e.g. is able to reduce short term congestions through *backpressure* [8]. On layer 5 certain actions should be initiated, if a transport connection of a desired quality cannot be achieved or sustained (e.g. through lessening the quality of service or by informing the initiating applications). It is obvious, that there is no perfect solution for load control currently. Each combination of algorithms exhibits certain advantages and disadvantages.

2.1. Acknowledgement Timers in Implicit Congestion Control

An acknowledgement timer can expire before the acknowledgement is received. The reason may be an inaccurately adjusted timer or an erroneous and delayed data transmission. The expired timer activates unnecessary mechanisms for error recovery and it initiates retransmissions of packets. Thus, the network load rises without transferring relevant data. A timer should at least cover the span of ***2 * propagation delay + receiver processing time*** (equals round trip time, RTT). On the other hand it should not run too long, so that an immediate reaction in case of error is possible. Timer values slightly exceeding the actual RTT are ideal.

Since RTTs change during a connection, adaptive timers are desirable. But if the RTT oscillation rate is higher than the adaptation rate of the timer, false alarms are likely [9]. Adjusting the timers using the SRTT (Smoothed RTT) method described in [1] no longer matches the rapid oscillation of high load phases and low load phases in today´s communication networks. Although we have made the SRTT mechanism sensitive for the measured RTT by using large values for the adaptation parameter (close to 1), the strategy cannot follow the RTT oscillation.

Various TCP implementations do not provide a correct mapping of an acknowledgement to the original packet or its duplicates, in case of retransmission. This phenomenon is called *acknowledgement ambiguity* and it results from the fact, that a TCP header does not indicate if an acknowledgement refers to the original transmission of a packet or to its retransmission. When this problem first was recognized, two different strategies have been evaluated: *measuring the first transmission* or *measuring the most recent transmission*. Measuring from the first transmission may cause SRTT to grow infinitely when there is loss on the network, whereas the second method only works if a timeout implies that previous transmissions almost certainly have been lost. Unfortunately this assumption is often false. Since the estimated RTT for the expiring of an acknowledgement timer normally is very close to the maximum possible RTT, most of the delayed acknowledgements will arrive just after the timer has expired. Since the actual timer has been reset just before, the new estimated RTT stabilizes at an unreasonably low value when measuring the most recent transmission. This behaviour inevitably leads to unnecessary retransmissions and wasting of much bandwidth. A possible solution is given by the Karn-algorithm [10]. To avoid RTT-ambiguity the last unequivocal round trip value is kept as a measure [10] and RTT measurements of retransmitted packets are ignored. A calculation of a new acknowledgement timer is only activated if the packet or a succeeding packet is acknowledged without an intervening retransmission. In addition to this technique Karn uses a back-off strategy which lengthens the acknowledgement timer by some factor before retransmitting data. This is necessary since without an acknowledgement there would be no new timing information available to calculate the actually valid timer value. To prevent timeouts from becoming too long, most implementations limit increases by this back-off strategy to an upper bound.

A better method to solve the ambiguity problem was introduced in [11]. It is proposed to place a timestamp in each data segment (*option* field), and the receiver has to reflect these timestamps back in ACK segments. Thus, an accurate RTT measurement of each ACK is possible. This strategy is already used in protocols as XTP or RTP [12]. The same option is also used in a procedure known as PAWS (Protect Against Wrapped Sequence) to prevent erroneous data delivery due to a combination of old duplicate segments and sequence number reuse at very high bandwidths [11].

2.2. Slow Start / Tri-S

To illustrate the correlation between RTT-measurement, acknowledgement timers and window based congestion control, results of an analysis of the *Slow-Start/Tri-S* mechanism [4] will be described. Results in [4], confirmed by our own tests show that the use of Slow-Start/Tri-S reduces the ***throughput oscillation*** compared to Slow-Start [1]. *Oscillation* describes the throughput development of the network near the optimal operating point (the so-called "*knee*"). A reduction of oscillation entails a reduction of the RTT-variance, making it possible to improve setting of the acknowledgement timer and reducing danger of false alarms. The optimal operating point is in a region where both a high throughput and a small round-trip delay is experienced. The knee is identified as the maximum of the power of the network which equals the maximum of the ratio of throughput and delay (cf. fig. 2.1).

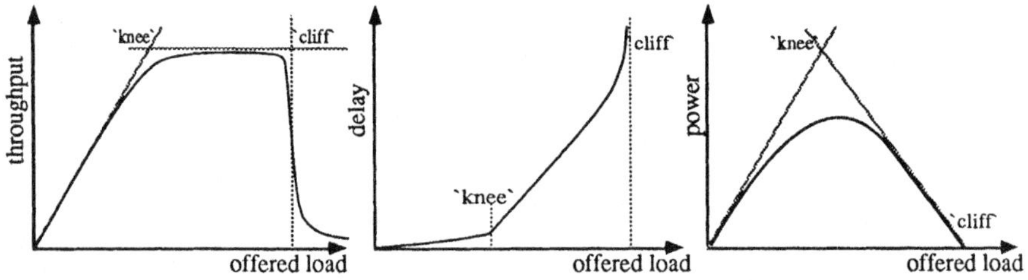

Fig. 2.1: Net-Power

Binary report methods (sender or relay system based) can only indicate whether or not an overload exists. If yes the load should be reduced, if no it should be increased. The adaptation algorithm is not able to decide how close the actual operating point is to the optimum and if it is useful to increase the offered load. This means, that binary methods increase the load until it becomes too large (except a previously determined maximum flow window is reached).

Tri-S is based on the idea to change the size of the flow-window only if the amount of network traffic changes significantly. This is the case if for instance an additional connection is created or an active connection is terminated. The *normalized throughput gradient (NTG)* serves as a determinator for the sender to compute the load development and to control the algorithm. The algorithm is divided into three phases:

- *initializing phase* with exponential window growth: each time an acknowledgement is received, the window size is increased by one basic adjustment unit (BAU) which is a user defined value that may correspond to a packet size. The initializing phase ends when the maximum window size specified by the end users has been reached. Thus, the initializing phase corresponds to the slow start phase of the Slow Start mechanism.
- *decrease phase* with decrease of the window to one BAU if an ack timer expires: Upon receiving the next acknowledgement NTG is checked: If it is larger than NTG_d (d for decrease) the window size is increased by one BAU, otherwise it enters the increase phase.
- *increase phase* with linear window growth to absorb released resources and probe the state of the bottleneck: The NTG is checked as soon as the accumulated value of BAU/(current window-size) is larger than one BAU. If NTG is less than NTG_i, the window size is decreased by one BAU, otherwise, nothing is done.

Compared to the Slow-Start algorithm *Tri-S* offers the possibility to maintain or even slightly decrease the window size. The shape of the throughput curve (fig. 2.1) up to the "*cliff*" is the basis for the computation of the gradient. The gradient is computed as follows (with W_n representing the window size at time n, and $T(W_n)$ representing the corresponding throughput):

$$TG(W_n) \quad := \quad \frac{T(W_n) - T(W_{n-1})}{W_n - W_{n-1}} \qquad \text{with } W_n = 0 \text{ and } W_n = \text{BAU}.$$

It is obvious, that the throughput gradient TG converges to 0 with increasing window size. The normalized throughput gradient ***NTG*** is computed as follows:

$$NTG(W_n) := \frac{TG(W_n)}{TG(W_1)}$$

According to [4] this value "varies approximately in the range [1;0]". In [7] a proof is introduced which qualifies the term *approximately*. This proof shows that the NTG computation can result in values outside this interval. Minor modifications to the algorithm eliminating these cases do not diminish the usefulness of Slow-Start/Tri-S [7]. With light network traffic the NTG will be close to 1, since the throughput grows proportional to the offered load. NTG will be close to 0 when transmitting on a saturated path, because in this case throughput cannot increase. Since the window size can change from window to window for at most one BAU

$$W_n - W_{n-1} = W_1 - W_0.$$

Thus, the normalized throughput gradient can alternatively be computed as follows:

$$NTG(W_n) = \frac{T(W_n) - T(W_{n-1})}{T(W_1)}$$

The throughput can be computed as $T(W_n) = W_n / D_n$. D_n is corresponds to the RTT of the actual transmitted BAU. Thus, NTG results in:

$$NTG(W_n) = \frac{\frac{W_n}{D_n} - \frac{W_{n-1}}{D_{n-1}}}{\frac{W_1}{D_1}}$$

Normalization to the first measured RTT (close to the pure propagation delay) should be understood as a method for avoiding effects, which are caused by various RTT-measurements.

2.3. Simulation Model and Test Scenario

To validate our simulation model and to test the modified Slow-Start/Tri-S, we have configured our model according to [4]. Asynchronous discrete event driven simulation is used for studying a simplified model of the underlying network in combination with Tri-S. The analysis is executed using the **ATLAS** tool (**A**nalysis **T**ool for **L**ocal **A**rea Network **S**imulation) which has been developed at the Technical University of Aachen [13]. Fig. 2.2 shows the queueing model of the network. The number of intermediate nodes can be varied. In the presented test scenarios the number of intermediate nodes is 5. Three scenarios have been simulated. *Scenario 1* simulates one end-to-end connection on a network without any other traffic, *scenario 2* simulates two end-to-end connections with different path lengths partly sharing a path and *scenario 3* simulates two end-to-end connections fully sharing the path.

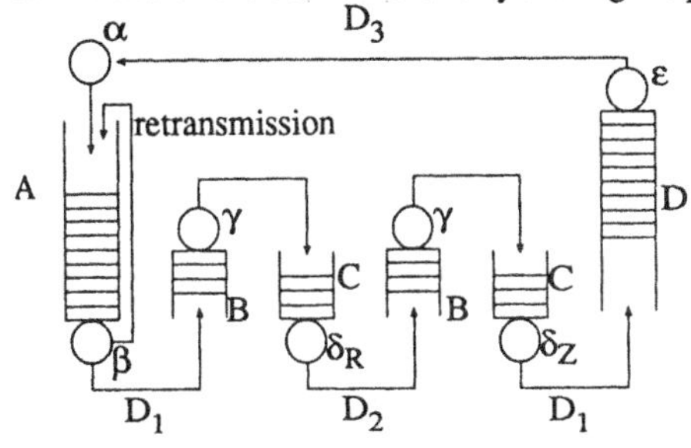

α	:	packet-rate
β	:	send-rate
γ	:	service-time
δ_R	:	send-rate to relay
δ_Z	:	send-rate to receiver
ε	:	service rate
D_1	:	delay sender to relay and relay to receiver
D_2	:	delay relay to relay
D_3	:	delay receiver to sender
A-D	:	queues

Fig. 2.2: Network model

In the intermediate systems incoming packets are served with different server strategies, depending on the simulation run. In one test, packets are served via **FCFS**, and in case of overflow either the incoming (**Drop-Tail**) or a random packet from the input queue is discarded (**Random Drop**). A separate logical queue runs in the other test and is maintained for each source-destination pair in order to obtain fairness among users. The single queues are served in round-robin order, using Fair Queueing [14,15].

To achieve unequivocal RTT-measurements the local transmission time is put in the header of each data packet and retransmitted in the corresponding acknowledgement. If this TCP extension is not implemented, the Karn-Algorithm is used as in most existing implementations. As an alternative to the method usually applied in TCP (Smoothed RTT, SRTT [1]), the acknowledgement timer has been determined by doubling the actual RTT in our simulations.

Four basic combinations to determine the RTT and the timer setting were examined in our simulations in order to test their influence on the Slow Start/Tri-S scheme:

A using timestamps for RTT calculation
acknowledgement timer: 2 * RTT

B using Karn algorithm for RTT calculation
acknowledgement timer: SRTT

C using Karn algorithm for RTT calculation
acknowledgement timer: 2 * RTT

D using timestamps for RTT calculation
acknowledgement timer: SRTT

Table 2.1 depicts the static input parameters for all of the executed simulation runs.

Parameter	Value
α	5000 kbit/sec
β	1000 kbit/sec
γ	1 msec (mean value)
δ_R	500 kbit/sec
δ_Z	1000 kbit/sec
ε	5000 kbit/sec
D_1	0 msec
D_2	50 msec
D_3	205 msec
A	85 packets
B	15 packets
C	15 packets
D	85 packets
simulation-period	180 seconds

Table 2.1: Input parameters

2.4. Simulation Results

The results of the simulations comparing Slow Start and Tri-S (cf. fig. 2.3) during the start phase of a connection illustrate the main difference between the two algorithms. Slow Start opens the connections by increasing the window until a packet loss occurs. It then enters consecutive cycles of slow start and linear increase phases. Tri-S follows the same policy for the connection start up. Since in the depicted example an NTG > 0.5 has been selected, a packet loss occurs. The decrease phase is entered and NTG calculation activated. After an exponential window growth Tri-S enters the increase phase with linear window growth.

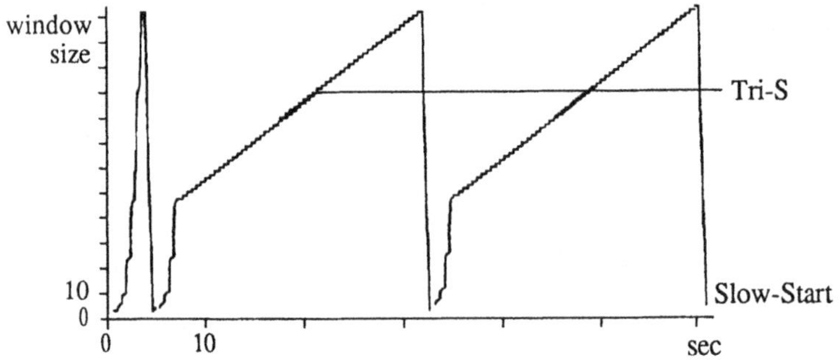

Fig. 2.3: Window size, Tri-S vs. Slow Start

Other simulation runs show that using Tri-S with the Karn algorithm leads to high oscillation of window size. The longer a retransmission delay, the worse the cooperation between Karn algorithm and Tri-S. This correlation results in an NTG value of one, which indicates the transmission path would be free; thus leading to a further exponential growth of the flow window.

The more reliable the RTT measurement, the easier the adjusting of the timer. Since the exact measurement of RTTs using time stamps is feasible with a relatively simple extension (option field) we have used this TCP extension (option field) in the simulation runs (**A** and **D**). A calmer throughput oscillation compared to the other strategies is obvious (cf. fig. 2.4).

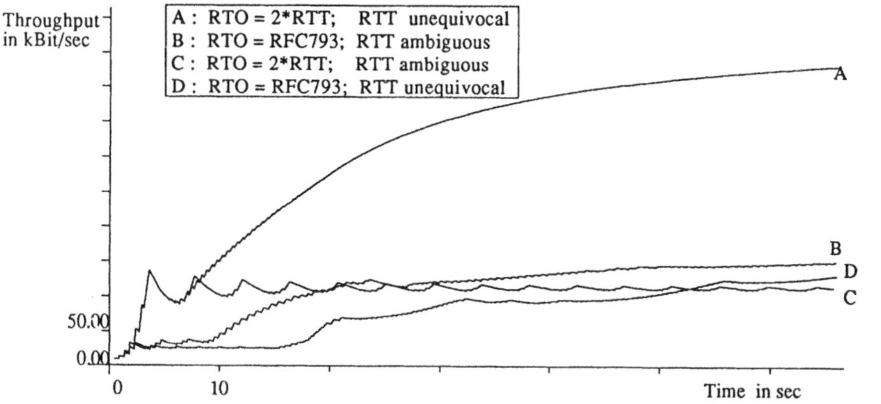

Fig. 2.4: Throughput of Tri-S

In fig. 2.5 the throughput development of Tri-S (scenario 2) with RTT measurements using time stamps and different queueing and service policies in the intermediate nodes is depicted. Obviously, the most complicated method (in the sense of implementation costs), *Fair Queueing (FQ)*, achieves the best long-term network efficiency regarding the sum of the two throughput curves compared with *Random Drop (RD)* and *Drop Tail (DT)*. Fair Queueing is a very robust method that can achieve an equal distribution of resources between different sources even if the Karn algorithm is used for acknowledgement timer calculations. However, the use of separate queues does not ensure the fairness among users from the same source. Fairness among those users can only be achieved if separate queues for each source-destination pair are maintained.

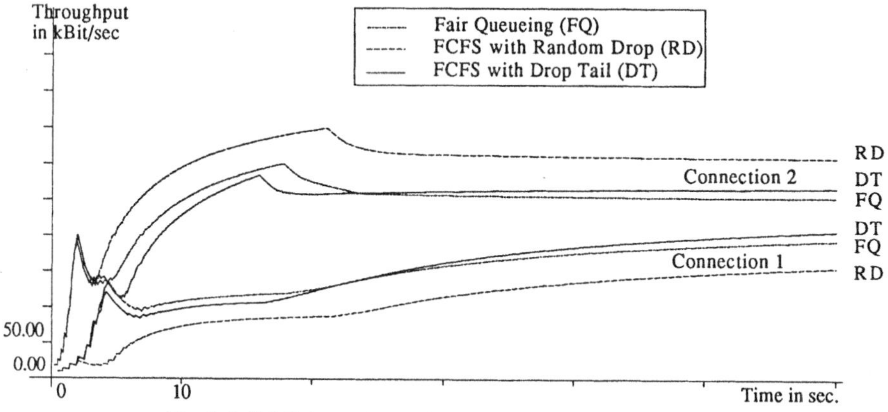

Fig. 2.5: Tri-S throughput with different service disciplines

The use of Fair Queueing shows, that efficiency maximization of one sender does not lead to an optimal use of the network. A global efficiency optimum can not be achieved from the sum of all local optima. To operate a network at optimal level, the senders have to be prepared to operate at a value below their individual optimum. To achieve a fair allocation of resources the intermediate gateways have to become active. The use of Random Drop entails only a few advantages. It is even a disadvantage for a connection with long RTT (connection 1), which shares a (part of a) path with a connection that has a short RTT (connection 2).

2.5. Concluding Remarks

In a large number of simulation runs we have studied the impact of RTT measurement and timer adjustment on congestion control. Using Slow Start / Tri-S, that the continuing measurement of RTTs is essential for the performance of the algorithm. Tri-S helps to eliminate periodic packet loss and reduces RTT variation. More research about the usefulness of implicit timer based congestion control schemes has to be done to decide, if explicit congestion control schemes are an alternative to implicit window based methods. In recent and future network architectures rate and window based load control mechanisms have to coexist. The window based schemes (based on local timer decisions) are used for buffer place allocation whereas rate control is used for the real-time adjustment of resources between different connections.

3. Multicasting and Concentration

The term **multicast** [16] is defined as a concurrent data transmission to a restricted set of recipients which is identifiable by a single (group) address. Hughes [17] calls this service a *multicast transmission.* The term **concentration** [6] is defined as the transmission of arbitrarily sized messages from a set of hosts to a single host. Hughes calls this service a *multicast reception.* We call this kind of communication also *transaction oriented multicast* since one transport entity sends a request to a group of receivers which all have to respond.

The increasing dissemination of high-speed networks gives rise to another development: **multi-media communications**. Although this term is mainly used as a synonym for either interactive audio and video communication or for exchange of documents including e.g. some graphics or spreadsheets, true multi-media does actually mean the integration of the above.

Putting it all together, we end up with the perception that a future communication system should (inter alia) be able to support on one hand the efficient distribution of multi-media information from one source to a large and widely dispersed number of recipients (**multicast**) and on the other hand the reliable and efficient transmission of information from a set of hosts to one recipient host (**concentration**). This imposes a number of problems on the system. The remainder of this chapter addresses some of these problems and recommends the use of timers to improve the performance of the algorithms. Paragraphs 3.1, 3.2 and 3.3 illustrate the currently most promising transfer mechanism to support "semi reliable" multicast, the bucket algorithm [6]. Paragraph 3.4 covers the problem of large RTT variation to the different receivers in a multicast group and on the one hand introduces threshold values for RTT to detect "**bad receivers**" and on the other hand improves the reception at bad receivers by forming "**local groups**". Chapter 3.4 ends with suggestions how to enhance the reliability of multicasting. In the final paper simulation results on the use of the different mechanisms will be integrated.

3.1. XTP Bucket Algorithm

XTP [16] multicast is based on the ***bucket algorithm*** which is able to accommodate semi reliable and best-effort semantics. In contrast to other concepts sender and receivers build one group. By setting the MULTI-bit in the packet header a sender can indicate that he addresses an arbitrarily large receiver group. The sender does not have to establish *n* unicast connections but replaces all the unicast connections by a single multicast transmission. This is only feasible on top of a transmission medium that defines a broadcast or multicast facility.

The receivers are passive - they do not send data and only respond when prompted. As with unicast, the multicast transmitter sets a Status **REQ**uest bit (**SREQ**) in an out-going CNTL packet in order to get receivers to issue an ACK CNTL. A bucket which is a collection of control information is created whenever the multicast sender transmits an SREQ. The time at which a request was issued is measured by the XTP *sync* counter, which is incremented by one each time a packet is transmitted from the context. Each in-coming CNTL packet contains an *echo* value that reflects the value of *sync* as known at the receiver. The *sync/echo* mechanism enables in-coming CNTL packets to be sorted into buckets by ´age´. CNTL packets of the same bucket are coalesced by recording the minimum of the flow parameters, the byte-based sequence number (*rseq*) that indicates the highest consecutive sequence number received without error at the issuing receiver, and the maximum of the round trip time estimate.

3.2. XTP Damping and Slotting

To reduce multicast overhead, the number of duplicated control messages generated in response to an SREQ or in case of errors should be decreased using efficient acknowledgement and buffer control strategies. If error control is not turned off by setting the NOERR bit a multicast-receiver generates a CNTL-packet. In the current version such a CNTL is a REJECT packet requesting a go-back-n retransmission. REJECTs are not exclusively addressed to the multicast sender but to the whole group. Consequently, the other group members recognize the control packet and can dequeue their retransmit demand if the incoming control packet records the same or a smaller missed sequence number (**DAMPING**). This mechanism lessens the number of superfluous control packets flowing to the transmitter. To make the damping method feasible each receiver imposes a random delay before sending its CNTL packet (**SLOTTING**). This technique allows time for CNTL packets to arrive that might damp the out-going CNTL packet. A problem when using slotted damping is that it must be compatible with the bucket algorithm, it must take care that slot delay is less than the lifetime of a bucket.

Disadvantages to the proposed damping mechanism result from its fragility due to timing considerations that may vary widely over disparate environments and from the case that the ACKs are anonymous. Thus, the resulting multicast algorithm is not completely reliable. The probability that a given receiver will "fall behind" (that means, the sender will drop some data from a retransmission buffer before all the receivers have got it), is the probability that a sent data packet is lost/corrupted and that the whole number of ACKs in a row are lost from every receiver that missed that packet (this number is implementation dependent and gives the number of buckets which will be created before dropping). To achieve a completely reliable multicast, the ACKs may not be anonymous. The group of receivers must be known (using a reliable group membership mechanism not defined within XTP) and data buffered for retransmission may not be dropped until every receiver in the group has indicated that it has received the data.

3.3. XTP Multicast Addressing

XTP does not define how to assign a multicast group address. It is assumed that address assignments are made by an external mechanism as e.g. agreement between peers or a dedicated address management protocol. Multicast addresses may be a special Internet Class D address, or a locally-defined, 32-bit direct address. Direct addressing is always available for multicast (or non-multicast) use. The usual restriction applies: direct addressing is intended for closed environments where peer conventions and knowledge guarantee correct assignment and interpretation of addresses. Systems using direct addressing must define a mapping between MAC addresses and direct XTP address values.

3.4 Illustrative Example

The following fig. 3.1 illustrates a concrete example of the bucket mechanism, where a simple DATA and CNTL packet exchange is shown. The scenario consists of one sender and a receiver pool. The diagram depicts the sender´s input buffers, the two buckets and the multicast packet exchange. After initializing a multicast transfer a sender maintains a bucket list with a life time of a single bucket which corresponds to a value close to the maximum estimated RTT. The horizontal thick lines indicate the time boundaries for the creation or recycling of buckets. Specific problems illustrated in fig. 3.1 are e.g. the occurrence of redundant retransmissions like DATA(100..399, sync=5,6) and the difficulty to find the right trade-off between reliability and efficiency. The latter depends on the two time values, maximum RTT and bucket life-time. Another problem is the enability of routers to filter CNTL packets with similar context information which cross a router during one RTT (e.g. CNTL(echo=2,rseq=200,dseq=100) and CNTL(echo=2,rseq=200,dseq=200)).

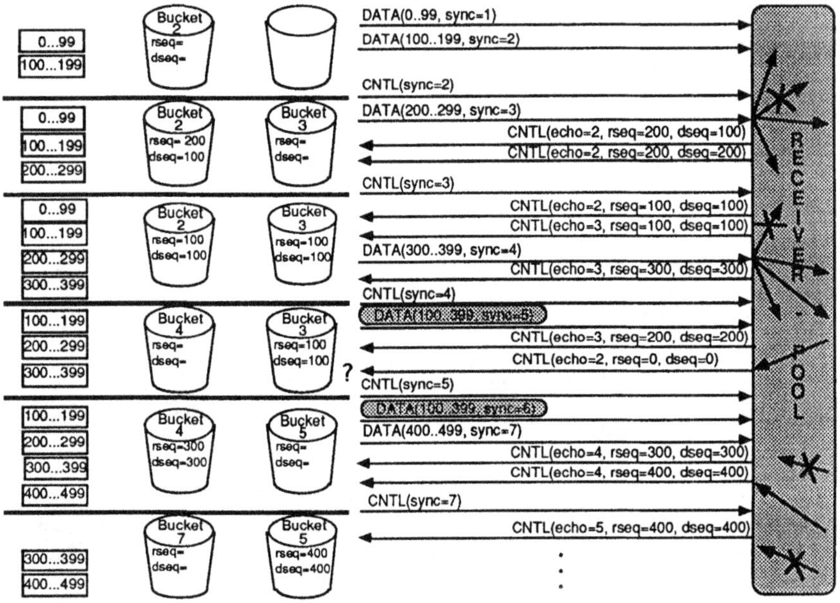

Fig. 3.1: XTP Bucket Algorithm

3.4. XTP Deficiencies and Enhancements

3.4.1. Ignoring 'Bad Receivers' using Local Timers

One drawback of the XTP multicast strategy using damping and slotting is the following: *Multicast receivers may be located at a large number of networks. At the same time, different load conditions may exist on the networks. Hence, there may be large differences in propagation delay which give rise to large differences between the RTT values reported to the multicast sender. Thus, receivers on overloaded LANs can considerably influence the overall performance of a multicast session.*

We propose to determine a threshold value $\mathbf{RTT_{max}}$ (maximum RTT value indicating the slowest receiver). Upon the session start a default value defined by the implementation can be used. Subsequently, new values substitute the default $\mathbf{RTT_{max}}$. These values are obtained from the mean values of the measured RTTs ($\mathbf{RTT_{actual}}$) during one bucket lifetime. This operation will take place at the end of the actual bucket period. The basic rules we are propose to detect *bad* receivers are given in the following pseudo code:

```
if ( RTTactual < RTTmax )
{   if ( RTTactual > RTTmean )
    {       RTTmax = RTTactual;
            the address of the station which has sent the packet is stored in a bad receivers' list ;
    }
    else
```

```
            if ( originating station of both RTTmax and RTTactual values is the same )
                    RTTmean = RTTactual;
}
else
{    if ( RTTactual > RTTmean )
     {      if ( originating station of both RTTmax and RTTactual values is the same )
            {       ignore the new value and put the station address into the bad receivers' list;
                    this receiver is getting harmful for the multicast performance when remaining
                    member of the group;
            }
            else
                    if ( already marked as a bad receiver ) ignore it ;
                    else put the station address into the bad receivers'list;
     }
     elseRTTmean = RTTactual;
}
```

The adoption of this procedure should not compromise the effectiveness of the retransmission strategy in environments where fast delivery is preferred to total integrity of data.

3.4.2. Establishing Local Groups

To improve behaviour of receivers located at overloaded subnetworks, an additional procedure can be adopted. We propose use of specific CNTL packets (not provided in the last release of XTP) with small ttl-values. Having chosen these values appropriately, only group members at e.g. the same subnetwork will receive such packets, since they will very soon be discarded by the network. This leads to local groups only known to the closest router. We also propose to collect control packets from the members of this group and to forward them to the multicast sender in a single composite control packet. To achieve this functionality a collector only needs to set up its address filters appropriately and some logic to coalesce control packets. From the multicast sender´s point of view data transfer is considerably simplified. The sender establishes a connection with the group and receives a fixed number of control packets (one per group) upon each SREQ. This strategy will widen the bottleneck at the source host´s network interface.

3.4.3. Unicast CNTL instead of Multicast CNTL

The usefulness of multicasting CNTL-packets is very questionable in global heterogeneous networks. For this scenario we propose to use unicast CNTL packets only sent to the transmitter rather than to the whole multicast group or to a local group, respectively. This implies to shift damping and slotting facilities from the end system to the routers. Routers collect the incoming control information and provide a slotted damping algorithm similar to the collector functionality described in the previous section. The problem of this procedure will be the appropriate adjustment of slotting intervalls in order to collect the optimal number of CNTL packets and to avoid congestion.

3.4.4. New Packet Format for Reliable Multicast

To implement a k-reliable multicast service, it must be assured that k members of the group have received the data. The transmitter must be able to distinguish between the single receivers. Therefore, we integrate a new field into the header of CNTL packets, identifying the individual addresses of the senders.

3.4.5. Relay Race based Endsystem Slotting

To make the slotting mechanism more practical we suggest to provide a relayed slotting algorithm. The receivers which are closest to the sender have to wait the longest time interval before creating an acknowledgement. Otherwise in most cases the closest receivers will have created their acknowledgement before an effective damping can start.

4. Conclusion and Further Work

Within this paper, we have considered two aspects of high performance communication: congestion control and multicasting. In particular, we propose the use of timers to manage both problem areas. In a large number of simulation runs, the interrelation between RTT estimation, timer adjustment and sophisticated transport system services for multicast and congestion control has been studied. The results may be summarized as follows:

Slow Start/Tri-S provides a method to approach the optimal throughput operating point in a more careful way compared to conventional Slow Start. The feasibility highly depends on the calculation of the RTT of the first packet in the Slow Start phase. Since this value should give a rather accurate estimation of the pure propagation delay on the network, we assign the highest priority to the first packet. Thus queueing delay can be minimized. Continuing measurement of RTTs via time stamps and an appropriate calculation of acknowledgement timers lead to improved throughput oscillation, reduction of packet loss and eliminates the preemption of the packet stream. To avoid unfairness caused by the fact that two users have different timeout periods, we suggest the use of Fair Queueing in intermediate nodes.

Drawbacks of the XTP bucket algorithm to support multicast error control are its enability to support a reliable service since the senders cannot identify active receivers, and its bad performance resulting from the inefficient handling of delayed receivers. Therefore we make recommendations to provide individual addresses to the sender, to eliminate "bad receivers" and to build local groups by using short controlling timers for performance enhancement.

The next step of our work will be concentrated on more detailed evaluations on the interrelation between window based and rate based control schemes and their influence on the achievable network power. Especially the area of heterogeneous wide area networks is more difficult to estimate than local area environment. This fact is true for congestion control as well as for multicast services. Currently we are working on a formal specification of the enhanced bucket algorithm which we first want to implement in an Ethernet-FDDI-Ethernet test scenario.

References

[1] V. Jacobson: *Congestion Avoidance and Control*, SIGCOMM '88 SYMPOSIUM Communications Architectures & Protocols, Stanford, California, August 16-19, 1988

[2] Allison Mankin, K.K. Ramakrishan: *Gateway Congestion Control Policies*, Internet-Draft, July 1989

[3] S.J. Golestani: *Congestion-Free Communication in High-Speed Packet Networks*, IEEE Transactions on Communications, Vol. 39, No. 12, December 1991

[4] Z. Wang, J. Crowcroft, *A New Congestion Control Scheme: Slow Start and Search (Tri-S)*, Computer Communication Review, Vol.21, No.1, Januar 1991

[5] K.K. Ramakrishan, R. Jain: *A Binary Feedback Scheme for Congestion Avoidance in Computer Networks with a Connectionless Network Layer*, SIGCOMM '88 and ACM CCR, Vol. 18, No. 4, 1988

[6] *XTP Protocol Definition*, Revision 3.6, Protocol Engine Incorporated, January 1992

[7] Wolfgang Mers: *Ratenbasierte versus fensterbasierte Lastkontrolle*, Diplomarbeit am Lehrstuhl für Informatik IV, RWTH Aachen, August 1992

[8] R. Jain: *Myths About Congestion Management in High-Speed Networks*, DEC-TR-726, Oktober 1990

[9] L. Zhang: Why TCP Timers don't work well, ACM CCR, Vol. 16, No. 4, 1986 and SIGCOMM'86

[10] P. Karn, C. Partridge: *Improving Round-Trip Estimates in Reliable Transport Protocols*, Computer Communication Review, Vol. 17, No. 5, 1987 and SIGCOMM'87

[11] V. Jacobson, R. Braden, D. Borman: *TCP Extensions for High Performance*, Internet Draft, 02/14/1991

[12] H. Schulzrinne: *A Transport Protocol for Audio and Video Conferences and other Multiparticipant Real-Time Applications*, Internet Draft, October 27, 1992

[13] P. Davids: ATLAS: *Analysis-Tool for Local Area Network Simulation*, Version 5.0, Lehrstuhl für Informatik IV, RWTH Aachen, January 1992

[14] J. Nagle: *On Packet Switches with Infinite Storage*, IEEE Transactions on Communications, Vol. 35, No. 4, pp. 435-438, 04/1987

[15] J. Nagle: *Congestion Control in TCP/IP Networks*, RFC 896, June 1984

[16] K. Paliwoda: *Transactions involving multicast*, ACM CCR, Vol.11, No.6, pp. 313-318, December 1988

[17] L. Hughes: *Multicast response handling taxonomy*, ACM CCR, Vol.12, No.1, pp.39-46, February 1989

Integrated Broadband Communication Networks
and Services (C-18) / V.B. Iversen (Editor)
Elsevier Science B.V (North-Holland)

On the Limit Behaviour of the ATM Traffic within a Network

M. D'Ambrosio and R. Melen*

CSELT, Via G. Reiss Romoli 274, 10148 Torino (Italy) - Tel.: +39 11 2285111 - Fax: +39 11 2285069 - E_mail: DAMBROSIO@CSELT.STET.IT

Abstract.
The paper is concerned with the ATM traffic characterization within the network. Most of the work performed up to now has studied the effects of the traffic on the access multiplexer and the first switch of an ATM network; various source models were assumed to generate the ATM traffic. So, while the performances of a single switch node have been exhaustively examined, the statistical behaviour of the traffic modified as it crosses the network has not been thoroughly analyzed yet. The paper, through an analysis of a network of cascaded queues, shows that limit distributions exist in the statistical behaviour of the traffic streams and in the queue performances. The first modelling step consists in deriving the exact interdeparture time distribution for the cells of a reference-connection arriving to the output queue of a switch node with a general interarrival (GI) time distribution and multiplexed with a background traffic stream, then the analysis is iterated through a long sequence of cascaded output queues, until the interdeparture time distribution converges to a limit one. Simulations show that the analitycal results are more than accurate at each stage of the network under the hypothesis of independent queues. The study shows that the queue performances at the equilibrium point are always better than the M/D/1 case. The limit distributions depend only on the connection bandwidth and on the background traffic behaviour; the initial characteristics of a connection (burst length distributions and burst interarrival time distributions) do influence only the convergence speed, not the limit distributions.

Keyword Codes: C.2.0; C.4
Keywords: Computer-Communication Networks, General; Performance of Systems

1. INTRODUCTION

ATM switches have been intensively studied in the last years with respect to their performance in order to meet the grade of service required by a wide variety of B-ISDN services. So the characterization of the ATM traffic is a relevant topic for making appropriate architectural choices in resource allocation and control mechanisms. Most of the studies in the field of performance evaluation of ATM networks were performed under the hypothesis of a uniform and uncorrelated traffic model. But the ATM traffic cannot always be described in terms of Bernoulli or Poisson sources and uncorrelated cells uniformly distributed in the network as sometimes accepted. Although the switch performance has been studied using alternative traffic models (the superposition of periodic and modulated periodic arrival processes have been considered at the sources, and bursty streams have been usually described with on-off or fluid-flow models [4]), very little work has been done to obtain a more accurate characterization of ATM traffic within the trunk network [1,4,5,7,12,13,14]. The fact that at least one half of the traffic arriving at a local switch comes from the network has an effective impact on the overall switch performance. Thus it is necessary to characterize

---

*Present address: Politecnico di Milano, Dipartimento di Elettronica, v. Ponzio 34/5, 20133 Milano (Italy) - Tel.: +39 2 23993608 - Fax: +39 2 23993413

the statistical behaviour of the traffic as it is modified across the network: the present paper is concerned with this topic. A simple analytical model is proposed: it allows to characterize the interdeparture process for the cells belonging to a given connection passing through a long sequence of cascaded switch output queues. The model shows how the statistical characteristics of the streams change stage after stage till to reach a stationary state. A limit behaviour is reached within the network in a number of stages depending on the number of connections and their initial characteristics. Furthermore it can be shown by intuitive reasons that the worst case limit traffic is given by a Poisson arrival process, so that the performance of a M/D/1 queueing system can be assumed as a pessimistic bound for the output queues inside the network at the equilibrium point. This result needs to be carefully evaluated in order to determine the minimum number of multiplexing stages that a real ATM traffic should pass before its statistical behaviour allow better performance than a Poisson process.

The paper is organized as follows. In section 2 preceding work is reviewed. Section 3 introduces the analytical model. The derivation of the interdeparture time distribution is performed in section 4. Numerical results are given in section 5 and validated by simulations. In section 6 the outcomes are discussed. Then, section 7 concludes the paper with suggestions for further research on the subject.

2. PRECEDING STUDIES

Only a few papers are focussed on the topic of traffic characterization within the ATM trunk network. Most of them are based on simulation approaches. Norros [4,13] has studied the burst expansion phenomenon in a long sequence of cascaded multiplexer queues. The length of a burst increases as the stream goes through the queues decreasing its burstiness (see figure 1). The same effects have been noted in more exhaustive simulation experiments [5,12]. In [5] a network of crossbar switches with even 50 or 100 stages has been simulated. The network model was simplified using a cascade of NxN switch modules with output queueing. The traffic was generated according to Bernoulli and Poisson processes, or with bursty on-off sources. The streams leaving a switch were fed into the next after being uncorrelated by shifting them in time. The results have shown that a Bernoulli traffic can be considered almost invariant as it goes through the network. When a bursty traffic is assumed, the distribution curves change stage after stage approaching the Bernoulli curves. In each stage bursts are interleaved and the correlation between successive packets on the output lines is reduced. The burstiness on the input lines is greater than the output burstiness, until the traffic reaches the steady state. Since the switch size in the experiment is finite, the equilibrium traffic seems to be characterized by a Bernoulli process. The connections were however not modelled in that experiment: the model has not taken into account the correlation between two successive bursts belonging to the same virtual connection; all the bursts were independently routed through the network.

Analytic approaches are rare and general [2,4] and often limited to the traffic analysis on the output lines of a single switch [1,5]. A notable exception exists in the work of Ohba et al. [14]. They have derived the interdeparture time distribution and the waiting time distribution for the cells of a stream having a general interarrival (GI) time distribution, multiplexed with a background traffic with Bernoulli arrivals of cells in batches and interrupted Poisson processes. A tandem network with four cascaded output queues has then been examined: at each node a constant background traffic (Bernoulli and IPP) is kept, while a GI-stream feeded in the first queue crosses the network. The results show that the interdeparture time distributions become less bursty than the interarrival time distributions at each node, and the mean waiting time decreases.

Now, what happens when a traffic stream goes through a large number of switching stages? Given the above results, the existence of a fixed-point traffic reached in the network becomes a intriguing hypothesis. Using an alternative approach, the present paper numerically calculates the equilibrium interdeparture time distribution to which a GI-stream connection tends as goes through the nodes. With respect to the paper by Ohba et al. [14], here the emphasis is on the limit behaviour of the traffic and on the surprising queueing performances.

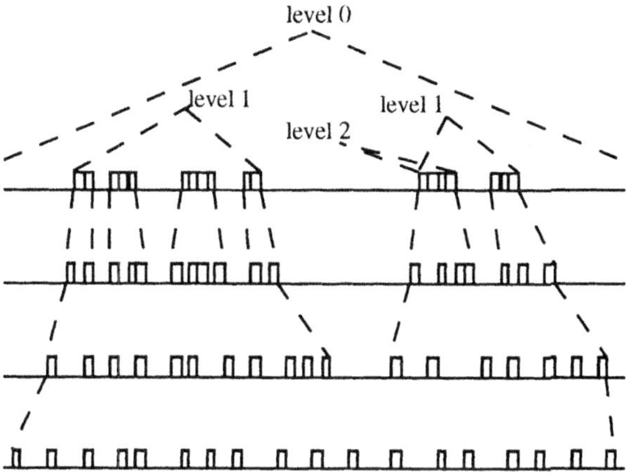

Fig. 1. Burst expansion phenomenon in a single connection.

3. THE MODEL

In the paper the attention will be focussed on a single output queue of an ATM switch. It will be assumed that the cell length is fixed, and the time is slotted with slot size being equal to a cell transmission time. Cells coming from a number of sources and destined to the same output link arrive simultaneously at the beginning of a time slot and are randomly stored in the buffer. The first cell in the queue is then immediately served, by the end of that time slot (see figure 2). So, the ATM switch is modelled as a discrete-time single-server queue.

An ATM connection carried by an input link sends its cells to the switch according to its interarrival time distribution. Usually a connection is characterized by a certain degree of burstiness. Its interarrival time distribution can be described with the composition of two or more sub-distributions: for instance, a connection can be characterized by the interarrival time distribution relative to the cells belonging to the same burst, and by a second distribution relative to the interarrival time of the first cell of a burst (that is the distribution of the silence period duration). The burstiness of a connection can be defined as the ratio between peak bandwidth and the average bandwidth (or the ratio between the overall mean cell interarrival time and the mean cell interarrival time during a burst). Another parameter characterizing an ATM connection is its burst length distribution. Note that bursts can be nested at different levels (like in the first line of figure 1) and, for each level, different interarrival time distributions can occur. However, if the interest is not on the burst phenomenon, a single general distribution suffices. All these features should be taken into account in modelling a single connection.

The model chosen to characterize a single connection according with the above observations is the following (see figure 2): in a certain time-slot x, the stream source is said to be in state (t) when the first cell arrival to the ATM output queue from that source occurs in time-slot x+t. So, state (0) corresponds to an effective cell arrival, and the steady state probability $P_{(0)}$ turns out to be the average offered load of the stream. Let A[(x),(t)] be the transition probability from state (x) to state (t) in one time-slot (with $x,t \in \{0,1,2,...\}$). Then the transition probability A[(0),(t)] is the probability that, after a cell arrival, the next one occurs after t+1 time slots. Let $f_i(t)$ $(1 \le i \le D, t \ge 1)$ be the i-th function of a set of $D \ge 1$ discrete general

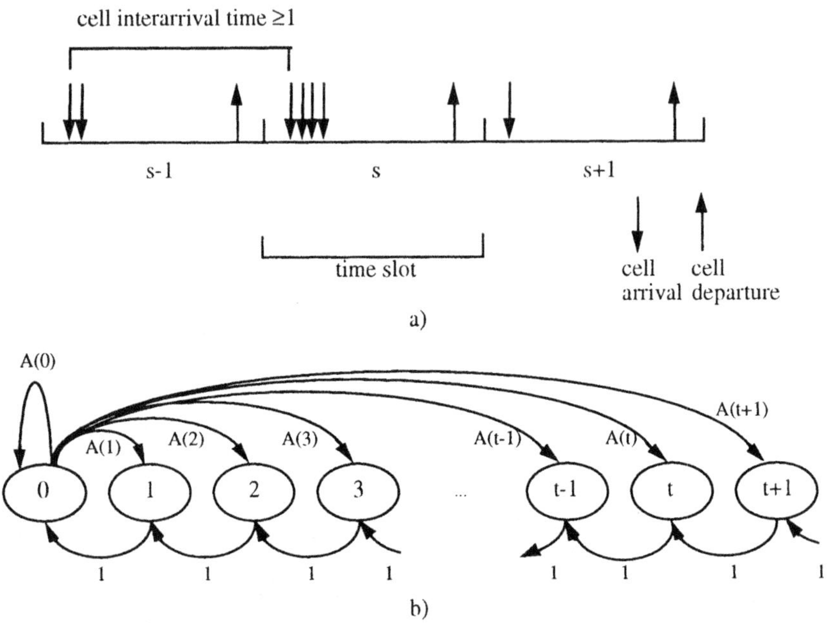

Fig. 2. Time-slotted model (a) and state model (b) for a single connection.

interarrival time distribution functions, and let q_i ($\sum_{i=1}^{i=D} q_i=1$) be the respective probability for that function to be applied after a cell arrival. The transition probabilities are then:

$$A[(x),(t)] = \begin{cases} 0 & \text{if } x \neq 0 \wedge x \neq t+1 \\ 1 & \text{if } x = t+1 \\ \sum_{i=1}^{i=D} q_i I_i(t+1) & \text{if } x = 0 \end{cases} \qquad x,t \in \{0,1,2,\ldots\} \tag{1}$$

Note that, if D>1, the composition of the interarrival time distributions by the set of probabilities q_i implies that the number of cells at each level is distributed according to a geometric distribution; however, setting D=1 the transition probability A[(0),(t)] can be associated to any general interarrival time distribution I(t). The model is useful to characterize interarrival time distributions by combining elements which have intuitive significance (interarrival times, burst length) into a composite distribution. In the following, let A[(0),(t)] be replaced by A(t) and, as the emphasis is put on the burst phenomenon, let D be set to 2 unless explicitly stated. This assumption allows a correct modelling of a bursty connection with only one level; the source appears characterized by on-off periods of activity. Let S be the random variable representing the interarrival time of a cell after the end of a burst and let $I_S(t)$ its density function, that is the probability distribution of the silence period between two successive bursts. During a burst, let B be the random variable for the interarrival time, with density function $I_B(t)$, and let L be the random variable for the burst length (distributed according to a geometric law); assume p=1/E[L] where E(L) is the expectation for L; p is the probability that the next cell will not belong to the current burst; therefore it is:

$$q_B = 1 - p$$
$$q_S = p \tag{2}$$

and in the following the transition probabilities for a bursty connection can be considered as:

$$A[(x),(t)] = \begin{cases} 0 & \text{if } x \neq 0 \wedge x \neq t+1 \\ 1 & \text{if } x = t+1 \\ pI_S(t+1) + (1-p)I_B(t+1) & \text{if } x = 0 \end{cases} \qquad x,t \in \{0,1,2,\dots\} \tag{3}$$

The equilibrium state equations for a single connection characterized by the above transition probabilities are (see figure 2):

$$\begin{cases} P_{(1)} = P_{(0)} \cdot (1 - A(0)) \\ P_{(t+1)} = P_{(0)} \cdot \left(1 - \sum_{s=0}^{t} A(s)\right) & t \geq 0; \end{cases} \tag{4}$$

solving the system, the steady state probabilities are given as:

$$\begin{cases} P_{(0)} = \dfrac{1}{(1-p)\alpha + p\beta} = \dfrac{1}{1+\gamma} \\ P_{(t+1)} = \dfrac{1 - \sum_{s=0}^{t} A(s)}{(1-p)\alpha + p\beta} = \dfrac{1 - \sum_{s=0}^{t} A(s)}{1+\gamma} & t \geq 0; \end{cases} \tag{5}$$

where α=E[S], β=E[B] and γ=E[A(t)]. Note that the $I_S(t)$ and $I_B(t)$ are defined for $t \geq 1$, while A(t) is defined for $t \geq 0$ so that a 1 is added to γ in the denominator.

Now suppose that a reference connection (R-connection) enters the output queue of a

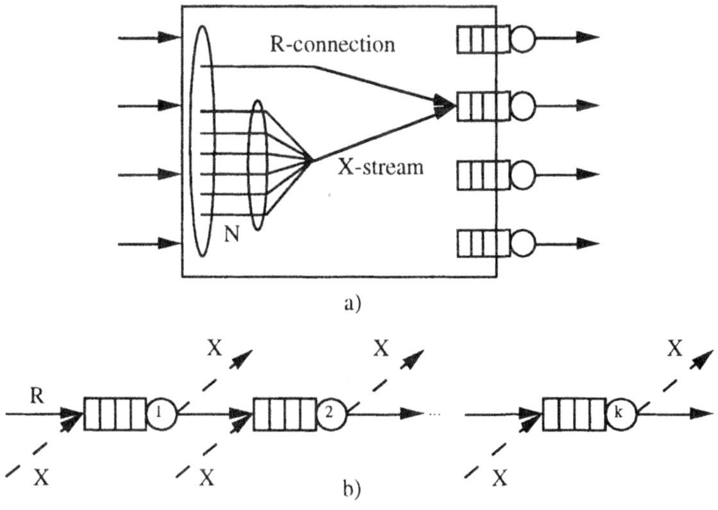

Fig 3. a) Model of a switch node with output queues.; the X-stream is made of N individual streams coming from the inputs, and forming the background traffic for the Reference-connection; b) A R-connection goes through a network of cascaded output queues; the background traffic X is multiplexed with the reference connection at each stage.

switch node; in the queue it will be merged with N other connections (see figure 3a). A random selection policy for service will be applied among the cells simultaneously arrived. The traffic stream due to those N connections, named X-stream, if N is large enough can be well approximated with a Poisson stream (P-stream) with mean λ ($\lambda<1$), or with any other traffic memoryless stream (for instance the arrival process due to N Bernoulli traffic sources characterized by a binomial distribution, named in the following B-stream). Let a_k be the probability of having $k\geq 0$ simultaneous cell arrivals from the X-stream.

Now, the system in a certain slot x can be described by the state (t,n), meaning that the first cell arrival from the R-connection occurs at time-slot x+t, and the number of cells waiting in the queue at the end of time x is n (without counting the cell just serviced). The steady state distribution of the system can be calculated by iteration with the following system of balance equations:

$$\begin{cases} P_{(0,n)} = \sum_{m=0}^{n} \left(P_{(1,m)} + P_{(0,m)}A(0)\right) a_{n-m} & n \geq 0 \\ P_{(t,0)} = \left(P_{(t+1,0)} + P_{(0,0)}A(t)\right)(a_0 + a_1) + \left(P_{(t+1,1)} + P_{(0,1)}A(t)\right) a_0 & t > 0 \\ P_{(t,n)} = \sum_{m=0}^{n+1} \left(P_{(t+1,m)} + P_{(0,m)}A(t)\right) a_{n-m+1} & t,n > 0 \end{cases} \tag{6}$$

The queue length distribution PQ(n) can be derived as:

$$PQ(n) = \sum_{t=0} P_{(t,n)} \tag{7}$$

To make the analysis computationally tractable, the cell interarrival times of the R-connection and the queue length must be bounded. As there can be infinite interarrival times, the tail of the time distribution A(t) is truncated and the tail probabilities are swept to the truncation point. Moreover, the buffer size of the output queue is assumed quite large, so that the cell loss probability is made negligible. Buffer sizes of 100-200 cells and maximal interarrival times equal to 200-400 slots have been considered in the numerical examples, according to the offered load and to the stream burstiness.

4. THE INTERDEPARTURE DISTRIBUTION

As the focus is on the modifications of a stream passing through a cascade of queues (see figure 3b), the interdeparture time of a R-connection from an output queue become essential to the analysis. In fact, modelling the interarrival process at queue i+1 with the interdeparture time distribution at stage i makes possible to iterate the analysis indefinitely.

The derivation of the interdeparture time distribution of a R-connection can be carried out using a transient analysis of the queue behaviour. Consider that in a certain slot s the R-connection is about to send a cell (R-cell); therefore the R-connection has just entered state (0); in the same slot, other $j\geq 0$ cells arrive from the background X-stream; at that time the R-cell is preceeded in the queue by n other cells, where n is comprehensive of the cells waiting in the buffer before the R-cell arrival and a few of the j cells from the X-stream (a random selection policy for service is applied among the cells simultaneously arrived).

Note that:

• the probability that the queue length is x just before the cell arrivals in slot s, given that the R-connection enters state (0), is:

$$\frac{P_{(0,x)}A(0) + P_{(1,x)}}{P_{(0)}} \tag{8}$$

• if the R-cell arrives simultaneously with j cells from the X-stream, k ($0\leq k\leq j$) cells will be stored in the buffer before the R-cell, with probability:

$$\frac{1}{j+1} \tag{9}$$

• The interdeparture time t_D between two successive R-cells is related to their interarrival time t_A as follows:

$$t_D = t_A - n + m \qquad (10)$$

where n is the number of cells waiting for service seen by the first R-cell before itself, and m is the number of cells in the buffer seen by the second R-cell before itself.

Assume that, after the first R-cell arrival at time s, the next arrival from the R-connection occurs at slot s+t. Define the following probabilities:

- $E_{s+t}(n,m)$ is the probability that the first R-cell sees n cells waiting for service before itself in the queue at time s, and the queue length at the end of slot s+t is m;

- $F_{s+t}(n,m)$ is the probability that the first R-cell sees n cells waiting for service before itself in the queue at time s, and the next R-cell arriving in s+t sees m cells waiting for service before itself.

- $G_{t_A}(t_D)$ is the probability that a couple of successive R-cells with an interarrival time t_A experiences at the output an interdeparture time t_D.

The probabilities above are all stationary, so it is possible to omit s from their expressions. $E_t(n,m)$ and $F_t(n,m)$ can be recursively calculated starting with $E_0(n,m)$ and using (8) and (9):

$$E_0(n,m) = \sum_{x=0}^{n} \frac{P_{(0,x)}A(0) + P_{(1,x)}}{P_{(0)}} \cdot \frac{a_{m-x}}{m-x+1} \qquad n \le m \qquad (11)$$

$$\begin{cases} E_{k+1}(n,0) = E_k(n,0)(a_0 + a_1) + E_k(n,1)a_0 \\ E_{k+1}(n,m) = \sum_{x=0}^{m+1} E_k(n,x)a_{m-x+1} \end{cases} \qquad k \ge 0 \qquad (12)$$

$$F_t(n,m) = \sum_{x=0}^{m} E_{t-1}(n,x) \sum_{j=m-x}^{\infty} \frac{a_j}{j+1} \qquad t \ge 1 \qquad (13)$$

The probability $G_{t_A}(t_D)$ can be obtained by (10) as:

$$G_{t_A}(t_D) = \sum_{n=\max(0,t_A-t_D)}^{\infty} F_{t_A}(n, t_D - t_A + n) \qquad t_D, t_A \in \{1,2,3,\ldots\} \qquad (14)$$

Then the interdeparture time distributions $I'_i(t)$ corresponding to the component interarrival time distributions $I_i(t)$ are:

$$I'_i(t) = \sum_{s=1}^{\infty} G_s(t) I_i(s) \qquad t \in \{1,2,3,\ldots\} \qquad (15)$$

and the transition probabilities of the R-connection at the following output queue in the network will be therefore:

$$A'[(x),(t)] = \begin{cases} 0 & \text{if } x \ne 0 \wedge x \ne t+1 \\ 1 & \text{if } x \ne 0 \wedge x = t+1 \\ \sum_{i=1}^{i=D} q_i I'_i(t+1) & \text{if } x = 0 \end{cases} \qquad x,t \in \{0,1,2,\ldots\} \qquad (16)$$

Note that, unfortunately, a correlation effec exists between consecutive queues in a real network. This effect can be however neglected as far as the queue length distribution and the interdeparture time distribution at the equilibrium point are concerned; this has been shown by comparing two different network model: the former under the hypothesis of independent queues, the latter maintaining the correlation between consecutive queues. The simulation results under the hypothesis of independent queues are very close to the analogous results in the case of correlated queues; the analytical results, very accurate in the former case, can be considered good approximations in any case.

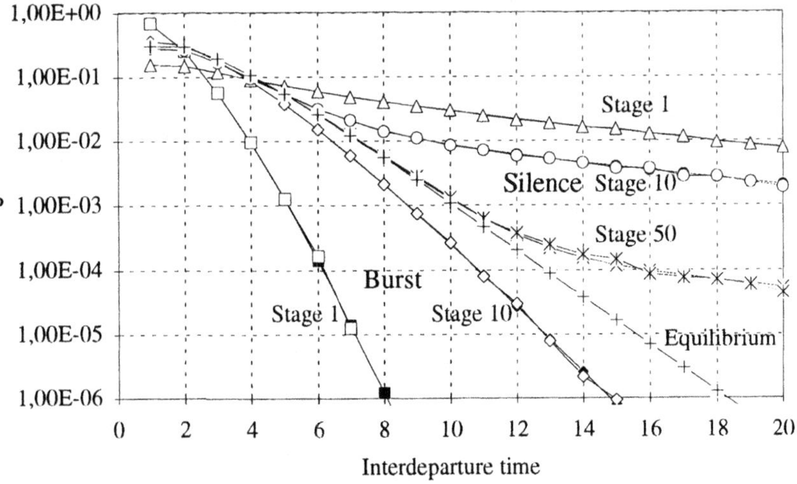

Fig. 4. Cell interdeparture time distribution during bursts and during silences (R-stream with load 0.4 and average burst length 5; total load 0.8).

5. NUMERICAL RESULTS

In the numerical examples, the reference connection was initially either a bursty one or a Bernoulli source. When a bursty connection was assumed, it was characterized by its normalized bandwidth (varying in [0.1;0.7]), its mean burst length L (2.5,5,10) and its interarrival time distributions $I_B(t)$ and $I_S(t)$ (usually geometric distributions at the first stage). A Poisson stream has been assumed as the background traffic X, but a few experiments have been carried out adopting Bernoulli streams or a second R-connection identical to the first one (using an extension of the above analytical method which has not been described here); in the paper the reported numerical results only refer to the case of a Poisson stream interfering with the R-connection, because the other cases show better performances. The total traffic obtained merging the R-connection and the X-stream was set to 0.8, although other load conditions have been tested.

The analysis has been validated by simulations. Figures 4-5 compare the simulation results (under the hypothesis of independent queues) with the analysis. The cell interdeparture time distributions during the burst transmissions and the silence periods can be seen in figure 4, while figure 5 shows the queue length distribution. It can be noted that all the simulation results are exactly predicted by the analytical distributions even after 50 stages; they actually cannot be distinguished at the first stages.

The numerical analysis shows that the interdeparture time distributions $I_B(t)$ and $I_S(t)$ both converge to the same distribution, after a number of stages depending on the burstiness, the burst length and the traffic load conditions. The limit distribution varies on the X-stream characteristics, but it is not influenced by the initial R-connection features. Figure 6 reports the limit queue length distributions. Note that the tail of the length distribution of a M/D/1 queue (with load 0.8) is always higher than the others. The correlation introduced by the existence of a connection improves the average performance.

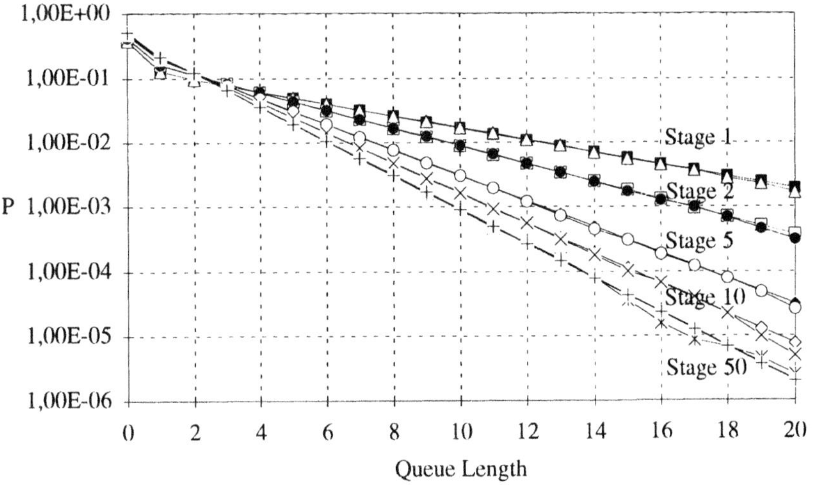

Fig. 5. Queue length distribution: simulation and analysis (R-stream with load 0.4 and average burst length 5; total load 0.8).

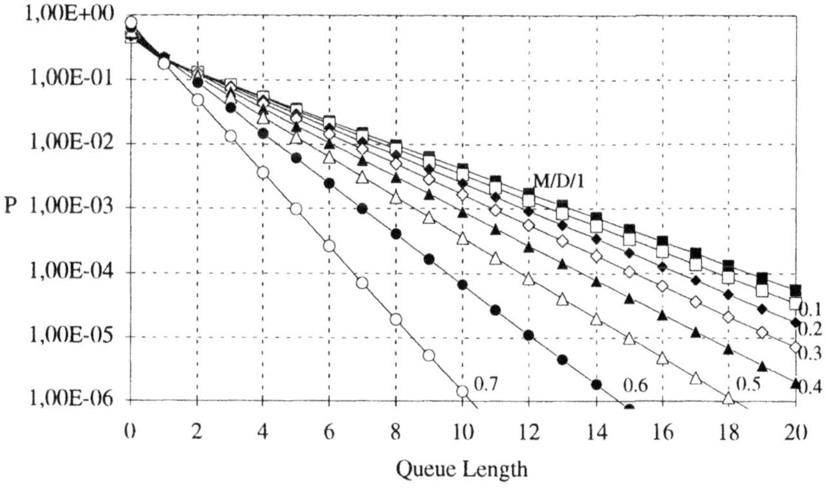

Fig. 6. Queue length distribution at the equilibrium point compared with the M/D/1 queue distribution (R-stream with load 0.1-0.7; average burst length 5; total load 0.8).

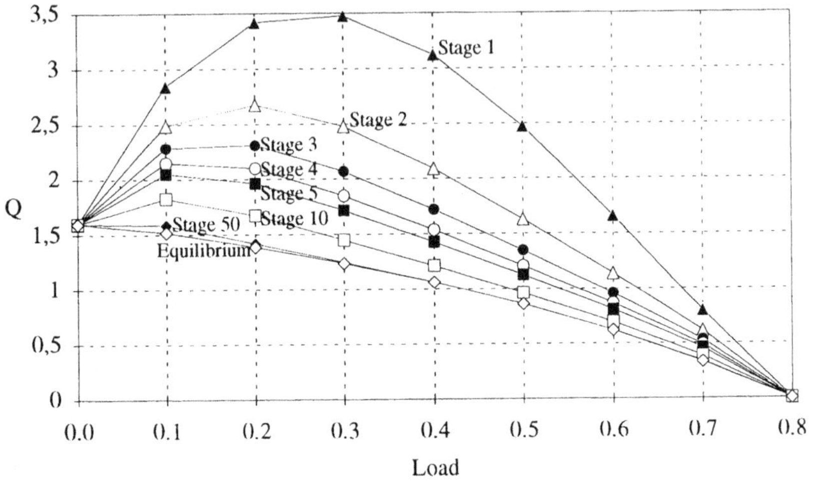

Fig. 7. Mean queue length vs. R-stream load at different stages (R-streams with average burst length=5; total load=0.8)

The graph of figure 7 gives the mean queue length against the R-connection load at different stages. A similar graph can be obtained for the variance. Note that, as the R-connection load decreases to 0, the limit of both the mean and the variance of the queue length approach the values of a M/D/1 (with load 0.8 the mean is 1.6). Note that the worst performances at the first stage occur around a load of 0.2-0.3, but after a few stages both the mean and the variance fall under the M/D/1 threshold. Note also that, as the X-stream load shrinks, the queue length goes down to 0 because the cell contention at the queues stops.

As far as the convergence speed is concerned, it must be noted that the process is faster for connections characterized by a high load, and slower for light connections. The burst length also influences the convergence speed. The limit of the mean interdeparture time is equal to $\frac{1}{\lambda}$ where λ is the R-connection load. The stage threshold (that is the first stage number) after which the queue performances become better than the M/D/1 case is higher as the burstiness increases and the R-connection load decreases. However, fixed the total load, as the R-connection load approaches 0, its effect on the queue gets lower and the queue performance becomes close to the M/D/1.

6. REMARKS

The analysis shows that a unique limit interdeparture time distribution exists for traffic streams going through a network of cascaded switch nodes. The limit interdeparture time distribution of a reference-connection is characterized as follows:

- all the cells belonging to the same R-stream have an identical limit interdeparture time distribution; so, at the equilibrium point there is no difference in the interdeparture process of correlated (that is belonging to the same burst) and uncorrelated cells; the resulting burstiness is therefore 1; however the limit behaviour cannot be correctly represented by a Bernoulli stream, because only the tail of the limit interdeparture time distribution is a straight line in a logarithmic scale graph (see figure 4, Equilibrium curve);

• the limit interdeparture time distribution depends on the connection bandwidth and on the behaviour of the background traffic; R-connections characterized by different initial interarrival distributions but identical load intensities converge to the same limit distribution with a given fixed background traffic. The limit changes as the background traffic characteristics vary.

• the initial burstiness, burst length and interarrival time distribution of a R-connection only influence the convergence speed, not the limit behaviour.

As far as the queue performance is concerned, the results show that the burstiness strongly affects only the initial multiplexing stages. Note that the queueing performances at the equilibrium point are always better than the M/D/1 case, which appears to be the worst case. With a total traffic load equal to 0.8 and R-connections with average burst length 5, less than 15 stages are enough to go under the M/D/1 threshold for R-connections with load larger than 0.2, and less than 5 stages are enough for R-loads larger than 0.4. A consistent number of stages are needed for R-loads lower than 0.1 though. Nevertheless, as the R-load decreases, the mean queue length and the variance approach the M/D/1 values (see figure 7). However, as the burst length increases, the convergence process becomes longer.

Note that in real systems there are more than a single connection multiplexed on a link; the worst case for the queue performance at a switch output occurs when all the incoming connections come from separate links, because the variance of simultaneus cell arrivals is the higher, resulting in a large contention effect. This means that the limit behaviour of a traffic composed of a finite number of connections is better than the case with a single R-connection multiplexed with a P-stream (which represents the aggregate of an infinite number of separate connections). This fact is consistent with the following property shown for a switch node in [11]: the performance under any nonuniform traffic distribution from inputs to outputs is better than or close to the performance under the uniform traffic distribution. So the equilibrium lines of figure 7 must be considered upper bounds for every kind of initial traffic conditions (with total load 0.8). Note moreover that in figure 7 the curve at stage 10 can be considered better than or very close to the M/D/1 values. Thus, it seems possible that a few multiplexing stages (~10) are sufficient to reach queue performances very close to the M/D/1 for a wide variety of traffic conditions.

An interesting study case is the characterization of the expansion phenomenon of two consecutive cells not bound to any connection, but following the same path in the network. This case can be seen as the expansion phenomenon of a connection when its load λ tends to zero. It was said above that the mean interdeparture time tends to be $1/\lambda$ at the equilibrium and therefore the interdeparture time between two cells of a permanent connection is bound to its load. If this is true also in the limit case $\lambda=0$, the expansion phenomenon never stops for successive cells which belong to temporary connections. This fact is not immediately intuitive and difficult to justify. There are arguments asserting the exact opposite [4] and there is no proof showing that the expansion phenomenon continues ad infinitum or stops at a certain limit. An analogy which could explain the phenomenon could be seeing the merging of two or more streams like a diffusion process of their particles into each other. This process is renewed at each multiplexing stage, causing the two consecutive cells to be separated indefinitely. The first cell causes a queue perturbation influencing the queue behaviour at the arrival of the second cell. Note also that the second cell can never surpass the first one, so that the queue seen by the second cell cannot be considered independent from the queue seen by the first. Moreover the diffusion process speed depends on the burst length (and on the carried load): if the burst length is large (and/or the carried load is low), the expansion phenomenon still continues but it cannot be perceived at the higher levels of the process in figure 1. Thus it is statistically true that, for a connection of duration 10 minutes, the distance between the first cell and the last increases, but this effect cannot be experimentally observed.

7. CONCLUSIONS

In this paper the limit behaviour of an ATM traffic within the network has been examined. By deriving the exact interdeparture time distribution for the cells of a connection arriving to the output queue of a switch node with a general interarrival time distribution and multiplexed with a background traffic stream, the analysis has been iterated through a long sequence of cascaded output queues, until the interdeparture time distribution converge to a limit one. Therefore a limit behaviour for an ATM traffic exists. Moreover it has been shown that the queue performances at the equilibrium point are always better than the M/D/1 case. The convergence speed to the limit depends on the initial characteristics of the traffic at the sources, on the background traffic and on the total traffic load. It seems however that a few multiplexing stages (~10) are sufficient to reach queue performances very close to the M/D/1 for a wide variety of traffic conditions. Therefore it appears reasonable that other studies would be done on the subject. An extension of this work coud be considering a background traffic composed of a finite number of connections comprising both traffic generated by bursty sources and traffic at the equilibrium. The resulting queue performance should be a better approximation for the sizing of the queues inside the network, avoiding an oversize of the buffers and allowing an effective cost saving with respect to the previous models with bursty sources at the first stage.

REFERENCES

1. F.Bonomi, S.Montagna, R.Paglino: ``Busy Period Analysis for an ATM Switching Element Output Line", Proceedings of INFOCOM'92, 1992, pp.544-551.

2. JP.J.Burke: ``The output of Queueing Systems", Oper. Research, vol.4, pp.699-704, 1956.

3. J.S.-C.Chen, T.E.Stern: ``Throughput Analysis, Optimal Buffer Allocation, and Traffic Imbalance Study of a Generic Nonblocking Packet Switch", IEEE Journal On Selected Areas In Communications, vol. 9,no. 3, April 1991.

4. COST 224: ``Performance Evaluation and Design of Multiservice Networks", COST 224 Final Report 1991, sections 6.1-4, 9.1-2.

5. M. D'Ambrosio: ``Prestazioni di strutture di commutazione ATM. Parte II: caratterizzazione del traffico all'interno della rete di connessione", in Italian, CSELT internal report RR 90.598, 1990.

6. M.D'Ambrosio, R.Melen: ``Performance Analysis of ATM Switching Architectures: a Review", CSELT Technical Reports, vol. XX, no. 3, June 1992, pp.265-281.

7. B.S. Davie, "ATM network modeling and design for bursty traffic", Bell Communications Research, July 1989.

8. M.Hluchyi, M.Karol: ``Queuing in High-Performance Packet Switching", IEEE J. on Sel. Areas in Comm., Vol.6, No.9, Dec. 1988.

9. C.Kruskal, M.Snir, A.Weiss: ``The Distribution of Waiting Times in Clocked Multistage Interconnection Networks", IEEE Trans. on Computers, vol.C-37, No.11, Dec. 1988, pp.1337-1352.

10. S.Liew, K.Lu: ``Comparison of Buffering Strategies for Asymmetric Packet Switch Modules", IEEE Journal on Selected Areas in Communications, Vol. 9, no. 3, April 1991.

11. S.Liew, T.Lee: ``A Fundamental Property for Traffic Management in ATM Networks", to appear in INFOCOM'93.

12. A. Massari, ``Simulazioni di reti ATM multinodo", in Italian, CEFRIEL -Politecnico di Milano, technical report RI 91081, 1991.

13. I. Norros, ``Deformation of a traffic stream passing through several ATM multiplexers", Technical Research Centre of Finland - Telecommunication Laboratory, doc 224TD(91), April 1991.

14. Y.Ohba, M.Murata and H.Miyahara: ``Analysis of Interdeparture Processes for Bursty Traffic in ATM Networks", IEEE J. on Sel. Areas in Comm. vol.9, no. 3, April 1991.

15. S.-Q.Li ``Performance of a Non-Blocking Space Division Packet Switch With Correlated Input Traffic", Proceedings of Globecom'89, 49.1, Dec.1989.

Integrated Broadband Communication Networks
and Services (C-18) / V.B. Iversen (Editor)
Elsevier Science B.V (North-Holland)

Time dependent behaviour and large deviations analysis for a virtual path in a B-ISDN

Suzanne P. Evans

Dept. of Mathematics and Statistics, Birkbeck College, University of London, Malet St., London WC1E 7HX, United Kingdom.

This paper considers a B-ISDN system operated using the technique of virtual paths. The time dependent behaviour of a single virtual path carrying a mix of services is examined using fluid approximations, weak convergence and the theory of large deviations. Results are obtained concerning the occurrence of overload, mean time between overload events and time taken to move from near equilibrium to overload, if overload occurs. There are implications for monitoring, control and resource management. The model has discontinuous transition rates at the boundary. Such processes pose problems for the application of fluid approximations and large deviations theory, particularly when, as here, behaviour at the boundary is of interest. These problems are examined in the paper.

Keyword Codes: C.2.0; C.2.3
Keywords: Computer-Communication Networks, General; Network Operations

1. INTRODUCTION

This paper is concerned with characterizing time dependent behaviour in a B-ISDN operated using the technique of virtual paths. We consider the implications for monitoring, control and resource management.

In Evans (1990a & 1990b) transient behaviour was analysed using a spectral expansion, a sum of weighted exponential terms in which the eigenvalues of the process generator appear in the exponents. Under conditions chiefly concerned with the separation between eigenvalues, the weights are bounded and dominant terms may be equated with dominant exponentials.

In this paper we look at various limiting results for time dependent behaviour in heavy traffic and apply large deviations theory to examine both the distribution and frequency of rare excursions far away from expectation, and the way such deviations occur. The network is modelled at the call level and the use of virtual paths provides the mechanism for allocating resources between groups of calls assigned to different virtual paths. Call admission rules on each virtual path are based on effective bandwidths associated with different call types. The stationary properties of this model are analysed in Evans (1989 & 1991) and transient behaviour on a single virtual path is discussed in Evans (1990a & 1990b). Results in this paper also refer to time dependent behaviour on a single virtual path and are based on large deviations theory. As we shall see a full large deviations analysis is not straightforward since we are dealing with a process whose transition rates are discontinuous across certain boundaries; the theory is less well developed for such processes. If the virtual path is in statistical equilibrium well away from the boundaries this causes no difficulty. However we are often interested in the behaviour of virtual paths where the capacity provided is well matched to the capacity required, or where the virtual path is overloaded. The system then spends a significant amount of time 'close' to the overload boundary. In this case we use some results on time scale separation in overload and look at the fast time-scale free capacity process and the averaged process.

The structure of the paper is as follows. In Section 2 we present the theory and two simple examples that are relevant later in the context of our virtual path model. Section 3.1 introduces the virtual path model, together with some of the heavy traffic results used in its analysis. Large deviations results for this model are derived in Section 3.2, and the theoretical difficulties at the capacity boundary are discussed in Section 3.3.

2. LARGE DEVIATIONS FOR PROCESSES

2.1. General theory for multidimensional birth-death processes

Following Weiss (1986) and Dupuis *et al* (1991) we consider multidimensional birth-death processes on the S-dimensional lattice Z^S and allow the system to become large in a well-defined sense. With appropriate scaling we then give a generalization of the law of large numbers for these processes. This is Theorem 1. Consider Z^S and suppose that we have a finite collection of step directions $\mathbf{e}_j$, $j = 1, \ldots, J$ of the form $\mathbf{e}_j^T = (e_j^1, e_j^2, \ldots, e_j^S)$ where each component $e_j^s \in \{0, +1, -1\}$. Let $\mathbf{n}(t)$ be the Markov Process on the state-space V with transition rates given by $q(\mathbf{n}(t), \mathbf{n}(t) + \mathbf{e}_j) = \alpha_j(\mathbf{n}(t))$. Suppose now that the system becomes large in the following way. Consider a sequence of processes $\mathbf{n}^N(t)$, indexed by some parameter N that increases without limit. The scaled process $\mathbf{z}^N(t) = \mathbf{n}^N(t)/N$ is a process in R^S moving in jumps $\mathbf{e}_j/N$ with associated transition rates $q(\mathbf{z}^N(t), \mathbf{z}^N(t) + \mathbf{e}_j/N) = \alpha_j^N(\mathbf{z}^N(t))$, where the α_j^N are defined on R^S. There exist functions α_j, also defined on R^S, such that $N^{-1}\alpha_j^N(\mathbf{z}) \to \alpha_j(\mathbf{z})$ as $N \to \infty$, so that

$$\alpha_j^N(\mathbf{z}) = N\alpha_j(\mathbf{z}) + o(N) \quad j = 1, \ldots, J \tag{1}$$

We shall assume henceforth that:

(i) The $\alpha_j(\mathbf{x})$ are bounded and Lipschitz continuous on R^S.

(ii) An $\alpha_j(\mathbf{x})$ that is nonzero at one point is uniformly bounded away from zero everywhere.

The process $\mathbf{z}^N(t)$ has an associated generator $\mathcal{L}^N$, an operator on real valued functions,

$$\left(\mathcal{L}^N f\right)(\mathbf{x}) = \sum_{j=1}^{J} \alpha_j^N(\mathbf{x}) \left[f(\mathbf{x} + N^{-1}\mathbf{e}_j) - f(\mathbf{x})\right] \tag{2}$$

We can use Taylor's theorem to expand the differences $f(\mathbf{x} + N^{-1}\mathbf{e}_j) - f(\mathbf{x})$ if we restrict the f to C^n. If we then ignore terms of $O(N^{-1})$, or smaller, then, from results (1) and (2), what remains is itself the generator of a Markov process, specifically the unique deterministic process $\mathbf{z}^\infty(t)$ which solves the system of ordinary differential equations

$$\frac{d}{dt}\mathbf{z}^\infty(t) = \sum_{j=1}^{J} \alpha_j(\mathbf{z}^\infty(t))\mathbf{e}_j \tag{3}$$

with a given initial condition $\mathbf{z}^\infty(0) = \mathbf{z}_0$. Note that (i) and (ii) ensure the existence and uniqueness of the solution to (3). Thus we expect $\mathbf{z}^\infty(t)$ to be a continuous *fluid approximation* to $\mathbf{z}^N(t)$ for large N. The following theorem formalizes this and provides the promised generalization of the law of large numbers to the processes under consideration.

Theorem 1. (a version of Kurtz(1978)). If the $\alpha_j(\mathbf{x})$ satisfy (i) and (ii), let $\mathbf{z}^\infty(t)$ be the unique solution to the system of differential equations (3). Suppose $\mathbf{z}^N(0) = \mathbf{z}^\infty(0)$, then for each $T > 0$, $\varepsilon > 0$ there exist positive constants $C_1 > 0$, $C_2 > 0$ such that

$$P\left(\sup_{0 \le t \le T} |\mathbf{z}^N(t) - \mathbf{z}^\infty(t)| > \varepsilon\right) \le C_1 e^{-NC_2}$$

Inferences about the *equilibrium* behaviour of the system can be drawn from the fixed points of equations (3), namely the solutions to $\sum_{j=1}^J \alpha_j(\mathbf{z}^\infty(t))\mathbf{e}_j = 0$. Roughly speaking, if the fluid approximation $\mathbf{z}^\infty(t)$ converges as $t \to \infty$ then the stochastic process $\mathbf{z}^N(t)$ reaches equilibrium. A precise version of this statement is proved in Mitra & Weiss (1989).

Note that taking the second term in the Taylor expansion of $\left(\mathcal{L}^N f\right)(\mathbf{x})$ gives the generator of a diffusion approximation with the fluid approximation as the centring term.

The following large deviations theorems are concerned with the probabilities and paths associated with rare excursions of $\mathbf{z}^N(t)$ far from $\mathbf{z}^\infty(t)$. Consider first the function:

$$h(\mathbf{x},\mathbf{y}) = \sup_{\boldsymbol{\theta} \epsilon R^S} (<\boldsymbol{\theta},\mathbf{y}> \; - H(\mathbf{x},\boldsymbol{\theta})) \tag{4}$$

$$\text{where } H(\mathbf{x},\boldsymbol{\theta}) = \sum_{j=1}^{J} \alpha_j(\mathbf{x}) \left(e^{<\boldsymbol{\theta},\mathbf{e}_j>} - 1\right) \tag{5}$$

is the logarithm of the moment generating function of our process and $h(\mathbf{x},\mathbf{y})$ is its Legendre transform. (See, for example, Bucklew (1990), pp 65—67). The function $H(\mathbf{x},\boldsymbol{\theta})$ is convex in $\boldsymbol{\theta}$ and continuous in $\mathbf{x}$, while the function $h(\mathbf{x},\mathbf{y})$ is convex in $\mathbf{y}$, continuous in $\mathbf{x}$ and nonnegative. For any piecewise absolutely continuous function $\mathbf{r}(t)$ in R^S define

$$I_{t_1}^{t_2}(\mathbf{r}) = \int_{t_1}^{t_2} h(\mathbf{r}(t),\mathbf{r}'(t))\,dt$$

$I_{t_1}^{t_2}$ is known as the action functional, or the Cramer functional. Clearly it is nonnegative. Theorems 2 and 3 which follow are found in Weiss (1986).

Theorem 2. If $\mathbf{r}(t)$ in R^S is a piecewise absolutely continuous function with $\mathbf{z}^N(t_1) = \mathbf{r}(t_1)$, then for any $\varepsilon > 0$ and $t_2 \in [t_1,\infty)$

$$P\left(\sup_{t_1 \le t \le t_2} \left|\mathbf{z}^N(t) - \mathbf{r}(t)\right| < \varepsilon\right) = \exp\left(-NI_{t_1}^{t_2}(\mathbf{r}) + o(N)\right)$$

The proof of Theorem 2 follows a standard pattern. An upper bound is proved via the martingale generalization of an exponential Chebychev inequality, and a lower bound using a change of measure argument whereby the path $\mathbf{r}(t)$ is made to appear "typical".

Theorem 3 follows fairly easily from Theorems 1 and 2. It concerns the occurrence and frequency of rare excursions of the process from a neighbourhood of equilibrium to some unlikely point or set of points. Such rare events E turn out to occur approximately as a Poisson process with parameter $\exp(-NI^*(E))$, where $I^*(E)$ is the smallest value of the action functional $I_{t_1}^{t_2}(\mathbf{r})$ over paths $\mathbf{r}(t)$ which cause the event E to occur. An event E is regarded as a set of piecewise, absolutely continuous paths $\mathbf{r}(t)$ such that any time translate of a path in E also belongs to E. For each $\mathbf{r}(t)$ in E, there are numbers $t_1$and

t_2, $-\infty \leq t_1 < t_2 < \infty$, such that $\mathbf{r}(t)$ is defined on the interval $[t_1, t_2]$. If $\mathbf{z}^N(t)\epsilon E$ then we say that E occurs at $t_2(\mathbf{z}^N)$. An *admissible* event satisfies further regularity conditions (see Weiss (1986). Events considered in this paper are all admissible.

Assume that the equations (3) defining $\mathbf{z}^\infty(t)$ have exactly one fixed point $\overline{\mathbf{z}}$, which is stable. Assume also that the process $\mathbf{z}^N(t)$ is irreducible and ergodic so that it makes sense to refer to $\overline{\mathbf{z}}$ as the equilibrium point.

Set $I^*(E) = \inf_{\mathbf{r}\epsilon E,\, \mathbf{r}(t_1)=\overline{\mathbf{z}}} I_{t_1(\mathbf{r})}^{t_2(\mathbf{r})}(\mathbf{r})$

It can be shown that the infimum is attained for a wide class of events.

Theorem 3. Let E be an admissible event with $I^*(E) > 0$ and suppose E is defined only on the behaviour of a path outside a neighbourhood of equilibrium. Then:

(a) The times at which E first occurs after $\mathbf{z}^N(t)$ has been near equilibrium are distributed approximately exponentially, with mean μ_N, where $\mu_N = \exp(NI^*(E) + o(N))$. Thus E occurs approximately as a Poisson process with parameter $\exp(-NI^*(E))$.

(b) If there is a unique (up to time translation) path $\mathbf{r}^*\epsilon E$ such that $I_{t_1(\mathbf{r}^*)}^{t_2(\mathbf{r}^*)}(\mathbf{r}^*) = I^*(E)$ and $\mathbf{r}^*(t_1(\mathbf{r}^*)) = \overline{\mathbf{z}}$, and if we translate $\mathbf{r}^*$ so that $t_2(\mathbf{r}^*) = T_2$, then for $T > 0$, $\varepsilon > 0$

$$\lim_{N\to\infty} P\left(\sup_{T_2 - T \leq t \leq T_2} \left|\mathbf{z}^N(t) - \mathbf{r}^*(t)\right| < \varepsilon \;\middle|\; \text{event occurs at } T_2 \right) = 1$$

2.2. The one-dimensional level crossing event

In this paper we are concerned with the multidimensional birth-death process used to model the call level process on a virtual path. Nevertheless the general solution to the one-dimensional level crossing problem is easy to obtain (see Bucklew (1990)) and is relevant to multidimensional overload events. Consider a series of general one-dimensional birth-death processes, indexed and scaled by the large parameter N. The step directions are just $e_1 = +1$ and $e_2 = -1$ and we write the transition rates as $q(z^N, z^N + 1/N) = N\lambda(z^N) + o(N)$ and $q(z^N, z^N - 1/N) = N\mu(z^N) + o(N)$. Consider the level crossing event E consisting of paths $r(t)$ on $[t_1, t_2]$ such that $r(t) < C$ for $t\epsilon[t_1, t_2)$ and $r(t_2) \geq C$. Here $C > \nu(r) = \lambda(r)/\mu(r)$, the unique equilibrium point. From (4) and (5)

$$h(r, r') = \sup_{\theta \epsilon R} \{\theta r' - H(r, \theta)\} \tag{6}$$

$$\text{where } H(r, \theta) = \lambda(r)\left(e^\theta - 1\right) + \mu(r)\left(e^{-\theta} - 1\right) \tag{7}$$

is convex in θ and continuous in r. It is easy to show that the supremum occurs at $\theta = \theta^*$

$$\text{where } \theta^* = \log\left[r' + \sqrt{(r')^2 + 4\lambda(r)\mu(r)} \Big/ 2\lambda(r)\right] \tag{8}$$

$$\text{so } h(r, r') = r' \log \frac{r' + \sqrt{(r')^2 + 4\lambda(r)\mu(r)}}{2\lambda(r)} + \lambda(r) + \mu(r) - \sqrt{(r')^2 + 4\lambda(r)\mu(r)} \tag{9}$$

Fixing the time of the event at $t = 0$, we solve the following variational problem. Find

$$\inf_{r,T\epsilon G} \int_T^0 h(r, r')\, dt \qquad \text{where } G = \{r, T : -\infty \leq T < 0,\ r(T) = \text{equilibrium},\ r(0) = C\}$$

Since h does not explicitly depend on t, the Euler equation for this problem is

$$h(r,r') - r'(\partial/\partial r')h(r,r') \;=\; \text{constant} \tag{10}$$

The endpoint T is free, so there is a transversality condition:

$$h(r(T),r'(T)) \;-\; r'(T)\,(\partial/\partial r')h(r(T),r'(T)) \;=\; 0 \text{ giving}$$

$$h(r,r') \;-\; r'(\partial/\partial r')h(r,r') \;=\; 0 \quad T \le t \le 0 \tag{11}$$

By definition $h(r,r') = \theta^* r' - H(r,\theta^*)$ so that $\frac{\partial h}{\partial r'} = \theta^*$ and Euler's equation becomes

$$H(r,\theta^*) = 0 \quad T \le t \le 0 \tag{12}$$

giving θ^* as a function of r only. From Euler's equation as given in (11) we have

$$\int_T^0 h(r,r')\,dt \;=\; \int_T^0 r'\frac{\partial h}{\partial r'}(r,r')\,dt \;=\; \int_{r(T)}^{r(0)} \frac{\partial h}{\partial r'}\,dr$$

$$\text{whence} \quad \inf_{r\in G} \int_T^0 h(r,r')\,dt \;=\; \int_{r(T)}^{r(0)} \theta^*(r)\,dr \quad \text{from (12)} \tag{13}$$

where $\theta^*(r)$ solves (12). From (7) it is easy to show that $H(r,\theta^*) = 0$ if and only if

$$\theta^*(r) \;=\; \log(\mu(r)/\lambda(r)) \tag{14}$$

$$\text{Thus } I^*(E) \;=\; \int_{r(T)}^{r(0)} \log(\mu(r)/\lambda(r))\,dr \tag{15}$$

from which we obtain the Poisson rate, $\exp(-NI^*(E))$, at which the event E occurs. The optimum trajectory $r^*(t)$ is obtained by equating two expressions for θ^*. Equation (8) gives the general expression for θ^* and equation (14) gives the expression for θ^* derived from Euler's equation. It follows easily that the optimum trajectory must satisfy

$$r' \;=\; \mu(r) - \lambda(r) \;\text{ on } [T,0] \text{ with } -\infty \le T < 0 \text{ and } r(0) = C. \tag{16}$$

This is just the negative of the fluid equations. Thus the most probable manner in which the system moves from equilibrium to the point of overload C is the time reversal of the most probable path from C to equilibrium; overload is reached by the rates $\lambda(r)$ and $\mu(r)$ interchanging temporarily. The result can also be derived by a reversibility argument.

We look now at a variant on the above problem in which we are concerned with evaluating the probability that the system moves from empty up to the overload point C in a given time T. Consider the following variational problem.

$$\text{Find} \quad \inf_{r\in G} \int_0^T h(r,r')\,dt \quad \text{where } G = \{r(t) :\; r(0) = 0,\;\; r(T) = C\}$$

together with the unique path $r^*(t)$, if it exists, at which the minimum is achieved. The Euler equation for this variational problem is

$$h(r,r') - r'(\partial/\partial r')h(r,r') \;=\; \text{constant} \tag{17}$$

In this problem neither of the endpoints is free and we cannot set the constant equal to zero. In fact it turns out to be negative and we anticipate this result by setting it equal to $-K$, where $K > 0$ and depends on T. Thus Euler's equation reduces to $H(r, \theta^*) = K$ and θ^* is a function of r only.

$$\text{Thus} \quad \inf_{r \in G} \int_0^T h(r, r')\, dt = \int_0^C \theta^*(r)\, dr - KT \tag{18}$$

where $H(r, \theta^*(r)) = K$ determines $\theta^*(r)$. It is easy to see that r' must be larger in this constrained case than in the unconstrained case. It follows from equation (8) that $\theta^*(r)$ must also be greater than in the unconstrained case. From (16), and the expression for $H(r, \theta)$ given in (7), it is easy to show that

$$\theta^*(r) = \log\left[\left(\lambda(r) + \mu(r) + K + \sqrt{(\lambda(r) + \mu(r) + K)^2 - 4\lambda(r)\mu(r)}\right) \Big/ 2\lambda(r)\right] \tag{19}$$

Clearly K must be positive to make $\theta^*(r)$ larger. It may be determined from the optimal path $r^*(t)$. Equations (8) and (9), and Euler's equation $H(r, \theta^*) = K$, give us

$$\sqrt{(r')^2 + 4\lambda(r)\mu(r)} = \lambda(r) + \mu(r) + K \tag{20}$$

and we need to solve (20), subject to $r(0) = 0$ and $r(T) = C$. Equation (20) yields

$$r' = \sqrt{[\mu(r) - (\lambda(r) - K)]^2 + 4\lambda(r)K} \tag{21}$$

and for given $\lambda(r)$ and $\mu(r)$ equation (21) may be solved, with the two boundary conditions determining K and the constant of integration.

As an example consider Erlang's model, an M/M/N/N queue with call arrival rate λ and mean call holding time μ^{-1}. We let the capacity, N, become large and consider the 'number of calls in progress' process scaled by N. We now have $\lambda(r) = \lambda$, $\mu(r) = \mu r$ and we consider some level $C \leq 1$, where, for simplicity, we assume that $\nu = \lambda/\mu < C$ so that rate discontinuities at the boundary $r = 1$ do not need to be considered.

Setting $R = \mu r - (\lambda - K)$ in (21), we have $R' = \sqrt{R^2 + 4\lambda K}$ which can easily be solved by separation of variables, yielding $\sinh^{-1} R \big/ 2\sqrt{\lambda K} = t + D$, where D is an arbitrary constant. The boundary conditions $r(0) = 0$ and $r(T) = C$ determine the constants K and D and the unique optimal path $r^*(t)$, $0 \leq t \leq T$, is given by

$$r^*(t) = \nu(1 - e^{-\mu t}) + K(T)(e^{\mu t} - 1) \quad \text{where} \quad K(T) = \left(C - \nu(1 - e^{-\mu T})\right) \Big/ \left(e^{\mu T} - 1\right)$$

Then, from (18) and (19), $I_0^T(r^*)$ is easily evaluated as

$$-K(T)T - C \log 2\lambda + \int_0^C \log\left[\sqrt{(\mu r - \lambda)^2 + 2(\mu r + \lambda)K(T) + K(T)^2} + \lambda + \mu r + K(T)\right] dr$$

$$\text{and} \quad P\left(\sup_{0 \leq t \leq T} \left|z^N(t) - r^*(t)\right| < \varepsilon\right) = \exp(-N I_0^T(r^*) + o(N)) \quad \text{from Theorem 2.} \tag{22}$$

These results are quoted in Mitra & Weiss (1988) for the case $C = 1$, and with $\mu = 1$. (The derivation is omitted there due to lack of space). We do not prove it here but assert that $P(z^N(T) = C \,|\, z^N(0) = 0)$ is precisely the probability given in (22), with $r^*(t)$ as the overwhelmingly probable path.

In the next section we introduce the call level model for a virtual path carrying a mix of services. We shall make use of the results of Section 2 to examine its time dependent behaviour and the occurrence of excursions to sets of rare states. We shall examine in more detail the effect of having discontinuous transition rates at the boundary.

3. THE VIRTUAL PATH MODEL AND LARGE DEVIATIONS ANALYSIS

We begin this section with a short review of our call level model for a virtual path (VP) in a broadband network, together with some of the heavy traffic results used in its analysis. Similar descriptions may be found in Evans (1990b) & (1991).

3.1. The model and heavy traffic results

We consider a VP of capacity M and carrying a mix of S services. Each call of service type $s = 1, \ldots, S$ is assumed to have a capacity requirement of a_s, where the a_s may represent "effective capacities". Calls of type s arrive as a Poisson stream of rate λ_s. We assume the simplest possible call admission algorithm. An arriving call of type s is accepted if, at its arrival time t, the free capacity $m(t)$ lies in the set $\mathcal{M}_s = \{m : m \in Z_+,\ m \geq a_s\}$. If accepted it occupies a capacity a_s for the call holding time, where holding times are exponentially distributed, with mean μ_s^{-1}. Let $\nu_s = \lambda_s \mu_s^{-1}$. The resulting process for $\mathbf{n} = (n_1, \ldots, n_S)^T$, the calls-in-progress vector, is a multidimensional birth-death process with state space $\mathcal{S}(M) = \{\mathbf{n} : \mathbf{a}^T\mathbf{n} \leq M\}$ where $\mathbf{a}^T = (a_1, \ldots, a_S)$.

Consider the limiting regime in which VP capacity and call arrival rates are increased without limit, while ratios are held roughly constant. We have a sequence of VPs indexed by N such that the Nth VP has capacity $M(N)$ and arrival rates $\lambda_s(N)$. We consider a limit in which $N^{-1}\lambda_s(N) \to \lambda_s$ and $N^{-1}M(N) \to M$ as $N \to \infty$. Let $\mathbf{n}(N) = (n_1(N), \ldots, n_S(N))^T$, where for the present we omit dependence on time. For our large deviations analysis we shall scale by N and consider the sequence $\mathbf{z}^N = \mathbf{n}(N)/N$, but first we consider a centred $\mathbf{n}(N)$, scaled by $N^{1/2}$, and examine the central limit theorem of Kelly (1986) and Hunt & Kelly (1989). The centring term is obtained as follows.

Let $\nu_s(N) = \lambda_s(N)\mu_s^{-1}$ so that $N^{-1}\nu_s(N) \to \nu_s$ as $N \to \infty$. Let $B^*(N)$ be the minimal solution in $[0,1)$ to $\sum_{s=1}^{S} a_s \nu_s(N)[1 - B(N)]^{a_s} \leq M(N)$. Then $\mathbf{x}^*(N)$, with elements $x_s^*(N) = \nu_s(N)[1 - B^*(N)]^{a_s}$, represents the most probable state of the Nth VP, where $B^*(N)$ is thought of as the blocking probability for a single unit of capacity. The sequence $\{B^*(N)\}$ has a limit B^* as $N \to \infty$ and the form of the central limit theorem depends on B^*, and one other condition if $B^* = 0$. Let $x_s^* = \nu_s(1 - B^*)^{a_s}$ for each s. Three cases may be distinguished. The VP is termed *supercritical* if $B^* > 0$ so that $\mathbf{a}^T\mathbf{x}^* = M$. If $B^* = 0$, then $\mathbf{x}^* = \boldsymbol{\nu}$. Then the VP is termed *critical* if $B^* = 0$ and $\mathbf{a}^T\mathbf{x}^* = M$, and *subcritical* if $B^* = 0$ and $\mathbf{a}^T\mathbf{x}^* < M$. Kelly's central limit theorem is for the vector $\mathbf{u}(N)$, where

$$u_s(N) = \left[n_s - \nu_s(N)(1 - \overline{B}(N))^{a_s}\right] \Big/ N^{1/2}$$

and $\overline{B}(N)$ is just $B^*(N)$ if the VP is supercritical and zero otherwise. The details of the theorem applied to this model are given in Evans (1989). Briefly, in the *subcritical* case when $\mathbf{a}^T\boldsymbol{\nu} < M$, we obtain the S-dimensional multivariate normal distribution $N(\boldsymbol{\nu}, \text{diag}(\nu_s))$. In the *critical* case, if we assume a convergence rate of $O(N^{-1/2})$ as $N^{-1}\left(M(N) - \mathbf{a}^T\boldsymbol{\nu}(N)\right)$ goes to zero, we obtain a truncated version of the previous distribution. Finally, in the *supercritical* case, we have a collapse to an $(S-1)$- dimensional multivariate normal distribution obtained by conditioning $N(\mathbf{x}^*, \text{diag}(x_s^*))$ on $\mathbf{a}^T\mathbf{x}^* = M$.

Kelly(1986) also obtains weak convergence results for the spare capacity variable m. In the supercritical case he shows that m is of $O(1)$ and has a geometric distribution $f(m) = D(1-B^*)^m$, where D is a normalizing constant for the set of feasible m. If units of capacity are chosen so that any nonnegative integer is a feasible m, then $D = B^*$.

These are stationary results. We are interested in the evolution over time of the stochastic process $\mathbf{n}(N,t)$, where the parameter t is included to emphasize the dependence on time. For large N we look for limiting results for a scaled version of the process, taking explicit account of the presence of the boundary defined by $\mathbf{a}^T\mathbf{n}(N) = M(N)$. The results are proved formally in Hunt (1990). Briefly the essential feature near the boundary is the time scale separation between the process $\mathbf{z}^N(t) = N^{-1}\mathbf{n}(N,t)$, which operates on an $O(1)$ time scale, and the free capacity process $m(N,t)$ which operates on an $O(N^{-1})$ time scale. The main result in Hunt (1990) is a functional law of large numbers. For each s we have $\mathbf{z}^N \to \mathbf{z}^N + N^{-1}\mathbf{e}_s$ at approximately rate $N\lambda_s\mathcal{X}_{\mathcal{M}_s}$ and $\mathbf{z}^N \to \mathbf{z}^N - N^{-1}\mathbf{e}_s$ at approximately rate $N\mu_s z_s^N \mathcal{X}_{\{\mathbf{z}^N : z_s^N \geq N^{-1}\}}$, where $\mathcal{X}_{\mathcal{M}_s}$ is the indicator for the set $\{\mathbf{z}^N : m(N,t) = M(N) - N\mathbf{a}^T\mathbf{z}^N \geq a_s\}$. Now suppose $\mathbf{z}(t)$ represents the state of the VP, suitably normalized, at some time t. We let $m_{\mathbf{z}(t)}$ be the Markov process on $Z_+ \cup \infty$, where $m \to m + a_s$ at rate $\mu_s z_s(t)$, $m \to m - a_s$ at rate $\lambda_s\mathcal{X}_{\{m \geq a_s\}}$ and ∞ is an absorbing state. Thus $m_{\mathbf{z}(t)}$ represents the free capacity process normalized so that transitions occur on a time scale of $O(1)$. On this time scale the process $\mathbf{z}^N(t)$ is regarded as fixed. Let $\tilde{\pi}$ be the equilibrium distribution for $m_{\mathbf{z}(t)}$ if it exists, and define $\pi_s(t)$ by

$$\pi_s(t) = \begin{cases} \tilde{\pi}(m \geq a_s) & \text{if } m_{\mathbf{z}(t)} \text{ is positive recurrent and } \mathbf{a}^T\mathbf{z}(t) = M \\ 1 & \text{otherwise} \end{cases}$$

Thus $\pi_s(t)$ represents the probability that a call of type s arriving at time t will be accepted. When we consider the process $\mathbf{z}^N(t)$ on *its* natural time scale, and let $N \to \infty$, we expect the probabilities $\pi_s(t)$ to replace the indicator variables $\mathcal{X}_{\mathcal{M}_s} \equiv \mathcal{X}_{\{m(N,t) \geq a_s\}}$, where $m(N,t)$ is the original free capacity process with unnormalized time scale. From Hunt (1990a) we have the following:

Theorem 4. If $\mathbf{z}^N(0)$ converges weakly to $\mathbf{z}^\infty(0)$ as $N \to \infty$ then $\mathbf{z}^N(\cdot)$ converges weakly to $\mathbf{z}^\infty(\cdot)$, where $\mathbf{z}^\infty(\cdot)$ is the unique (continuous) solution to the equations

$$\tfrac{d}{dt} z_s^\infty(t) = \lambda_s \pi_s(t) - \mu_s z_s^\infty(t) \quad s = 1, \ldots, S$$

Notice that $\pi_s(t)$ is a function of $\mathbf{z}^\infty(t)$, not just of $z_s^\infty(t)$, except away from the boundary, when $\mathbf{a}^T\mathbf{z}^\infty(t) < M$ and

$$\tfrac{d}{dt} z_s^\infty(t) = \lambda_s - \mu_s z_s^\infty(t) \quad s = 1, \ldots, S \tag{23}$$

When $\mathbf{a}^T\mathbf{z}^\infty(t) = M$ the equations are dependent. Note that we have lost the information about $O(N^{1/2})$ fluctuations in $\mathbf{n}(N,t)$, although the result should give us the centring term in a possible diffusion limit.

3.2. Large deviations analysis of VP model — subcritical case

In this section, we shall restrict ourselves to examining the subcritical case for which $\mathbf{a}^T\boldsymbol{\nu} < M$. In this case equations (23) have the unique fixed point $\boldsymbol{\nu} = (\nu_1, \ldots, \nu_S)^T$, which is the centring term in the central limit theorem for the subcritical case. We shall consider the event E consisting of paths $\mathbf{r}(t)$ in R^S that start at or near the equilibrium point $\boldsymbol{\nu}$ and rise to reach the boundary $\{\mathbf{z} : \mathbf{a}^T\mathbf{z} = M\}$ at time $t_2(\mathbf{r})$. First we identify the function $h(\mathbf{r}, \mathbf{r}')$. Now

$$\begin{aligned} h(\mathbf{r},\mathbf{r}') &= \sup_{\boldsymbol{\theta}\,\epsilon\, R^S} \left\{ < \boldsymbol{\theta}, \mathbf{r}' > - \sum_{s=1}^{S} \left[\lambda_s \left(e^{\theta_s} - 1 \right) \right] - \sum_{s=1}^{S} \left[\mu_s r_s \left(e^{-\theta_s} - 1 \right) \right] \right\} \\ &= \sum_{s=1}^{S} \sup_{\theta_s \,\epsilon\, R} \left[\theta_s r_s' - H_s(r_s, \theta_s) \right] = \sum_{s=1}^{S} h_s(r_s, r_s') \end{aligned}$$

and we have S one dimensional maximizations identical to that analysed in section 2.2.

To formulate the associated variational problem we define the function $g : R^S \to R$ such that $g(\mathbf{z}) = \mathbf{a}^T\mathbf{z} - M$. Thus, $g(\mathbf{z}) \le 0$ for all $\mathbf{z}\,\epsilon\,\mathcal{S}(M)$ and only equals zero at overload. Fixing the time of overload event E at $t = 0$, we need to find

$$\inf_{\mathbf{r},T\epsilon G} \int_T^0 h(\mathbf{r},\mathbf{r}')\,dt \quad \text{where} \quad G = \{\mathbf{r}, T : -\infty \le T < 0,\ \mathbf{r}(T) = \boldsymbol{\nu},\ g(\mathbf{r}(0)) = 0\}$$

and, if it exists, the unique path $\mathbf{r}^*(t)$ in G at which the infimum is achieved.

From standard calculus of variations theory, any extremal path $\mathbf{r}(t)$ must satisfy $\partial h/\partial r_s - \frac{d}{dt}\partial h/\partial r_s' = 0$, $\forall\, s$ and since $h(\mathbf{r}, \mathbf{r}')$ is separable and does not depend on t, we have $h_s(r_s, r_s') - r_s'\,(\partial/\partial r_s')\,h_s(r_s, r_s') = 0$, $\forall\, s$ since the end-point T is free. From Section 2.2, it follows that the optimum trajectory $\mathbf{r}^*$ must satisfy

$$r_s' = \mu_s r_s(t) - \lambda_s \quad \forall\, s, \tag{24}$$

that $h(\mathbf{r}^*, (\mathbf{r}^*)') = \sum_{s=1}^{S} (r_s^*)' \log(\mu_s r_s^* / \lambda_s) = \sum_{s=1}^{S} (r_s^*)' \log(r_s^* / \nu_s)$ (25)

and $\inf_{\mathbf{r},T\epsilon G} \int_T^0 h(\mathbf{r},\mathbf{r}')\,dt = \int_T^0 \sum_{s=1}^{S} (r_s^*)' \log(r_s^*/\nu_s)\,dt = \sum_{s=1}^{S} \int_{\nu_s}^{r_s^*(0)} \log(r/\nu_s)\,dr$ (26)

It is easy to show that any solution to equations (24) is a global minimal path. To complete our solution we need to identify $\mathbf{r}^*(0)$, the endpoint at overload, where $g(\mathbf{r}^*(0)) = \mathbf{a}^T\mathbf{r}^*(0) - M = 0$. Since $\frac{\partial g}{\partial t} = 0$, the following transversality equations hold for $\mathbf{r} = \mathbf{r}^*$.

$$h(\mathbf{r},\mathbf{r}')\,\partial g/\partial r_s - \partial h/\partial r_s' \left(\textstyle\sum_{v=1}^{S} r_v' \partial g/\partial r_v \right) = 0 \quad s = 1, \ldots, S$$

Now clearly $\frac{\partial g}{\partial r_s} = a_s$ and for $\mathbf{r} = \mathbf{r}^*$ we have $h(\mathbf{r}, \mathbf{r}') = \sum_{v=1}^{S} r_v' \log(r_v/\nu_v)$ from (25) and $\partial h/\partial r_s' = \theta_s^* = \log(r_s/\nu_s)$ (cf (14)). Thus the transversality equations become

$$a_s \sum_{v=1}^{S} r_v' \log(r_v/\nu_v) - \log(r_s/\nu_s) \sum_{v=1}^{S} a_v r_v' = \sum_{v \ne s} r_v' \left(a_s \log(r_v/\nu_v) - a_v \log(r_s/\nu_s) \right) = 0 \tag{27}$$

An obvious solution to equations (27) is to set $a_s \log(r_v^*/\nu_v) = a_v \log(r_s^*/\nu_s)$ for all $s \ne v$, whence, $(r_s^*/\nu_s)^{1/a_s} = (r_v^*/\nu_v)^{1/a_v}$. Then, at $t = 0$

$$r_s^*(0) = \nu_s \delta^{a_s} \quad s = 1, \ldots, S \tag{28}$$

where the constant δ in (28) is given by the unique solution to $g(\mathbf{r}^*(0)) = 0$. That is

$$\sum_{s=1}^{S} a_s \nu_s \delta^{a_s} = M \tag{29}$$

The resulting end-point $\mathbf{r}^*(0)$, determined by (28) and (29), will be denoted by $\mathbf{z}^*$. It is the overload point of maximum probability as can be shown by maximizing the exact product-form equilibrium distribution of $\mathbf{n}$ subject to $\mathbf{a}^T\mathbf{n} = M$. It can be also be shown that this is the only solution to equations (27) that also satisfies $g(\mathbf{r}^*(0)) = 0$.

From equations (24) subject to $r_s(T) = \nu_s$ and $r_s(0) = \nu_s \delta^{a_s}$ we see that

$$r_s^*(t) = \nu_s \left(\delta^{a_s} - 1\right) e^{\mu_s t} + \nu_s \quad s = 1, \ldots, S \quad \text{with } T = -\infty \tag{30}$$

We have $I^*(E) = \inf_{\mathbf{r}, T \epsilon G} \int_T^0 h(\mathbf{r}, \mathbf{r}')\, dt$, for the overload event E, and from (26)

$$I^*(E) = \sum_{s=1}^{S} \int_{\nu_s}^{r_s^*(0)} \log \frac{r}{\nu_s}\, dr = \sum_{s=1}^{S} \left(r \log \frac{r}{\nu_s} - r \right) \Bigg|_{\nu_s}^{r_s^*(0)} = M \log \delta - \sum_{s=1}^{S} \nu_s \left(\delta^{a_s} - 1\right)$$

where δ satisfies (29). Therefore the event E occurs approximately as a Poisson process with mean time between occurrences of about $\exp(NI^*(E))$. We may also calculate an approximation to the time it takes to move from *near* the equilibrium point $\boldsymbol{\nu}$ along the path $\mathbf{r}^*(t)$ to reach overload at $\mathbf{z}^*$. (Since $T = -\infty$ it will take an infinite amount of time to reach $\mathbf{z}^*$ starting from $\boldsymbol{\nu}$.) As we traverse the extremal path $\mathbf{r}^*$, the value of the function g moves from $g(\boldsymbol{\nu}) = \mathbf{a}^T\boldsymbol{\nu} - M < 0$, to overload at $g(\mathbf{z}^*) = 0$. Let us suppose that the system starts from a point $\hat{\mathbf{z}}$ on $\mathbf{r}^*$ such that $g(\hat{\mathbf{z}}) = g(\boldsymbol{\nu}) + \varepsilon$, where $\varepsilon > 0$ is small. Then $\mathbf{a}^T\hat{\mathbf{z}} = \mathbf{a}^T\boldsymbol{\nu} + \varepsilon$ and $\hat{\mathbf{z}} = \mathbf{r}^*(\hat{t})$ for some time $\hat{t}$. It follows from (30) that $\hat{t}$ is defined as the solution to $\sum_{s=1}^{S} a_s \nu_s \left(\delta^{a_s} - 1\right) e^{\mu_s \hat{t}} = \varepsilon$ and that $-\hat{t}$ is the time required for the system to reach overload from $\hat{\mathbf{z}}$, the point close to $\boldsymbol{\nu}$, given that overload is reached.

We may also obtain results for the probability that the system moves from empty up to the most probable overload point $\mathbf{z}^*$ in a given time T. The problem of finding

$$\inf_{\mathbf{r} \epsilon G} \int_0^T h(\mathbf{r}, \mathbf{r}')\, dt, \quad \text{where} \quad G = \{\mathbf{r}(t) : \ \mathbf{r}(0) = \mathbf{0}, \ \mathbf{r}(T) = \mathbf{z}^*\}$$

and the unique $\mathbf{r}^*$ at which the minimum is achieved, separates into S independent, one-dimensional problems of the type considered in Section 2.2. Thus

$$r_s^*(t) = \nu_s(1 - e^{-\mu_s t}) + K_s(T)(e^{\mu_s t} - 1) \quad 0 \le t \le T \tag{31}$$

$$\text{where } K_s(T) = \frac{\nu_s \delta^{a_s} - \nu_s(1 - e^{-\mu_s T})}{e^{\mu_s T} - 1} = \frac{\nu_s \left(\delta^{a_s} - (1 - e^{-\mu_s T})\right)}{e^{\mu_s T} - 1} \tag{32}$$

We can now evaluate $I_0^T(\mathbf{r}^*)$ as

$$\begin{aligned} I_0^T(\mathbf{r}^*) &= -T \sum_{s=1}^{S} K_s(T) - \sum_{s=1}^{S} z_s^* \log 2\lambda_s \\ &+ \sum_{s=1}^{S} \int_0^{z_s^*} \log \left[\sqrt{(\mu_s r - \lambda_s)^2 + 2(\mu_s r + \lambda_s) K_s(T) + K_s(T)^2} + \lambda_s + \mu_s r + K_s(T) \right] dr \end{aligned}$$

where $z_s^* = \nu_s \delta^{a_s}$. It follows that $P\left(\sup_{0\le t\le T}\left|\mathbf{z}^N(t) - \mathbf{r}^*(t)\right| < \varepsilon\right) = \exp(-NI_0^T(\mathbf{r}^*) + o(N))$. This completes our analysis for the call level VP model when the subcritical case applies.

3.3. Time dependent analysis of VP model — critical and supercritical cases

In the critical and supercritical cases the problems of discontinuous statistics at the overload boundary are important. The first order fluid approximation needs to be adapted using Theorem 4, and second order approximations are needed to reflect both stationary and time-dependent behaviour accurately. For the VP model accurate approximations for stationary behaviour are given in Evans (1991), with second order approximations required in the critical case. For Erlang's model, discussed in Section 2.2, Mitra & Weiss (1988) look at time dependent behaviour and the probability that the system moves from empty to full in a given time T. They obtain first and second order approximations via asymptotic expansions of the Laplace transforms. The first order results are special cases of the second order results and are consistent with the first order large deviations results, also quoted in the paper. For the supercritical case, and possibly for the critical case, the second order terms are required to give a good approximation.

Returning to our model for a VP carrying a mix of services, we consider the supercritical case and adapt the fluid approximation that appears in Theorem 1 using timescale separation at the boundary. From Theorem 4, $\mathbf{z}^N(.)$ converges weakly to the unique solution to equations $\frac{d}{dt}z_s^\infty(t) = \lambda_s \pi_s(t) - \mu_s z_s^\infty(t)$ with given initial condition. The fast moving, free capacity process has the geometric distribution given by Kelly (1986), and so $\pi_s(t) = (1 - B^*)^{a_s}$ giving $\frac{d}{dt}z_s^\infty(t) = \mu_s\left(\nu_s(1 - B^*)^{a_s} - z_s^\infty(t)\right)$. These equations have a stable, fixed point at $\mathbf{z}^*$ where $z_s^* = \nu_s(1 - B^*)^{a_s}$. Thus for both supercritical and critical cases, the fixed point of the appropriate fluid equations lies on the boundary. To examine excursions from it we need to investigate the *time dependent* behaviour of the fast moving free capacity process. For Erlang's model, the free capacity process is just an M/M/1 queue and Mitra & Weiss (1988) give a second order asymptotic result for the probability of returns to the boundary. Further analysis is needed for the VP model.

Large deviations theory in the presence of discontinuous statistics is still developing. For recent advances see Dupuis *et al* (1991) and Dupuis & Ellis (1992), but note that only first order terms are available. This may not be sufficient here.

4. CONCLUSION

This paper extends the analysis of a call level model of a virtual path in a B-ISDN, to examine time dependent behaviour and the occurrence of rare events. The load on the system may be subcritical, critical or supercritical. In each case we have a first order fluid approximation to the process as it evolves over time, taking account of behaviour at the overload boundary by averaging over the fast moving free capacity process there. For a subcritical load, large deviations theory gives us the distribution and frequency of the rare overload event. We can identify the overwhelmingly probable path to overload from near equilibrium and estimate the time taken to trace it. We also develop formulae for the probability that the system is full at some time T, if it is empty at time zero.

For the critical and supercritical cases the discontinuities in transition rates at the overload boundary become important. To examine time dependent behaviour near the boundary, and excursions from the boundary, we need the *time dependent*, not the *average* behaviour of the fast moving free capacity process. Recent advances in large deviations theory for systems with discontinuous statistics also need to be followed up, but first order approximations do not appear to be very accurate in this area.

More detailed analysis is required, both for the call acceptance model used here, and for extensions such as 'trunk reservation'. For the free capacity process we need weak convergence results at all loads and asymptotic results for time dependent behaviour. Other approaches, perhaps via the related diffusion approximations, may be appropriate.

The importance of characterizing time dependent behaviour is apparent. In a system where overload is rare, its occurrence follows approximately a Poisson process with large mean interarrival time. However, when it occurs, the system can rise to overload very rapidly; it may not be possible to detect such a rise in time for any reactive control to be effective. We need to establish whether reactive controls are possible, and to identify from the qualitative behaviour the access and control policies most likely to be effective.

REFERENCES

Bucklew J. A. (1990), Large deviation techniques in decision, simulation and estimation, Wiley.

Dupuis P. Ellis R. S. & Weiss A. (1991), "Large deviations for Markov processes with discontinuous statistics, I: general upper bounds", Ann. Probab. **19**, pp 1280—1297.

Dupuis P. and Ellis R. S. (1992), "Large deviations for Markov processes with discontinuous statistics, II: random walks", Probab. Theory Relat. Fields **91**, pp 153—194.

Evans Suzanne P. (1989), " A mathematical model and related problems of optimal management and design in a broadband integrated services network", The Journal of the Australian Mathematical Society, Volume 31 (Series B), pp 150–175.

Evans Suzanne P. (1990a), "A quasi-stationary analysis of a virtual path in a B-ISDN network shared by services with very different characteristics", Computer Networks and ISDN Systems, Volume 20, Numbers 1–5, pp 391–399.

Evans Suzanne P. (1990b), "Performance and control in B-ISDN: transient behaviour and time scales", Paper No. 4.1, 7th ITC Specialists Seminar, New Jersey, U.S.A.

Evans Suzanne P. (1991), "Optimal bandwidth management and capacity provision in a broadband network using virtual paths", Performance Evaluation, Volume 13, pp 27–43.

Hunt P. J. (1990a), "Limit theorems for stochastic loss networks", Ph. D. thesis, University of Cambridge.

Hunt. P. J. and Kelly F. P. (1989), "On critically loaded loss networks", Adv. Appl. Prob. **21**, pp 831–841.

Kelly F. P. (1986), "Blocking probabilities in large circuit-switched networks", Adv. Appl. Prob. **18**, pp 473–505.

Kurtz T. G. (1978), "Strong approximation theorems for density dependent Markov chains", Stoch. Proc. Appl. **6**, pp 223—240.

Mitra D. and Weiss A. (1988), "The transient behaviour in Erlang's model for large trunk groups and various traffic conditions", Proc. 12th ITC, Torino, pp 5.1B4.1—5.1B4.8.

Mitra D. and Weiss A. (1989), "The fluid limit of a closed queueing network with applications to data networks", preprint.

Weiss A. (1986), "A new technique for analyzing large traffic systems", Adv. Appl. Prob. **18**, pp 506—532.

Integrated Broadband Communication Networks
and Services (C-18) / V.B. Iversen (Editor)
Elsevier Science B.V (North-Holland)

On the broadening of cell bursts in an ATM network

Jorma Virtamo and Ilkka Norros [a]

[a]Technical Research Centre of Finland
Telecommunications Laboratory
P.O.Box 34, SF-02151 Espoo, Finland
Tel. +358 0 4561, Fax +358 0 4550115
E-mail: jorma.virtamo@vtt.fi

The expansion of a burst during its traversal through a number of multiplexing stages in an ATM network is analysed. The phenomenon is first described and demonstrated by simulation experiments. The expansion is caused by a self-induced load increase of the multiplexers by the burst, whereby the latter cells of the burst experience, on the average, longer delays than the first cell. An analytical study on the growth of the interval between the first and the second cell is presented. Assuming Poissonian background traffic, this interval forms a Markov chain, from stage to stage. The transition matrix of the chain is derived by calculating the delay distribution of the second cell conditioned on the delay of the first cell and the interval between the two on the input side of the multiplexer, and then unconditioning this with respect to the delay of the first cell which is distributed as the delay in an $M/D/1$ queue. A simplified analysis is also given for the growth of the average interval.

1. INTRODUCTION

The output processes of ATM multiplexers, i.e. the deformation of an individual cell stream passing through one or several multiplexers, have been studied relatively little. Brown and Simonian [1] considered the deformation of a periodic cell stream in a multiplexer where it is mixed with Poissonian background traffic. Roberts showed how this model, the so called $M + D/D/1$ queue, can be analyzed with the Beneš approach. This work is reported in Section 6.4 of [3]. Recently, van den Berg and Resing [4] found an approximate form of the limit distribution of the cell interarrival time when a periodic stream passes through an infinite number of multiplexing nodes with Poissonian cross traffic.

In all the studies mentioned above, an infinitely long stream was considered. In this paper, we study a slightly different problem where, instead of a stationary cell stream, a finite burst of cells is transmitted through a number of multiplexers, again loaded with Poissonian cross traffic. It turns out that the burst becomes, in the mean, longer in each multiplexer, and it is this broadening phenomenon that we are focusing at. Initial results from this work were reported in Section 9.2 of [3], which, as far as we know, was the first time this problem has been presented.

The problem is defined in Section 2. Section 3 presents an analysis of the growth of the distance between the two first cells of a burst. Numerical results along with a simplified

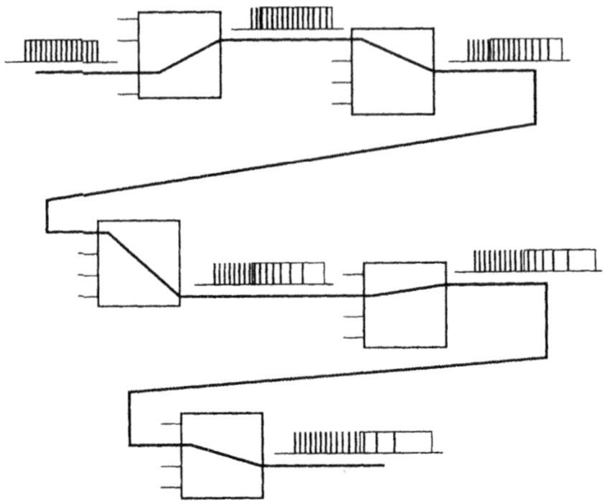

Figure 1. Burst broadening in a cascade of multiplexers. The rectangles show what the positions of the burst were if there were no queueing.

analysis of the mean interval expansion are given in Section 4. In the concluding Section 5 we discuss some situations where the broadening phenomenon may have practical significance.

2. THE PHENOMENON OF BURST BROADENING

The burst broadening we are considering is essentially a transient phenomenon which is caused by a self-induced load increase of the multiplexers: the first cell of the burst sees each multiplexer in a stationary state, loaded only by the cross traffic, but for the later cells the multiplexer is loaded by both the cross traffic and the burst itself. Therefore, the later cells experience longer delays, and the burst as a whole becomes longer in each stage, as is schematically depicted in Figure 1.

In contrast to the delay jitter of a cell in a continuous stream, this transient effect is linearly accumulative as a function of the number of passed multiplexing stages. The broadening starts at the leading edge of the burst. The intercell interval between the first and second cell begins first to grow. As it gets longer, the load jump effectively shifts towards the second cell, and the interval between the second and third cells begins to grow, etc. In this way a rarefaction wave eats its way into the burst. The linear accumulation of the burst broadening continues until the rarefaction wave has reached the end of the burst, whereafter the growth continues at a slower pace.

Figure 2 shows the result from the simulation of one realisation of the system when a burst of 21 cells with an initial intercell interval of 20 (cell transmission time is used as the unit of time) traverses across a network of 100 multiplexers, each of them loaded by a Poissonian background traffic of $\rho = 0.9$ (this large number of stages is, of course,

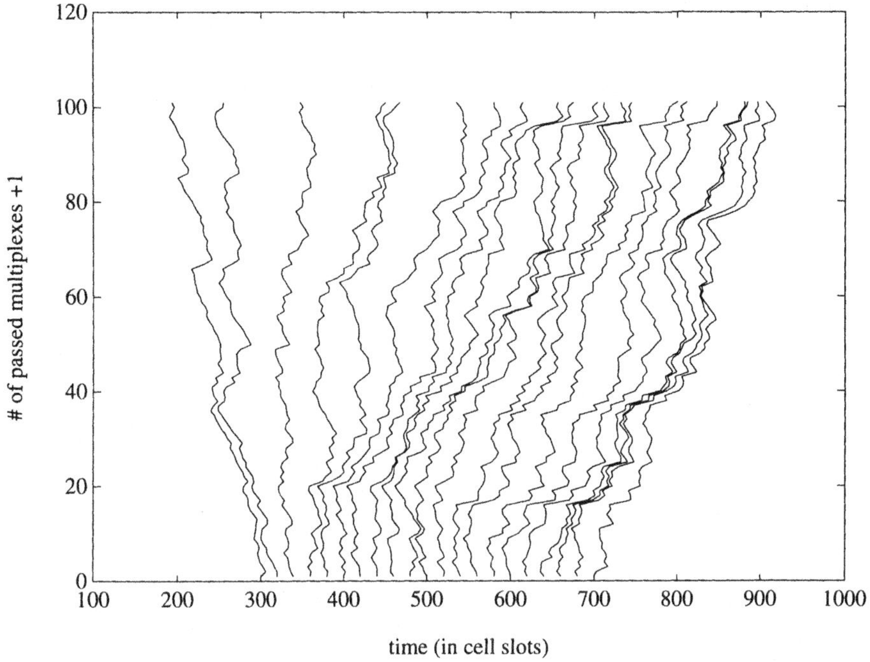

Figure 2. Deformation of a burst with an intercell interval of 20 and a background load $\rho = 0.9$.

unrealistic but is used just to demonstrate the effect). The linearly proceeding expansion phenomenon is clearly seen. A cleaner view of the phenomenon is given by Figure 3 where the same experiment was repeated 1000 times and mean positions of the cells after each multiplexer are shown (the theoretical mean arrival time of the first cell, as calculated from the mean delay in a $M/D/1$ queue, is used as time zero for each multiplexer). We also see that the expansion is strongest at the beginning of the burst as expected by the above discussion. In the following section we make a quantitative analysis of the growth of the interval between the first and second cells of the burst.

3. BROADENING OF THE LEADING EDGE: AN EXACT ANALYSIS

Our analysis is based on the results on the $M + D/D/1$ system presented in [2] and [3]. The $M + D$ arrival process consists of a superposition of Poisson traffic of intensity ρ and a deterministic stream with period D. As noted in [3] the delays of the cells of the periodic stream constitute a Markov chain. The probability of the delay V_i of cell i exceeding some

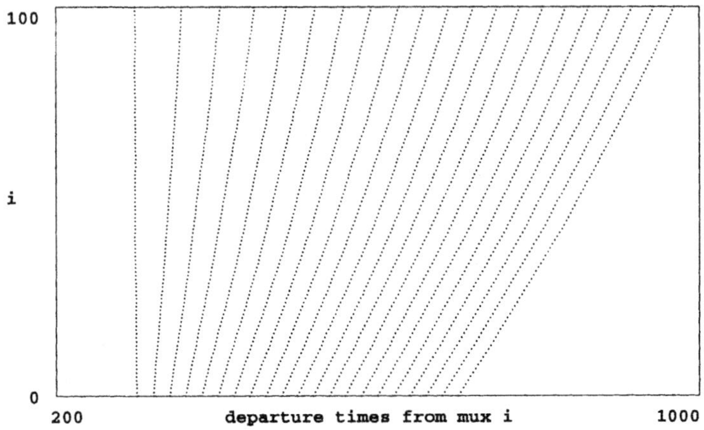

Figure 3. Mean burst expansion from 1000 simulation runs.

value x depends only on the delay V_{i-1} of the preceding cell (and, parametrically, on the intercell interval D_{in} at the input of the multiplexer). The basic results thus apply to any pair of cells in an arbitrary test stream (not necessarily a periodic one) superposed on the Poisson background traffic. We apply the results to cells 1 and 2 of the burst. Our main interest is now in the dependence of the intercell interval at the output, D_{out}, on the interval at the input, D_{in}. These intervals also form a Markov chain, from stage to stage.

In accordance with [3] we restrict ourselves to a slotted time system where the service can only start at instants $\ldots -2, -1, 0, 1, 2, \ldots$ (using the cell service time as the time unit). Arrivals of the test stream (specifically, cells 1 and 2) are also assumed to occur at these discrete instants. Then the delays V_1 and V_2, as well as the intercell intervals D_{in} and D_{out}, all take integer values.

Central to the analysis is the discrete transition matrix

$$t(m,n;\, i) = \Pr\{V_2 = n \mid V_1 = m, D_{in} = i\}, \tag{1}$$

where we have added the argument i to emphasize the dependence on the cell interval at the input. By the results of [3] and [2] we can calculate the transition matrix from the following set of formulae

$$t(m,n;\, i) = T(m,n-1;\, i) - T(m,n;\, i), \tag{2}$$

where

$$T(x,y;D) = \sum_{k=0}^{\infty} T_k(x,y;D)\,\frac{(\rho D)^k}{k!}\,e^{-\rho D} \tag{3}$$

and

$$T_k(x,y;D) = \begin{cases} 1, & \text{for } y < x+1+k-D \text{ or } y < 0, \\ Q_D^k(y), & \text{for } y \geq x+1+k-D, \end{cases} \tag{4}$$

with

$$Q_D^k(x) = \sum_{x<n\leq k} \binom{k}{n} \left(\frac{n-x}{D}\right)^n \left(1 - \frac{n-x}{D}\right)^{k-n} \frac{D-k+x}{D-n+x}. \tag{5}$$

Now, the interval D_{out} is related to D_{in}, V_1 and V_2 by

$$D_{out} = D_{in} + V_2 - V_1. \tag{6}$$

Thus, we get the conditional probability

$$S(i,j;m) = \Pr\{D_{out} = j \mid D_{in} = i, V_1 = m\} = t(m, m+j-i; i), \tag{7}$$

which represents the transition probability from the cell interval on the input side to that on the output side, given that cell 1 has experienced a delay of m time units.

The delay of the first cell is independent of the interval between cells 1 and 2, and its distribution is given by that of a cell coming into a (discrete time) $M/D/1$ queue (the independence is not true for later cells, which makes their analysis more complicated). By denoting this distribution by $\pi(m)$,

$$\pi(m) = \Pr\{V_1 = m\}, \tag{8}$$

we get an unconditional transition matrix $S(i,j)$,

$$S(i,j) = \Pr\{D_{out} = j \mid D_{in} = i\} = \sum_m S(i,j;m)\pi(m). \tag{9}$$

Let $p^k(i)$ denote the probability of cell interval at the input of stage k being equal to i and let p^k be the distribution vector

$$p^k = (p^k(0),\ p^k(1), \ldots). \tag{10}$$

Then the distributions p^k evolve according to the Markov chain

$$p^{k+1} = p^k \cdot S. \tag{11}$$

4. NUMERICAL RESULTS

We have calculated numerically the transition matrix S for a system with the background (Poisson) load of $\rho = 0.8$. The size of the matrix we used was 30×30.

Figure 4 gives the row averages of the transition matrix as a function of the row number, i.e. the mean intercell interval at the output of a stage as a function of the interval (deterministic) at the input. This figure clearly shows the expansion tendency of the interval (all the points lie above the diagonal). The expansion is strongest for very short intervals; there is a kind of repulsion between the cells.

Figure 5 shows the cell interval distribution after 1, 3 and 10 stages, given that the interval at the input of stage 1 was 5 cell service times (corresponding to a total load of

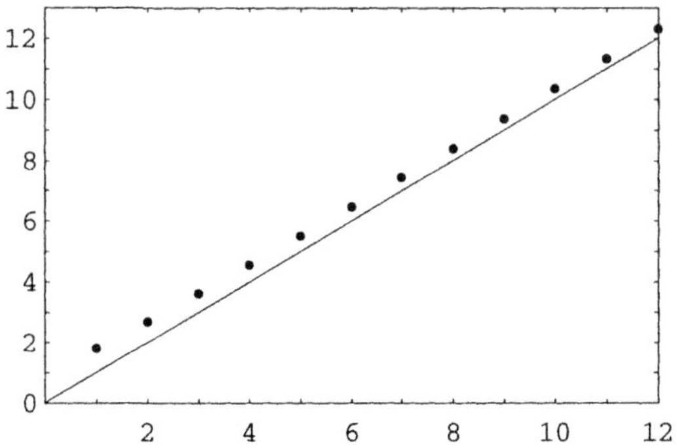

Figure 4. Mean intercell interval at the output as a function of the interval at the input. The diagonal is drawn for reference only.

$0.8 + 1/5 = 1$ during the burst). Besides of the spreading of the distribution after each additional stage, the growth of the mean intercell interval can again be noted. In Figure 6 this is shown more explicitly, i.e. the mean intercell interval is plotted against the number of traversed multiplexing stages.

Finally, Figure 7 gives a simplified analysis of the expansion of the mean. It is based on the observation that the input-to-mean-output relation in Figure 4 is very close to linear. If it were perfectly linear then, irrespective of the distribution of the input, the mean interval at the output would only depend (deterministically) on the *mean* input interval. The progression of the mean values is then represented by the steps between the diagonal and the input-to-mean-output curve of Figure 2, which has been approximated by a parabola (we are relaxing the slotted time requirement). How well this approximation works is shown by the dots which represent the exact input-output relation of Figure 6 (each dot in Figure 7 has a y-value equal to that of a dot in Figure 6 and an x-value equal to the y-value of the preceding dot in Figure 6). The true expansion is a little faster than that predicted by the simplified model. Due to the small quadratic term in the input-to-mean-output curve the mean output has an extra positive term reflecting the dependence on the variance of the input distribution.

5. Concluding remarks

We have described the broadening phenomenon of a finite burst of cells traversing across a network of ATM multiplexers and, in particular, given an analysis of the growth of the first intercell interval. The notable feature in this phenomenon is its additive nature; while the expectation of the jitter of a given cell in an infinite stream of cells is zero, the expectation of the expansion of the transient burst grows in proportion to the number of traversed multiplexers.

The practical significance of the phenomenon depends on a number of parameters. The most important ones are the background load level of the multiplexers and, in particular,

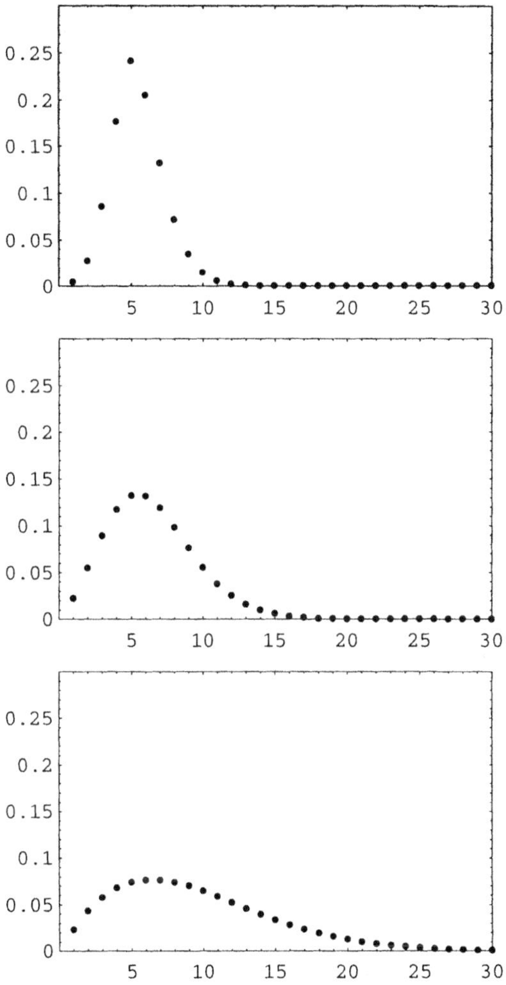

Figure 5. Distribution of the intercell interval after 1, 3 and 10 multiplexing stages. Initial interval 5 and background load 0.8.

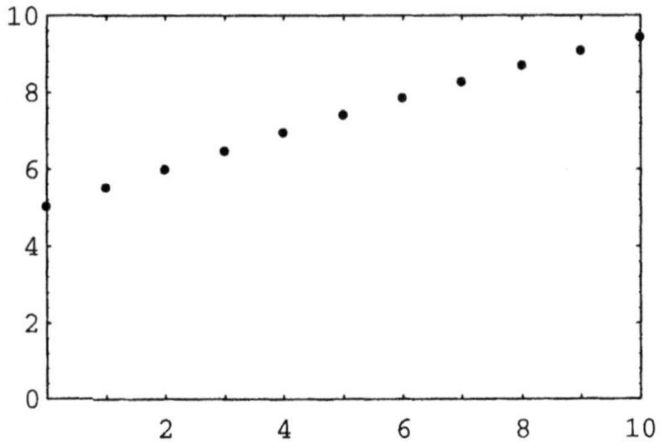

Figure 6. Mean intercell interval as a function of the number of traversed multiplexers. Initial interval 5 and background load 0.8.

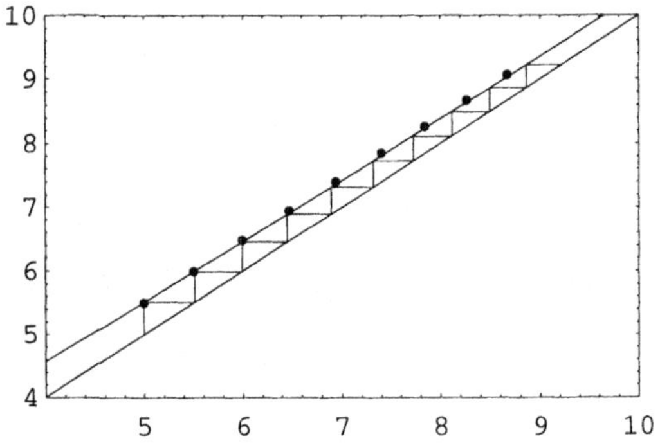

Figure 7. A simplified analysis of the mean expansion. The mean output interval of each stage depends almost solely on the mean input interval, leading to the expansion described by the steps. The dots represent the exact input-output relation of Figure 7.

the total load during the burst. If these are close to 1, the expansion can be quite significant. The burst can even temporarily overload the multiplexer. In such a situation a burst scale queue builds up with a dramatic increase in the burst length (or the buffer overflows). But even with more modest load levels the broadening can become noticeable if the number of multiplexing stages is large.

One of the undesirable consequences of the broadening phenomenon is that it makes it more difficult to make reliable rate measurements within the network for congestion control purposes. Because of the burst broadening phenomenon a rate measurement based on the observation of a few first cells leads to an underestimation of the actual rate of the burst.

REFERENCES

1. P. Brown and A. Simonian. Perturbation of a periodic flow in a synchronous server. In *Performance '87, Brussels*, 1987.
2. J.W. Roberts. Jitter due to an ATM multiplex — application to peak rate policing. Technical Report 242TD(89)035, COST, 1989.
3. J.W. Roberts, editor. *Performance Evaluation and Design of Multiservice Networks. COST 224 Final Report.* CEC, 1992.
4. J.L. van den Berg and J.A.C. Resing. Tandem queues with crossing interference. Technical Report 242TD(92)040, COST, 1992.

Integrated Broadband Communication Networks
and Services (C-18) / V.B. Iversen (Editor)
Elsevier Science B.V (North-Holland)
1994 IFIP

Analysis of ATM Switching Modules with Channel Grouping in a Bursty–Source Environment

BART STEYAERT and HERWIG BRUNEEL

University of Ghent, Laboratory for Communications Engineering,
Sint–Pietersnieuwstraat 41, B–9000 Gent, Belgium.

ABSTRACT.

In this paper, a method for analyzing the buffer behavior of a discrete–time queue with bursty arrivals and multiple servers is presented. Specifically, an ATM switching element with a finite number of independent and identical bursty traffic sources, and a multiple number of transmission lines assigned per destination is considered. The burstiness of the sources is captured by representing the cell arrival process on one particular input line of the switch as a first–order Markov chain. Arriving cells are assumed to be routed to their destinations in a uniform and independent way. A correlated–arrivals queueing model for the switch is obtained, which is analyzed in this paper by purely analytical means. As a result, an explicit expression for the probability generating function of the system contents is derived. This leads to closed–form expressions for the mean system contents and an accurate analytic approximation for the tail probabilities of the system contents.

Keyword Codes : G3.
Keywords : Mathematics of Computing, Probability and Statistics; (*) ATM (Asynchronous Transfer Mode) Networks; (*) Performance Evaluation.

1. SYSTEM DESCRIPTION.

In the past couple of years, discrete–time queueing models have become very popular in the performance analysis of ATM switching elements with output queueing. In many studies, the cells are assumed to be generated according to an uncorrelated (i.e., Bernoulli [1–3]) process by each of the sources. It has been observed however, that in practice, the arrival stream of the cells tends to be bursty, i.e., of a correlated nature. Therefore, several attempts have been made to incorporate bursty (nonindependent) cell arrival processes in the queueing analysis of buffers, both for single–server queues [4–7] and for multiple–server queues [8]. The discrete–time queueing models considered in these papers are analyzed by mainly analytical means (i.e., a generating–functions approach), and they have in common that they are applicable for ATM multiplexers, meaning that no routing mechanism of arriving cells to one of various output buffers (each corresponding with one possible destination) is taken into account. In this paper, we present a method for analyzing the buffer behavior of an ATM switching element

The second author wishes to acknowledge the support of the Belgian National Fund for Scientific Research (NFWO). The support of Alcatel Bell Mfg. Co. is also greatly acknowledged.

with first–order Markovian arrivals on the inlets, and independent uniform routing of the arriving cells to their destinations (which is a reasonable assumption for a variety of architectures for a self–routing switching network, as was pointed out in [9]), where multiple servers are assigned per destination. The case of single–server output queues has been treated in [10].

As in most studies concerned with ATM–related discrete–time queueing models, the time axis is divided into slots of fixed length, where one slot suffices for the transmission of exactly one ATM cell. Cells enter the switch via one of its *input links* (sources), and are then routed to one of the *output links*, where they are temporarily buffered in a designated *output queue* to await transmission of the cells that arrived before them in this particular output queue and that are not yet sent. In this paper, we are concerned with multiple–server output queues, i.e., each output queue (corresponding with one single destination) has a constant number of transmission lines (denoted by the integer $c \geq 1$)

In the remainder of this paper, N denotes the number of input and output links of a switch. Cell arrivals on different input links are assumed to be statistically independent, as they originate from different sources. It is assumed that the routing of cells from the input to the output links is performed in a uniform and independent way, i.e., each arriving cell is routed to a tagged output queue with probability c/N (since there are N/c possible destinations), independently of the routing of other cells.

Throughout this paper, the arrival process on one particular input link will be modeled as a first–order Markov chain. Let us now define an *active period* on an input link as a number of consecutive slots during which a cell arrival occurs. Similarly, a *passive period* is a number of consecutive slots during which there is no cell arrival on this input link. A slot belonging to an active (passive) period on one particular input link will be referred to as an active (passive) slot, and, similarly, the input link will be referred to as an active (passive) input link. In the following, we will assume that the lengths of both active and passive periods are geometrically distributed random variables with parameters α and β respectively. Note that this assumption implies the above mentioned Markovian nature of the arrival processes on the individual inlets. An uncorrelated Bernoulli arrival law on each input link corresponds with $\alpha + \beta = 1$.

It is now clear that in our model, the cell arrival process on an inlet alternates between two possible states, say A_1 and A_2 (where A_1 corresponds to the active, and A_2 to the passive periods). Therefore, if we concentrate on one tagged input link, then the state of the input link can be described by a 2–dimensional first–order Markov chain. If we denote by $[a_i(j)]$ the 2x2 transition matrix describing the transitions between the 2 possible states at the slot boundaries separating consecutive slots, then this matrix fully describes the arrival process one each input link, and due to the previous definitions and assumptions, it is given by

$$[a_i(j)] = \begin{bmatrix} a_1(1) & a_1(2) \\ a_2(1) & a_2(2) \end{bmatrix} \triangleq \begin{bmatrix} \alpha & 1-\alpha \\ 1-\beta & \beta \end{bmatrix} . \tag{1}$$

If we denote by $a_{i,k}$,$1 \leq i \leq 2$, the number of input links in state A_i during slot k, then, due to the two–state model previously described, $A(x_1,x_2)$, the generating function describing the number of input links in state A_i during an arbitrary slot, is given by

$$A(x_1,x_2) \triangleq \lim_{k\to\infty} E[x_1^{a_{1,k}} x_2^{a_{2,k}}] = (\sigma x_1 + (1-\sigma)x_2)^N , \tag{2}$$

where E[.] denotes the expected value of the quantity between brackets, and

$$\sigma \triangleq (1-\beta)/(2-\alpha-\beta) \tag{3}$$

is the probability that an input link is in state A_1 during an arbitrary slot. Note that, since each input link is either in state A_1 or A_2 during any slot, we always find that $a_{1,k} + a_{2,k} = N$, (i.e., the number of inlets of the switch) for any value of k.

Furthermore, we assume that if an input link is in state A_i during slot k, then we denote by $f_i(z)$, $1 \le i \le 2$, the routing function corresponding to state A_i, which is the probability generating function describing the number of cell arrivals in an output queue during a slot on an input link in state A_i. With the previous definitions, we find that

$$f_1(z) = 1 + (z-1)c/N \qquad (4.a)$$
$$f_2(z) = 1 \ , \qquad (4.b)$$

since exactly one cell is generated on an input link in state A_1 during a slot, which is routed uniform and independently to one of the the N/c output queues, and no cells are generated on an input link which is in state A_2 during a slot. Although in our model for the ATM switch $f_1(z)$ and $f_2(z)$ satisfy (4.a,b), the analysis in the remainder of this paper holds for general $f_1(z)$ and $f_2(z)$ (that could be used, for instance, to describe a non–uniform routing mechanism of arriving cells to their destinations), as will become clear later.

2. DERIVATION OF THE SYSTEM EQUATION.

We now focus attention on one particular output queue, and we denote by v_k the random variable describing the system contents just after slot k (i.e., at the beginning of slot k+1). This is the number of cells present in the output queue at the beginning of slot k+1, including those that will be transmitted during this slot. Furthermore, let us denote by e_k the number of packet arrivals during slot k in the output queue. Since the buffer has c severs providing the transmission of packets at a rate of one packet per slot, the equilibrium condition requires that the mean number of packet arrivals during an arbitrary slot, σc, must be strictly less than c, which implies that $\sigma<1$.

Our final goal is to find an expression for the steady–state probability generating function of the system contents at the beginning of a slot. If we choose $(a_{1,k},a_{2,k},v_k)$ as the set of random variables describing the system during slot k, then, due to the arrival model described in Section 1, it is clear that this set forms a three–dimensional Markov chain that fully describes the evolution of the system contents from slot–to–slot. Note that, due to the property that $a_{1,k} + a_{2,k} = N$, one of the first two random variables in the set $(a_{1,k},a_{2,k},v_k)$ could be omitted, and still we would obtain a proper (two–dimensional) state description of the system. However, for reasons that will become clear later in this paper, we prefer to maintain both random variables $a_{i,k}$, $1 \le i \le 2$, in our state description. Therefore, defining the joint the joint probability generating function of these random variables as

$$P_k(x_1,x_2,z) \triangleq E[x_1^{a_{1,k}} x_2^{a_{2,k}} z^{v_k}] \ ,$$

and using the system equation $v_k = (v_{k-1} - c)^+ + e_k$ (where $(.)^+ \triangleq \max(.,0)$), we obtain

$$P_k(x_1,x_2,z) = E\left[z^{(v_{k-1}-c)^+} \prod_{i=1}^{2} (x_i f_i(z))^{a_{i,k}}\right] \ . \qquad (5)$$

With the characteristics of the traffic model described in Section 1 and using some standard queueing–theoretical techniques, we find that

$$P_k(x_1,x_2,z) = E\left[z^{(v_{k-1}-c)^+} \prod_{i=1}^{2} A_i(x_1,x_2,z)^{a_{i,k-1}}\right] . \tag{6}$$

where

$$A_i(x_1,x_2,z) \triangleq \sum_{j=1}^{2} a_i(j)\, x_j\, f_j(z) . \tag{7}$$

Expression (6) can then be transformed into

$$P_k(x_1,x_2,z) = z^{-c} \{P_{k-1}(x_1,x_2,z)$$
$$+ E\left[(z^c - z^{v_{k-1}}) \prod_{i=1}^{2} A_i(x_1,x_2,z)^{a_{i,k-1}} \middle| v_{k-1}<c\right] \mathrm{Prob}[v_{k-1}<c]\} .$$

When the system has reached its steady state, this becomes independent of k, and we thus obtain an equation for the steady–state probability generating function $P(x_1,x_2,z)$

$$P(x_1,x_2,z) \triangleq \lim_{k\to\infty} P_k(x_1,x_2,z) = z^{-c} \{P(A_1(x_1,x_2,z),A_2(x_1,x_2,z),z)$$
$$+ Q(A_1(x_1,x_2,z),A_2(x_1,x_2,z),z)\} , \tag{8.a}$$

where

$$Q(x,y,z) \triangleq \sum_{j=0}^{c-1} (z^c - z^j) \sum_{i=0}^{N} x^i\, y^{N-i}\, \mathrm{Prob}[a_1=i, a_2=N-i, v=j] . \tag{8.b}$$

In this expression, the random variable v denotes the system contents at the beginning of an arbitrary slot, while a_1 and a_2 denote the number of input links in state A_1 respectively A_2 during the previous slot. Equations (8.a,b) define a functional equation for the joint probability generating function of the system contents at the beginning of an arbitrary slot and the number of input links in state A_1 and A_2 during the previous slot, in terms of a number of probabilities still to be determined. From this equation, all the information concerning the buffer behavior can be extracted. In the next section, we will describe a technique for solving this equation for $P(x_1,x_2,z)$. It will also become clear then how the unknowns that still appear in (8.b) can be computed.

3. SOLVING THE FUNCTIONAL EQUATION.

If we denote by

$$a_{ij}(z) \triangleq a_i(j)\, f_j(z) \quad , \quad 1 \leq i,j \leq 2 , \tag{9.a}$$

then combining (9.a) and (7) reveals that

$$A_i(x_1,x_2,z) = \sum_{j=1}^{2} a_{ij}(z)\, x_j \quad , \quad 1 \leq i \leq 2 . \tag{9.b}$$

Defining $[A_i(x_1,x_2,z)]$ as the 2x1 column matrix with $A_i(x_1,x_2,z)$ on the i–th row, $[x_j]$ as the 2x1 column matrix with x_j on the j–th row, and $[a_{ij}(z)]$ as the 2x2 matrix with $a_{ij}(z)$ on the intersection of the i–th row and the j–th column, then (9.a,b) can be written as $[A_i(x_1,x_2,z)] = [a_{ij}(z)]\, [x_j]$. Let us now introduce the set of functions, for $n \geq 2$ defined recursively as

$$A_{i,n}(x_1,x_2,z) \triangleq A_i(A_{1,n-1}(x_1,x_2,z),A_{2,n-1}(x_1,x_2,z),z) \quad , \quad 1 \le i \le 2 \quad , \qquad (10.a)$$
$$A_{i,1}(x_1,x_2,z) \triangleq A_i(x_1,x_2,z) \quad , \qquad 1 \le i \le 2 \quad . \qquad (10.b)$$

First of all, since $A_i(x_1,x_2,z)$, $1 \le i \le 2$, is a linear function of x_1 and x_2, $A_{i,n}(x_1,x_2,z)$ will also be a linear function of x_1 and x_2, which means that this function can be written as

$$A_{i,n}(x_1,x_2,z) \;=\; a_{i1,n}(z)\,x_1 \;+\; a_{i2,n}(z)\,x_2 \quad , \quad 1 \le i \le 2 \quad ,$$

for some functions $a_{ij,n}(z)$, still to be determined. Using the matrix representation of the above formula consecutively, starting from (10.b), we obtain

$$[A_{i,n}(x_1,x_2,z)] \;=\; [a_{ij}(z)]^n\,[x_j] \quad , \quad n \ge 1 \quad , \qquad (11)$$

due to the definition (10.a) and (9.b).

For each value of z, we now denote by $\lambda_i(z)$, $1 \le i \le 2$, the 2 eigenvalues of the matrix $[a_{ij}(z)]$, by $[w_{ij}(z)]$ a 2x2 matrix for which the i–th row is a left row eigenvector of $[a_{ij}(z)]$ with eigenvalue $\lambda_i(z)$, and by $[u_{ij}(z)]$ the inverse matrix of $[w_{ij}(z)]$; it can be easily shown that $[u_{ij}(z)]$ is a 2x2 matrix for which the j–th column is a right column eigenvector of $[a_{ij}(z)]$ with eigenvalue $\lambda_j(z)$. In the following, it will become clear that it is preferable to define the left row eigenvectors of $[a_{ij}(z)]$ such that

$$w_{i1}(z) + w_{i2}(z) \;=\; 1 \quad , \quad 1 \le i \le 2 \quad . \qquad (12)$$

The elements of $[w_{ij}(z)]$ (and, consequently, its inverse $[u_{ij}(z)]$) are hereby uniquely determined. Specifically, due to equations (1–3), it can be shown that for the model considered here, the eigenvalues $\lambda_1(z)$ and $\lambda_2(z)$ of $[a_{ij}(z)]$ are given by

$$\lambda_{1,2}(z) = \frac{\alpha f_1(z)+\beta \pm \left((\alpha f_1(z)+\beta)^2 - 4(\alpha+\beta-1)f_1(z)\right)^{1/2}}{2} \quad , \qquad (13.a)$$

(with the agreement that $(\rho.\exp\{\iota\theta\})^{1/2}=\rho^{1/2}\exp\{\iota\theta/2\}$ for all $\rho>0$, $0 \le \theta < 2\pi$) where $\lambda_1(z)$ corresponds with the plus–sign in the numerator of the above expression (note that $\lambda_1(1)=1$). Furthermore, the following expressions for $[w_{ij}(z)]$ and $[u_{ij}(z)]$ can be obtained:

$$[w_{ij}(z)] \;=\; \begin{bmatrix} \dfrac{(1-\beta)\,f_1(z)}{\lambda_1(z)-(\alpha+\beta-1)f_1(z)} & \dfrac{\lambda_1(z)-\alpha f_1(z)}{\lambda_1(z)-(\alpha+\beta-1)f_1(z)} \\[2ex] \dfrac{(1-\beta)\,f_1(z)}{\lambda_2(z)-(\alpha+\beta-1)f_1(z)} & \dfrac{\lambda_2(z)-\alpha f_1(z)}{\lambda_2(z)-(\alpha+\beta-1)f_1(z)} \end{bmatrix} \qquad (13.b)$$

$$[u_{ij}(z)] = \begin{bmatrix} \dfrac{(\lambda_1(z)-(\alpha+\beta-1)f_1(z))(1-\alpha)}{(\alpha f_1(z)-\lambda_1(z))(\lambda_2(z)-\lambda_1(z))} & \dfrac{(\lambda_2(z)-(\alpha+\beta-1)f_1(z))(1-\alpha)}{(\alpha f_1(z)-\lambda_2(z))(\lambda_1(z)-\lambda_2(z))} \\[2ex] \dfrac{\lambda_1(z)-(\alpha+\beta-1)f_1(z)}{\lambda_1(z)-\lambda_2(z)} & \dfrac{\lambda_2(z)-(\alpha+\beta-1)f_1(z)}{\lambda_2(z)-\lambda_1(z)} \end{bmatrix} \quad . \qquad (13.c)$$

Finally, in view of the property $[a_{ij}(z)] = [u_{ik}(z)]\,[\lambda_k(z)]\,[w_{kj}(z)]$, equation (11) can be transformed into

$$[A_{i,n}(x_1,x_2,z)] \;=\; [u_{ik}(z)]\,[\lambda_k(z)^n]\,[w_{kj}(z)]\,[x_j] \quad ,$$

which becomes, for $1 \le i \le 2$

$$A_{i,n}(x_1,x_2,z) = \sum_{k=1}^{2} u_{ik}(z)\, \lambda_k(z)^n \sum_{j=1}^{2} w_{kj}(z)\, x_j \triangleq \sum_{k=1}^{2} u_{ik}(z)\, \lambda_k(z)^n\, W_k(x_1,x_2,z) \quad . \quad (14)$$

Let us now focus attention on the functional equation (8.a). With the definitions (10.a,b), it is not difficult to see that (8.a) can be transformed into

$$z^{-(n-1)c}\, P(A_{1,n-1}(x_1,x_2,z),A_{2,n-1}(x_1,x_2,z),z) = z^{-nc}\{P(A_{1,n}(x_1,x_2,z),A_{2,n}(x_1,x_2,z),z) + Q(A_{1,n}(x_1,x_2,z),A_{2,n}(x_1,x_2,z),z)\} \quad ,$$

for all $n \geq 2$. Summing (8.a) and this equation for consecutive values of n and using the definitions of P(.) and Q(.) and the fact that $a_1+a_2=N$, then yields

$$P(x_1,x_2,z^N) = P\left(\frac{A_{1,n}(x_1,x_2,z^N)}{z^{nc}},\frac{A_{2,n}(x_1,x_2,z^N)}{z^{nc}},z^N\right) + \sum_{i=1}^{n} Q\left(\frac{A_{1,i}(x_1,x_2,z^N)}{z^{ic}},\frac{A_{2,i}(x_1,x_2,z^N)}{z^{ic}},z^N\right) \quad .$$

Let us now consider values of z for which

$$|\lambda_k(z^N)| < |z|^c \leq 1 \quad , \quad 1 \leq k \leq 2 \quad .$$

In the following section, we will prove that this is the case for all values of z on the unit circle and different from 1, i.e., for the set of values $\{z : |z|=1 \text{ and } z\neq 1\}$. Then, in view of expression (14) and taking the limit for $n\to\infty$, the sum in the right hand side of the above expression converges, and we find

$$P(x_1,x_2,z^N) = \sum_{i=1}^{\infty} Q\left(\frac{A_{1,i}(x_1,x_2,z^N)}{z^{ic}},\frac{A_{2,i}(x_1,x_2,z^N)}{z^{ic}},z^N\right) \quad ,$$

where we have also used the property that $P(0,0,z) = 0$. Let us define the functions $v_{im}(z)$, $0 \leq i,m \leq N$, as

$$\left[\sum_{k=1}^{2} u_{1k}(z)\, x_k\right]^i \left[\sum_{k=1}^{2} u_{2k}(z)\, x_k\right]^{N-i} \triangleq \sum_{m=0}^{N} v_{im}(z)\, x_1^{m}\, x_2^{N-m} \quad , \quad 1 \leq i \leq N \quad , \quad (15.a)$$

meaning that

$$v_{im}(z) = \sum_{j=(i+m-N)^+}^{\min(i,m)} \begin{bmatrix} i \\ j \end{bmatrix} \begin{bmatrix} N-i \\ m-j \end{bmatrix} u_{11}(z)^j u_{12}(z)^{i-j} u_{21}(z)^{m-j} u_{22}(z)^{N-i-m+j} \quad , \quad (15.b)$$

where $\begin{bmatrix} k \\ \ell \end{bmatrix} = \frac{k!}{\ell!(k-\ell)!}$. Using this definition and equations (8.b) and (14), and setting $x_1=x_2=1$, then in view of (12), we find the following expression for $V(z^N) \triangleq P(1,1,z^N)$:

$$V(z^N) = \sum_{m=0}^{N} \sum_{j=0}^{c-1} (z^{cN} - z^{jN}) \sum_{i=0}^{N} v_{im}(z^N) \sum_{\ell=1}^{\infty} \left[\frac{\lambda_1(z^N)}{z^{\ell c}}\right]^{\ell m} \left[\frac{\lambda_2(z^N)}{z^{\ell c}}\right]^{\ell(N-m)} p_i(j) \quad , \tag{16}$$

where $p_i(j) \triangleq \text{Prob}[a_1=i, a_2=N-i, v=j]$. Working out the sum for ℓ in the above formula, we thus finally find

$$V(z) = \sum_{m=0}^{N} \frac{\lambda_1(z)^m \lambda_2(z)^{N-m}}{z^c - \lambda_1(z)^m \lambda_2(z)^{N-m}} \sum_{j=0}^{c-1} (z^c - z^j) \sum_{i=0}^{N} v_{im}(z)\, p_i(j) \quad . \tag{17}$$

This expression for $V(z)$, the probability generating function of the system contents at the beginning of an arbitrary slot still contains the unknowns $p_i(j)$. In the next section, it will become clear how these quantities can be computed.

4. CALCULATION OF THE $p_i(j)$'s.

We first prove the following lemma :

LEMMA :

$|\lambda_i(z)|<1$ for all $\{z : |z|=1 \text{ and } z\neq 1\}$, $1 \leq i \leq 2$.

PROOF :

Since $[u_{ij}(z)]$ is a matrix containing right column eigenvectors of $[a_{ij}(z)]$, we find that

$$\sum_{j=1}^{2} a_k(j)\, f_j(z)\, u_{ji}(z) = \lambda_i(z)\, u_{ki}(z) \quad , \quad 1 \leq i,k \leq 2 \quad .$$

For a given value of i and z, if we now choose k such that $u_{ki}(z)$ satisfies $|u_{ki}(z)| \geq |u_{ji}(z)|$, for $1 \leq j \leq 2$ (i.e., $u_{ki}(z)$ is the element of $\{u_{1i}(z), u_{2i}(z)\}$, $1 \leq i \leq 2$, with the largest modulus), then, the following inequality holds :

$$|\lambda_i(z)| \leq \sum_{j=1}^{2} a_k(j)\, |f_j(z)| \left|\frac{u_{ji}(z)}{u_{ki}(z)}\right| \leq \sum_{j=1}^{2} a_k(j)\, |f_j(z)| < \sum_{j=1}^{2} a_k(j) = 1 \ , \quad 1 \leq i \leq 2 \quad .$$

for all $\{z : |z|=1 \text{ and } z\neq 1\}$, where we have used the property that since $f_1(z)$ is a probability generating function and not a constant, $|f_1(z)| < 1$ if $\{z : |z|=1 \text{ and } z\neq 1\}$. ∎

We now consider the case $z=1$. Since the underlying Markov chain of the arrival process is irreducible and aperiodic, the transition matrix $[a_i(j)]$ has one eigenvalue equal to 1, (say $\lambda_1(1)=1$, in accordance with (13.a)), and the other eigenvalue satisfies $|\lambda_2(1)|<1$ (see e.g. [11], pp. 13–24). Then combination of this observation with the previous lemma leads to

$$\left|\lambda_1(z)^m \lambda_2(z)^{N-m}\right| < 1 \quad \text{for all } \{z : |z|=1\} \quad ,$$

except if $z=1$ and $m=N$. Using Rouché's theorem, we thus find that the denominators in expression (17) for $V(z)$ each have c zeros inside the unit disk $\{z : |z|<1\}$ if $m\neq N$; if $m=N$, then, apart from $z=1$, we find $c-1$ zeros inside the unit disk. A similar reasoning as above can then be made to prove that each of the equations

$$z - \left(\lambda_1(z)^m \lambda_2(z)^{N-m}\right)^{1/c} \exp\{2\pi k\iota/c\} = 0 \quad , 0 \le k \le c-1 \quad , \tag{18}$$

(where ι is the imaginary unit, and $(\rho.\exp\{\iota\theta\})^{1/c} \triangleq \rho^{1/c}\exp\{\theta\iota/c\}$ for all $\rho>0$, $0 \le \theta < 2\pi$) has exactly one solution inside the unit disk $\{z : |z|<1\}$, if $m \neq N$ or $k \neq 0$; $z=1$ is a solution of (18) if $m=N$ and $k=0$. The calculation of these roots using a Newton–Rhapson scheme, or, even simpler, a numerical substitution algorithm, has shown to converge quite fast, and is not memory nor CPU–time consuming.

Finally, if we denote by z_{mk} the root of equation (18) found for each possible set of values for $0 \le m \le N$ and $0 \le k \le c-1$, then, since V(z) is analytic inside the unit disk, it is clear that these quantities must also be zeros of the numerator of the corresponding term in expression (17) for V(z) (this is already the case for z=1, the root found for m=N and k=0). This leads to the following set of equations for the unknowns $p_i(j)$:

$$\sum_{j=0}^{c-1} (z_{mk}^{\ c} - z_{mk}^{\ j}) \sum_{i=0}^{N} v_{im}(z_{mk})\, p_i(j) = 0 \quad , \tag{19}$$

that must be satisfied for all possible sets of values for m and k, except if m=N and k=0. Furthermore, setting z=1 in (17), we then find the normalization condition

$$\sum_{j=1}^{c-1} (c-j)\, \mathrm{Prob}[v=j] = c(1-\sigma) \quad , \tag{20}$$

where we have used the property that $\lambda_1(1)=\sigma c/N$ (where primes denote derivatives with respect to the argument), and $v_{im}(1)=1$ if m=N. Equations (19) and (20) form a set of linear equations for the unknowns $p_i(j)$. It can be verified that in general, the number of unknowns, as well as the number of equations in (19) and (20) are equal to (N+1)c, and it has been observed that this set of equations has a unique solution

5. PERFORMANCE MEASURES OF THE SYSTEM CONTENTS.

In this section, we will derive closed–form expressions for two important performance measures of the system contents, being the mean system contents and the tail distribution of the system contents.

5.1. Mean Value.

A straightforward method for calculating the mean system contents, is, of course, taking the first order derivative of expression (17) for V(z) in z=1. However, this will lead to complicated equations, that will not always be easy to evaluate. Therefore, we prefer a different approach. Let us start from equations (8.a,b), and consider values of (x_1,x_2,z) for which

$$x_i = u_{i1}(z) \quad , \quad 1 \le i \le 2 \quad . \tag{21}$$

Choosing these values of (x_1,x_2) for a given value of z, then, since $A_i(x_1(z),x_2(z),z) = \lambda_1(z)u_{i1}(z)$, $1 \le i \le 2$, equation (8.a) can be transformed into

$$P(u_{11}(z),u_{21}(z),z) = \frac{\lambda_1(z)^N\, Q(u_{11}(z),u_{21}(z),z)}{z^c - \lambda_1(z)^N} \quad . \tag{22}$$

Note that $u_{11}(1) = u_{21}(1) = 1$, due to expression (13.c). We can obtain a closed–form formula for the mean system contents by taking the first order derivative of (22) in z=1. Applying de l'Hospitals rule and using the property that $u_{j1}(1)=1$, $1 \leq j \leq 2$, we then find

$$E[v] = V'(1) = \frac{Q_1(1) + Q_2(1) - c(c-1) + N\lambda_1''(1) + N(N-1)\lambda_1'(1)^2}{2(c - N\lambda_1'(1))} + \sigma c \quad , \qquad (23.a)$$

where $P(x_1,x_2,1)$ is given by equation (2), and where we have used the relations

$$u_{11}'(1) = \frac{c}{N}\frac{\alpha+\beta-1}{2-\alpha-\beta}(1-\sigma) \quad , \quad u_{21}'(1) = -\frac{c}{N}\frac{\alpha+\beta-1}{2-\alpha-\beta}\sigma \qquad (23.b)$$

$$Q_1(1) = \sum_{j=0}^{c-1} (c(c-1)-j(j-1)) \sum_{i=0}^{N} p_i(j) \qquad (23.c)$$

$$Q_2(1) = 2 \sum_{j=0}^{c-1} (c-j) \sum_{i=0}^{N} (i.u_{11}'(1) + (N-i)u_{21}'(1))\, p_i(j) \ . \qquad (23.d)$$

The quantity $\lambda_1'(1)$ in (23.a) is equal to $\sigma c/N$, while $\lambda_1''(1)$ can be obtained by taking the second order derivative of expression (13.a) for $\lambda_1(z)$, for z=1. This yields

$$\lambda_1''(1) = 2\left[\frac{c}{N}\right]^2 \frac{\alpha+\beta-1}{2-\alpha-\beta}\sigma(1-\sigma) \ . \qquad (23.e)$$

Using expressions (23.a–e), the calculation of the mean system contents is straightforward, once the boundary probabilities $p_i(j)$ have been calculated.

5.2. The Tail Distribution of the System Contents.

Another important performance measure of interest is the tail distribution of the system contents, that can be used as a criterion for dimensioning purposes. It has been observed in many cases (see e.g. [2], [3], [10] and [12]) that approximating the tail distribution of the system contents by a geometric form is quite accurate, if the zeros of the denominator of V(z) have a different modulus, which is, in general, the case. For the model considered here, this observation will be confirmed in Section 6 through comparison with simulation results, for some values of the parameters.

Approximating the distribution of the random variable v by a geometric form corresponds to approximating V(z) by

$$V(z) \cong \frac{C}{z - z_0} = -z_0^{-1} C \sum_{n=0}^{\infty} (z/z_0)^n \ ,$$

where we are particularly interested in sufficiently large values of n. In the above expression for V(z), z_0 is the pole of V(z) with the smallest modulus (which, of course, lies outside the unit disk), and the terms that have been neglected correspond to the poles of V(z) with modulus larger than z_0. It is obvious that these terms converge faster to zero than the one in the above formula. The pole z_0 is a real and positive quantity, and it has been observed that it is the solution of $z-\lambda_1(z)^{N/c} = 0$ outside the unit disk with the smallest modulus. The above approximate expression for V(z) then leads to the following approximation for the tail probabilities of the system contents :

$$\text{Prob}[v=n] \cong -C\, z_0^{-n-1} \quad , \quad n \geq 0 \ , \tag{24.a}$$

and this approximation improves for increasing values of n

Furthermore, using the residue theorem, the quantity C in (24.a) can be shown to be equal to

$$C = \lim_{z \to z_0} (z-z_0)V(z) \quad .$$

Using expression (17) for V(z), this becomes

$$C = \frac{\lambda_1(z_0)^N}{c z_0^{c-1} - N\lambda_1'(z_0)\,\lambda_1(z_0)^{N-1}} \sum_{j=0}^{c-1} (z_0^c - z_0^j) \sum_{i=0}^{N} u_{11}(z_0)^i\, u_{21}(z_0)^{N-i} p_i(j) \ . \tag{24.b}$$

The quantities $\lambda_1(z_0)$, $\lambda_1'(z_0)$, $u_{11}(z_0)$ and $u_{21}(z_0)$ can be easily calculated from (13.a,c).

6. NUMERICAL RESULTS.

Let us define the parameter K as the ratio of the mean length of an active period (or a passive period) in the correlated–arrivals case versus the uncorrelated–arrivals case, i.e., $K = (1-\sigma)/(1-\alpha)$, or, equivalently, $K = \sigma/(1-\beta)$. It is clear that the couple (K,σ), as well as the couple (α,β), fully describes the first–order Markov cell arrival process on each of the input links : σ, as defined in (3), denotes the fraction of active slots on an input link and thus is a measure for the relative lengths of the active and

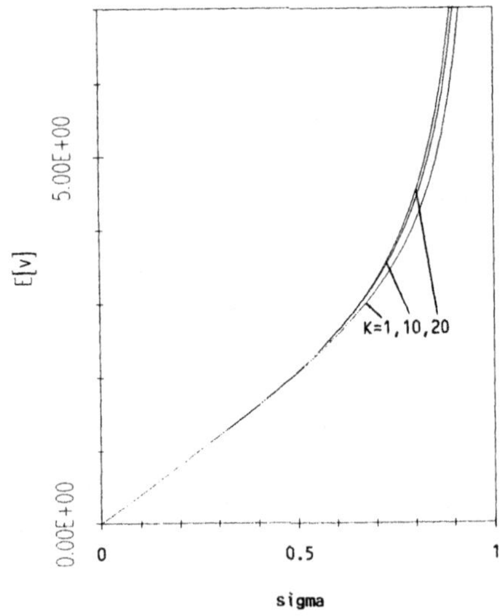

Fig. 1 : Mean system contents versus σ, for N=16, c=4, K=1,10,20.

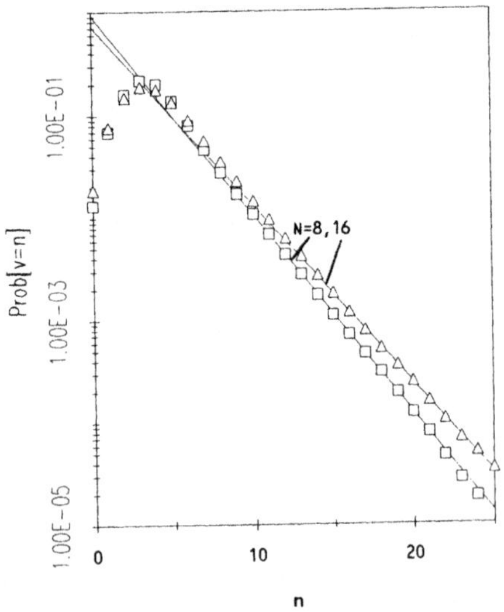

Fig. 2 : Prob[v=n] versus n, for N=8,16, c=4, σ=0.8, K=5, analytic (solid line) versus simulation (dots).

passive periods, while K characterizes the absolute lengths of these periods and thus is a measure for the degree of correlation in the arrival process (K=1 characterizes uncorrelated arrivals).

First of all, in Fig. 1, the mean system contents is plotted versus σ, for N=16, c=4, and K=1,10,20. It is obvious that higher values of K (i.e., more "burstiness" in the cell arrival process) lead to higher values of the mean system contents, although the differences observed here are not extremely large.

Furthermore, in Fig. 2, the comparison has been made between simulation results (dots) and the values predicted by equations (24.a,b) (solid lines) of Prob[v=n] versus n, for N=8,16, c=4, σ=0.8 and K=5, and these results match extremely well, especially in the low probability area. It may therefore be concluded that for the model for the ATM switch currently under study, approximating the tail distribution of the system contents by a geometric form, is an extremely good approximation.

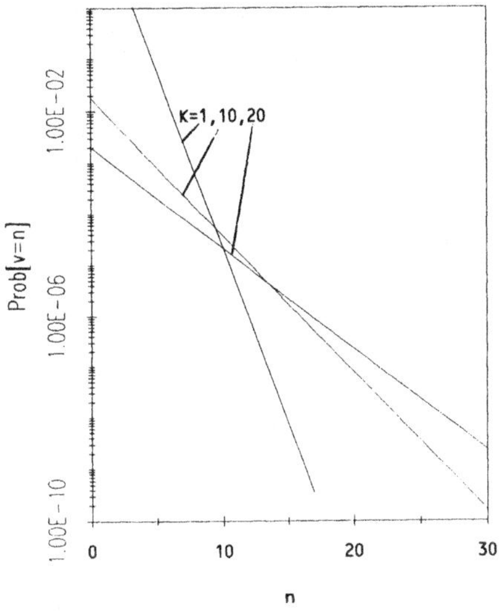

Fig. 3 : Prob[v=n] versus n, for N=16, c=4, σ=0.5, K=1,10,20.

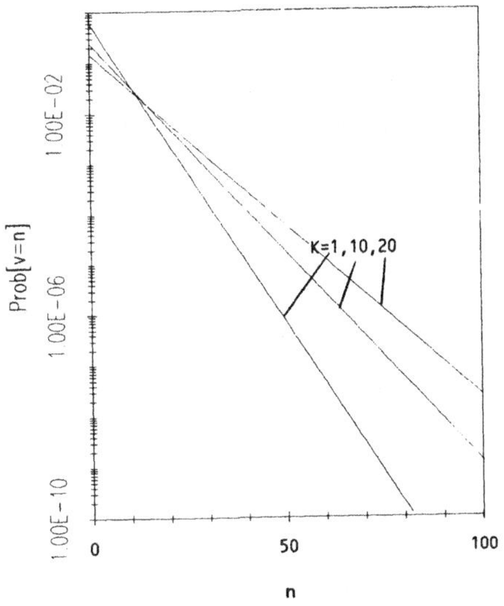

Fig. 4 : Prob[v=n] versus n, for N=16, c=4, σ=0.9, K=1,10,20.

Finally, in Figs. 3 and 4, the geometric approximation for the tail distribution of the system contents is plotted versus the buffer size, for N=16, c=4, K=1,10,20 and σ=0.5 and 0.9 respectively. As could be expected, the buffer requirements considerably increase for increasing values of K (i.e., increasing correlation in the cell arrival process), and fixed values of N, c and σ. Note that in spite of the uniform and independent routing of arriving cells to their destinations, the impact of the correlation in the cell arrival process on the buffer behavior cannot be neglected.

7. CONCLUSIONS.

In this paper, we have presented a method for analyzing a multiserver discrete–time queue with correlated arrivals that can be used to model an ATM switching element with bursty sources and independent uniform routing, and multiple transmission lines assigned per destination. The model has been analyzed according to a generating–functions approach, and this has resulted in explicit expressions for the probability generating function, mean, and a geometric–tail approximation for the probability mass function, of the system contents. It has been observed that, although the routing of the cells to their destinations is assumed to be uniform and independent, the correlation in the cell arrival process has a far from negligible influence on the buffer requirements of the switching element.

REFERENCES.

[1] M.J. Karol, M.G. Hluchyj and S.P. Morgan, "Input versus output queueing on a space–division packet switch", *IEEE Trans. Comm.*, vol. COM–35 (1987), pp. 1347–1356.

[2] E. Desmet, B. Steyaert, H. Bruneel and G.H. Petit, "Tail distributions of queue length and delay in discrete–time multiserver queueing models, applicable in ATM networks", *Proceedings of ITC 13, Copenhagen, June 1991,* vol. *Queueing, Performance and Control in ATM*, pp. 1–6.

[3] H. Bruneel and B. Steyaert, "Buffer requirements for ATM switches with multiserver output queues", *Electron. Lett.*, vol. 27 (1991), pp. 671–672.

[4] A.M. Viterbi, "Approximate analysis of time–synchronous packet networks", *IEEE J. Sel. Areas in Comm.*, vol. SAC–4 (1986), pp. 879–890.

[5] H. Bruneel and I. Bruyland, "Performance study of statistical multiplexing in case of slow message generation", *Proceedings of ICC '89, Boston, June 1989,* pp. 951–955.

[6] H. Bruneel, "Queueing behavior of statistical multiplexers with correlated inputs", *IEEE Trans. Comm.*, vol. COM–36 (1988), pp. 1339–1341.

[7] Y. Xiong and H. Bruneel, "Performance of statistical multiplexers with finite number of inputs and train arrivals", *Proceedings of INFOCOM '92, Firenze, May 1992.*

[8] San–Qi Li, "A general solution technique for discrete queueing analysis of multimedia traffic on ATM", *IEEE Trans. Comm.*, vol. COM–398 (1991), pp. 1115–1132.

[9] G.H. Petit, A. Buchheister, A. Guerrero, P. Parmentier, "Performance evaluation methods applicable to an ATM multi–path self–routing switching network", *Proceedings of ITC 13, Copenhagen, June 1991,* vol. *Teletraffic and Datatraffic in a period of change*, pp. 917–922.

[10] B. Steyaert and H. Bruneel, "Analytical study of the buffer contents in an ATM switch with Markovian arrivals", *Proceedings of ISTN–92, Beijing, September 1992*, pp. 9–12.

[11] J.J. Hunter, "Mathematical Techniques of Applied Probability, Volume 2, Discrete Time Models : Techniques and Applications", *Academic Press, 1983.*

[12] C.M. Woodside and E.D.S. Ho, "Engineering calculation of overflow probabilities in buffers with Markov interrupted service", *IEEE Trans. Comm.*, vol. COM–35 (1987), pp. 1272–1277.

Integrated Broadband Communication Networks
and Services (C-18) / V.B. Iversen (Editor)
Elsevier Science B.V (North-Holland)

A Refined Approximation of a Finite Capacity Polling System under ATM Bursty Arrivals

Y. Frank Jou[a], Arne A. Nilsson[a], and Fuyung Lai[b]

[a] Center for Communications and Signal Processing, Department of Electrical and Computer Engineering, North Carolina State University, Raleigh, N.C. 27695-7914

[b] IBM, V57/B660 P.O. Box 12195, Research Triangle Park, N.C. 27709, U.S.A.

Abstract

This paper is concerned with both the mean delay and the probability of cell loss that bursty arrivals incur in an ATM switching system which can be modeled as a finite capacity polling system with nonexhaustive cyclic service. The arrival process to each input port of the system is assumed to be bursty and is modeled by an Interrupted Bernoulli Process (IBP). A practical polling system with finite capacity, as the one we deal with here, does not lend itself to an exact solution. In this paper, we introduce an effective approach to provide an analytical approximation. This approach is validated extensively by comparing it against simulation results under different configurations. It is shown that both the mean delays and the cell loss probabilities obtained from this analysis provide highly accurate estimates.

Keyword Codes: B.4.4; C.2.1; C.4
Keywords: Performance analysis; Network Architecture; Performance of Systems

1 INTRODUCTION

The emerging needs for high speed communications and the promise of the technologies to support these services in an integrated fashion have generated a lot of interest in research, development, and standardization of broadband integrated networks. Among the transport and switching techniques for B-ISDN, the Asynchronous Transfer Mode (ATM) technique has shown to be the most promising solution. In this paper, we consider the mean delay and cell loss probability that bursty arrivals incur in an ATM switch architecture as shown in figure 1 [1]. This ATM switch architecture is constructed by connecting self-routing switching modules (SRMs) in a three-stage link configuration which is called a multi-stage self-routing network (MSRN). Each stage of MSRN consists of eight self-routing switching modules. Each module is an 8×8 crossbar switch which has a finite buffer associated with each crosspoint. The cells in each buffer are transmitted in a cyclic order. In each stage of the MSRN, there is no blocking mechanism in the service given that the destination queue is full at the next stage.

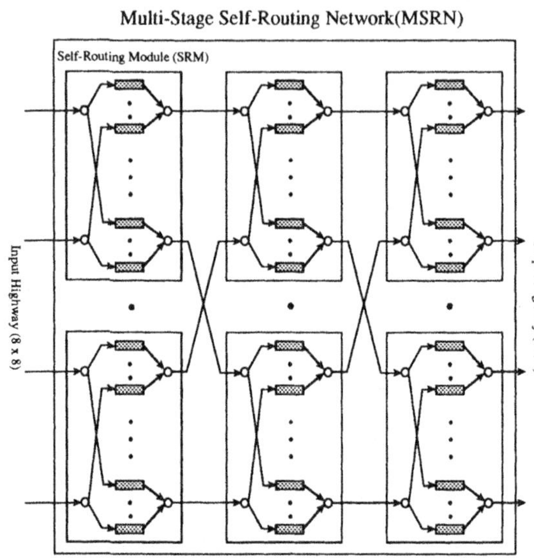

Figure 1: ATM Switching Architecture.

To study the performance of this SRM, we model it by a polling system with cyclic service. Excellent surveys of results on polling systems with extensive lists of references have been given by Takagi in [2]. In most of these investigations, the input processes are assumed to be Poisson, and the queues of the polling system are assumed to have infinite capacity. In order to include more realistic modeling elements in the class of polling systems, we consider bursty arrival processes as inputs into the polling system, which has a finite buffer capacity.

In this paper, we use a queueing model to compute the mean delay and cell loss probability that cells incur in the finite capacity polling system. This model will assume symmetric traffic load, zero switchover time, and limited-1 service. In Section 2 we describe in detail the model we propose. This model will require analysis of the queue length distribution of the polling system and a multiple urn model with uniform occupancy probability which are presented in Sections 3 and 4, respectively. In Section 4, we assume that each queue (urn) can accommodate any number of cells up to its capacity with equal probability. The cell loss probability obtained from this approach overestimates, and the mean delay underestimates the simulation result. To improve accuracy, in Section 5, we solve an $IBP/D/1/L$ with vacation queueing model to obtain its queue length distribution. By assuming the independence among the queues, we then take this queue length distribution as a basis to compute the occupancy probability of all the possible configurations. This occupancy probability enables us to refine the aggregate queue length distribution and provide accurate performance measures. Numerical results validated by computer simulations are given in Section 6. Finally, Section 7 presents our conclusions.

2 MODEL DESCRIPTION

In this section we describe the arrival process, and the proposed queueing model.

2.1 Arrival process

Since most of the traffic sources that an ATM network supports are bursty, a Poisson process may no longer be suitable for describing the network traffic. For instance, interactive data and compressed video generate cells at a near-peak rate for a very short period of time. Immediately, following a near peak rate such a source may become inactive, thus generating no cells. With this scenario, the usual approximation of arrival process by a Poisson process will fail to capture the bursty nature of input traffic and may result in a quite dramatic error in the performance estimation. Kuehn[3] has shown that the system behavior is much more sensitive to arrival processes than to service process. Therefore, we propose to use the IBP which is the generalization of the Interrupted Poisson Process in a discrete time system.

A simple IBP is governed by a Markov chain with two states, an active state and an idle state. The duration of stay in these two states are geometrically distributed. Arrivals occur in a Bernoulli fashion with parameter α when the process is in the active state. No arrivals occur if the process is in the idle state. Given that the process is in the active state (or idle state) at slot i, it will remain in the same state in the next slot $i+1$ with probability p (or q), or will change to the idle state (or active state) with probability $1-p$ (or $1-q$). During the active state, a slot contains a cell with probability α. Here we assume α equals 1 which will generate the most bursty traffic. A complete description on IBP is given in [4].

2.2 Queueing model

Based on the structure of SRM, we evaluate its performance by using a multiqueue system as shown in figure 2. Under the assumption of symmetric traffic, the arrival processes to the multiqueue system will be characterized also as IBPs with the same parameters of p and q as in the original arrival processes. However, since every original arrival process branches to eight possible outlets, the parameter α of the arrival processes to the polling system will only be one eighth.

Instead of considering the queue lengths of the multiqueue system individually, we will call the distribution of the total number of cells in the polling system as the aggregate queue length distribution. In order to obtain this aggregate queue length distribution, it is necessary to consider the blocking effect due to finite buffer space.

We model the queues in the multiqueue system as multiple urns which have the same limited capacity. Given the number of cells waiting in the system, it is assumed that the occupancy of each queue is independent from each other and the cells are uniformly distributed in any queue. With this model, we can compute the weighting of the cell occupancy configuration which could cause cell loss and then establish the transition

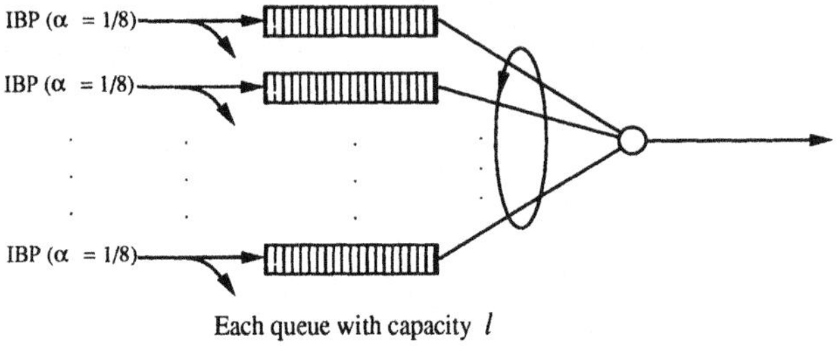

Figure 2: Multiqueue System Served by a Single Server.

matrix which allows us to compute the aggregate queue length distribution. After this distribution is obtained, the mean delay and cell loss probability will follow readily.

Notice that given R cells in the system, the occupancy of these cells in reality will more likely be evenly distributed among these queues because in general the server will visit the queues with longer queue sizes more frequently than the queues with shorter queue sizes. Therefore, given the number of cells in the polling system exceeding a single queue capacity, the occupancy configuration which has at least one full queue is less likely to occur. Hence, the assumption of uniform occupancy will give us a conservative estimate which can serve as an upper bound for cell loss probability of the polling system.

In order to provide better approximation, we refine the uniform assumption such that the probability for a queue to accommodate a certain number of cells is according to the queue length distribution of this single queue. Based upon this refined assumption, we propose a queueing model, $IBP/D/1/L$ with vacation, to obtain the queue length distribution of a particular queue in the polling system. This queue length distribution will serve as a basis for computing the weighting of all the possible configurations. The total weighting of all the cell occupancy configurations can be found by a technique originated from the closed queueing model. We use the sum of these weightings as a normalization factor to modify the weighting of the configurations which could cause cell loss. This refined weighting will be incorporated in the steady state equations to compute the aggregate queue length distribution which provides us with highly accurate performance measures.

3 AGGREGATE QUEUE LENGTH DISTRIBUTION

In our model, we assume that arrivals can only occur at the beginning of each slot and departures leave the system at the end of each slot. This arrangement is called an early arrival system according to Hunter [5]. During a slot period, one cell may arrive on each input link, and one cell may be transmitted given that the system is not empty. The state change, from active to idle or vice versa, only occurs at the slot point.

Next, the cell arrival process is analyzed for the cells arriving from all inputs in a slot time. We assume that there are N queues in the multiqueue system. The number of active input links K and the number of arrival cells M in a unit time can both vary from zero to N. If the number of active input links is k, then m cells will arrive in a unit time with probability

$$P_{M|K}(m|k) = \binom{k}{m} \alpha^m (1-\alpha)^{k-m}.$$

The state transition probability of having k' active lines in a slot given k active lines in the previous slot is given by

$$P_{K'|K}(k'|k) = \sum_{j=0}^{k} \binom{k}{j} p^j (1-p)^{k-j} \binom{N-k}{k'-j} q^{N-k-k'+j}(1-q)^{k'-j}. \quad (1)$$

In order to describe the blocking effect, we extend the concept introduced in [6] and define a conditional probability $P(m"|m', qsize)$ as

$P\{$ *m" cells accepted* | *m' cells arrived and qsize cells in the system before arrivals* $\}$.

This probability will be further described and computed by a multiple urn model which is discussed in the next section.

Now, we define a two dimensional state variable (K, Q) such that the queue length becomes Q as the result of having M' cells arrive and $M"$ cells accepted in a slot given that K input lines are active. The state probability $P_{K,Q}(k, qsize)$ can be obtained by a numerical solution of the following steady state equation:

$$P_{K,Q}(k', qsize') = \sum_{k=0}^{N} \sum_{m'=0}^{k'} \sum_{qsize=0}^{Q_m} P_{K,Q}(k, qsize) P_{K'|K}(k'|k) P_{M|K}(m'|k') P(m"|m', qsize"), \quad (2)$$

where Q_m is the total capacity of the multiqueue system, and $\sum_{k=0}^{N} \sum_{qsize=0}^{Q_m} P_{K,Q}(k, qsize) = 1$. In equation (2), $qsize"$ and $qsize'$ are given by $qsize" = max(qsize - 1, 0)$ and $qsize' = m" + qsize"$, respectively. From $P_{K,Q}(k, qsize)$, we can sum over K and find the queue length distribution $P_Q(qsize)$. Since we have assumed zero switchover time in the system, the mean output rate λ_{out} can be determined as

$$\begin{aligned} \lambda_{out} &= 1 - P_Q(0) \\ &= \lambda_{in}\,(1 - P_{loss}). \end{aligned}$$

Therefore, the cell loss probability P_{loss} is obtained as

$$P_{loss} = 1 - \frac{\lambda_{out}}{\lambda_{in}}.$$

After we compute the mean queue length L, the mean delay W can be determined by using Little's result as

$$W = \frac{L}{\lambda_{out}}.$$

4 MULTIPLE URN MODEL

In order to compute the conditional probability $P(m"|m', qsize)$ defined in the last section, we use a multiple urn model to find $T_{n,r,l}$, the total number of ways to place r indistinguishable balls into n distinguishable boxes, given that the capacity of each box is limited to l. We have

$$T_{n,r,l} = \sum_{i=0}^{n}(-1)^i \binom{n}{i} \binom{n+r-i(l+1)-1}{n-1}.$$

Also, define $C_{n,r,l}$ as the number of ways of having at least one full box for this multiple urn model. If the total number of balls r is less than l, $C_{n,r,l}$ equals zero. For $kl \leq r < (k+1)l$, it can be shown that

$$C_{n,r,l} = \sum_{i=1}^{k} \binom{n}{i} \left(T_{n-i,r-i\times l,l} - C_{n-i,r-i\times l,l}\right), \tag{3}$$

where $C_{n,r,l}$ is obtained recursively. Notice that the expression in the summation in equation (3) denotes the number of ways to *have exactly i full boxes*. For further derivation, the readers are referred to [7].

According to the definition of $P(m"|m', qsize)$, we now obtain the conditional probability as

$$P(m"|m', r) = \frac{\sum_{i=0}^{\lfloor \frac{r}{l} \rfloor} \frac{\binom{N-i}{m"}\binom{i}{m'-m"}}{\binom{N}{m'}} \binom{N}{i} \left(T_{N-i,r-i\times l,l} - C_{N-i,r-i\times l,l}\right)}{T_{N,r,l}}.$$

5 REFINEMENT OF THE APPROXIMATION

From the computations in Sections 3 and 4, we find out that the cell loss probabilities obtained from the analytical model overestimate the results obtained from simulation. On the other hand, the mean delays tend to underestimate. These phenomena are due to the assumption of equal probability among the configurations of occupancy in the multiple urn model. Based upon this observation, we solve an $IBP/D/1/L$ with vacation queueing model and formulate the derivation to compute the normalization factor in order to refine the approximation as we discussed in Section 2.2.

5.1 $IBP/D/1/L$ with vacation

If we focus on a specific queue in the polling system, the queueing model of this particular queue can be identified as an $IBP/D/1/L$ with vacation where L denotes the capacity of this queue. To solve this queueing model, we need the aggregate queue length distribution of the polling system in order to characterize the vacation time distribution of this single queue. The aggregate queue length distribution of the polling system was obtained by using the algorithm implemented in Sections 3 and 4. The vacation time, or to be more precise, the cycle time distribution defined by $c(k)$ which refers to the time between two immediate visits of the server can be computed by the equations in Section 4 as follows.

Define $D_{i,j}$ as the number of configurations which have i empty queues given that there are j cells in the system. Or equivalently, we can define $D_{i,j}$ as the number of configurations which have i queues full of holes given that there are $N \times L - j$ holes in the system. Therefore, $D_{i,j}$ can be expressed as

$$D_{i,j} = \binom{N}{i} \left(T_{N-i,(N\times L-j)-i\times L,L} - C_{N-i,(N\times L-j)-i\times L,L}\right).$$

The cycle time distribution, $c(k)$, can be obtained as

$$c(k) = \sum_{j=1}^{N\times L} \frac{D_{N-k,j} P_Q(j)}{T_{N,j,L}} + 1_{\{k=1\}} P_Q(0),$$

where $1 \le k \le N$.

We now proceed to solve the queueing model of $IBP/D/1/L$ with vacation. The basic techniques used in this derivation are similar to those used in [8]. The steady state distribution of queue length is computed by the embedded Markov chain approach. Cells arrive at the system in accordance with an IBP with parameters α, p, and q. There is a single server whose service time is deterministic of one slot. If an arriving cell finds exactly L cells in the system, then this cell leaves the system without being served.

The system will be examined at time epochs $\{t_0, t_1, \cdots\}$ of service completion or vacation termination. The state space of the system is $\{R_i, \xi_i, \eta_i\}$, defined as follows. R_i is the number of cells in the system at the embedded point t_i. If the point t_i is a vacation termination instant, i.e., the point when the server starts to scan this particular queue, then $\xi_i = 0$; otherwise $\xi_i = 1$. The state of the arrival process at the embedded point t_i is denoted by η_i, where η_i is 1 if the IBP is in its active state or 0 if it is in the idle state. The state transitions occur at cell departure instants or vacation termination instants. Let $X_{r,\eta}$ and $Y_{r,\eta}$ represent the limiting probability distributions defined as follows

$$\begin{aligned} X_{r,\eta} &\equiv \lim_{i\to\infty} P_r\{R_i = r, \xi_i = 1, \eta_i\} && 0 \le r \le L-1, \\ Y_{r,\eta} &\equiv \lim_{i\to\infty} P_r\{R_i = r, \xi_i = 0, \eta_i\} && 0 \le r \le L. \end{aligned}$$

Due to the deterministic service and early arrival assumption, we have $X_{r,\eta} = Y_{r+1,\eta}$, $0 \le r \le L-1$. Therefore, we can set up the steady state equations by focusing only on $Y_{r,\eta}$.

Define $g(j,n|i,k)$ as the probability of having n slots in state 1 among k transition slots with the arrival process starting at state i and ending at state j between two immediate vacation termination instants. This probability can be obtained recursively by the following equations:

$$\begin{aligned} g(0,n|0,k) &= q\, g(0,n|0,k-1) + (1-p)\, g(1,n|0,k-1), \\ g(1,n|0,k) &= (1-q)\, g(0,n-1|0,k-1) + p\, g(1,n-1|0,k-1), \\ g(0,n|1,k) &= q\, g(0,n|1,k-1) + (1-p)\, g(1,n|1,k-1), \\ g(1,n|1,k) &= (1-q)\, g(0,n-1|1,k-1) + p\, g(1,n-1|1,k-1), \end{aligned}$$

where the initial conditions are given by

$$\begin{aligned} g(0,0|0,1) &= q, & g(0,1|0,1) &= 0, \\ g(1,0|0,1) &= 0, & g(1,1|0,1) &= 1-q, \\ g(0,0|1,1) &= 1-p, & g(0,1|1,1) &= 0, \\ g(1,0|1,1) &= 0, & g(1,1|1,1) &= p. \end{aligned}$$

We further define $G(j,r|i,k)$ as the probability of having r cells arriving in k transition slots with the arrival process starting at state i and ending at state j between two immediate vacation termination instants. The probability $G(j,r|i,k)$ can be expressed in terms of $g(j,n|i,k)$ as

$$G(j,r|i,k) = \sum_{n=0}^{N} \binom{n}{r} \alpha^r (1-\alpha)^{n-r} g(j,n|i,k).$$

When the system is under steady state, the limiting probability distribution $Y_{r,\eta}, 0 \le r \le L-1$ satisfies the following equations:

$$Y_{r,0} = Y_{0,0}H(0,r|0) + Y_{0,1}H(0,r|1) + \sum_{k=0}^{r} (Y_{k+1,0}H(0,r-k|0) + Y_{k+1,1}H(0,r-k|1)), \quad (4)$$

$$Y_{r,1} = Y_{0,0}H(1,r|0) + Y_{0,1}H(1,r|1) + \sum_{k=0}^{r} (Y_{k+1,0}H(1,r-k|0) + Y_{k+1,1}H(1,r-k|1)), \quad (5)$$

and

$$\begin{aligned} Y_{L,0} &= Y_{0,0} \sum_{r=L}^{N} H(1,r|0) + Y_{0,1} \sum_{r=L}^{N} H(1,r|1) \\ &+ \sum_{k=0}^{L-1} \left[Y_{k+1,0} \sum_{r=L}^{N} H(1,r-k|0) + Y_{k+1,1} \sum_{r=L}^{N} H(1,r-k|1) \right], \quad (6) \\ Y_{L,1} &= Y_{0,0} \sum_{r=L}^{N} H(1,r|0) + Y_{0,1} \sum_{r=L}^{N} H(1,r|1) \\ &+ \sum_{k=0}^{L-1} \left[Y_{k+1,0} \sum_{r=L}^{N} H(1,r-k|0) + Y_{k+1,1} \sum_{r=L}^{N} H(1,r-k|1) \right], \quad (7) \end{aligned}$$

where the transition probability $H(j,r|i)$ is given by

$$H(j,r|i) = \sum_{k=1}^{N} G(j,r|i,k)c(k),$$

which represents the probability of having r arrivals between two vacation termination instants. The arrival process at these two embedded points is in the state i, and j, respectively. Combining the steady state equations (4) and (5) together with the normalization equation

$$\sum_{r=0}^{L} \sum_{\eta=0}^{1} (X_{r,\eta} + Y_{r,\eta}) = 1,$$

we can easily solve $X_{r,\eta}$ and $Y_{r,\eta}$. (For simplity, we assume $X_{L,\eta} = 0$.) The queue length distribution P_{IBP_Q} is followed readily as

$$P_{IBP\_Q}(r) = \sum_{\eta=0}^{1}(X_{r,\eta} + Y_{r,\eta}), \qquad 0 \le r \le L.$$

5.2 Computing the normalization factor

We recognize that the discrepancy of the results obtained from Sections 3 and 4 is due to the assumption that each position in a queue has the same probability to be occupied. In other words, the queue length distribution under this assumption is $1/L$ uniformly. This is of course not the case in reality. In order to refine this assumption, we use the queue length distribution $P_{IBP\_Q}(r)$ found in Section 5.1 as the probability for a particular queue to accommodate r cells. Given R cells in the polling system, the stationary distribution of the system state $\underline{\mathbf{r}} = (r_1, r_2, \cdots, r_N)$ is equal to

$$\begin{aligned} P(\underline{\mathbf{r}}) &= P(r_1, r_2, \cdots, r_N) \\ &= \frac{1}{G}\prod_{i=1}^{N} P_i(r_i), \qquad \sum_{i=1}^{N} r_i = R \end{aligned} \tag{8}$$

where the states of the various queues are assumed to be independent. $P_i(r_i)$ in equation (8) is given by $P_{IBP\_Q}(r)$. This setup is exactly the same as the case in the closed queueing network except that we have finite capacity in each queue, i.e. $r_i \le L$, rather than $r_i \le R$ in the general cases. Following the same spirit, we extend the derivation given in [9] to compute the normalization factor $G_N(R)$ recursively. In particular, we have

$$\begin{aligned} G_1(R) &= P_1(R) && 0 \le R \le L, \\ G_2(R) &= \sum_{r_2=max(0,R-L)}^{min(L,R)} P_2(r_2)G_1(R-r_2) && 0 \le R \le 2L, \\ &\vdots \\ G_N(R) &= \sum_{r_N=max[0,R-(N-1)\times L]}^{min(L,R)} P_N(r_N)G_{N-1}(R-r_N) && 0 \le R \le N \times L. \end{aligned}$$

This recurrence relation allows $G_N(R)$ to be computed in $O(NR^2)$ steps.

The above equations are subject to the constraint of having L spaces in each queue. In order to refine the conditional blocking probability in Section 3, we need another set of normalization factor $\overline{G_N}(R)$ which gives us the total weighting of the configurations that none of the N queues is full. $\overline{G_N}(R)$ follows the same derivation but is subject to the constraint of $L-1$ buffer spaces. When $R > N \times (L-1)$, we simply let $\overline{G_N}(R)$ be zero.

Define $P_{full}(K, R)$ as the probability of having R cells in the polling system and K out of N queues full. Therefore, we have

$$P_{full}(0, R) = \frac{\overline{G_N}(R)}{G_N(R)},$$

$$P_{full}(K,R) = \frac{\binom{N}{K} G_K(K \times L)\overline{G_{N-K}}(R - K \times L)}{G_N(R)}, \qquad 1 \le K \le N.$$

The refined conditional blocking probability $P(m"|m',R)$ can be expressed as

$$P(m"|m',R) = \sum_{K=0}^{\lfloor \frac{R}{L} \rfloor} \frac{\binom{K}{m'-m"}\binom{N-K}{m"} P_{full}(K,R)}{\binom{N}{m'}}.$$

This blocking probability can be inserted into equation (2) in Section 3 to obtain the desired performance measures.

6 NUMERICAL RESULTS

In this section, we examine several configurations where the presented approximations are compared against the simulation results. The performance measures, mean delay, and cell loss probability are affected by the buffer capacity of the multiqueue system and the burstiness of the arrival processes. In terms of queue capacity, we present three cases where the buffer sizes are 4, 8, and 16, respectively.

In order to show the effect of the arrival burstiness, we vary the squared coefficient of variation (C^2) of the arrival processes from 1, to 20, to 200. These three kinds of burstiness represent three typical cases. When C^2 equals 1, we can regard this arrival process as being smooth. The burstiness of voice is represented by the case where C^2 equals 20. We use $C^2 = 200$ for the burstiness of data traffic.

Figures 3, 4, and 5 show the mean delay times that the arrivals incur under different burstinesses when the queue capacities are 4, 8, and 16, respectively. The worst case is where C^2 equals 200 and the arrival rate is 0.9. In this worst case, the relative errors between the analytical results and the simulations are 3.2%, 5.0%, and 5.1% when the queue capacities are 4, 8, and 16, respectively. It is clearly shown that the analysis follows the simulation closely. Comparing these three figures, we can see that for the larger queue capacity the bursty effect becomes more obvious.

The impact of the burstiness toward the cell loss probability can be seen in figures 6, 7, and 8 where the buffer capacities vary from 4 to 16. From these figures, we see that the analysis traces the simulation nicely when the queue capacities are 4 and 8. With buffer capacity of 16, the analytic results could be twice as high as the simulations. From our experiments, the larger the queue capacity is, the higher the discrepancy. However, a benefit of this analysis is that it overestimates the cell loss probability and provides us with a conservative approximation.

7 CONCLUSION

A realistic polling system with finite capacity does not lend itself to an exact analysis. In this paper, we presented an effective approach to provide an analytical approximation.

As for the arrival processes, we also take into account the effect of bursty arrivals which is an essential feature in the ATM environment.

It is shown that the analytical model works very well with a wide range in both the burstiness and the traffic load of the arrival process as well as the queue capacity of the polling system. The analysis is also computationally effective. The major portion of the computation time of this analysis is devoted to solving the steady state equations in order to obtain the aggregate queue length distribution. It is our experience that the speedup of the analysis over simulation on average is more than two orders of magnitude.

The strength of this analysis is that it is able to provide accurate performance measures in a short period of time. Therefore, it can be applied to decide the system specifications in order to meet the performance requirements under ATM environment without running the simulations.

REFERENCES

[1] K. Hajikano, K. Murakami, E. Iwabuchi, O. Isono, and T. Kobayashi, "Asynchronous transfer mode switching architecture for broadband ISDN - multistage self-routing switching," *IEEE International Conference on Communications*, vol. 2, pp. 911–915, 1988.

[2] H. Takagi, "Queueing analysis of polling models: an update," *Stochastic Analysis of Computer and Communication Systems*, pp. 267–318, H. Takagi (editor), Elsevier Science Publishers B.V. (North–Holland) Amsterdam, 1990.

[3] P. J. Kuehn, "Multiqueue systems with nonexhaustive cyclic service," *B.S.T.J.*, vol. 58, no. 3, pp. 671–698, 1979.

[4] A. A. Nilsson, F.-Y. Lai, and H. G. Perros, "An approximate analysis of a bufferless N × N synchronous Clos ATM switch," *Proceedings, Canadian Conference on Electrical and Computer Engineering*, pp. 39.1.1–39.1.4, 1990.

[5] J. J. Hunter, *Mathematical Techniques of Applied Probability, Vol. 2, Ch. 9.* New York, NY: Academic Press, Inc., 1983.

[6] A. Ganz and I. Chlamtac, "Tractable analytical models of demand assignment protocols in networks with arbitrary buffer capacity," *IEEE Transactions on Communications*, vol. 40, no. 5, pp. 926–939, 1992.

[7] Y. F. Jou A. A. Nilsson and F.-Y. Lai, " Approximate Analysis of an ATM Switching System with Bursty Arrivals and Finite Capacity," *Proceedings of IFIP Workshop on Modelling and Performance Evaluation of ATM Technology*, Elsevier Science Publishers B.V. (North–Holland) Amsterdam, 1993.

[8] Tony T. Lee, "M/G/1/N queue with vacation time and limited service discipline," *Performance Evaluation*, vol. 9, pp. 181–190, 1989.

[9] E. Gelenbe and I. Mitrani, *Analysis and Synthesis of Computer Systems, Ch. 3.* New York, NY: Academic Press, Inc., 1980.

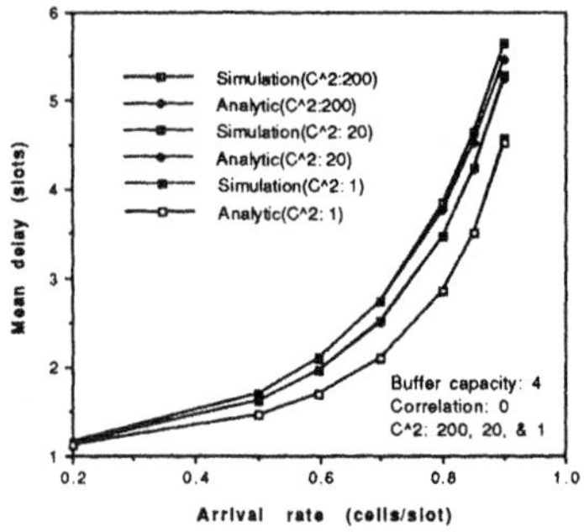

Figure 3: Mean delays incurred in the polling system when queue capacity = 4.

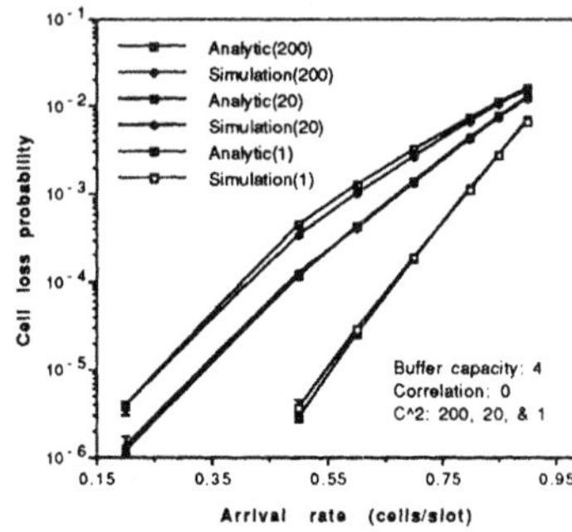

Figure 6: Cell loss probabilities incurred in the polling system when queue capacity = 4.

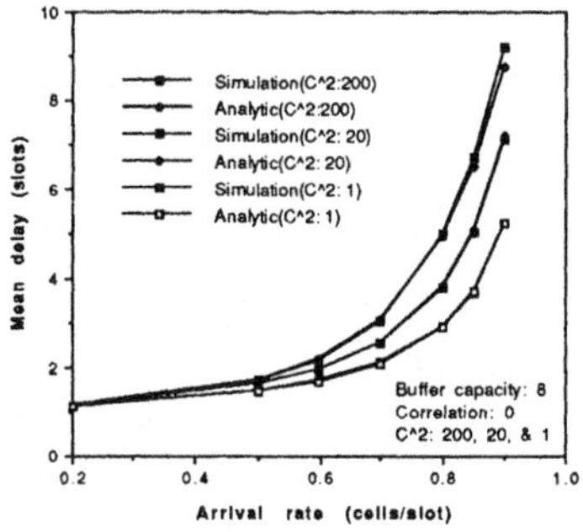

Figure 4: Mean delays incurred in the polling system when capacity = 8.

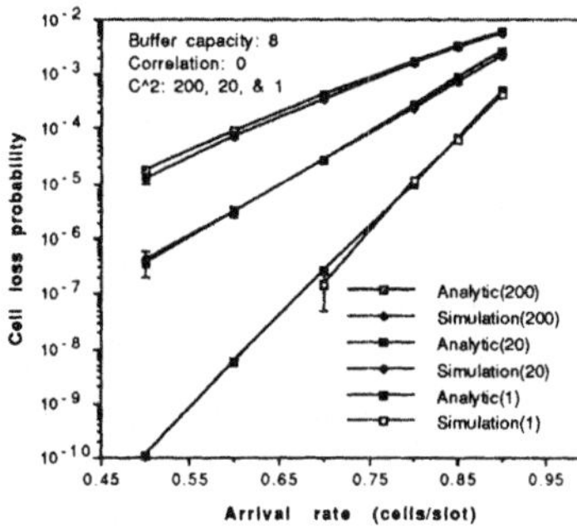

Figure 7: Cell loss probabilities incurred in the polling system when capacity = 8.

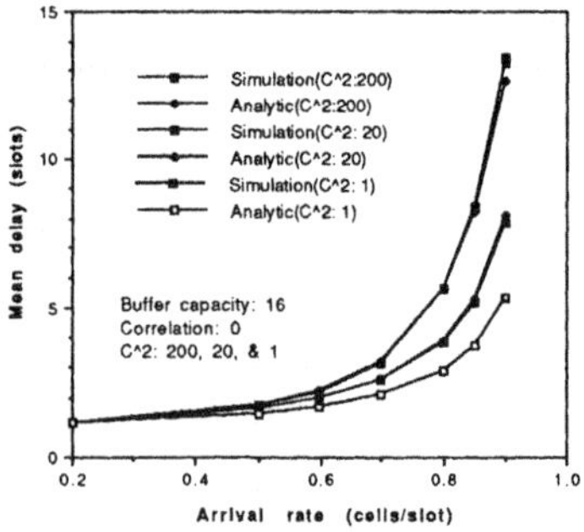

Figure 5: Mean delays incurred in the polling system when queue capacity = 16.

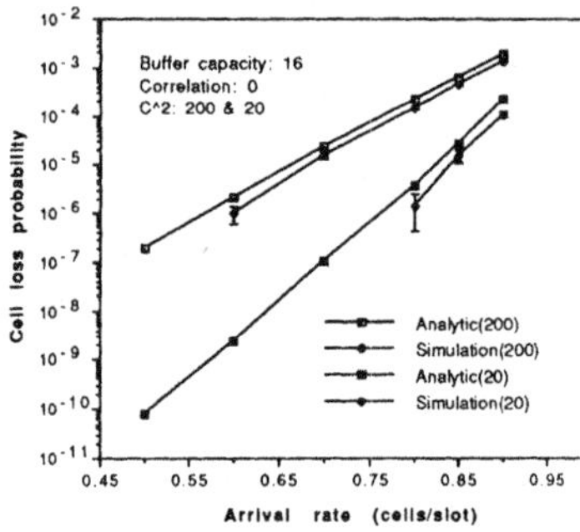

Figure 8: Cell loss probabilities incurred in the polling system when capacity = 16.

Integrated Broadband Communication Networks
and Services (C-18) / V.B. Iversen (Editor)
Elsevier Science B.V (North-Holland)

AN INFORMATION MODEL FOR THE QOS MONITORING OF AN ATM SWITCH:
From Analysis to Implementation

Spiridon ARSENIS - Noëmie SIMONI - Simon ZNATY

TELECOM Paris, Dept Réseaux
46 rue Barrault, 75634 Paris Cedex 13, FRANCE
E-mail: arsenis@res.enst.fr

Abstract

The information model is the conceptual support of every application, including network supervision and management. To obtain a unified view of the network resources, considering the complexity and heterogeneity of networks, it is important to base the management on a generic and powerful model. This paper presents a generic information model of network elements applied to the administration of an ATM switch. This hierarchical and modular model follows the guidelines of the different network management standards. Moreover, we have used an object oriented methodology. Based on that model, the ATM switch model has been implemented using an object-oriented programming language.

Keyword codes: C2.3; D2.2
Keywords: ATM, B-ISDN, ATM Switch, Information Model, Quality of service, Object Orientation.

1. Introduction

For the efficient realization of a network, among the different network services, network administration has an important status, offering the control, coordination and monitoring of network resources. Therefore its integration is required at the conception phase of the network. The conception phase defines the initial and threshold values using the results of the performance models.
But afterwards, how can we guaranty the Quality of Service during the connection?
How can we have knowledge on the events resulting of a service degradation ?
In fact, the ATM switch is the basic network element of B-ISDN, since it has access to as well as supports a great variety of services including conversational, messaging, retrieval as well as broadcasting services, each of which have different performance requirements. To satisfy the demand of all these B-ISDN services, a powerful management scheme is required in order to provide dynamic monitoring, control and management.
This paper introduces an information model providing the management abstraction of all the elements needed to compute, monitor, control and manage

all services in order to obtain all the necessary information to monitor the switch's behavior during the information transfer phase.
Indeed, our purpose is to track the behavior of the switch at this phase and manage the alarms generated by exceeding the threshold values of the QoS parameters. To define all the manageable objects, we used a generic service model, and an object method. The *genericity* of the model is imposed by the necessity to obtain a unified view of the network resources considering the complexity and heterogeneity of networks. Our service model has been developed in parallel with the model proposed by the CCITT.
For the analysis, design and implementation of the information model of an ATM switch, following the object approach, the consideration and definition of a number of conception steps is required. We are going to follow this order for our presentation.
In the following section of the paper, we will analyze the functions of an ATM switch for defining the service that has to be provided by the switch, which allows us to define later in the second section the QoS criteria and their associated traffic parameters supported by our information model. In the third section we present the abstraction of our generic information model using the elements mentioned at the first two sections. We present the object methodology that permits us to start with the definition of the information model so to derive the application of administration of the switch. The last section deals with the implementation of this model.

2. Functions of the switch

The ATM switch is the network element charged with the realization of connections, based on a sophisticated allocation scheme that guaranties the required QoS of each connection. It is also responsible for making profit of the flexible routing mechanism of ATM, introduced by the structure of the cells.
We stress in our model the "transport" part of the switch ATM **[1]**, which is responsible for the correct transportation of cells from one ATM incoming channel to another ATM outgoing channel preserving the QoS specifications negociated during the connection's establishment.
At the ATM switch level we can identify the following operations need to be managed:

- *Buffering of the cells:* The cells arriving at the switch should be buffered while waiting for their turn to be forwarded. Depending on the approach, we can deal with input buffers (only at the level of input controllers), output buffers, combined input/output buffers or crosspoint buffers (located at the crosspoints of the matrix). This function is associated with our model's parameter of Cell Loss.

- *Analysis and translation of the header of each cell:* The header of the cells arriving at the switch is controlled for errors. The values of VPI and VCI, carried in the header, are read and translated to other values according to the switching table. This function is associated with our model's parameters of Cell Error, and the counting of different types of cells arriving and leaving the switch.

The *switching table,* which is responsible for the routing function, is structured in six fields. The first three (number of the port receiving the cell, identifier of the VP entering through that port, identifier of VC included within this VP) identify the inlink , the link through which the cell has arrived to the switch. The last three fields (number of the port emitting the cell, identifier of the VP leaving that port, identifier of VC included within this VP) identify the outlink from which the cell will leave the switch.

- *Commutation of space of cells*: The transportation of cells from an incoming logical ATM channel (out of N incomming channels) to an outgoing logical ATM channel, chosen from a number of outgoing channels, is described in the switching table. (The cell is guided through the switch to an output link and leaves the switch containing the new values of VCI, VPI.). This function is associated with our model's parameter of Cell Misinsertion.

- *Multiplexion/concentration:* The number of links entering a switch may be different to the number of links leaving the switch. The inverse operation (expansion/demultiplexing) is also thinkable **[1].**
These functions are performed on several types of cells, as **[9]** :

- *Idle* cells only at the physical level;
- *Assigned* cells which contain useful user information;
- *Unassigned* cells for the synchronization of the sender and the receiver;
- *Metasignaling* cells for the negotiations of VCI;
- *General broadcast* cells destinated to all terminals of the User Network Interface (UNI);
- *Physical layer maintenance* cells transporting maintenance information to the physical layer

The classification of cells is based on the value of the Payload Type field (PT) **[9,I.361]** as seen in the following scheme.

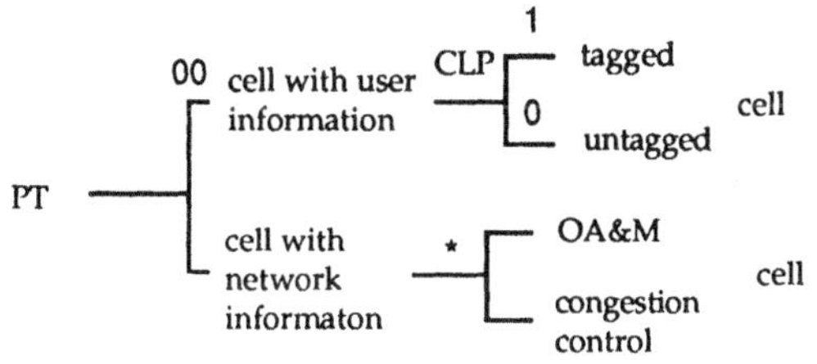

* To distinguish the type of cells with network information, further treatment of the information field is required.

3. Quality of service

QoS can be seen as a set of criteria that possibly affect each other. By this set, we are able to monitor the behavior of a network and tune the values of certain parameters to improve these QoS criteria according to the environment constraints and the user requirements [3].

3.1 Quality of service parameters

The inspection of network performance at the level of the ATM switch concerns the following parameters:

- *Cell error*: in the case of a detection of an errored cell at the Physical layer (PL).
- *Cell loss*: in the case of discarding one or more cells due to completion of the available buffer space.
- *Cell misinserted*: in the case of a logical error causing that the values of VCI and VPI in the cell header do not correspond to any value stored into the switching table.
- *Cell delay*: It measures the delay introduced to the network by the ATM switch. It comprises of a switching delay, depending on the time needed for internal packet transfer through the hardware and a queueing delay, depending on the time needed for the cells to wait in the buffers [6]. The mechanism to measure cell delay is based on the OA&M cells.

The monitoring of these parameters provides us the following statistical information [7]:

$$CER = \frac{\text{number of cells with error}}{\text{Total number of cells sent}}$$

$$CLR = \frac{\text{number of cells lost}}{\text{Total number of cells sent}}$$

$$CIR = \frac{\text{number of cells misinserted}}{\text{Total number of cells sent}}$$

3.2 Trafic descriptors

A fundamental characteristic of ATM is the way of establishing its connections. Connections are established after a negociation on the expected characteristics for a connection based on the QoS required for a service. These specific characteristics have to be monitored during the duration of the connection in order to further prevent congestion and/or service degradation. Usage parameter's control is part of the CCITT framework for traffic management controls.
The number of traffic parameters which need to be monitored in order to be able to describe the desired QoS has not been yet defined. In our approach we consider three parameters based on the analysis in [2].

1. *Connection peak cell transfer rate*: The minimum inter-cell arrival time between two consecutive cells;
2. *Connection burst transfer rate*: minimum time allowed for the arrival of B cells, where B is the maximum expected burst length
3. *Connection average bandwidth*: number of cells expected to arrive during a fairly long time.

4. The information Model

4.1 Introduction

Adopting an object oriented approach is useful for developing an integrated network-operation environment. In this approach, a unified view of the network resources is essential and is represented by our network information model [8]. Following this point of view a generic network element information model has been defined which is used for the representation of any network element. This model identifies the resource objects that exist in a network and their related attribute types, events, actions, and behaviors. It also provides a foundation for understanding the inter-relationships between these resources and attributes, and may in turn, promote uniformity in dealing with the various aspects of managing these resources and attributes.
The CCITT also proposes a generic information model based on 35 object classes[10]. The considered viewpoints are: the network element level, the network level and the service level. Our generic model is based on 8 object classes that are applicable to four abstract levels, corresponding to the levels at which decision making must be done: physical, data link, network and application[11]. It is applicable to any network element. The network and service level are obtained using aggregations and recursiveness rules.
Concerning the service, we have focused here on a particular service, namely the QoS.

4.2 Modeling Issues

A network is a system for providing services by connecting network elements. Topologically, it can be represented by a set of nodes and links. These nodes and links are recognized as managed entity types. A link connects two nodes and a node terminates one or multiple links[8][3].
The main question remains:What aspects of a network element should be managed and which service provided?
Part of the answer by the architecture element. An architecture element is a manageable software or hardware resource supporting the whole or a part of the function of a network element.
For the hardware part, we can use a description of the hardware components of the network element as buffers, processors etc.
The software part can be divided into an *entity* containing all the operational parameters, a part containing the *connection table,* dealing with the existing

connections with the other network elements and the *SAP* (Service Access Point) through which we access the network element[11].
A network element must provide a service. To be able to monitor, control and manage any network element we have defined traffic and QoS entities.
At last a service entity, models the services that the network element provides to the layer above. In these entities we can find parameters related to [3]:

- the *availability* of the object, defined as the time during which the service is available to the user;
- its *reliability* corresponding to the response time and actual flow throughtput from the user's point of view ;
- its *transit delay* that we have already presented;
- Its *capacity* in terms of throughput.

To tackle the informational, behavioral and computational complexities of such a model, we have defined an object oriented approach. This approach provides three differents views for the description of the application but all are required for complete description. These three views are:

- A static view (the static structure of the model), describing the objects in the system and their relationships;
- A dynamic view (the object's behavior);
- The functional view (the object's functions) describing the data transformations.

This object oriented approach assists during the analysis, design and implementation steps.
We have chosen this object approach because this bottom-up concept provides mechanisms to ensure that all data needed will be taken into consideration.
The analysis step leads toward the definition of the inheritance tree and of the object model.
We have arrived to the presentation of the inheritance tree (figure 1) following the guidelines of the definition of the information model (ISO 10165-1) and the guidelines of the definition of managed objects (GDMO - ISO 10165-4) [4]. In fact the graphical presentation (inheritance tree) we provide for our model is the graphical representation of the textual description of the managed objects (GDMO).
The object model (figure 1) is provided by the object oriented methodology. It is defined as the reorganization of the inheritance tree by introducing the aggregation relationships between the object classes. In the object model, which represented by the object diagram, nodes are object classes and arcs are the generalization/specialization and aggregation relationships between classes. This static (object) view of the model is the key to easily pass from the inheritance tree to the containment tree (figure 1-2).
The conceptual (design) step leads toward the construction of the containment tree (figure 2).
The containment tree is represented by a graph whose nodes are object instances and whose arcs are containment relationships between instances. We can find all

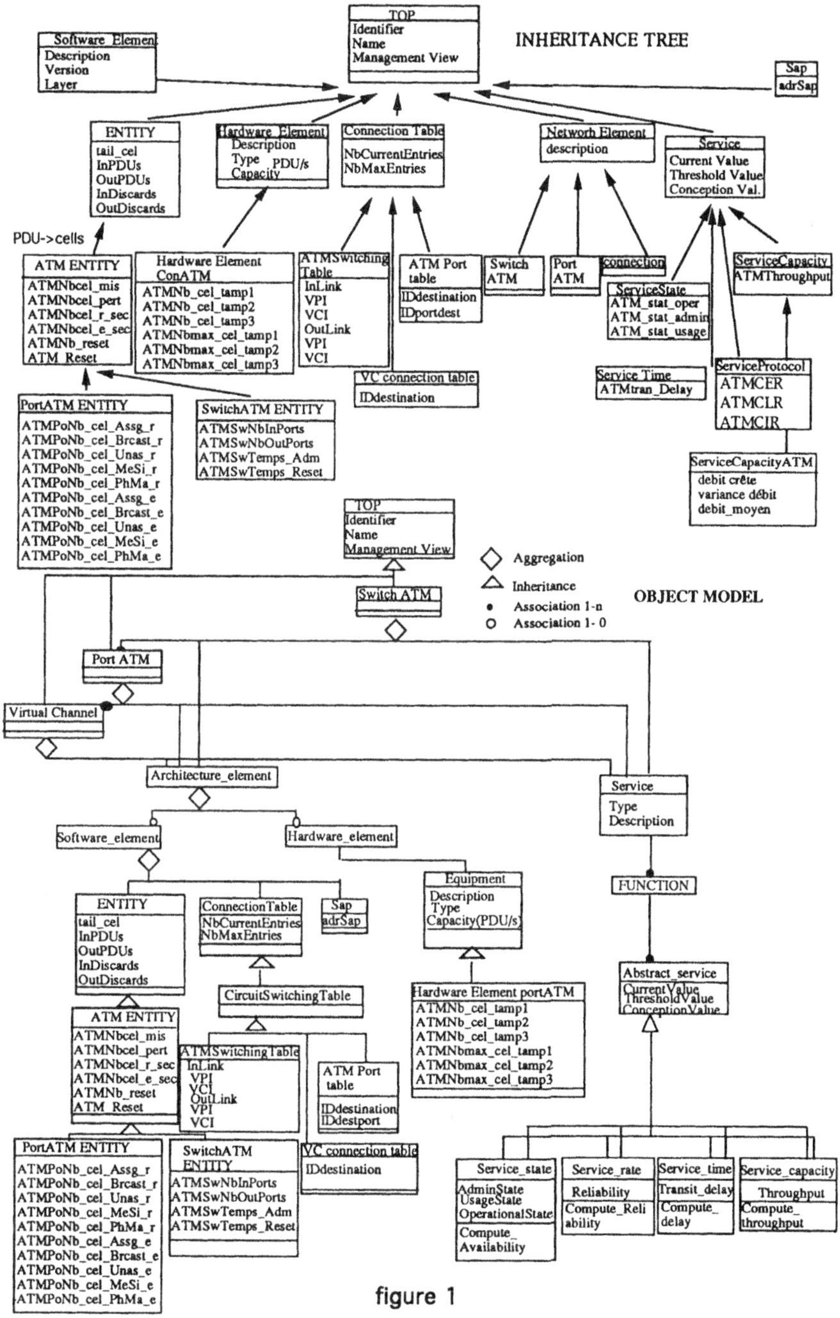
INHERITANCE TREE
TOP
Identifier
Name
Management View
Software Element
Description
Version
Layer
Sap
adrSap
ENTITY
tail_cel
InPDUs
OutPDUs
InDiscards
OutDiscards
Hardware Element
Description
Type PDU/s
Capacity
Connection Table
NbCurrentEntries
NbMaxEntries
Networh Element
description
Service
Current Value
Threshold Value
Conception Val.
PDU->cells
ATM ENTITY
ATMNbcel_mis
ATMNbcel_pert
ATMNbcel_r_sec
ATMNbcel_e_sec
ATMNb_reset
ATM_Reset
Hardware Element ConATM
ATMNb_cel_tamp1
ATMNb_cel_tamp2
ATMNb_cel_tamp3
ATMNbmax_cel_tamp1
ATMNbmax_cel_tamp2
ATMNbmax_cel_tamp3
ATMSwitching Table
InLink
VPI
VCI
OutLink
VPI
VCI
ATM Port table
IDdestination
IDportdest
Switch ATM
Port ATM
connection
ServiceState
ATM_stat_oper
ATM_stat_admin
ATM_stat_usage
ServiceCapacity
ATMThroughput
VC connection table
IDdestination
Service Time
ATMtran_Delay
ServiceProtocol
ATMCER
ATMCLR
ATMCIR
ServiceCapacityATM
debit crête
variance débit
debit_moyen
PortATM ENTITY
ATMPoNb_cel_Assg_r
ATMPoNb_cel_Brcast_r
ATMPoNb_cel_Unas_r
ATMPoNb_cel_MeSi_r
ATMPoNb_cel_PhMa_r
ATMPoNb_cel_Assg_e
ATMPoNb_cel_Brcast_e
ATMPoNb_cel_Unas_e
ATMPoNb_cel_MeSi_e
ATMPoNb_cel_PhMa_e
SwitchATM ENTITY
ATMSwNbInPorts
ATMSwNbOutPorts
ATMSwTemps_Adm
ATMSwTemps_Reset
OBJECT MODEL
TOP
Identifier
Name
Management View
Aggregation
Inheritance
Association 1-n
Association 1- 0
Switch ATM
Port ATM
Virtual Channel
Architecture_element
Service
Type
Description
Software_element
Hardware_element
FUNCTION
ENTITY
tail_cel
InPDUs
OutPDUs
InDiscards
OutDiscards
ConnectionTable
NbCurrentEntries
NbMaxEntries
Sap
adrSap
Equipment
Description
Type
Capacity(PDU/s)
CircuitSwitchingTable
Abstract_service
CurrentValue
ThresholdValue
ConceptionValue
ATM ENTITY
ATMNbcel_mis
ATMNbcel_pert
ATMNbcel_r_sec
ATMNbcel_e_sec
ATMNb_reset
ATM_Reset
ATMSwitchingTable
InLink
VPI
VCI
OutLink
VPI
VCI
ATM Port table
IDdestination
IDdestport
Hardware Element portATM
ATMNb_cel_tamp1
ATMNb_cel_tamp2
ATMNb_cel_tamp3
ATMNbmax_cel_tamp1
ATMNbmax_cel_tamp2
ATMNbmax_cel_tamp3
PortATM ENTITY
ATMPoNb_cel_Assg_r
ATMPoNb_cel_Brcast_r
ATMPoNb_cel_Unas_r
ATMPoNb_cel_MeSi_r
ATMPoNb_cel_PhMa_r
ATMPoNb_cel_Assg_e
ATMPoNb_cel_Brcast_e
ATMPoNb_cel_Unas_e
ATMPoNb_cel_MeSi_e
ATMPoNb_cel_PhMa_e
SwitchATM ENTITY
ATMSwNbInPorts
ATMSwNbOutPorts
ATMSwTemps_Adm
ATMSwTemps_Reset
VC connection table
IDdestination
Service_state
AdminState
UsageState
OperationalState
Compute_Availability
Service_rate
Reliability
Compute_Reliability
Service_time
Transit_delay
Compute_delay
Service_capacity
Throughput
Compute_throughput

figure 1

the elements already defined inthe inheritance tree. The presentation of the objects is based this time on the containment relationships between the objects. It also reflects a dependence of existence between an object and its superior managed object **[3]**.
The organization of the structure, the fact that it follows the guidelines mentioned above as well as an object-oriented methodology, makes our model powerful, guiding it from the abstract genericity to a well-organized inheritance and thereby avoiding the duplication of common object attributes, operations and notifications presented in similar classes. Thus our model supports reusability, extensibility, efficiency and flexibility.

4.3 Modeling of an ATM Switch

The ATM switch management model represents the vision of OSI management for this network resource. The switch model is the abstraction of the real equipment representing its properties as they are seen from the OSI point of view.
The ATM switch is a particular case of Network Element. It is placed in the inheritance tree, just below the object NETWORK ELEMENT, from whom it inherits the description. It also inherits from the object TOP the identifier, the name and the management view.
The objects placed at the first level after TOP, are the generic objects describing any network element and in our case the switch, the ports and the connections. In the top-down direction of the tree, we place the objects describing the particularities of these network elements, representing thereby subclasses.
The notion of recursiveness is present as the switch's component ports and connections have the same structure as the switch.
The cells arriving at the switch are buffered while waiting for their turn to be forwarded. This buffering occurs at the port's (link) level. The counter ATMNb cel tamp existing at this level, gives an indication of the number of cells buffered at a specific instance.
The entity class provides us the information concerning the operational parameters at three levels of the network element: the different events dealing with the cells (number of arriving cells (ATMPoNb cel r), lost (ATMPoNbcel pert), misinserted etc) and also the events concerning the network element (number of times that the switch or the port has been reinitialised (ATMSwNb reset etc). At connection and port levels we can find information dealing with the different types of cells arriving or leaving the switch.
The values of the operational parameters dealing with the cells at the level of each particular port are obtained with aggregation of the corresponding parameters of all the connections of each particular port. For the switch we consider the aggregation of all the ports.
Our model considers three tables, one for each of the three levels mentioned above:

- The ATMSwitching Table responsible for routing cells;

- The ATMPort table and the VC connection table providing the identifier of the switch to which the virtual path or the virtual channel is destinated.

The service class models the different services that the switch provides:

- The availability, service state, represents the instantaneous condition of availability and operability of the associated resource from management's point of view. It is composed of three factors:

  - Operability: gives an indication whether or not the resource is physically installed and works;
  - Usage: whether or not the resource is actively in use at a specific instant, and if so, whether or not it has spare capacity for additional cells at that instance;
  - Administration: permission to use or prohibition against using the resource, imposed through the management services.

For each one of these three states an automat is used providing all the specific events associated with the resource and causing specific transitions from a state value to the other. The combination of these three states results in the management state which shows the relationship (dependance) between the three states. For example the usage state value is always idle if the operational state value is disable or the administrative state value is locked[5].

- The capacity, service capacity, in terms of throughput at the level of the switch and in terms of connection's burst transfer rate, connections's peak cell transfer rate and connection's average bandwidth at the level of ports and connections;
- The reliability, service protocol, monitoring the QoS parameters in terms of Cell Lost Rate, Cell misInserted rate;
- The Transit delay, service time, dealing with the transit delay;

This project makes part of the project for the development of a multi-network (FDDI, DQDB, ATM, TOKEN RING, ETHERNET) demonstrator carried out at the Département Réseaux of Télécom Paris.

5. Implementation

Our model is specified using object oriented analysis and object oriented design. It is implemented with an object oriented language. This approach has permitted us to preserve the objects and classifications identified during analysis to enrich them during design and then to implement them directly in code. Our framework for implementation consists of an object oriented language, Eiffel, and some object oriented databases management systems facilities.
The use of the recommendations defining the object behavior through the system management functions (SMF) [5] has provided us a formal description of the operations and the notifications applied on the managed objects, the state management functions, the attributes for representing relationships and the

alarm reporting function.
Thereby, it was possible to specify the requirements for the dynamic view. It consisted of defining scenarios and examining them to identify all the external or internal events, building state diagrams for each class with non trivial behavior and showing the events that the object receives and sends.
- Concerning the Object Management Functions [OSI 10164-1] applied on the objects, a set of nine functions has been standardized. Also four types of notifications are generated by the switch's object:

Object creation Notification;
Object deletion Notification;
Object change name notification;
Attribute value change notification.

These notifications are structured according to the recommended format whose parameters are identical to M-EVENT-REPORT service [CMIS ISO/IEC 9595].
The choise to decide which one of the above functions is suitable for applying them to each of our objects was left open.
- The State Management Function [OSI 10164-2] provides us a formal description of the state graphs, describing the behavior of the resources on a reasonable level of abstraction.
- The Alarm Reporting Function [OSI 10164-4] classifies the different faults occuring within a system and formalizes the report of these events, providing their degree of severity. The mechanism of monitoring the error rates, which is also given, is very important, for the early detection of faults before significant effects could be felt by the user.

6. Conclusion

This paper presents a generic, hierarchical and modular information model for network elements used for the administration of a network element. This model allows us to describe any network element. In this paper we have demonstrated its application on the ATM switch. This model follows the guidelines of the different standards and thereby an object oriented methodology. Its structure is characterized by a well-organized inheritance. This way our model supports reusability, extensibility, efficiency and flexibility. Moreover, the use of the object approach has permitted us to distinguish the dynamic aspect relative to the object as well as the functional aspect relative to the application. The steps have guided us from the analysis to the implementation.

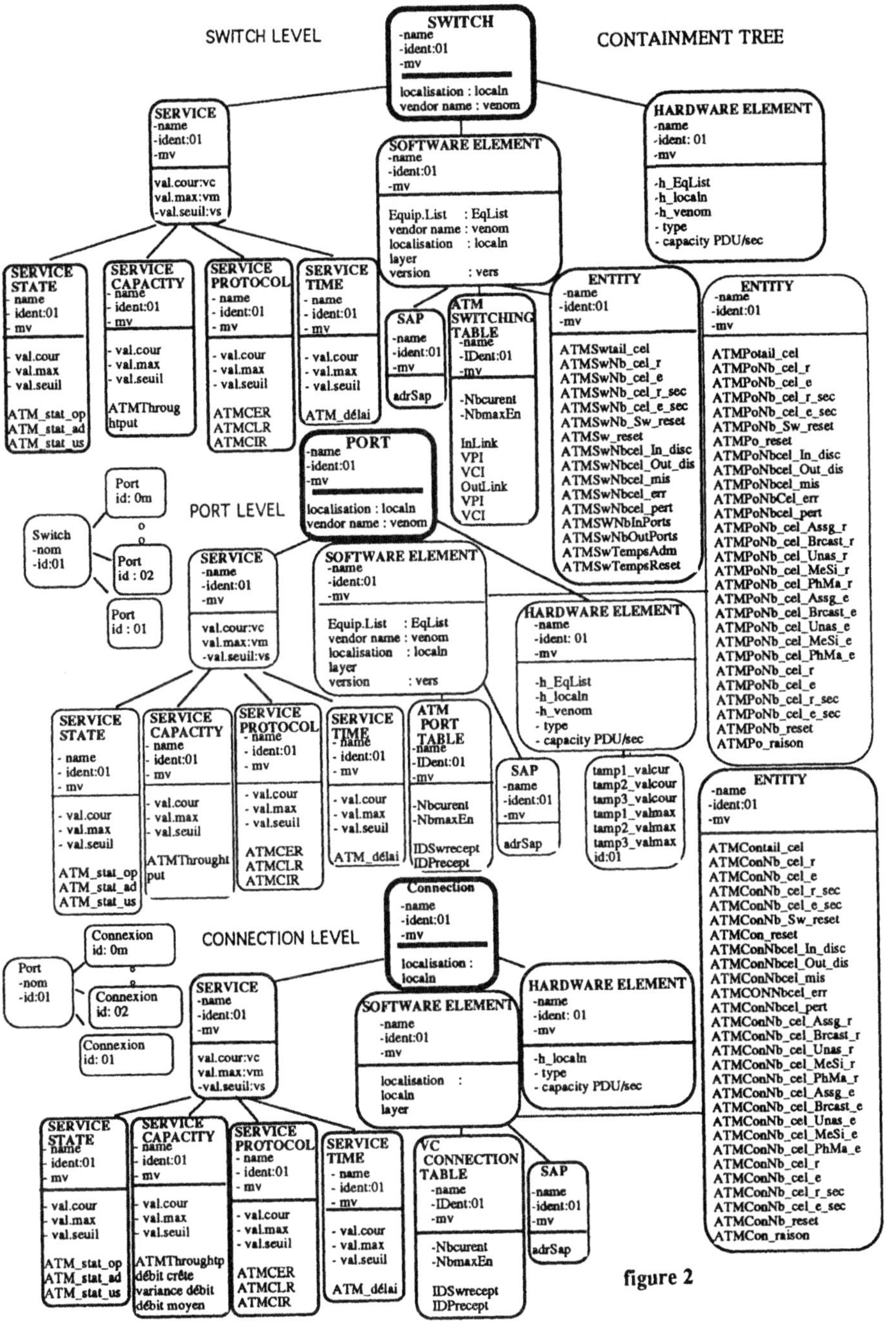
SWITCH LEVEL
CONTAINMENT TREE
SWITCH
-name
-ident:01
-mv
localisation : localn
vendor name : venom
SERVICE
-name
-ident:01
-mv
val.cour:vc
val.max:vm
-val.seuil:vs
SOFTWARE ELEMENT
-name
-ident:01
-mv
Equip.List : EqList
vendor name : venom
localisation : localn
layer
version : vers
HARDWARE ELEMENT
-name
-ident: 01
-mv
-h_EqList
-h_localn
-h_venom
- type
- capacity PDU/sec
SERVICE STATE
- name
- ident:01
- mv
- val.cour
- val.max
- val.seuil
ATM_stat_op
ATM_stat_ad
ATM_stat_us
SERVICE CAPACITY
- name
- ident:01
- mv
- val.cour
- val.max
- val.seuil
ATMThroughtput
SERVICE PROTOCOL
- name
- ident:01
- mv
- val.cour
- val.max
- val.seuil
ATMCER
ATMCLR
ATMCIR
SERVICE TIME
- name
- ident:01
- mv
- val.cour
- val.max
- val.seuil
ATM_délai
SAP
-name
-ident:01
-mv
adrSap
ATM SWITCHING TABLE
-name
-IDent:01
-mv
-Nbcurent
-NbmaxEn
InLink
VPI
VCI
OutLink
VPI
VCI
ENTITY
-name
-ident:01
-mv
ATMSwtail_cel
ATMSwNb_cel_r
ATMSwNb_cel_e
ATMSwNb_cel_r_sec
ATMSwNb_cel_e_sec
ATMSwNb_Sw_reset
ATMSw_reset
ATMSwNbcel_In_disc
ATMSwNbcel_Out_dis
ATMSwNbcel_mis
ATMSwNbcel_err
ATMSwNbcel_pert
ATMSWNbInPorts
ATMSwNbOutPorts
ATMSwTempsAdm
ATMSwTempsReset
ENTITY
-name
-ident:01
-mv
ATMPotail_cel
ATMPoNb_cel_r
ATMPoNb_cel_e
ATMPoNb_cel_r_sec
ATMPoNb_cel_e_sec
ATMPoNb_Sw_reset
ATMPo_reset
ATMPoNbcel_In_disc
ATMPoNbcel_Out_dis
ATMPoNbcel_mis
ATMPoNbCel_err
ATMPoNbcel_pert
ATMPoNb_cel_Assg_r
ATMPoNb_cel_Brcast_r
ATMPoNb_cel_Unas_r
ATMPoNb_cel_MeSi_r
ATMPoNb_cel_PhMa_r
ATMPoNb_cel_Assg_e
ATMPoNb_cel_Brcast_e
ATMPoNb_cel_Unas_e
ATMPoNb_cel_MeSi_e
ATMPoNb_cel_PhMa_e
ATMPoNb_cel_r
ATMPoNb_cel_e
ATMPoNb_cel_r_sec
ATMPoNb_cel_e_sec
ATMPoNb_reset
ATMPo_raison
PORT LEVEL
PORT
-name
-ident:01
-mv
localisation : localn
vendor name : venom
Switch
-nom
-id:01
Port
id: 0m
Port
id : 02
Port
id : 01
SERVICE
-name
-ident:01
-mv
val.cour:vc
val.max:vm
-val.seuil:vs
SOFTWARE ELEMENT
-name
-ident:01
-mv
Equip.List : EqList
vendor name : venom
localisation : localn
layer
version : vers
HARDWARE ELEMENT
-name
-ident: 01
-mv
-h_EqList
-h_localn
-h_venom
- type
- capacity PDU/sec
SERVICE STATE
- name
- ident:01
- mv
- val.cour
- val.max
- val.seuil
ATM_stat_op
ATM_stat_ad
ATM_stat_us
SERVICE CAPACITY
- name
- ident:01
- mv
- val.cour
- val.max
- val.seuil
ATMThroughtput
SERVICE PROTOCOL
- name
- ident:01
- mv
- val.cour
- val.max
- val.seuil
ATMCER
ATMCLR
ATMCIR
SERVICE TIME
- name
- ident:01
- mv
- val.cour
- val.max
- val.seuil
ATM_délai
ATM PORT TABLE
-name
-IDent:01
-mv
-Nbcurent
-NbmaxEn
IDSwrecept
IDPrecept
SAP
-name
-ident:01
-mv
adrSap
tamp1_valcur
tamp2_valcour
tamp3_valcour
tamp1_valmax
tamp2_valmax
tamp3_valmax
id:01
ENTITY
-name
-ident:01
-mv
ATMContail_cel
ATMConNb_cel_r
ATMConNb_cel_e
ATMConNb_cel_r_sec
ATMConNb_cel_e_sec
ATMConNb_Sw_reset
ATMCon_reset
ATMConNbcel_In_disc
ATMConNbcel_Out_dis
ATMConNbcel_mis
ATMCONNbcel_err
ATMConNbcel_pert
ATMConNb_cel_Assg_r
ATMConNb_cel_Brcast_r
ATMConNb_cel_Unas_r
ATMConNb_cel_MeSi_r
ATMConNb_cel_PhMa_r
ATMConNb_cel_Assg_e
ATMConNb_cel_Brcast_e
ATMConNb_cel_Unas_e
ATMConNb_cel_MeSi_e
ATMConNb_cel_PhMa_e
ATMConNb_cel_r
ATMConNb_cel_e
ATMConNb_cel_r_sec
ATMConNb_cel_e_sec
ATMConNb_reset
ATMCon_raison
CONNECTION LEVEL
Connection
-name
-ident:01
-mv
localisation :
localn
Port
-nom
-id:01
Connexion
id: 0m
Connexion
id: 02
Connexion
id: 01
SERVICE
-name
-ident:01
-mv
val.cour:vc
val.max:vm
-val.seuil:vs
SOFTWARE ELEMENT
-name
-ident:01
-mv
localisation :
localn
layer
HARDWARE ELEMENT
-name
-ident: 01
-mv
-h_localn
- type
- capacity PDU/sec
SERVICE STATE
- name
- ident:01
- mv
- val.cour
- val.max
- val.seuil
ATM_stat_op
ATM_stat_ad
ATM_stat_us
SERVICE CAPACITY
- name
- ident:01
- mv
- val.cour
- val.max
- val.seuil
ATMThroughtp
débit crête
variance débit
débit moyen
SERVICE PROTOCOL
- name
- ident:01
- mv
- val.cour
- val.max
- val.seuil
ATMCER
ATMCLR
ATMCIR
SERVICE TIME
- name
- ident:01
- mv
- val.cour
- val.max
- val.seuil
ATM_délai
VC CONNECTION TABLE
-name
-IDent:01
-mv
-Nbcurent
-NbmaxEn
IDSwrecept
IDPrecept
SAP
-name
-ident:01
-mv
adrSap

figure 2

References

[1] M. de Prycker: Asynchronous transfer mode, solution for broadband ISDN. Ellis Horwood Limited, 1991.
[2] Henry Gilbert, Osama Aboul-Magd and Van Phug: "Developing a Cohesive Traffic Management Strategy for ATM Networks", IEEE Communication Magasine, October 1991.
[3] Noëmie Simoni, Simon Znaty. "QoS: From Definition to Management". 4th IFIP Conference on High Performance Networking, Liège, Dec 92.
[4] ISO/IEC 10165 : Information Processing Systems - Open Systems Interconnection - Management Information Services - Structure of Management Information. Aug 91.
Part 1 : Management Information Model (ISO/IEC JTC1/SC21 N5253)
Part 2 : Definition of Management Information (ISO/IEC JTC1/SC21 N4867)
Part 4 :Guidelines for the Definition of Managed Objects (ISO/IEC JTC1/SC21 N4852)
[5] ISO/IEC 10164 : Information Processing Systems - Open Systems Interconnection - Management Information Services - Systems Management
Part 1 : Object Management Function (ISO/IEC JTC1/SC21 N4067)
Part 2 : State Management Function (ISO/IEC JTC1/SC21 N4068)
Part 3 : Relationship Management Function (ISO/IEC JTC1/SC21 N4069)
Part 4 : Alarm Reporting Function (ISO/IEC JTC1/SC21 N4070)
Part 5: Event Report Management Function (ISO/IEC JTC1/SC21 N4071)
Part 6 : Log Control Function (ISO/IEC JTC1/SC21 N4063)
Part 7 : Security Alarm Reporting Function (ISO/IEC JTC1/SC21 N4064)
[6] Noëmie Simoni, Simon Znaty: "An Object Oriented Methodology For Realisation and Management of a High Speed Gateway", AIP for Network Management. Paris, April '93, IFIP Conference on MAN & LAN Management.
[7] Tadahiro Yokoi, Yutaka Yamamoto, Yoshihosa Fujii and Toshiko Betchaku, "Performance Design Method for ATM Networks and Systems". NTT REVIEW, vol. 4 n° 4 July 1992.
[8] H.Yamaguchi, S.Isobe, T.Yamaki and Y.Yamanaka, "Network Information Modeling for Network Management", NOMS'92 IEEE.
[9] CCITT Recommendations
I.113 Vocabulary of terms for Broadband Aspects of ISDN
I.121 Broadband Aspects of ISDN
I.150 BISDN ATM Functional Characteristics
I.211 BISDN Service Aspects
I.311 BISDN General Network Aspects
I.321 BISDN Protocol Reference Model and its Application
I.327 BISDN Network Functional Architecture
I.361 BISDN ATM Layer Specification
I.413 BISDN User-Network Interface
[10] CCITT Draft Recommendation, M3100, Generic Network Information Model, 1992.
[11] Noëmie Simoni, Simon Znaty: "Interconnection of High-Speed Data Networks: Contribution of Dynamic Control and QoS", Computer Networks and ISDN Systems 25 (1992), North-Holland.

Integrated Broadband Communication Networks
and Services (C-18) / V.B. Iversen (Editor)
Elsevier Science B.V (North-Holland)

High speed data services for LAN interconnection in Norway

Berit Svendsen[a]

[a]Norwegian Telecom Research, P.O Box 83, N-2007-Kjeller, Norway

Abstract

An increasing number of companies all over the industrialized world have already installed or plan to install local area networks (LANs). Due to the bandwidth gap in LANs and wide area networks, it is expected that the demand for high speed data services will come from the LAN users and that data services will constitute an important proportion of future broadband traffic. A lot of work has been done regarding technical solutions for LAN interconnection. We will, on the other hand, focus on market demand for LAN interconnection. The intention with the study described in this article is to give a status of the market situation for LAN installations in Norway. Our impression is that LAN implementation and use of telecommunications is highly advanced in Norway and in the other Nordic countries. The goal of our study was to (i) find the total number of LAN installations, (ii) estimate the total market for interconnection of geographically separated LANs, and (iii) find the use of datacommunication services offered by Norwegian Telecom for LAN interconnection. We report the results from this study and describe how it is used for planning of new data services for LAN interconnection in Norway.

Keyword Codes: C.2.5; K.4.0; C.2.0
Keywords: Local Networks, Computers and Society; General, Computer-Communication Networks; General

1. INTRODUCTION

All over the industrialized world, an increasing number of companies have already installed or are going to install LANs (Local Area Networks) to meet the demand for data communications at the main-office and at branch-offices. For companies that have offices geographically separated, there is a move towards exchanging information between the LANs in different offices. Up to now there has been a gap in the bandwidth offered on a traditional LAN (Ethernet with 10 Mbit/s and Token Ring with 4 and 16 Mbit/s) and the bandwidth offered by a public network operator (up to 2 Mbit/s) over a wide area. The introduction of FDDI with 100 Mbit/s bandwidth is further increasing the gap in bandwidth. FDDI can be used both as a backbone for slower LANs such as Ethernets and Token Rings, and as a high-speed LAN with workstations directly attached to the FDDI network.

Due to the gap in bandwidth in the LANs and in the wide area networks, it is expected that the demand for high speed data services in public networks will probably come from the LAN users. This demand will further increase with the emergence of more powerful personal computers and workstations and high-performance applications in a wide area such as distributed processing and distributed databases. It is therefore expected that data services will constitute an important proportion of future broadband traffic.

There is a series of technical solutions for LAN interconnection both in a metropolitan area and in a wide area. But technical solutions is one thing, market demand for LAN interconnection is another. The intention of the study described in this article is to give a status for LAN implementations in Norway. We think that the status does not differ so much from the status in other industrialized countries. In this article, you will find that we divided the study into three different tasks:

- find the total number of LAN installations sold
- estimate the total market for interconnection of geographically separated LANs
- find the use of data services offered by Norwegian Telecom for LAN interconnection.

We approached the LAN vendors to find the total number of LAN installations in Norway and to estimate the total market for LAN interconnection (find the total number of remote bridges and routers sold). This study is described in chapter 3. To find the use of the data services offered by Norwegian Telecom, we approached 50 companies in Norway. This study is described in chapter 4. Chapter 5 will describe how we used the results to plan new data services in Norway for LAN interconnection.

Before we present the results from the study, it is very important to define what we mean by LAN interconnection. Figure 1 describes what we mean by LAN interconnection and LAN to LAN communication. LAN to LAN communication is between two geographically separated LANs interconnected via remote routers/bridges over a wide area network owned by a public operator. Hence, accessing a remote host from a LAN via a gateway is not what we define as LAN interconnection. Figure 1 shows that the data services offered by the public network operator is only part of the total network configuration.

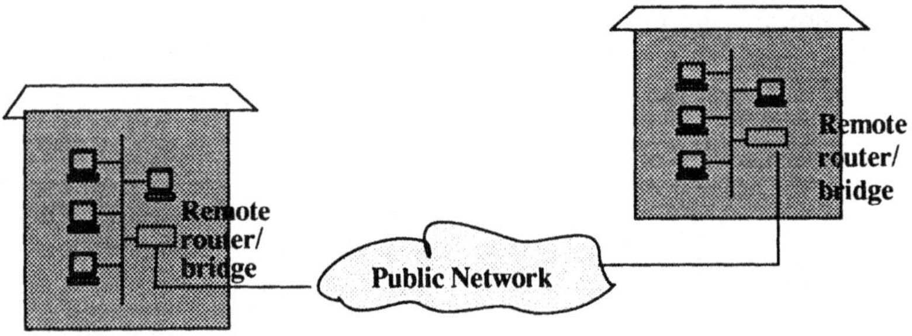

Figure 1. LAN interconnection

2. LAN, BRIDGES AND ROUTERS

Local Area Networks or LANs are a recognition of the need for people to use data, and as a by-product, to pass data from one person to another. People connected to a LAN can, in addition to communicating with each other, have access to the following functions:

- sharing of expensive peripherals such as printers and external communication ports,
- centralized storing of data and programs,
- centralized backup functions.

PC LAN operating systems are software that enables PCs attached to local area networks to communicate with the network server (where the data and programs can be stored), the peripherals and with other PCs on the network.

Ethernet is today a market leader, but Token Ring is also widely used. Typical characteristics of LANs based on these standards are limited geographical coverage (an Ethernet cable can be extended to 2.5 kilometres), high bandwidth (4-16 Mbit/s) and low error rate. The FDDI standard expands the concept of local networking to include an area with a radius of about 50 kilometres and increasing in bandwidth to 100 Mbit/s.

Both bridges and routers are necessary to expand connectivity and to allow people via a PC or workstation to communicate with one another or with network resources not located on a LAN of which one is a primary member. A bridge is used to link two technologically similar networks to one another. Two Ethernets can be linked with a bridge (it is worth noting that it is also possible to bridge an Ethernet and a Token Ring network). A bridge forwards packets between networks that are destined for the other network (operates on layer 2 according to the OSI-model). Broadcast packets are forwarded to the other networks. A router operates on layer 3 according to the OSI-model. A router forwards packets between networks, but in contrast to a bridge does not forward broadcasting messages. Bridges and routers can be either local and/or remote depending upon their ability to interface with wide area networks. Bridges and routers with no wide area interfaces are local bridges and routers. Remote routers and bridges have wide area interfaces (e.g. G.703, V.35, V.36). Figure 1 shows a remote router/bridge directly attached to another remote router/bridge via wide area network.

3. STUDY AMONG NORWEGIAN LAN-EQUIPMENT VENDORS

The international estimates and figures mentioned are from International Data Corporation and Yankee Group International.

The installed base of PC LAN operating systems in Norway is about 30,000 at the end of 1992. This is a rapidly increasing number. (The number of LANs that use Unix and Network File System is not included in this number). The number indicates how many PC LAN operating systems have been sold, but does not mention how many of these LANs are interconnected. To get an answer to this question, we have to look at the total number of remote bridges and routers sold to interconnect geographically dispersed LANs. About 150,000 remote routers and bridges have been sold worldwide. 62% of this number have been sold in USA, and the rest (38%) in the rest of the world. 2% of the total number (3,200 units) have been sold in Norway; which is a high number for such a small country (4 million inhabitants).

Cisco is the leading remote router vendor and Vitalink is the leading remote bridge vendor both on a world-wide basis and in Norway.

The information can indicate how many LANs that are interconnected in Norway today. The study among 50 companies (chapter 4) shows that the data communication pattern for a company today is star-based where all the branch offices have connection to one main office. If we take this into consideration and also assume one remote router or bridge per LAN (this is not always true, a bridge and router can have several ports or interfaces to the wide area network), we have an indication that 3,200 LANs in Norway can be interconnected via remote bridges/routers. This is only a small amount (11%) of the total number of LAN operating system installations sold (30,000). But the figure shows that despite the fact that Norway is a small country, 2% of the total number of remote bridges/routers sold are installed in Norway. An explanation why so many remote bridges and routers are sold in Norway may be that many companies in Norway use advanced datacommunication technology, and furthermore that Norway's regional policy has made it possible for many companies to be located both in the cities and in the regions. IDC confirms in (8) that LAN implementation and use of telecommunication is highly advanced in Norway (and also in the other Nordic countries).

One important aspect is that all the remote bridges in Norway are sold with interfaces for the leased line service and most of them for 64 kbit/s (less than 10% are sold with interfaces for the 2 Mbit/s service). Routers are sold both for leased line and X.25 service (none of the routers are sold with interfaces for 2 Mbit/s leased line service).

4. STUDY AMONG NORWEGIAN COMPANIES

There are about 200,000 companies in Norway, but only approximately 1,000 of these companies have more than 100 employees. Typically for Norway is a large amount of small companies, but relatively few large ones.

Common for all the companies in the study are that they all have LANs to meet the demand for communication internally and maybe also externally. A very important question is which companies want to communicate with each other? A company needs to communicate with:

a. other parts of the corporation if the company belongs to a corporation

b. cooperating companies

c. suppliers and vendors.

d. public administration

Security aspects are crucial to many companies. A company will often only allow companies in group a) to get access to its LAN. As for communication with companies in the other groups, electronic mail is a good alternative.

The 50 companies in the study are all among the largest companies in Norway, and some of them have much more than 100 employees. The 50 companies employ 156,700 persons and had a total turnover of 423,911 mill NOK in 1991 (U.S.$70.65 mill). Another common factor for the companies (the first is that they all have LANs) is that they have a hierarchical structure with main office and branch offices geographically spread all over Norway. The companies are among the largest in Norway within the following areas:

- Bank and insurance
- Industry (oil, production, engineering)
- Trade
- Transport
- Public authorities (local administrations and Government institutions)
- Service sector

Figure 2 describes a typical communication network for a bank or an insurance company. The solution is based on hierarchical terminal networks and a central IBM-compatible host machine located in the company's main office. All bank and insurance applications, data about the customers, policies, account data and information about the employees are stored in the IBM-compatible host machine.The branch offices need the information stored in the IBM-compatible host machine, so the demand for data communication is met by using communication lines over a wide area that belongs to a public operator. The company has started to install Token Ring local networks in the branch offices and this replaces the well-known terminals and terminal controllers. The gateway in the local area network will act as an IBM-compatible terminal controller and the PCs in the local area network will emulate IBM-compatible terminals. The local area network is also used for text editing. Typically, the data traffic exchanged between the main office and branch offices are transactions from a terminal/terminal-controller (or a PC/gateway) to a central host machine. It consist of short inquiries (about 200 characters) from terminals to main machine and data for a half or full screen picture (about 1000 characters) as answers from main machine to terminals. IBMs SNA is a dominant network architecture in bank and insurance companies.

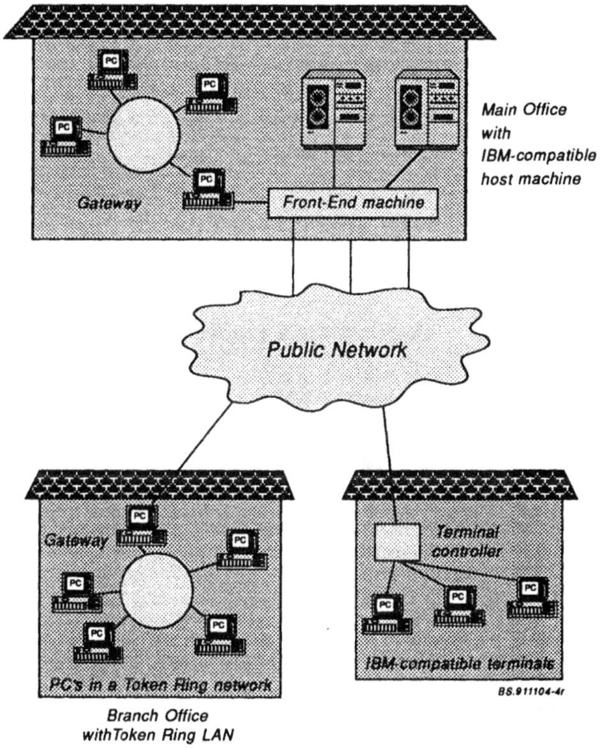

Figure 2. Typical datacommunication solution for bank and insurance companies

There are foreign and Norwegian oil companies in Norway. The study showed that the trend among these companies is to move away from their SNA networks to internetworks building up local area networks in main office and branch offices and connect them with bridges/routers over a public network, see figure 3. Like the bank and insurance companies, oil companies store account data and data about the employees and customers in the host machines in the main office. The companies have no plans about moving this information from the host machine. Therefore, it is still a need for the SNA network. The study showed that the companies either plans to maintain two separate networks (the SNA network and the internet) or have both the SNA traffic and the internet traffic in the same network. For the latter approach, the SNA traffic is packaged within a routable protocol (typically TCP/IP) and passed through the routers and the internetwork. The oil companies use VAX- and Unix-machines for technical applications like reservoir modelling and simulations, seismic data analysis and applications for exploration and production of oil. Important suppliers to the oil companies are engineering companies. Documentation regarding platform design and steel-calculations are today handed over on paper or as documents that can be scanned in, and later stored on optical disks. These documents cannot be exchanged electronically, because the companies use different CAD tools. Electronic mail, files and transactions are exchanged between the local area networks in branch offices and main office. The files are often large. Decnet and Unix (TCP/IP) is often used for technical applications, while local area networks with network operating systems Novell Netware and Lan Manager are being used for administrative applications. For communication between main office and the production platforms, 64 kbit/s and 2 Mbit/s permanent digital circuits via the NORSAT-A satellite system is used by several oil companies.

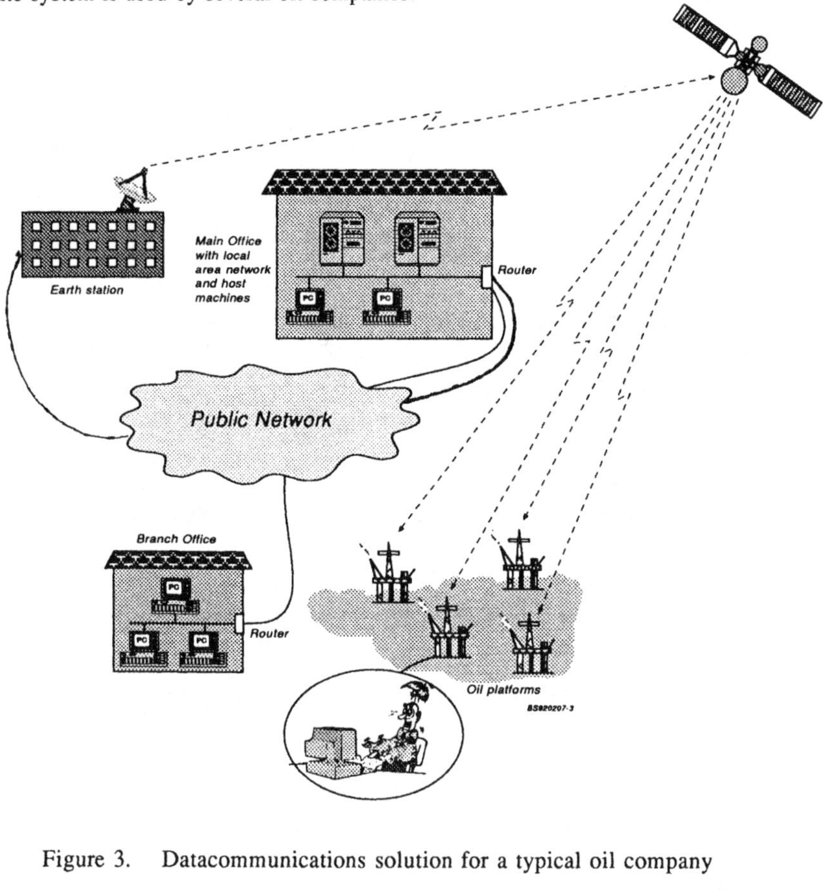

Figure 3. Datacommunications solution for a typical oil company

NOSIP (Norwegian OSI Profile) is a guideline for buying and using uniform solutions for data exchanging in Government institutions or between Government institutions. TCP/IP solutions shall be used if OSI-solutions are not available. Therefore, we find a large base of TCP/IP users in the Government institutions. Some of the Government institutions (Police, Norwegian Telecom, Health Service, and so on) differ from the other business companies in that they have a lot more branch offices spread ut all over the country. One of the institutions in the study has one headquarter and 450 branch offices. Information about customers, personnel, account data and so on is stored in a main machine in the headquarters. Local area networks in branch offices are used for administrative applications. Government institutions have a large demand for information. Some information can be electronically read out from different databases that belong to the Government (national register, criminal register, driver licence register and so on). Electronic mail, files and transactions are exchanged between the local area networks in branch offices and main office.

5. USE OF DATA SERVICES OFFERED BY NORWEGIAN TELECOM

Figure 4 shows the development of the total number of subscriptions for the data services offered by Norwegian Telecom. Our Datex service (X.21 switched service) is offered from 600 bit/s to 9600 bit/s. Our leased line service is being from 1200 bit/s to 2 Mbit/s and our X.25 service is offered from 2400 bit/s to 64 kbit/s. X.28 subscriptions to the X.25 service is not included in the total number of subscriptions for the Datapak service shown in figure 4.

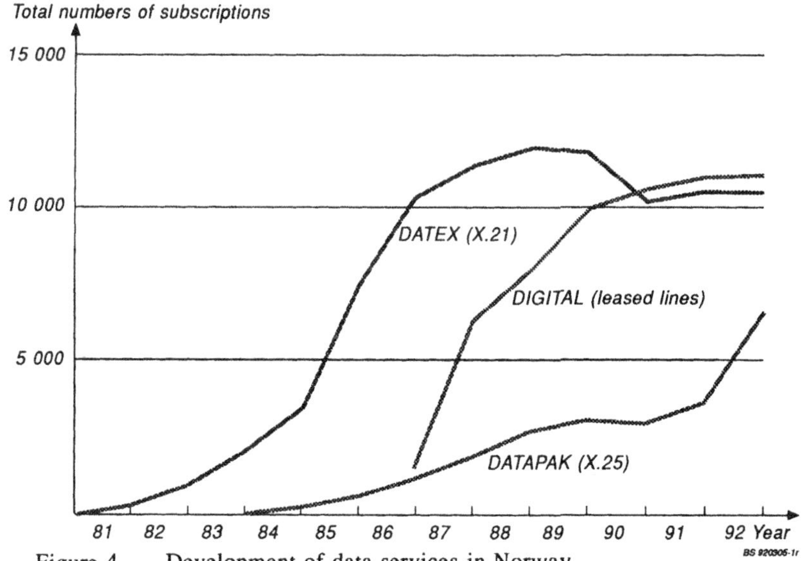

Figure 4. Development of data services in Norway

The trend in Norway regarding use of our "high speed" data services, 2 Mbit/s and 64 kbit/s leased line and 64 kbit/s X.25 service, is that very few large companies (small total number of subscribers) have a large share in the total number of subscriptions. For the 64 kbit/s X.25 service, 7 subscribers out of total number of 74 subscribers (9.5%) have 85 subscriptions out of total number of 182 subscriptions (47%). Table 1 shows this relationship for the X.25 service and the 64 kbit/s and 2 Mbit/s leased line service in 1992 in Norway.

Table 1 Relationship between subscribers and subscriptions for the data services in 1992 in Norway.

Data Service	%share in number of subscribers	%share in number of subscriptions
2 Mbit/s leased line	5	51
64 kbit/s leased line	3	44
64 kbit/s X.25 service	9.5	47

LAN interconnection as defined in figure 1 can be found within industry, trade business (very little), service sector and Government institutions. The other businesses do not have LAN interconnection. Some of the companies in the study without LAN interconnection have "stand alone" LANs, others use the LAN to get access to a remote main machine via a gateway, others have traditional SNA networks or they use minimachines for datacommunication. Both the study among LAN equipment-vendors and among companies show that very few 2 Mbit/s leased lines are being used for LAN interconnection at the time being (very few remote bridges and routers are being sold with interface for 2 Mbit/s leased line service). The study showed that most of the 2 Mbit/s leased lines are used for PBX interconnection.

91% of the total number of subscriptions used for LAN interconnection among the 50 companies in the study is the X.25 service, while the rest (9%) is the leased line service. Typically, the X.25 service is used by Government institutions, while the leased line service is being used by all the other businesses (industry, trade, etc.) The reason why Government institutions alone use so many trunks belonging to the X.25 service is that some of the Government institutions are rather large with one main office and over 400 branch offices geographically spread all over Norway. All the branch offices are connected to the main office using the X.25 service.

6. PLANNING OF NEW DATA SERVICES

In this chapter, we will describe how we used the results in the study to plan new data services for LAN interconnection in Norway.

The debate on the respective advantages of the connectionless and connection-oriented data services is not a new one, and has often underlined the gap between the data processing world and the telecommunication world. But which service is best suited for LAN interconnection today? Is it a connection-oriented data-service like Frame Relay or is it a connectionless service like CBDS/SMDS? In (10) is indicated that a connectionless datagram service seems predisposed toward a decentralized enterprise with a very large number of sites that communicate on an unpredictable basis. Connection-oriented data services will usually excel in the case of medium to heavy traffic densities among well-defined communication partners (2). A connectionless service also raises some problems:

- Congestion situations may easily occur in the networks.
- It is not easy to charge on the basis of the information sent, delivery is not guaranteed.

With these assumptions in mind, what is the preferable data service to offer in Norway for LAN interconnection? The study among 50 companies shows that the data communications pattern today is star-based where all the branch offices have connection to one main office. The branch offices need access to the data (account data, customer data, employee data etc.) often stored in a computer in the main office. Much terminal traffic is still exchanged between main office and branch offices. The study also shows that it is not generally today to have direct datacommunication between branch offices, and that distributed processing is not commonly used at the time being. The demand for high speed data services are not evident right now. The communication is point-to-point between predefined communication partners (main office to branch office). A very small amount of 2 Mbit/s is today being used for LAN interconnection, but a larger amount of our 64 kbit/s leased line service is being used for LAN interconnection. The study also shows that there is a demand for a wide-area data-service, not just within a metropolitan area.

We therefore suggest that for LAN interconnection, a connection oriented service will be the best at present. A connection oriented service like Frame Relay is best suited for LAN interconnection today because the study shows that LAN communication is between fixed partners belonging to the same company. There is also a need for a data service with a bandwidth between 64 kbit/s and 2 Mbit/s. Therefore, Norwegian Telecom plans to offer Frame Relay as a service for LAN interconnection in 1993.

The demand for more bandwidth than frame relay and leased line services can offer today (2 Mbit/s) will probably first come from oil companies. They are planning to use more distributed processing with applications such as exploration simulation in new oil and gas field. Special offers regarding price and bandwidth to these companies is also possible.

References

1. Taylor, S A. Frame transport systems. *IEEE Communications Magazine,* 30, (3), 66-70, 1992

2. Peker, C A. To connect or not to connect (data communications network planning) *Telephony,* 218, (24), 36-42, 1990.

3. International Data Corporation Norge. *Server marketplace 1989 - 1995: an analysis of the Norwegian marketplace.* Oslo, IDC, 1990.

4. Godø, H, Netland, B, Orøy, K. *Storbrukere av datakommunikasjon i Norge.* Kjeller, Teledirektoratets forskningsavdeling, 1987. (TF-rapport R50/87.)

5. NOSIP-Nytt. *Norsk OSI Profil for offentlig forvaltning.* Oslo, Statskonsult, 1991.

6. Network review: users find their voice. *CommunicationsWeek International,* (84), 22-35, 1992

7. Svendsen, B. *Norske bedrifters behov for LAN-LAN kommunikasjon.* Kjeller, Teledirektoratets forskningsavdeling, 1992. (TF-rapport R10/92.)

8. International Data Corporation Norge. *Local area network marketplace: multi client study 1992*. Oslo, IDC, 1992.

9. International Data Corporation. *Analysis of the LAN internetwork market.* Framingham, Mass., IDC 1992.

10. Thinking strategically about LAN interconnect. *Business Communication Review* (22), 10-12, 1993

Integrated Broadband Communication Networks
and Services (C-18) / V.B. Iversen (Editor)
Elsevier Science B.V (North-Holland)

Buffering and Packet Loss in the DQDB to ATM Interworking Unit

Sam Manthorpe
Laboratoire de Télécommunications (TCOM), EPFL, CH-1015 Ecublens, Switzerland

Buffering and Packet Loss in the DQDB to ATM Interworking Unit

Abstract

This paper describes methods for reducing the connectionless packet loss rate in the Distributed Queue Dual Bus (DQDB) to Asynchronous Transfer Mode (ATM) Interworking Unit (IWU). Results from simulation studies are presented to show the relationship between the buffering available in the IWU and the reduction in cell and packet loss probabilities. Further performance studies compare the performance of the Lost Packet Screening scheme, the Pushout Buffering scheme and the Datagram Acceptance Control scheme for further reducing the packet loss rate. It is shown that the packet loss rate can be reduced by one order of magnitude. The implementational complexity of the three approaches is also discussed.

1. INTRODUCTION

The Distributed Queue Dual Bus (DQDB) protocol [2,12] provides a high bit rate Media Access Control (MAC) Sublayer service for connectionless traffic and has a high degree of compatibility with the Asynchronous Transfer Mode (ATM) [3]. The use of ATM networks to interconnect DQDB based Metropolitan Area Networks (MANs) is an important evolutionary step in the introduction of Broadband Integrated Services Digital Networks (B-ISDN) [4]. Figure 1 shows the interworking scenario considered in this paper.

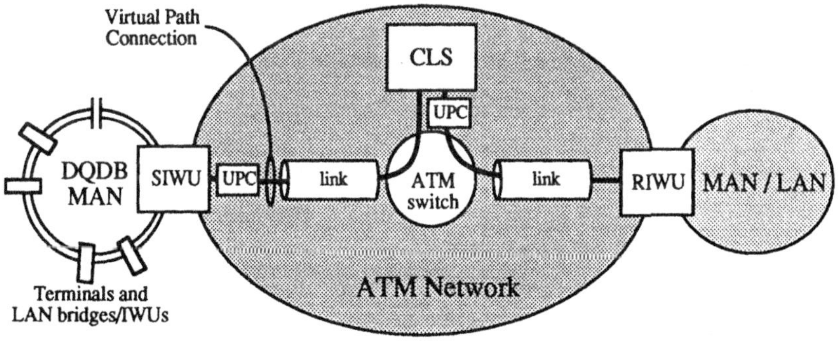

SIWU Sending Interworking Unit UPC Usage Parameter Control
RIWU Receiving Interworking Unit CLS Connectionless Server

Figure 1

Connectionless packets (at the IMPDU layer [12]) are divided into segments for transmission on the DQDB bus, and due to their highly compatible formats, these can be converted directly into ATM cells in the Sending Interworking Unit (SIWU) without the need for packet reassembly. This mode of interworking is known as *cell-to-slot interworking* (also *streaming mode*). A second possibility for interworking is *frame interworking*, where the packets are reassembled before protocol conversion. Cell to slot interworking offers many advantages over frame interworking (e.g. smaller delays, less buffer requirements, less bursty traffic [5]), but has the disadvantage of allowing "orphaned" cells to continue to pass through the IWU, even after the packet to which they belong has become useless due to the destruction of one or more of its segments. This paper examines methods of overcoming this drawback. Frame interworking is not considered in this paper, since the packet loss reduction techniques discussed are not relevant to this mode of operation.

There remains, a conflict between the connectionless traffic orientation of DQDB and the connection-oriented nature of ATM. The unpredictability of connectionless LAN traffic has been demonstrated in the much-referenced paper by Fowler and Leland [1], and while this is not a problem in the DQDB network, an ATM network *requires* a prediction of all traffic entering the network [13]. It is important that the cell loss probability in the ATM network be kept low, since one lost cell results in the loss of the whole packet to which it belongs, perhaps consisting of hundreds of cells. The effective information loss resulting from a single lost cell can therefore be rather large. The most significant source of cell loss in the considered interworking scenario will be that in the Usage Parameter Control (UPC), due to the difficulty in predicting the traffic parameters [14], as mentioned above. The UPC cell loss can actually be completely eliminated by placing a shaping buffer at the output of the SIWU, but cell loss will then occur when this buffer overflows. In this paper we discuss the dimensioning of this buffer and investigate three means of reducing the packet loss rate resulting from cell loss in the buffer.

Section 2 describes the model network adopted for the simulation studies. Section 3 describes the SIWU with shaping for the connectionless traffic and presents results showing its performance. Lost packet screening, the pushout and Datagram Acceptance Control (DAC) methods for reducing the packet loss rate in the shaper are presented in section 4, together with the results of a comparative performance study. Finally, section 5 draws some conclusions from the results of the preceding sections.

2. MODEL NETWORK

In this paper we consider the external traffic (destined for another network) arriving from a DQDB network and its behaviour within the SIWU. Simulation studies of a detailed model of the DQDB network (see figure 2a) and of a process-sharing queueing model (see figure 2b)

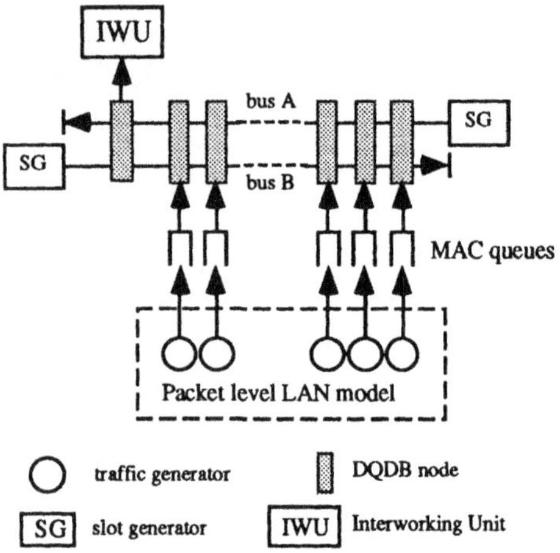

Figure 2a

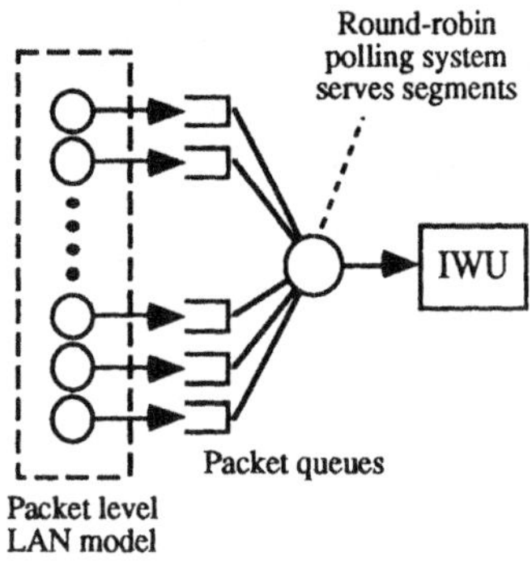

Figure 2b

[10] confirmed that the later provides a good approximation of the behaviour of the DQDB network. In order to reduce simulation times, the results presented in this paper were obtained through simulations employing a process-sharing queue to model the DQDB network. A "round-robin" service protocol was employed for the process-sharing queue.

Figure 3 shows the models for file transfer-like and interactive session-like traffic used in the simulation studies presented here. For the file transfer (FT) model, the lengths of the files (l_0, l_1,...) and the time between successive files (m_0, m_1,...) are geometrically and exponentially distributed, respectively. The files themselves are then broken down into fixed length packets, which access the DQDB network via a LAN bridge; the bitrate of the LAN is modelled by adding a constant time, representing the packet transmission time to the inter-packet times. Once a packet is ready for transmission on the DQDB network, it is broken down into segments, according to the DQDB protocol. The segments are queued for transmission on the bus via the MAC layer (the process-sharing queues). The interactive traffic is modelled as packets with fixed lengths and exponentially distributed inter-packet times. Two types of traffic load the DQDB network: internal and external traffic. The internal traffic is generated to serve as a background load that will influence the bus access delay (waiting time in the process-sharing queue) of the external traffic segments. Once the internal traffic is served by the PS queue, it is no longer required. The external traffic is the traffic that is routed to the SIWU and it is this traffic that is studied.

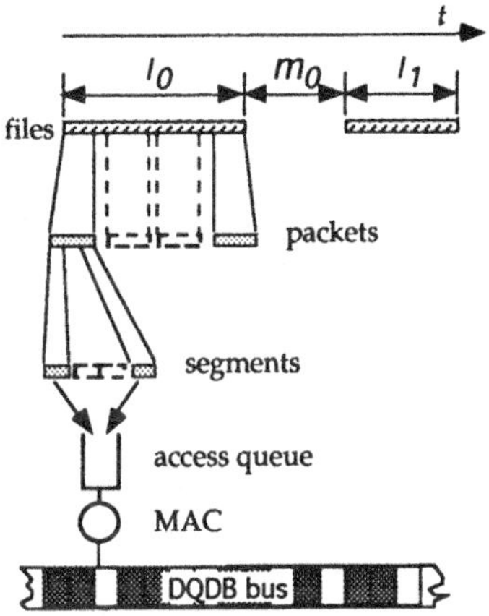

Figure 3

The detailed simulations of the DQDB network were performed using an extension to the DQDB simulator developed at the University of Paderborn [5] and the process-sharing model simulations were performed using a simulator developed at the Laboratoire de Télécommunications, Ecole Polytechnique Fédéral de Lausanne. Both simulators are based upon the ATLAS simulation tool [6], developed at the University of Aachen.

3. SHAPING FUNCTION

In the SIWU, connectionless traffic packets will be multiplexed onto Virtual Path Connections (VPCs), according to their destination address. In direct mode interworking, this VPC carries the traffic to a connectionless server (CLS), where the packets are demultiplexed, routed and then remultiplexed onto a new VPC, according to the next CLS or IWU on their respective routes. So, there will be at least one VPC leaving the SIWU for each logically attached CLS, with UPC being applied to each one. In fact, there may be more than one VPC for each SIWU to CLS link, each having different bandwidth allocations and UPC parameters to support different Quality of Service (QOS) requirements.

In order to guarantee zero cell loss between the SIWU and a CLS, the VPC bit rate must be dimensioned according to twice the DQDB bus rate (since there are two busses). Since this is rather an expensive solution, it is preferable to statistically multiplex the traffic onto a VPC whose capacity is less than that of the bus; this, however, introduces the possibility of cell loss at the UPC, which enforces this bitrate. To reduce this cell loss rate, a shaping function has been proposed.

The shaping function considered here is simply a buffer with a deterministic service rate. The only parameter currently defined by the CCITT to be controlled by the UPC function is the peak rate (we ignore any cell delay variation tolerance here) and so by setting the service rate of the shaper equal to the peak cell rate policed by the UPC, it can be guaranteed that there will be no cell loss in the UPC. Cell loss will still occur, but only as a result of overflow in the shaping buffer.

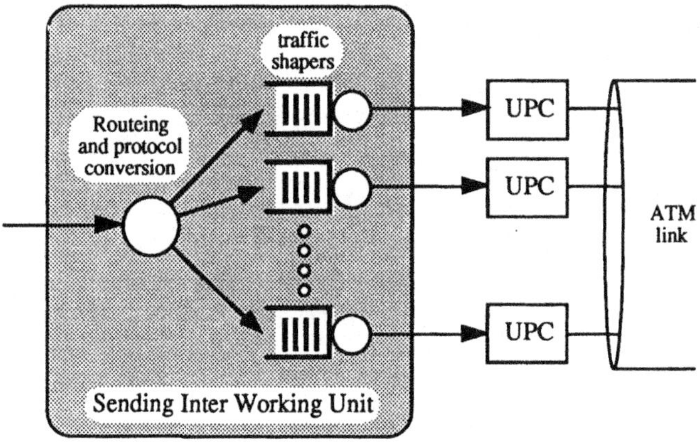

Figure 4

Although connectionless traffic is traditionally more loss sensitive than delay sensitive, new services and increasing computer performance are introducing more and more connectionless traffic that also has very strict delay requirements. This not only places a limit upon the size of the shaping buffer, but also on its load, in order to ensure the necessary loss statistics. Figure 4 shows a model of the SIWU with shaping functions applied to each VPC carrying connectionless traffic. In this paper, we consider only one shaper.

Graph 1 shows the actual cell loss rate and effective cell loss rate for a range of buffer sizes. The simulation scenario is described in table 1, below. For these simulations, the minimum inter-packet time allowed was increased to 0.964 ms, since this allows a higher shaper load, by reducing the burst lengths. (Note that, if the minimum interpacket time is very small (eg. 9.6 ms), and the shaper service rate more than one order of magnitude less than the DQDB bus rate, then the load on the shaper has to be very low and substantial buffering must be provided to keep the loss rate acceptable, since very long bursts of back to back segments may arrive). The actual cell loss rate is that measured in the shaper buffer, whereas the effective cell loss rate also represents the cells that become useless due to another cell from the same packet being lost; it is calculated by multiplying the packet loss rate by the packet size in cells.

LAN bus bitrate	10 Mbits/s	Shaper service rate	39.3kcells/s
Min. inter packet time	0.964 ms	Slot time	4.24 μs
Nº external sources (file transfer traffic)	60	Mean file length	20000 octets
Background load on DQDB network	25%	Background traffic	File-transfer

Table 1

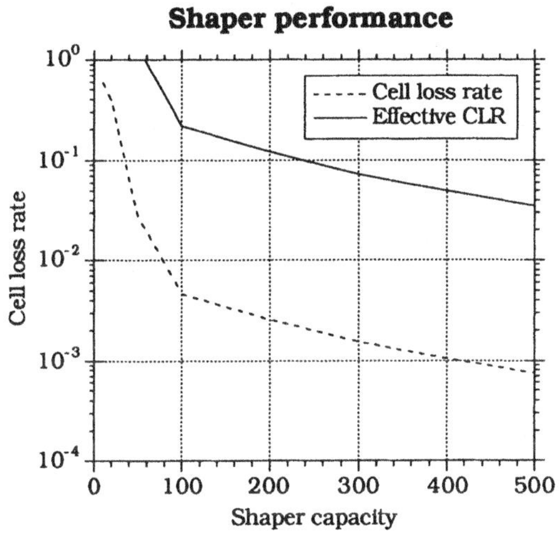

Graph 1

The actual cell loss curve can be divided into two parts: a rapidly falling part (occupancy < 100) and a low negative gradient part (occupancy > 100) (occupancy figures being approximate). The first part represents the cell scale queueing effects and the second part the burst scale queueing effects; this behaviour has been well documented in the literature, for example [7] and [8]. These results suggest that a significant reduction in cell loss can be obtained (of two orders of magnitude) by providing a shaping buffer with a capacity of around 100 cells to absorb cell-scale congestion (for this scenario). In fact, for low loads on the network, it is possible that large packets will arrive at the peak bitrate and so the buffer should be dimensioned to accomodate at least the maximum packet size of 210 segments. The long flat part to the curve shows that decreasing the cell loss probability in this region becomes more expensive in terms of the extra buffer places required and consequently the maximum delay that may be experienced. In this example, the maximum shaper delay (for 500 spaces) is 12.75 ms. For the remainder of the paper, a shaper capacity of 200 is assumed. The difference between the actual cell loss and the effective cell loss curves demonstrates the large potential for reducing the effective cell loss rate.

4. REDUCTION OF PACKET LOSS RATE

In this section, three methods for reducing the packet loss rate are examined: lost packet screening, the pushout buffering scheme and the Datagram Acceptance Control scheme.

4.1 Lost Packet Screening

The loss of a single cell effectively results in the loss of the whole packet to which it belongs. Lost packet screening is a simple modification to the basic buffer which ensures that if a packet is lost (for example, one of its cells is lost due to buffer overflow, or a next-cell timeout expires), then any cells subsequently arriving, which belonging to the same packet, are destroyed. Such cells are no longer useful, since their parent packet cannot be reassembled anyway, and so it is better to reduce the network load by discarding them. To implement Lost Packet Screening, it is necessary to keep a record of all the packets currently passing through the buffer (they can be indexed using the value held in the Message Identification (MID) field). When a cell arrives, before it is allowed into the buffer, the table is checked to see if its parent packet is still "alive". The scheme is illustrated in figure 5.

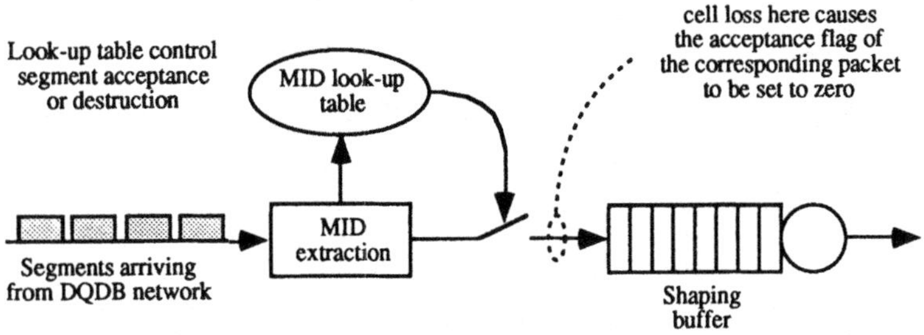

Figure 5

4.2 Pushout buffer

The pushout buffer takes the same idea a little further; it automatically removes from the buffer cells belonging to packets that have effectively become useless due to the loss of one or more cells. If a cell is lost in a buffer, then the probability that the immediately following cells will also be lost is greater than average. Due to the way that cells from different sources may be interleaved with each other, the consequences of a few lost cells are disproportionately severe; 10 lost cells could result in 10 lost packets. However, with the pushout buffer, as soon as a cell cannot be admitted to the buffer due to congestion, all the other cells in the buffer that belong to the same packet are removed, thus alleviating this congestion. The operation of the pushout buffer is illustrated in figure 6.

The implementation of such a buffer may be rather complex, especially considering the speed at which it must operate. It must be possible to remove up to 210 (maximum number of segments in onr DQDB IMPDU) cells from the buffer within one service time. Probably the only way to do this is to implement the buffer using a "linked list" approach, with the memory being used as a heap. Associated with each element in the queue is a pointer which points to the next element in the queue. When a new cell arrives, a free memory element is found (it's location in the memory is irrelevant), marked as being in use and then linked to the end of the queue by setting the pointer of the previous end-of-queue segment. In this way, cells can be removed easily from the buffer simply by changing pointers and no segment re-shuffling is required. There is, of course, an overhead in the memory required for the linking pointers and so it must be evaluated whether the reduction in packet loss obtained merits the extra complexity of the buffer.

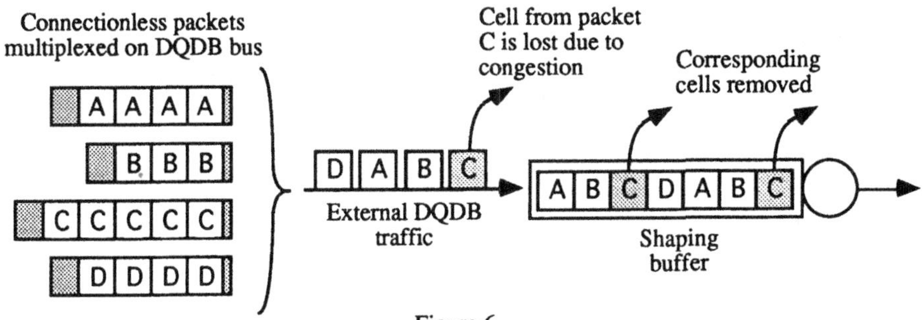

Figure 6

4.3 Datagram Acceptance Control

The DAC [11] performs an admission control on all arriving DQDB packets at the SIWU and is very similar to an approach described by Turner in [9]. The algorithm adopted here for the DAC differs slightly from that described in [11], but the principle of operation is the same. Whenever a BOM segment arrives, the imbedded packet length field is used to calculate whether the buffer will be able to support that packet. If there is sufficient capacity, then it is accepted, and this packet has then been *announced*. If

> $p(t)$ is the number of cells that have been announced, but not yet served at time t (called *pending* cells), and
>
> $l(t)$ is the shaper occupancy at time t,

then the number of pending cells that have been announced, but have not yet arrived at the shaper is given by $p(t) - l(t)$. The DAC avoids congestion in the shaper by predicting the maximum shaper occupancy in the future, l_{pred}, based upon knowledge of the currently announced packets. This can be done by assuming that the pending cells will arrive with a certain constant cell rate,

$$\frac{1}{\Delta}.s$$

where Δ is the time duration of 1 DQDB slot ($1/\Delta$ gives the maximum possible cell arrival rate) and s is a factor to control the tolerance of the prediction. If s=1, then the prediction is based on the worst case assumption that the cells arrive at the maximum possible rate. The prediction of the time it takes for all the cells pending at time t, to arrive is made using:

$$(p(t) - l(t)).\frac{\Delta}{s}$$

l_{pred} is then calculated as follows:

$$l_{pred} = p(t) - \frac{\Delta}{s}.\lambda[p(t) - l(t)] \quad (1)$$

Where l is the service rate of the shaper in units of cells/Δ. The *p(t)* term represents the segments already in the buffer plus those whose arrival has been announced and the term on the right is the number of segments that will be served between time t and the predicted arrival time of the last announced segment. The DAC decision is made by calculating l_{pred} with the proposed packet (ie. p(t) includes the packet length of the newly arrived packet). It will be accepted if and only if:

$$l_{pred} \leq \text{shaper capacity}$$

The algorithm is simple to implement, requiring only one variable, *p(t)* to be maintained. Whenever a packet is accepted, *p(t)* is updated according to the size of the packet and whenever a cell is served by the shaper, it is decremented by 1.

The performance of the DAC algorithm is heavily dependant on the value of s used. If s is too low, the algorithm will be too optimistic and not enough packets will be rejected, resulting in indiscriminate cell loss; on the other hand, if s is too high, the algorithm will be too pessimistic and reject packets when it is not necessary.

4.4 Controlling the DAC severity

Two model parameters that can affect the optimal value of s are the burstiness of the aggregate traffic arriving at the shaper (peak arrival rate divided by the mean arrival rate) and the background load in the DQDB network. Graph 2 shows how the performance of the DAC scheme depends on the value of s chosen, for two different example scenarios with different values for burstiness and background load. The parameters are given in table 2.

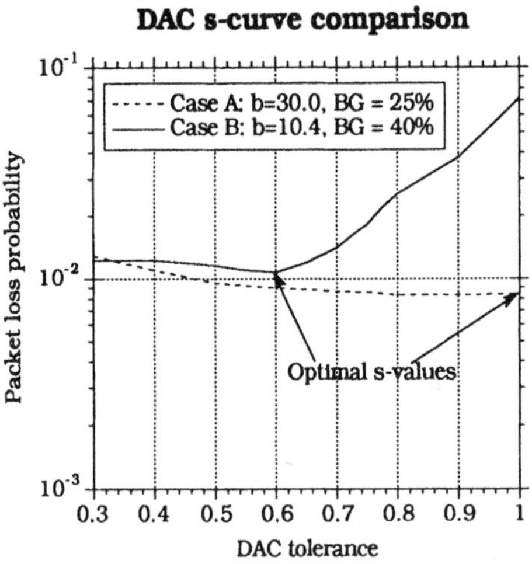

Graph 2

	Case A	Case B
Number of sources	60	20
Burstiness at shaper	30.0	10.4
Background load	25%	40%
Shaper load	20%	20%
Minimum inter packet time	$9.6 * 10^{-6}$	$9.635 * 10^{-4}$

Table 2

As can be seen, the optimal value of *s* changes considerably with the traffic conditions. This can be a problem for the DAC sheme, since if the value of *s* used is incorrect, then too many cells can be discarded and the packet loss rate is higher than it would be with the basic buffer. Furthermore, traffic the behaviour of LAN traffic is notoriously difficult to predict. One possibility would be to introduce a self-adjustment mechanism, which changes *s* on-line, based upon measurements of the arriving traffic.

4.5 Performance

The performance of the pushout shaper has been compared to that of a standard shaper through simulation. Graph 3 compares the performances of the standard buffer, the screened buffer, the screened, pushout buffer and the screened buffer with DAC, in terms of the packet loss rate (PLR). The DAC results are obtained using a value of $s = 0.8$, which was found to give good performance. The parameters of the simulation scenario are given in table 3 below.

Shaper capacity	200	Shaper service rate	39.3kcells/s
Min. inter packet time	0.964 ms	Slot time	4.24 µs
Nº external sources (file trasnfer traffic)	60	Mean file length	20000 octets
Background load on DQDB network	25%	Background traffic	File-transfer
LAN bus bitrate	10 Mbits/s		

Table 3

It can be seen that all three schemes, especially the pushout buffer and the DAC, can significantly reduce the packet loss rate. The increase in throughput obtainable decreases for lower packet loss probabilties. For a PLR of $10^{-2.15}$, for example, the DAC and pushout buffer offer an increase in throughput from 14% to 25%, whereas for PLR of less than 10^{-3}, the gain becomes smaller (extrapolating the curves would suggest an increase of 6% to 10%). From the point of view of packet loss *within* the shaper, the pushout buffer will always provide optimum performance, since it only destroys cells when it is absolutely necessary. However, this means that the pushout buffer often creates redundant, or "orphaned" cells, since the decision to destroy a packet may be taken after the first cells of that packet have already been transmitted. From the point of view of implementational complexity, the pushout buffer is considerably more complex than the DAC and screened buffer, since it requires a linked-list data structure in order to be able to remove multiple cells from any position in the buffer at high speed. In contrast, the DAC algorithm requires only one variable and has to perform a simple calculation for each BOM segment.

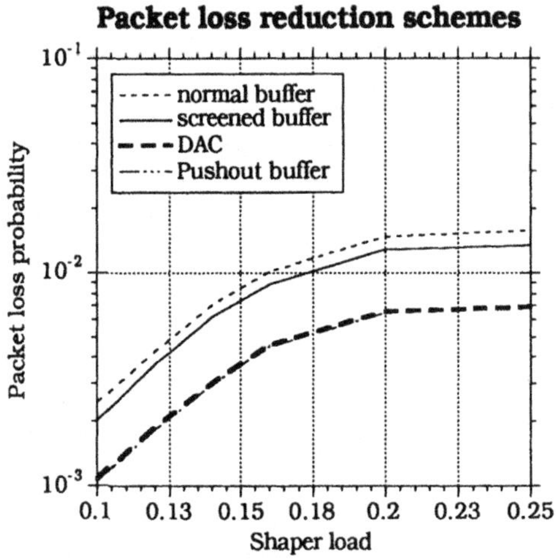

Graph 3

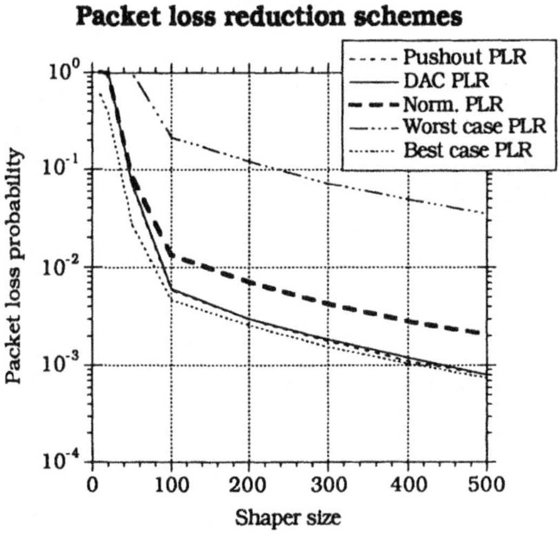

Graph 4

Graph 4 shows the performance of the DAC and pushout schemes for different buffer sizes. The parameters are the same as those in table 3, with a shaper load of 14%. Two additional curves are plotted to show the theoretical worst and best case packet loss rates. These are calculated from the cell loss rate; the worst case is when every cell loss falls in a different packet (PLR = number of cells lost/number of packets arrived) and the best case is when as many cell losses as possible occur in the same packet (PLR = cell loss rate).

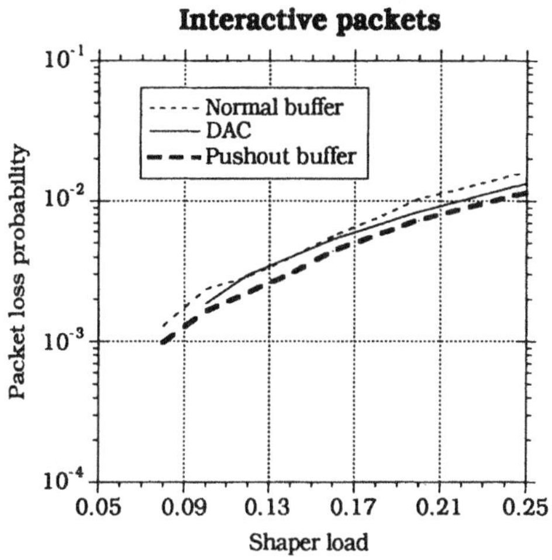

Graph 5

The algorithms do not always provide significant improvements in performance; graph 5 shows results from a scenario where the reduction in packet loss is negligable. Here, the interactive traffic model (see section 2) is used for the external traffic (the parameters are given below), resulting in less bursty cell loss and consequently less opportunity for reducing the packet loss rate.

Shaper capacity	200	Shaper service rate	39.3kcells/s
Min. inter packet time	9.6 μs	Slot time	4.24 μs
Nº external sources (interactive traffic)	60	External traffic packet lengths	4000 octets
Background load on DQDB network	25%	Background traffic	Interactive,

Table 4

CONCLUSION

This paper has shown that the cell loss rate for connectionless traffic at the SIWU and UPC may be significantly reduced by introducing a shaper function before the UPC. This buffer should be dimensioned to at least the maximum packet size of 210 cells. Three different schemes for reducing the packet loss in this shaper have been described: the screened buffer, the pushout buffer and the Datagram Acceptance Control scheme. A simulation based performance study revealed that both the pushout and DAC methods offer significant reductions in the packet loss rate. The screened buffer offers a more modest reduction in packet loss rate, but is very simple to implement and is therefore recommended. An advantage of the DAC is that it will always discard a *whole* packet and not leave orphaned cells to traverse the network, as is the case with the pushout buffer. The DAC is considerably simpler to implement than the pushout buffer and provides more or less the same performance. However, the choice of values for the DAC tolerance, *s*, is an important issue and the optimal value depends on variables that are difficult to predict. For unsuitable *s* values, the DAC can actually give worse performance than the basic buffer. Further work will address the problem of choosing a value for *s* and the use of a dynamic self-adjusting mechanism to control *s* based upon on-line traffic measurements.

ACKNOWLEDGEMENTS

The author would like to acknowledge the support of the RACE 2032 project, COMBINE and the useful input from the members of Workpackage 1. He would also like to acknowledge the help of the University of Paderborn for making their DQDB simulator available for use.

REFERENCES

[1] W.Leland, D.Wilson, "High Time-Resolution Measurement and Analysis of LAN Traffic: Implications for LAN Interconnection", IEEE INFOCOM 91.

[2] M.T. Le, R. Petty, "The IEEE 802.6 Metropolitan Area Network Distributed-Queue, Dual-Bus Protocol", Journal of Data & Computer Comms., Fall 1990, pp23-43.

[3] M. de Prycker, "Asynchronous Transfer Mode Solution for Broadband ISDN", Ellis Horwood Limited, 1991, ISBN 0-13-053513-3.

[4] T.D. Poelhekken, F.van den Dool, "Evolution of MANs towards B-ISDN", Proceedings of R1022 workshop on ATM Network Planning and Evolution, London, April 1991.

[5] I.Norros (editor), "Models for heterogeneous networks and their elements", COMBINE Deliverable 3, R2032/KTA/WP1/DS/P/003/a1, December 1992.

[6] M. Rümekasten, A. Calcagno, "ATLAS Reference Manual", release 5, University of Paderborn, Dept. Mathematics and Computer Science, December 1991.

[7] C.Rasmussen, J.H.Sørensen, K.S.Kvols, S.B.Jacobsen, "Source Independent Call Acceptance Procedures in ATM Networks", IEEE Journal on Selected Areas in Communications, Vol.9, No.3, April 1991.

[8] H. Kröner, "Statistical Multiplexing of Sporadic Sources - Exact and Approximate Performance Analysis", TELETRAFFIC AND DATATRAFFIC in a Period of Change, Proceedings of ITC-13, 1991, Copenhagen.

[9] J.S.Turner, "A proposed bandwidth and Congestion Control Scheme for Multicast ATM Networks", Washington University Technical Report WUCCRC-91-1, May 1991.

[10] J.Virtamo, S.Aalto, I.Norros, K.Kvols, S.Manthorpe, "Performance Aspects of Streaming and Message Modes of Interworking", Proceedings of Interworking 92, Bern November 1992.

[11] R.Vogt, U. Killat, J.Ottensmeyer, M. Rümekasten, "A Concept for Interconnecting DQDB MANs through ATM-based B-ISDN and Related Issues with respect to Simulation" , Proceedings of Interworking 92, Bern, November 1992.

[12] "Distributed Queue Dual Bus (DQDB) Subnetwork of a Metropolitan Area Network (MAN)", IEEE Standard 802.6, July 3 1991, ISBN 1-55937-093-9

[13] J.W. Roberts (editor), "Performance Evaluation and Design of Multiservice Networks", COST224 Final Report, October 1991, EUR 14152, ISBN 92-826-3728-X, pp 31-35.

[14] CCITT Recommendation I.371

Integrated Broadband Communication Networks
and Services (C-18) / V.B. Iversen (Editor)
Elsevier Science B.V (North-Holland)

ATM - N-ISDN 2 Mbit/s Interworking Unit

M. S. Nunes[a], P.D. Soerensen[b], M. Frieden[c], J. Gonçalves[d], M. Günay[e], H. Blaser[f], V. Tao[g]

[a]INESC, R. Alves Redol 9, 1000 Lisbon, Portugal
[b]JTAS, Sletvej 30, DK-8310, Arhus-Tranbjerg J, Denmark
[c]ASCOM TECH, Freiburgstrasse 370, 3018 Bern, Switzerland
[d]CET, Sítio das Palhas, 3800 Aveiro, Portugal
[e]MAZ, Karnapp 20, D-2100, Hamburg 90, Germany
[f]ASPA, Wallstrasse 22, 4002 Basel, Switzerland
[g]LEP, Avenue Descartes 22, F-94453, Limeil-Brevannes, France

Abstract

This paper presents the specifications of an ATM - N-ISDN 2 Mbit/s Interworking Unit being developed in RACE Project 2061 EXPLOIT. A description of different interworking scenarios between the B-ISDN and N-ISDN and the protocol reference model of the IWU is given, followed by an overview of the system requirements to be met by the IWU. The system structure of the IWU is presented and its main functional blocks are described. The external interfaces of the IWU are also specified, namely the 2 Mbit/s N-ISDN NNI interface and the ATM interface. Finally the synchronisation aspects of the different interworking scenarios are discussed.

Keyword Codes: C.2; C.2.1; C.2.2
Keywords: Computer-Communication Networks; Network Architecture and Design; Network Protocols

1. INTRODUCTION

It is generally recognised that the deployment of the B-ISDN will require a period of time as (public and/or private) network operators seek to find the most economical means of evolving to the B-ISDN. These evolutionary phases will need to be harmonised with the overall B-ISDN concepts ensuring the continued support of existing interfaces and services specified in CCITT recommendations. This requires in particular, the possibility of interconnection of N-ISDN terminals and services with the B-ISDN, what requires the definition and implementation of an Interworking Unit (IWU) between the two networks.

The purpose of the ATM - N-ISDN 2 Mbit/s IWU is to provide interworking between an ATM based network, the EXPLOIT testbed [1], and a N-ISDN 2.048 Mbit/s network, the SWISSNET [2]. The location of the IWU is shown in Figure 1.

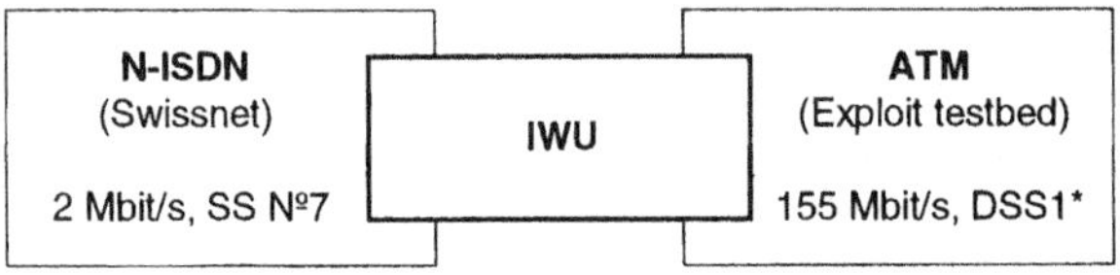

Figure 1. Environment of the ATM-2 Mbit/s-IWU

The EXPLOIT testbed provides 155 Mbit/s data transfer rate. At the NNI it is used a modified and enhanced version of the Digital Subscriber Signalling System (DSS1) including extensions specific to ATM is available, called DSS1* [3] in the following text.

The N-ISDN in the test environment (SWISSNET) provides a 2.048 Mbit/s transmission system comprising 32 time slots of 64 kbit/s. Time-slot 16 (D channel) is used for signalling and 0 for synchronisation. The remaining 30 time-slots (B channels) are used for user data transmissions. For signalling between N-ISDN and the IWU the Common Channel Signalling System Nº. 7 (SS Nº7) is used.

Additionally, interworking with the MEGACOM [4] network will also be provided, in order to test the emulation of n* 64 Kbit/s connections on a semi-permanent basis.

2. INTERWORKING SCENARIOS BETWEEN B-ISDN AND N-ISDN

A B-ISDN user with terminals supporting 64 Kbit/s based ISDN services would like to communicate with a N-ISDN user transparently. Consequently, interworking between both networks is required.

Two different interworking scenarios were identified:

1) Access to N-ISDN through semi-permanent point-to-point ATM connections, transparent to N-ISDN signalling.
2) Access to N-ISDN using B-ISDN signalling interworking

Figure 2 shows the reference configuration for scenario 1.

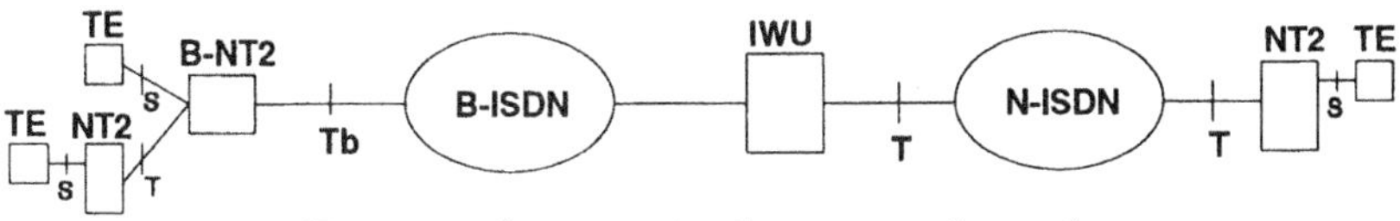

Figure 2. Scenario 1 reference configuration

In this scenario a basic/primary N-ISDN access will be emulated by a semi-permanent ATM connection, transparent to N-ISDN signalling. The semi-permanent ATM connection is established between the B-ISDN user and the IWU, using OAM procedures, so signalling is not required. The user has always a virtual channel connection between himself and the IWU.

The B-NT2 or the TA and the IWU only have to do the circuit emulation in AAL type 1 of the basic/primary access. The IWU is seen as a NT2 + NT1 to the basic N-ISDN access and a NT1 to the primary N-ISDN access, which means the N-ISDN exchange will always see the IWU as a primary N-ISDN access. The 64 Kbit/s based ISDN TE´s connected through B-ISDN to a specific IWU are subscribers of the N-ISDN exchange where the specific IWU is connected. Figure 3 shows the protocol stacks corresponding to this scenario.

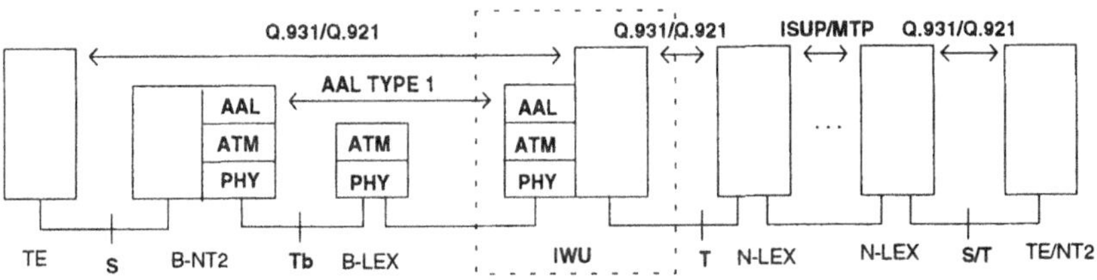

Figure 3a. Signalling transfer in U-plane (basic N-ISDN access)

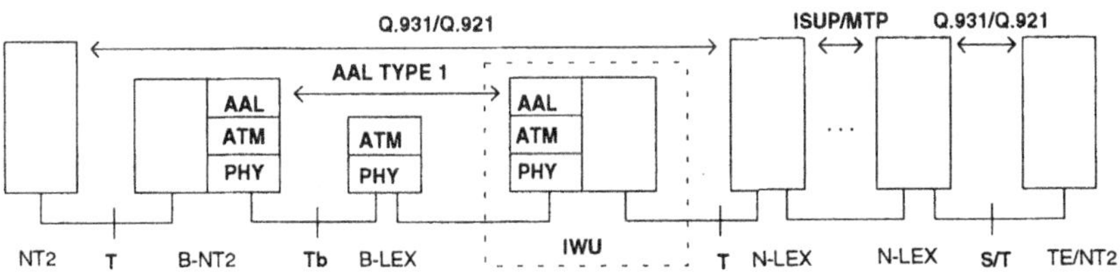

Figure 3b. Signalling transfer in U-plane (primary N-ISDN access)

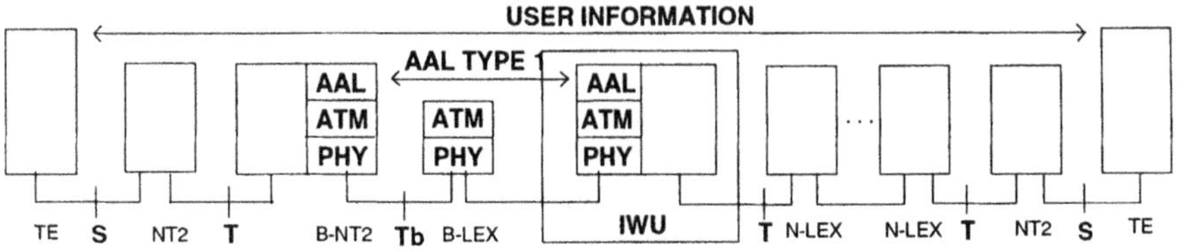

Figure 3c. User information transfer (interworking in the U-plane)

In scenario 2 the interworking is implemented using the B-ISDN signalling. The IWU must take into account the interworking of protocols between B-ISDN and N-ISDN. Figure 4 shows the reference configuration for scenario 2.

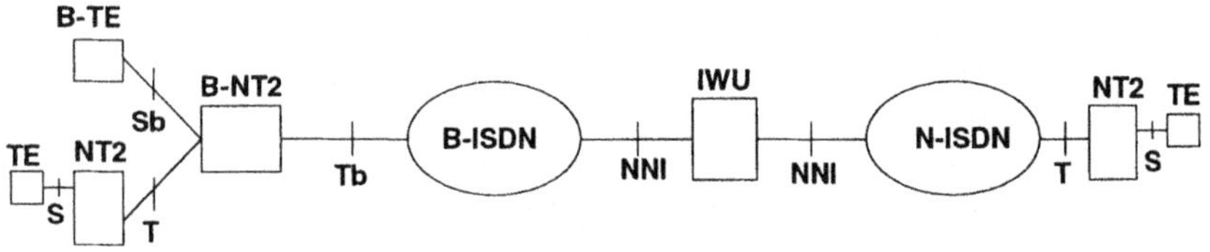

Figure 4. Scenario 2 reference configuration

Two different communications between B-ISDN users and N-ISDN users can occur:

i) B-TE supporting 64 Kbit/s based ISDN services <---> N-ISDN TE
ii) TE connected through B-NT2 or TA to B-ISDN <---> N-ISDN TE

In the first case the B-TE will access to B-ISDN using the broadband access protocols. An interconnection between networks is then necessary and the IWU has to do the interworking of the NNI signalling protocols from B-ISDN to N-ISDN and vice versa.

In the second case the B-NT2 or the TA has to do the interworking of the N-ISDN access protocols in B-ISDN access signalling protocols to connect the TE into the B-ISDN network. Then the IWU will do the same as in the first case to provide the communication between both networks. Figure 5 explains the interworking in this scenario.

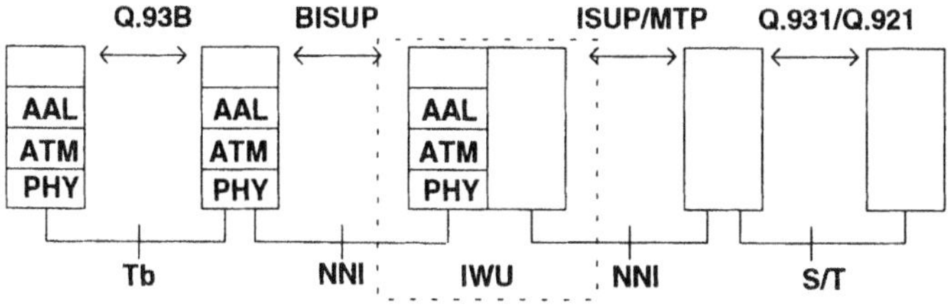

Figure 5a. Signalling transfer (interworking in the C-plane)

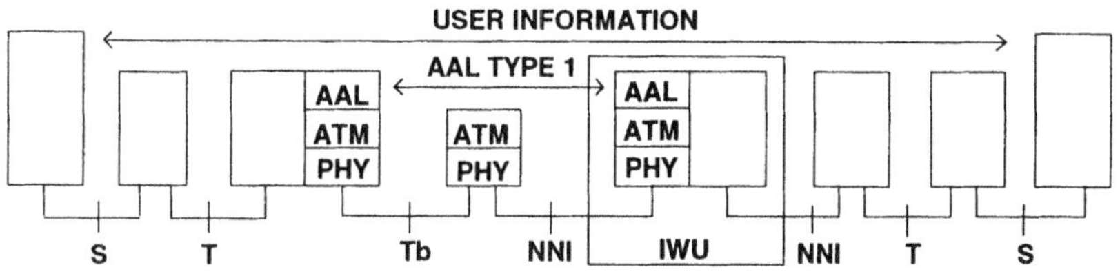

Figure 5b-1. User information transfer (interworking in the U-plane).
64 Kbit/s based N-ISDN TE connected to B-ISDN

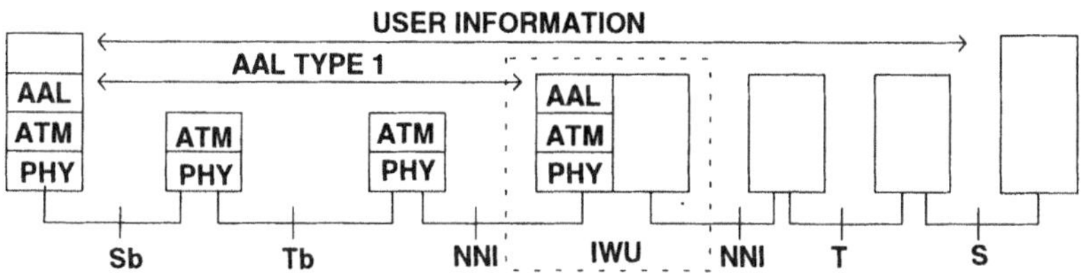

Figure 5b-2. User information transfer (interworking in the U-plane).
B-TE supporting 64 Kbit/s based ISDN service

A comparison between the different scenarios can now be made to select the most suitable for implementation.

If B-ISDN will be seen as an overlay network with few 64 Kbit/s based ISDN terminals connected, scenario 1 is more applicable. These terminals are seen as if they are directly connected to the N-ISDN network and the interworking is simplified. In scenario 1 if different TE´s connected to different B-ISDN users wish to communicate between them, they have to access the N-ISDN network first.

Scenario 2 is more powerful and flexible, however its functionality is more complex due to the interworking of signalling protocols. TE´s connected to B-ISDN are users of the B-ISDN network, so they can communicate between them without access to N-ISDN. A B-TE which supports 64 Kbit/s based ISDN services can communicate to a N-ISDN TE in the same way as it communicate to another B-TE within scenario 2.

Both scenarios will be implemented in RACE project EXPLOIT, the first one for interworking with MEGACOM [4], using n*64 kbit/s (n=1-30) circuits and 2.048 Mbit/s, the second one for interworking with SWISSNET [2], using only 64 kbit/s circuits.

3. FUNCTIONALITY OF THE IWU

Considering the different scenarios considered before, the IWU will have the following main functionalities:

- generation of the ATM cell stream for the EXPLOIT-testbed from the incoming N-ISDN data stream
- generation of the data stream for N-ISDN from the incoming EXPLOIT-testbed cell stream
- interworking between signalling systems
- managing the establishment and release of connections
- adaptation of the ATM system to the PDH 2.048 Mbit/s system and vice versa.

Regarding the cell stream generation the IWU has to perform the cell generation for each of the first order PDH time-slots in the following manner:

- mapping of circuit connections onto VCIs
- insertion of the VCI into the header
- insertion of the sequence number and sequence number protection into the SN and SNP fields of the AAL1 PDU
- filling the cell information fields with the corresponding PDH time-slot data.
- insertion of idle cells at the ATM interface, whenever necessary

The incoming cells from EXPLOIT testbed have to be evaluated. The information fields must be distributed to the corresponding time-slots in the first order PDH frame with a frequency of 8 KHz after disassembling. Regarding the data stream generation for the N-ISDN the IWU must control the following functions:

- sequence number checking according to CCITT recommendation I.363:
  - detection and correction of errors in the sequence number
  - in case of an uncorrectable error the cell will be discarded
  - when a cell is lost, the corresponding time-slot will be filled with a programmable pattern
- removing of header, SN and SNP respectively
- buffering a number of cells to compensate for variable network delay before starting playout into N-ISDN
- mapping of VCIs onto circuit numbers
- generation of a continuous bit stream from the cell information fields for the corresponding first order PDH time-slots.

To guarantee the separate processing of cells containing signalling information, a splitting of the cell stream is necessary. The insertion of signalling cells into the cell stream should also be guaranteed. In this context further requirements to the IWU with regard to the control of the cell stream follow:

- splitting of the cell stream: recognition and extraction of the cells containing signalling or OAM information and forwarding them to the processing unit
- inserting into the cell stream: insertion of cells containing signalling or OAM information by the processing unit.

As the signalling in the EXPLOIT testbed and in the N-ISDN are different, the IWU has to handle the SS Nº7 as well as the DSS1*, a modified and enhanced version of the DSS1 including extensions specific to ATM. Thus, the following requirements are to be met by the IWU:

- establishment and release of VC connections in the EXPLOIT testbed direction
  - reservation of the time-slot(s) for a circuit connections
  - requesting and reserving a VC from EXPLOIT testbed
  - updating of the translation table with the assigned VCI value
  - translation of the circuit number to the corresponding VCI according to the table entry
  - release of the VC and circuit connections
- establishment and release of circuit connections in the N-ISDN direction
  - reservation of the VCI, which is defined in setup
  - requesting and reserving time-slot(s) for a circuit connection
  - updating of the translation table
  - translation of the VCI to the corresponding circuit number according to the table entry
  - release of the VC and circuit connections

The necessary adaptation of the first order PDH-technique to ATM-technique and vice versa implies the following requirements:

- first order PDH side, outgoing data flow:
  - conversion: byte parallel to serial
  - bit clock generation for the serial bit-stream
  - filling the time-slots of the PDH-frame with the corresponding time-slot data
  - buffering in order to compensate for the transmission gaps created by not equidistantly spaced cell streams
  - filling the time-slots of a PDH-frame with a certain pattern when there is no user information for the corresponding time-slots
- first order PDH side, incoming data flow:
  - bit regeneration and bit clock recovery from the N-ISDN
  - recognition of the first order PDH frame start

- conversion: serial to byte parallel
- ATM side, outgoing data flow:
  - cell rate decoupling (insertion of idle cells)
  - delineation of data structure units for n x 64 Kbit/s circuits (n=2-32), based on insertion of "pointers" [5]
  - insertion of residual time stamp information for end-to-end synchronisation [5]
- ATM side, incoming data flow:
  - delineation of cells
  - evaluation of the HEC
  - detection of idle cells
  - detection of cell synchronisation loss
  - cell rate decoupling (discarding of idle cells)
  - recover of data structure units for n x 64 Kbit/s circuits (n=2-32), based on "pointers"
  - recover of residual time stamp information for end-to-end synchronisation

4. IWU PROTOCOL REFERENCE MODEL

The N-ISDN 2 Mbit/s IWU protocol reference model will be a combination of the B-ISDN and N-ISDN protocol reference models, as shown in figure 6.

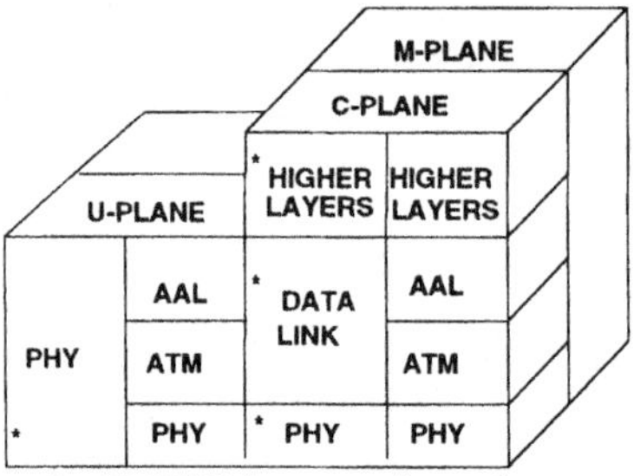

* - N-ISDN protocol stack

Figure 6. 2 Mbit/s IWU protocol reference model

The protocol reference model has a layered structure with 3 different planes: user plane, control plane and management plane

The user plane is responsible for the transfer of the user information from the B-ISDN network to the N-ISDN and vice versa, transparently to the user. As the IWU will perform n x 64 Kbit/s circuit emulation, the AAL type 1 protocol will be provided. The information flow in this plane is shown in the figure 7.

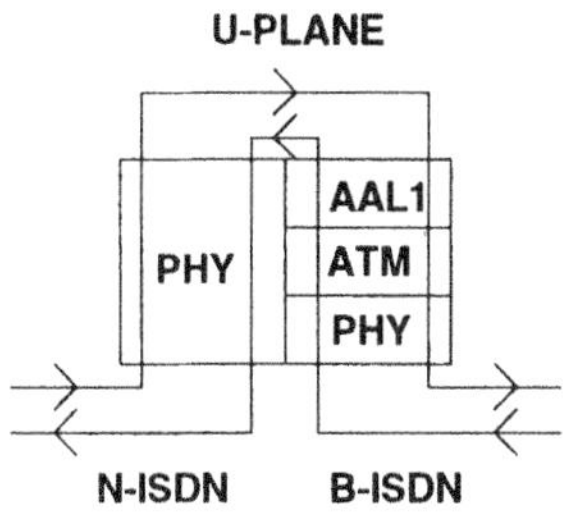

Figure 7. Interworking in the U-plane

The control plane deals with the signalling necessary to setup, supervise and release calls and connections. As we have two different networks with different signalling protocols, the IWU has to do the interworking of the protocols between both networks. From the control plane point of view the IWU can be seen as a signalling point of the both networks and additionally with a signalling interworking unit between the SPs. The information flow in the control plane is shown in figure 8.

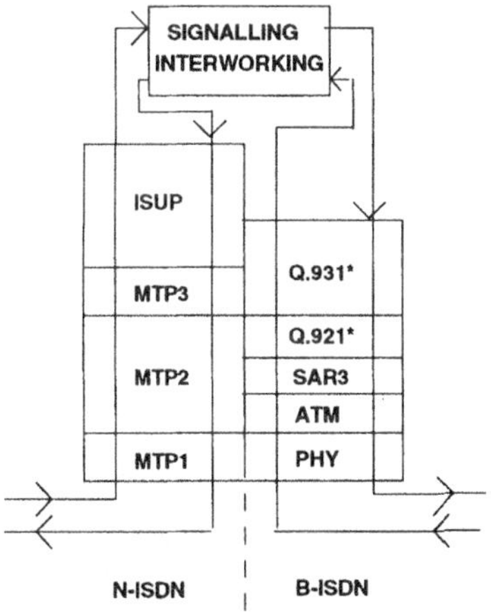

Figure 8. Interworking in the C-plane

5. SYSTEM STRUCTURE OF THE 2 MBIT/S IWU

The overall system structure and external interfaces of the ATM - 2 Mbit/s is shown in figure 9.

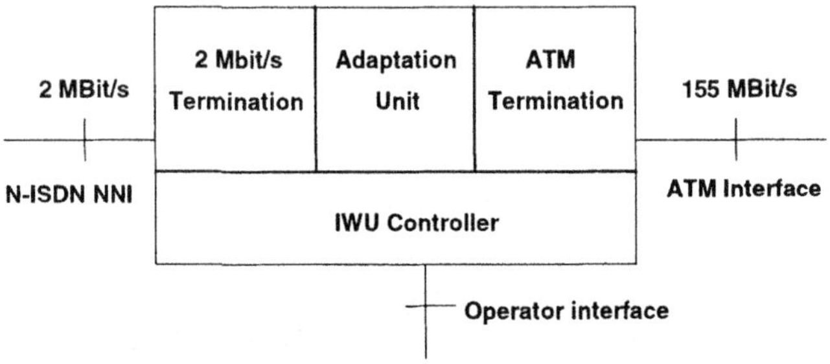

Figure 9. Overall system structure of the IWU.

The interface towards the N-ISDN 2 Mbit/s network is a standard G.703/G.704 (Blue book version) interface. The ATM interface is a partly non standard interface as the interface has to comply with the requirements defined in RACE project 1022. [6]. The operator interface is used for maintenance/control of the IWU. The interface can be used to show statistical information, manually setup and clearing of calls etc.

The ATM - 2 Mbit/s IWU is functionally divided into 4 main blocks:

- 2 Mbit/s termination
- Adaptation Unit
- ATM termination
- IWU Controller

The 2 Mbit/s termination implements the receiving and transmission line interface functions of a first order PDH system. The 2 Mbit/s termination conforms to the CCITT recommendation G.703/G.704. Clock and data recovery functions and the handling of the "low level" functions (MTP) of the SS Nº7 are implemented in this block.

The Adaptation performs the adaptation of the N-ISDN B-channels to ATM connections and vice versa. In the direction of the EXPLOIT testbed, the adaptation block provides the cell assembling of the B-channels into ATM cells. In the direction towards the synchronous 2 Mbit/s network, it provides the disassembling of the ATM cells into 64 kbit/s B-channels. In case of interconnection between two IWU, synchronisation information is also required between them.

The ATM termination implements the 155,52 Mbit/s interface which meets the requirements of the EXPLOIT testbed. The ATM termination is described in detail in [6]. The ATM interface will be connected to the Adaptation Unit via an internal ATM parallel interface which is based on 8 bit parallel data, a cell start and a byte clock.

The IWU Controller is mainly responsible for the control of the other blocks in the system and handling of the signalling interworking. The IWU Controller must implement "high level" parts of the SS Nº7 protocol, i.e. ISUP, the DSS1* and the interworking between the two different protocols. An operator terminal will be connected to the IWU Controller. The internal control interface will be a standard VMEbus.

6. SYNCHRONISATION

In the EXPLOIT project different interworking environments will be implemented, namely the interconnection between an IWU and a N-ISDN TA and the interconnection between two IWUs.

In the first case it is necessary to synchronise with the IWU the clock of the N-ISDN TE. As this terminal is connected to the N-ISDN TA developed in RACE Project 1022, we have to use the technique implemented in this TA which is based on the plesiochronous method [7]. This technique requires the use of a stable reference clock in the TA in order to reduce the frequency of occurrence of slips due to frequency mismatch between the public N-ISDN and the TA clocks.

In the second case two IWUs are interconnected through the EXPLOIT testbed, allowing the implementation of a more advanced synchronisation technique. The method chosen for this case is based on the Synchronous Residual Time Stamp (SRTS) method, as described in CCITT draft recommendation I.363 [5]. This technique was chosen because it is compliant with CCITT recommendations and meets the jitter specifications of rec. G.822/G.823 for the 2.048 Mbit/s plesiocronous digital hierarchy.

The SRTS method uses alternate bits of the CSI field [5] of consecutive cells of a virtual channel. To guarantee that the private N-ISDN receives permanently from the public network the SRTS information without interruption, a permanent virtual circuit with a bit-rate of 64 Kbit/s with a reserved VPI/VCI could be used to carry this information. This virtual channel is unidirectional and established between the IWU connected to the public N-ISDN to the other IWU connected to the private N-ISDN. Alternatively, the SRTS information could be sent in the user virtual circuit, performing in-band synchronisation. This second method has the advantage of saving bandwidth since no additional channel is required for synchronisation, being the preferred solution for implementation.

The use of the SRTS requires that a common network clock, e.g. 155.52 MHz, is available at both sides of the ATM network. This requirement is assured by the EXPLOIT testbed only if both IWUs are connected to the same ATM switch.

7. CONCLUSIONS

An ATM - N-ISDN 2 Mbit/s Interworking Unit being developed in Project 2061 EXPLOIT was described. Several interworking scenarios have been analysed, allowing to choose two of them for implementation, one using semi-permanent ATM connections, the other using B-ISDN signalling protocols to access the N-ISDN. The overall structure of the IWU was defined and the design of the different hardware and software modules has now started. For interconnection between IWUs different synchronisation methods will be implemented, one using the adaptive technique, the other the SRTS technique.

REFERENCES

1. External Specifications of the ATM - 2 Mbit/s Interworking Unit, R2061/EXP/SW2/DS/006/B2, November 1992.
2. Specification of the ISDN User Part for SWISSNET, H. Blaser, R2061-ASPA, May 1992.
3. Report on the Design of the R1022 Generic Control Functions. Appendix 4: Layer 3 Specification, M. Frieden, ASCOM, December 1991.
4. MEGACOM Technical Specification - User Interfaces, SWISS PTT, February 1992.
5. Draft Recommendation I.363, CCITT, June 1992.
6. ATM Layer Specification for RACE 1022 ATD Technology Testbed (RATT), PKI, April 1991
7. Report on the Use of generic Components in Terminal Adaptors, R1022/22/SW2/222/DS/C/204/B1, March 1990.

LIST OF AUTHORS

Aagesen, F.A.	261	Melteig, O.	287
Albanese, A.	201	Mitrou, N.M.	225
Arsenis, S.	407	Morland, L.	287
Baireuther, O.	67	Møller, A.M.	253
Basso, C.	119	Nilsson, A.A.	107,395
Bean, N.G.	241	Norros, I.	373
Blankers, P.F.C	313	Nunes, M.S.	441
Blaser, H.	441	Nørgaard, K.	11
Bruneel, H.	383	Pedersen, H.	11
Calvignac, J.	119	Perros, H.	107
D'Ambrosio, M.	349	Raatikainen, P.	177
Dittmann, L.	275	Reinhardt, W.	189
Ercole, D.	201	Resmini, L.	151
Evans, S.P.	361	Saito, T.	95
Falch, M.	163	Simoni, N.	407
Flinck, H.	177	Skouby, K.E.	163
Frieden, M.	441	Smit, T.A.	213
Girardi, G.	201	Soerensen, P.D.	441
Goncalves, J.	441	Sourbes, H.	119
Gruenenfelder, R.	303	Spaniol, O.	1
Guarneri, R.M.	151	Stainov, R.	189
Günay, M.	441	Staxén, P.	139
Händel, R.	131	Stevenson, D	107
Haugen, R.	23	Steyaert, B.	383
Heinrichs, B.	189,337	Stock, T.	45
Helvik, B.E.	287	Svendsen, B.	419
Henten, A.	163	Tao, V.	441
Jakobs, K.	337	Thomsen, A.	11
Jessen, K.E.	11	Tran-Gia, P.	45
Jou, Y.F.	395	Tuyt, J.	325
Lai, F.	395	Ullio, M.	201
Lanting, C.J.M.	151	van Till, J.W.J. baron	35
Koukos, A.K.	225	Villén-Altamirano, M.	77
Kvols, K.	275	Virtamo, J.	373
Mannsåker, B.	23	Zidbeck, J.	177
Manthorpe, S.	429	Znaty, S.	407
Melen, R.	201,349		

IFIP

The INTERNATIONAL FEDERATION FOR INFORMATION PROCESSING is a multinational federation of professional and technical organisations (or national groupings of such organisations) concerned with information processing. From any one country, only one such organisation – which must be representative of the national activities in the field of information processing – can be admitted as a Full Member. In addition a regional group of developing countries can be admitted as a Full Member. On 1 October 1993, 44 organisations were Full Members of the Federation, representing 65 countries.

The aims of IFIP are to promote information science and technology by:

- fostering international co-operation in the field of information processing;
- stimulating research, development and the application of information processing in science and human activity;
- furthering the dissemination and exchange of information about the subject;
- encouraging education in information processing.

IFIP is dedicated to improving worldwide communication and increased understanding among practitioners of all nations about the role information processing can play in all walks of life.

Information technology is a potent instrument in today's world, affecting people in everything from their education and work to their leisure and in their homes. It is a powerful tool in science and engineering, in commerce and industry, in education and adminstration. It is truly international in its scope and offers a significant opportunity for developing countries. IFIP helps to bring together workers at the leading edge of the technology to share their knowledge and experience, and acts as a catalyst to advance the state of the art.

IFIP came into official existence in January, 1960. It was established to meet a need identified at the first International Conference on Information Processing which was held in Paris in June, 1959, under the sponsorship of UNESCO.

Organisational Structure
The Federation is governed by a GENERAL ASSEMBLY, which meets once every year and consists of one representative from each Member organisation. The General Assembly decides on all important matters, such as general policy, the programme of activities, admissions, elections and budget.

IFIP ORGANISATION CHART

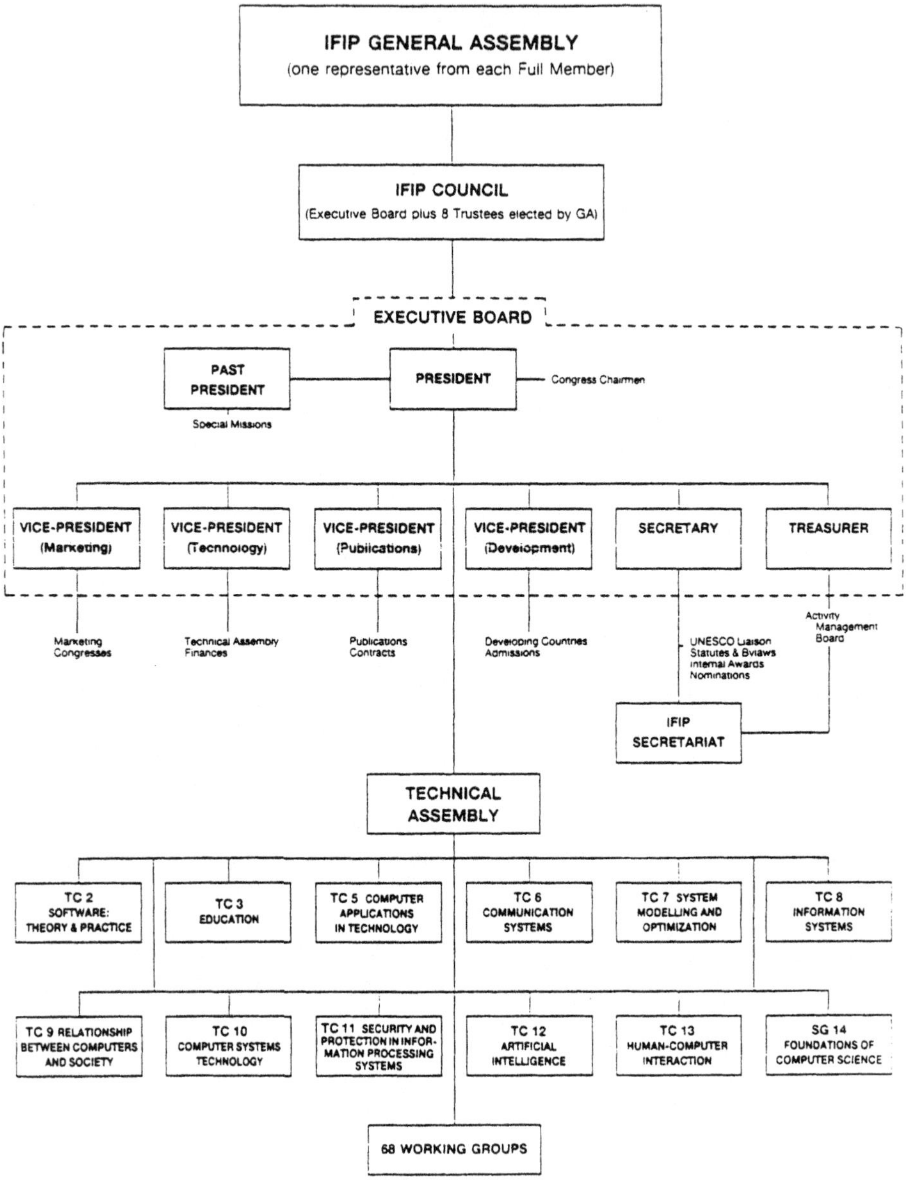

The day-to-day work of IFIP is directed by its Officers: the President, Vice-Presidents, Secretary and Treasurer, who are elected by the General Assembly and together constitute the EXECUTIVE BOARD.

The COUNCIL, consisting of the Officers and up to eight Trustees elected from the General Assembly, meets twice a year and takes decisions which become necessary between General Assembly meetings.

The headquarters of the Federation are in Geneva, Switzerland where the IFIP Secretariat administers its affairs.

For further information please contact:

IFIP Secretariat
attn. Mme. GWYNETH ROBERTS
16 Place Longemalle
CH-1204 Geneva, Switzerland
telephone: 41 (22) 310 26 49
facsimile: 41 (22) 781 23 22
Bitnet: ifip@cgeuge51

IFIP's MISSION STATEMENT

IFIP's mission is to be the leading, truly international, apolitical organisation which encourages and assists in the development, exploitation and application of Information Technology for the benefit of all people.

Principal Elements

1. To stimulate, encourage and participate in research, development and application of Information Technology (IT) and to foster international co-operation in these activities.
2. To provide a meeting place where national IT Societies can discuss and plan courses of action on issues in our field which are of international significance and thereby to forge increasingly strong links between them and with IFIP.
3. To promote international co-operation directly and through national IT Societies in a free environment between individuals, national and international governmental bodies and kindred scientific and professional organisations.
4. To pay special attention to the needs of developing countries and to assist them in appropriate ways to secure the optimum benefit from the application of IT.
5. To promote professionalism, incorporating high standards of ethics and conduct, among all IT practitioners.
6. To provide a forum for assessing the social consequences of IT applications; to campaign for the safe and beneficial development and use of IT and the protection of people from abuse through its improper application.

7. To foster and facilitate co-operation between academics, the IT industry and governmental bodies and to seek to represent the interest of users.
8. To provide a vehicle for work on the international aspects of IT development and application including the necessary preparatory work for the generation of international standards.
9. To contribute to the formulation of the education and training needed by IT practitioners, users and the public at large.

IFIP TRANSACTIONS

IFIP TRANSACTIONS is a serial consisting of 15,000 pages of valuable scientific information from leading researchers, published in 36 volumes per year. The serial includes contributed volumes, proceedings of the IFIP World Conferences, and conferences at Technical Committee and Working Group level. Mainstream areas in the IFIP TRANSACTIONS can be found in Computer Science and Technology, Computer Applications in Technology, and Communication Systems.

From 1993 onwards the IFIP TRANSACTIONS are only available as a full set.

IFIP TRANSACTIONS A:
Computer Science and Technology
1992: Volumes A1–A19
1993: Volumes A20–A40
1994: Volumes A41–A61
ISSN 0926-5473

IFIP Technical Committees that are involved in IFIP TRANSACTIONS A
Software: Theory and Practice (TC2)
Education (TC3)
System Modelling and Optimization (TC7)
Information Systems (TC8)
Relationship Between Computers and Society (TC9)
Computer Systems Technology (TC10)
Security and Protection in Information Processing Systems (TC11)
Artificial Intelligence (TC12)
Human-Computer Interaction (TC13)
Foundations of Computer Science (SG14)

IFIP TRANSACTIONS B:
Applications in Technology
1992: Volumes B1–B8
1993: Volumes B9–B14
1994: Volumes B15–B19
ISSN 0926-5481

IFIP Technical Committee that is involved in IFIP TRANSACTIONS B
Computer Applications in Technology (TC5)

IFIP TRANSACTIONS C:
Communication Systems
1992: Volumes C1–C8
1993: Volumes C9–C16
1994: Volumes C17–C26
ISSN 0926-549X

IFIP Technical Committee that is involved in IFIP TRANSACTIONS C
Communication Systems (TC6)

IFIP TRANSACTIONS FULL SET: A, B & C
1992: 35 Volumes. US $ 1892.00/Dfl. 3500.00
1993: 35 Volumes. US $ 2100.00/Dfl. 3885.00
1994: 36 Volumes. US $ 2277.00/Dfl. 4212.00
The Dutch Guilder prices (Dfl.) are definitive. The US $ prices mentioned above are for your guidance only and are subject to exchange rate fluctuations. Prices include postage and handling charges.

The volumes are also available separately in book form.

Please address all orders and correspondence to:

ELSEVIER SCIENCE B.V.
attn. PETRA VAN DER MEER
P.O. Box 103, 1000 AC Amsterdam
The Netherlands
telephone: 31 (20) 5862602
facsimile: 31 (20) 5862616
email: p.meer@elsevier.nl

Printed in Dunstable, United Kingdom

91898095R00268

Printed in Dunstable, United Kingdom

91898089R00228